Human Resource Management in the Digital Economy:

Creating Synergy between Competency Models and Information

Susana de Juana-Espinosa
University of Alicante, Spain

José Antonio Fernández-Sánchez
University of Alicante, Spain

Encarnación Manresa-Marhuenda
University of Alicante, Spain

Jorge Valdés-Conca
University of Alicante, Spain

Information Science
REFERENCE

Managing Director: Lindsay Johnston
Senior Editorial Director: Heather Probst
Book Production Manager: Sean Woznicki
Development Manager: Joel Gamon
Development Editor: Michael Killian
Acquisitions Editor: Erika Gallagher
Typesetters: Chris Shearer, Lisandro Gonzalez
Print Coordinator: Jamie Snavely
Cover Design: Nick Newcomer, Greg Snader

Published in the United States of America by
Information Science Reference (an imprint of IGI Global)
701 E. Chocolate Avenue
Hershey PA 17033
Tel: 717-533-8845
Fax: 717-533-8661
E-mail: cust@igi-global.com
Web site: http://www.igi-global.com

Library of Congress Cataloging-in-Publication Data

Human resource management in the digital economy: creating synergy between competency models and information / Susana de Juana-Espinosa ... [et al.], editors.
p. cm.
Text in English and Spanish.
Includes bibliographical references and index.
Summary: "This book demonstrates how to implement e-management and competency models in companies, offering perspectives on the impact of integrated e-human resource policies and provides recommendations for addressing the shift from traditional human resource policies to new perspectives"--Provided by publisher.
ISBN 978-1-61350-207-5 (hbk.) -- ISBN 978-1-61350-208-2 (ebook) -- ISBN 978-1-61350-209-9 (print & perpetual access) 1. Personnel management. 2. Organizational effectiveness. I. Juana-Espinosa, Susana de, 1975-
HF5549.H79155 2012
658.3--dc23

2011047367

British Cataloguing in Publication Data
A Cataloguing in Publication record for this book is available from the British Library.

Editorial Advisory Board

List of Reviewers

Drew Rodgers, *Oslo University College, Norway*
Vicente Sabater Sempere, *University of Alicante, Spain*
Sandra Mª Sánchez Cañizares, *University of Córdoba, Spain*
Cristina Santandreu-Mascarell, *Polytechnic University of Valencia, Spain*
Teresa Torres-Coronas, *Universitat Rovira i Virgili, Spain*
Mireia Valverde, *Universitat Rovira i Virgili, Spain*
Sharna Wiblen, *University of Sydney, Australia*

Table of Contents

Detailed Table of Contents

facilitate time- and space-independent interaction and collaboration. The chapter closes with a brief discussion of implications for HR managers and professionals and avenues for future research.

Human resources are considered the company's most valuable resource. For that reason, organizations require a Human Resource Management (HRM) that provides the adequate work force within the new global economy. The psychological contract includes solid agreements between company and its employees, beyond the written contract, specifying their contributions, expectations, beliefs, promises, and obligations between both parties. Their management requires the definition of the concept of psychological contract, the analysis of its main characteristics and contents; and the identification of its stages of development. Electronic Human Resource Management (e-HRM) introduces a way of implementing HRM strategies through Web-based tools, improving the psychological contract management. Information phase, Intranet, and internal electronic mail have an important role in socialization stage. The consolidation of psychological contract (maintenance phase) is favoured by Intranet, business-to-employees, internal electronic mail, database, videoconference, and groupware. Finally, in breach phase, intranet, electronic mail, database, videoconference, and groupware are important E-HRM.

The aim of this chapter is to assess the impact of different organizational factors on the success of e-learning programs, in terms of both self-reported satisfaction and the level of learning. Hence, this study adds to the analysis of the efficacy of e-learning models from an organizational perspective by providing some useful insights, which may help to improve decision-making related to employee's continuing education and satisfaction. This simulation, using a bivariate ordered probit model, shows that economic and indirect economic incentives play a key role in augmenting the level of both satisfaction and learning. This analysis also considers how efficacy of learning programs may thus improve by linking the human resources development policy with results obtained in e-learning courses.

People are an essential source on the competitiveness of organizations. Numerous authors have recognized the importance of organizational commitment and job satisfaction as key attitudes related with work and influencing aspects like rotation, absenteeism or productivity. However, a clearly delimited posture on the relationship between both constructs does not exist. It is because of this that the present work theoretically analyses several models regarding this matter. Subsequently a proposed model is estimated

It seems appropriate and coherent to start this study with a detailed description of the basic analysis unit in the human resource management model that constitutes the study object of this chapter: competency. With this aim in mind, the first section in this chapter is dedicated to three tasks the authors consider basic and introductory: (a) defining the concept of competency through the different approaches made by some of the most outstanding authors in this field; (b) dissecting that concept in its various elements or components so that it can be better interpreted; and finally (c) presenting a number of classifications thanks to which the treatment of certain competencies can be prioritised.

Web 2.0 is changing the rules, reducing costs of access to technology, increasing the volume of cyber social contact through networking, both social and professional, making possible that any information is just a click away. This evolution causes the birth of new requirements, including the adaptation of people and, because organizations are made up of people, changing their map skills. This chapter presents a theoretical model for defining and managing a presence plan on the Web, emphasizing in those individual Competencies 2.0. It is defined from a Delphi method, involving experts in the use and development of Web 2.0. The strengths of this model are two: 1) It is a model that attempts to explain new tasks resulting from magazines not covered in sufficient depth from an academic and scientific perspective, though, they are been studied and discussed in forums 2.0; and 2) It explains the relationship between individual and organizational competencies 2.0, so that the first one can act on to modify the latter. On the other hand, this chapter has certain shortcomings, mainly two: 1) It has not been empirically validated, although research is underway to rectify this deficiency, and 2) There is no academic scientific knowledge enough, probably due to the novelty of this issue.

In this chapter, the authors list different points of view about entrepreneurship and entrepreneurs. For this, they take into account the relationship between an entrepreneur and his or her environment. Thus, the chapter identifies the main competences that characterize an entrepreneur distinguishing between innate and/or learned competencies.

Cross-cultural knowledge management in the e-environment results from interactions with others from different cultures that require new understanding and interpretations. The authors find that few studies

Foreword

When the authors told me that they were writing a book on human resources management in the new economic context, I recognized that I thought it was an excellent idea, since it is known that the best managerial strategy will not get any result if it is not sustained and supported by the human factor.

The way of managing human resources in a company is central for the efficient realization of its activities, what will affect directly in its results, and in the possibilities of survival and success in this new international context.

Let us not forget that the companies of our days compete in a complex, turbulent, and more and more dynamic environment; what becomes indispensable is that people (with independence of their level in the hierarchical scale and in any organization type) are able to face the challenges that are outlined in their work in the short and large term.

In this context, it becomes necessary that the management of the firm, and more concretely, the management of human resources, must facilitate the knowledge, skills, and abilities to develop the potentials of the organizational members (working by teams or individually), and as a result, bring a competitive advantage to the company. In this way, human resources management must be able to transmit commitment and to involve people through the appropriate communication channels, and must help to solve the problems that will lead to profits for their organization. In summary, it's a key element able to add value to the company, to the employees and, hopefully, to society.

Therefore, the objective of this book is to contribute to the development in the company of the necessary knowledge, skills, and abilities to confront the challenge of the new economy with success. The language of the chapters is accessible, simple, and with a clear focus to any person. This permits to not only reach with a theoretical content, but also a real practical focus on the different aspects approached in each one of the chapters. In this sense, I specially recommend the text for its applicability through a selection of the empiric examples of the chapters to highlight the topics analyzed.

Juan Llopis-Taverner
University of Alicante, Spain

Juan Llopis-Taverner *has been the Dean of the Faculty of Economics and Business at University of Alicante, Spain since 2006. He is a Full Professor of Business Organization and graduated in Business Studies from University of Alicante (with honours; best student of the promotion). His Ph D Degree is from University of Alicante as well, and he was selected as the best Faculty Thesis of year 1991. He has authored or co-authored of 10 books, 15 chapters in books, 100 papers published in national and international academic and professional journals, 75 papers in national and international academic and professional work-*

shops and conferences on Management and Business topics, is member of 40 private research projects and 10 public research projects, 6 International editorial advisory boards, and Referee in 15 international business management journals. He has been the visiting professor at Universities of UK, United Arab Emirates, USA, Germany, Lithuania, Belgium, and China, and has 23 years of experience in lectures at the university level. He also has professional consulting experience for private firms and public institutions and is member of 25 evaluation committees of PhD dissertations of different universities.

Preface

The current times reflect the beginnings of a paradigm shift to a global economy in which the presence of the Internet is nothing short of pervasive. Businesses all around the world face major challenges related to the progressive (and many times unavoidable) incorporation of Information Technologies into their processes, using different tools and solutions. However, many organizations do not react to these environmental exigencies suitably. This problem often results in policies, practices, and strategies that are outdated. Furthermore, the profile of the employees has changed lately. Businesses have to consider their modern knowledge workers a set of competences rather than merely "people." Today's competency models assist in developing abilities like adaptability, learning, and knowledge creation and sharing. Since these models consider both subjective and objective skills together, they offer a much wider and more complex vision of a company's personnel. Managing by competencies leads to excellence in work performance, and increases the inimitability and uniqueness of the organization's human resources, enhancing their value as source of sustainable advantage. Therefore, competency based E-human resource practices can become the leverage that many companies need to survive and succeed without resorting to hard, cost-cutting based measures.

This book is a reference for both practitioners and academics for designing and implementing e-Management and competency models in companies. The publication's purpose is to show future Human Resource practitioners how to boost the added value of people as a company asset by means of an adequate strategy. It presents the impacts of integrated e-human resource policies, and provides some recommendations for addressing the shift from traditional Human Resource policies to a new economy perspective.

OVERVIEW

Chapter 1, "*Toward a Unifying Framework for Defining Internal Human Resource Flexibility: A Proposal Based on the Resource-Based View Approach*" by Inmaculada Beltrán Martín, Universitat Jaume I, Spain, discusses how a flexible workforce is emerging as a critical success factor to counteract certain organizational rigidities and to guarantee organizational competitiveness in challenging environments. This chapter provides a review of the relevant definitions and classifications of human resource (HR) flexibility that have appeared during recent years. Furthermore, the chapter presents a definition of internal HR flexibility based on the resource-based view approach. From this perspective, HR flexibility is defined as a multidimensional concept. Specifically, this chapter assumes that employees are flexible when they show intrinsic flexibility (i.e. they can easily move between tasks and roles), modification flexibility (i.e. they alter their skills and/or behaviors to adapt to new circumstances), and relational flexibility (i.e. they participate in collaborative activities).

Chapter 2, "*Outsourcing the HR Function in the new Economy: Literature, Facts and Research Agenda*" by Mireia Valverde, Universitat Rovira i Virgili, Spain, Sergi Romeu, Universitat Rovira i Virgili, Spain, and José Luís Gascó, University of Alicante, Spain, presents solutions for organisations searching for more flexibility in their quest towards competitiveness. One of the avenues towards flexibility is the outsourcing of some business processes, whether it is to achieve cost cutting or acquire expertise from external organisations. The HR function or some of its components have also been subjected to outsourcing for some time. However, the decision of what to outsource, how, when, and to whom in HR is not exempt from risks. This chapter reviews the existing literature on outsourcing the HR function that has addressed these questions and generates and agenda for research in order to orient researchers' efforts to reach more conclusive evidence about the practice of HR outsourcing.

Chapter 3, "*Talent Management: A New Perspective in Human Resource Management*," by Roberto Luna-Arocas, University of Valencia, Spain, underlines the powerful perspective of talent management, integrating practices from organizations and providing a scientific approach. Some previous results from different research are expressed in relation to strategy and organizational performance. Talent Management is clearly a related concept to high performance work system, with the difference being a more strategic approach. More than practices are goals for the organization to achieve excellence at work. So Talent Management is clearly linked with a strategic approach to the organization in order to attract, develop, evaluate, and retain professionals with high capacities.

Chapter 4, "*Ethics and Learning Organizations in the new Economy*," by Alexis Bañón, Polytechnic University of Valencia, Spain, Manuel Guillén, University of Valencia, Spain, and Ignacio Gil, Polytechnic University of Valencia, Spain, studies ethics and learning in the context of the new economy. Business organizations need to learn faster, and to maintain and to improve knowledge, producing creative solutions based on their knowledge, skills, and new technologies to develop a customer responsive culture in a more economic and efficient way. In order to achieve this, CEOs and Human Resource (HR) policies should potentially contribute to knowledge development by creating authentic learning organizations. The authors propose in this study that learning improvements in organizations are not just a matter of techniques or aptitudes, but also a matter of feelings, attitudes, and, above all, of the moral habits of their members. The authors strongly suggest complementing currently established conceptions of knowledge management and organizational learning through an explicit inclusion of ethics and ethical learning in organizations.

Chapter 5, "*E-Recruitment: The Move towards a Virtually Organized Recruitment Process*," by Anna B. Holm, Aarhus University, Denmark, explores E-recruitment, also known as online or Web-based recruitment, which is rarely discussed in research from an organizational perspective. The aim of this chapter, therefore, is to analyze and discuss the process of e-recruitment, its key constituents, and organizing principles, based on the results of a qualitative study conducted in 2008-2009, together with data from industrial reports, articles from practitioner magazines, and in-depth interviews. The chapter provides a summary of e-recruitment properties and a composite matrix of the overall elements of e-recruitment organizing. E-recruitment is viewed as a case of virtual organizing, i.e. the organization of processes and activities, which, via technology and human agents, facilitate time- and space-independent interaction and collaboration. The chapter closes with a brief discussion of implications for HR managers and professionals and avenues for future research.

Chapter 6, "*The Importance of Psychological Contract in Human Resource Management within the new Global Economy*," by Inocencia Mª Martínez-León, Technical University of Cartagena, Spain, highlights HR's role in the psychological contract between a company and its employees, beyond the written contract, specifying their contributions, expectations, beliefs, promises, and obligations between both parties. Their management requires the definition of the concept of psychological contract, the analysis of its main characteristics and contents, and the identification of its stages of development. Electronic Human Resource Management (e-HRM) introduces a way of implementing HRM strategies through Web-based tools, improving the psychological contract management. Information phase, Intranet, and internal electronic mail have an important role in socialization stage. The consolidation of psychological contract (maintenance phase) is favoured by Intranet, business-to-employees, internal electronic mail, database, videoconference, and groupware. Finally, in breach phase, intranet, electronic mail, database, videoconference, and groupware are important E-HRM.

Chapter 7, "*Strategic and Organizational Considerations Related to an E-Learning Model: A Case of Study*," by Nuria Calvo, University of A Coruña, Spain, and Paolo Rungo, University of A Coruña, Spain, aims to assess the impact of different organizational factors on the success of e-learning programs, in terms of both self-reported satisfaction and the level of learning. Hence, this study adds to the analysis of the efficacy of e-learning models from an organizational perspective by providing some useful insights, which may help to improve decision-making related to employee's continuing education and satisfaction. This simulation, using a bivariate ordered probit model, shows that economic and indirect economic incentives play a key role in augmenting the level of both satisfaction and learning. This analysis also considers how efficacy of learning programs may thus improve by linking the human resources development policy with results obtained in e-learning courses.

Chapter 8, "*Organizational Commitment and Job Satisfaction: An Empirical Analysis of Their Relationship in Private Teachers*," by Sandra Mª Sánchez Cañizares, University of Córdoba, Spain, and Fernando J. Fuentes García, University of Córdoba, Spain, offers a model for understanding the effect of employee's perceptions on an organization. People are an essential source for the competitiveness of organizations. Numerous authors have recognized the importance of organizational commitment and job satisfaction as key attitudes related with work, influencing aspects like rotation, absenteeism, or productivity. However, a clearly delimited posture on the relationship between both constructs does not exist. It is because of this that the present work theoretically analyses several models regarding this matter. Subsequently, a proposed model is estimated and validated using the methodology of structural equations modeling (SEM). This model confirms the influence of job satisfaction over organizational commitment and its moderating effect on the influence on the latter of various perceptual factors of the employee related to his/her work.

The economic situation in the last few years has forced companies to adjust their workforces and reduce hierarchical levels. They have also had to undertake overhead cost and expense adjustment programmes, including salary freeze and the reduction of training budgets, amongst others; all of it is within an environment where human capital has acquired special relevance. It has been shown that individuals represent enterprises' main asset, the element that can generate sustainable competitive advantages over time. Power has changed hands; it no longer lies in the traditional means of production, but in knowledge. And knowledge resides in people. Knowledge and people are the new engines of economy. The management of individuals, which is nothing but the management of their intangibles and the need to attract and retain talent, becomes especially important in this context. Commitment management is one of the most effective policies to achieve this aim. Managers must implement policies to commit employees to not only retain them, but also encourage them to do their best. The best tool to achieve this necessary commitment is internal communication. Chapter 9, "*The Role of Internal Communication in Commitment Management*," José Antonio Carazo, Capital Humano, Spain, aims to offer a conceptual reflection on commitment management and analyses the state of internal communication in Spanish companies. The purpose is to establish a link between the emphasis placed by companies on achieving their employees' commitment and internal communication as a tool that can help to reach that goal.

Chapter 10, "*Staff Restructuring in the New Economy*," José Antonio Fernández Sánchez, University of Alicante, Spain, and Encarnación Manresa Marhuenda, University of Alicante, Spain, tackles an important HR concern. In times of recession, survival for most businesses means to do more with fewer resources. It is necessary to encourage policies that will help a company to become more efficient, like staff restructuring. This policy encompasses many decisions, from laying people off to relocating employees to other positions as a consequence of the deployment of a more rational structure, which in turn answers to a change in strategy caused by the turn in the environment. This chapter explains the concept of staff restructuring in detail and offers several methods to carry it out. Afterwards, the potential problems that may arise from implementing this strategy are discussed, and, finally, some clues on how to avoid such problems, or, at least, mitigate their consequences, are presented. It is also made clear that a short-term approach in this regard will lead to a future lack of competitiveness, and therefore, endanger long-term survival.

Chapter 11, "*The Competency-Based Human Resource Management Model*," by Jorge Valdés Conca and Susana de Juana-Espinosa, University of Alicante, Spain, starts with a detailed description of the basic analysis unit in the human resource management model that constitutes the study object of this chapter: competency. With this aim in mind, the first section in this chapter is dedicated to three tasks the authors consider basic and introductory: (a) defining the concept of competency through the different approaches made by some of the most outstanding authors in this field; (b) dissecting that concept in its various elements or components so that it can be better interpreted; and finally (c) presenting a number of classifications thanks to which the treatment of certain competencies can be prioritised.

Chapter 12, "*Competencies 2.0: A Theoretical Model for Defining and Managing a Presence Plan on the Web*," by Jaime Izquierdo-Pereira, NewMahwah - The Social Xperts Lab, Spain, Carmen Avilés-Palacios, Universidad Politécnica de Madrid, Spain, Joaquín García-Alfonso, Universidad Politécnica de Madrid, Spain, and Manuel López-Quero, Universidad Politécnica de Madrid, Spain, presents a theoretical model for defining and managing a presence plan on the Web, emphasizing in those individual Competencies 2.0. It is defined from a Delphi method, involving experts in the use and development of Web 2.0. The strengths of this model are two: 1) It is a model that attempts to explain new tasks resulting from magazines not covered in sufficient depth from an academic and scientific perspective, though they have been studied and discussed in forums 2.0; and 2) It explains the relationship between individual and organizational competencies 2.0, so that the first one can act on to modify the latter. On the other hand, this chapter has certain shortcomings, mainly two: 1) It has not been empirically validated, although research is underway to rectify this deficiency, and 2) There is not enough academic scientific knowledge, probably due to the novelty of this issue.

In Chapter 13, "*Competence-Based Profile to Characterize Successful Entrepreneurs*," by Lourdes Canós-Darós and Cristina Santandreu-Mascarell, Polytechnic University of Valencia, Spain, the authors list different points of view about entrepreneurship and entrepreneurs. For this, they take into account the relationship between an entrepreneur and his or her environment. Thus, the chapter identifies the main competences that characterize an entrepreneur distinguishing between innate and/or learned competencies.

Chapter 14, "*Cross-Cultural Competences in the New Economy*," by Irene Martin-Rubio, Universidad Politécnica de Madrid, Spain, and Drew Rodgers and Erik Døving, Oslo University College, Norway, explores cross-cultural knowledge management in the e-environment results from interactions with others from different cultures that require new understanding and interpretations. The authors find that few studies have been conducted to address the objective of identifying the competences that are needed to promote learning and information processing in the 21st century. The authors' primary goal is to identify competences that will promote learning and information processing in global organizations. The chapter outlines a taxonomy of competences necessary for cross-cultural learning in the global e-environment based on Kolb's Experiential Learning Theory.

Chapter 15, "*Leadership Talent Development in the New Economy*," by José Manuel de Haro, AGBAR Group, Spain and José Antonio Carrión, Opem Consultants, Spain analyzes how it is possible to achieve the best results in the task of developing leaders. For this purpose, they first present, as a starting point, a framework based on strategic principles for managerial development that have to be assumed before taking action and, after which, they propose the operational means that allow them to go beyond what has so far constituted the content of the organizational practices in this field. Intentions are as follows: reducing the level of complexity perceived about the concept itself and about its tools; including the most relevant managerial development contents; presenting an orderly, consistent process, which can guide future interventions in this field. At present, and even though the authors think that the contribution made in this chapter will help introduce significant improvements in managerial development practices, there are still quite a few pending issues that will gradually have to be solved in order to make progress.

Chapter 16, "*Management and Leadership of Innovative Work Teams*," by Santiago Gutiérrez, Castilla-La Mancha University, Spain, focuses on the analysis of two factors; the characteristics of new work teams and the management and leadership tools required for these innovative teams, in order to generate greater effectiveness. Firstly, the authors present the differences between a work team and a group and the basic characteristics of innovative work teams (multifunctional teams, self-managing teams, virtual teams, and open-innovation teams). Then, the authors focus on the management systems of these teams with the aim of generating greater effectiveness. High-performance practices and management skills are presented as tools for increasing motivation and commitment to the company and its business culture. Finally, the authors analyzed the leadership and the effect of leadership has an impact on the team's results as well as on the company's innovativeness, making it an essential part of creating effective innovative teams.

For Javier Pagán Castaño, Portsmouth University, UK, and Dolores Garzón Benítez, Polytechnic University of Valencia, Spain, the authors of Chapter 17, "*Listening and Leadership*," this era is both characterized by continually evolving and changing market conditions and the relevance of knowledge to adapt to the new environment. In this chapter, the authors focus on three ideas: the connection between the company results and the leader's listening skills; people as the firm's most valuable resource; and therefore, internal communication as the key for success and full potential of the company. However, the chapter concludes that internal communication is scarce and not always effective since one way communication seems to be the most common form of interaction, even though leaders know the importance and value of their human resources.

The main goal of Chapter 18, "*Analysis of Gender Equality in Higher Management Levels: A Study Model*," by Nuria Calvo, University of A Coruña, Spain, María Bastida, University of Santiago de Compostela, Spain, and Jacobo Feás, University of Santiago de Compostela, Spain, is to undertake a critical analysis of the current situation concerning the equal treatment of female managers in Spain. In this chapter, the authors analysed the dynamics of business behaviour in order to understand why inequality of women managers persists for gender reasons in spite of the anti-discrimination measures recommended by the legislative framework in place. This analysis has allowed proposals for measures to be drawn up to be taken into account in designing human resources strategies, based on systems of management by competencies and assessment of managerial performance.

Susana de Juana-Espinosa
University of Alicante, Spain

José Antonio Fernández-Sánchez
University of Alicante, Spain

Encarnación Manresa-Marhuenda
University of Alicante, Spain

Jorge Valdés-Conca
University of Alicante, Spain

Acknowledgment

The slow and often interrupted evolution of this book has been a collaborative experience involving the efforts and responses of many people. We want to express our gratitude to our friends, family and colleagues, whose support and goodwill kept us going through the accomplishment of this edition, only a proportion of which we have space to acknowledge here.

We have learnt a great deal from those who have worked with us over the years and gratefully acknowledge our debt to them, especially our department colleagues and members of our research group. We would like to express our thanks to our publisher, IGI Global, and particularly Mike Killian and Joel Gamon, for leading and supporting us through this project. Our deepest thanks go also to our contributors, reviewers and members of the editorial advisory board, without whom this book would have never seen the light.

This book is dedicated to our new bundles of joy, born in 2011: Anaïs, Mario, Nico and Encarni (respectively).

Susana de Juana-Espinosa
University of Alicante, Spain

José Antonio Fernández-Sánchez
University of Alicante, Spain

Encarnación Manresa-Marhuenda
University of Alicante, Spain

Jorge Valdés-Conca
University of Alicante, Spain

December, 2011

Section 1
Global Issues in HRM

Chapter 1
Toward a Unifying Framework for Defining Internal Human Resource Flexibility:
A Proposal Based on the Resource-Based View Approach

Inmaculada Beltrán Martín
Universitat Jaume I, Spain

1. ABSTRACT

A flexible workforce is emerging as a critical success factor to counteract certain organizational rigidities and to guarantee organizational competitiveness in challenging environments .This chapter provides a review of the relevant definitions and classifications of human resource (HR) flexibility that have appeared during recent years. Furthermore, the chapter presents a definition of internal HR flexibility based on the resource-based view approach. From this perspective, HR flexibility is defined as a multidimensional concept. Specifically, this chapter assumes that employees are flexible when they show intrinsic flexibility (i.e. they can easily move between tasks and roles), modification flexibility (i.e. they alter their skills and/or behaviors to adapt to new circumstances), and relational flexibility (i.e. they participate in collaborative activities).

2. INTRODUCTION[1]

Tendencies such as the increasing spread of market globalization, new technological developments, the reduction of product life cycles and aggressive competition, are generating high levels of environmental changes and uncertainty for organizations of all types (Volberda, 1996; Sanchez, 1997). These circumstances require rapid responses through adaptations of organizational attitudes and capabilities, which lead to innovative management approaches and organizational methods (Bueno, 1996: 262).

Traditional sources of competitive advantages are changing and it is imperative to deploy new strategies to successfully compete under changing

DOI: 10.4018/978-1-61350-207-5.ch001

external conditions. In this context, flexibility is emerging as a competitive weapon that allows organizations to counteract current market evolution and competitive levels (Ahmed et al., 1996; Volberda, 1996). Flexibility is a broad concept that can refer to operational issues such as manufacturing flexibility, or to strategic decisions such as alterations in the organization's product-market combinations. All these factors are associated with the organization's efforts to adjust available means to external challenges. Regardless of the specific response adopted by organizations, it is broadly believed that environmental dynamism forces managers to pay increasing attention to the management of the organization's social issues (Wright and Snell, 1998).

Nowadays, organizations require a new type and level of contribution from their workforce. In order to successfully compete under dynamic conditions, people's performance of a fixed set of prescribed tasks is no longer considered adequate. Instead, competitive advantage comes from employees who are engaged in broad open-ended and interdependent roles (Campbell, 2000; Parker, 2000). The recognition of the importance of employee flexibility for modern organizations has led to a body of research on labor flexibility or human resource flexibility[2]. Broadly speaking, HR flexibility refers to the possibility of varying the quality and quantity of personnel to suit changes in the environment (Gouswaard et al., 2001; Peiró et al., 2002). Interest in these questions arose in the early 1980s and has expanded during recent decades, as can be seen from the development of large-scale studies (e.g. Brewster et al.'s, 1994 study of labor flexibility strategies in Europe) or from the appearance of academic journal special issues on human resource flexibility[3].

Due to the growing interest in people as a source of flexibility, it is worth analyzing whether HR flexibility has the same meaning for everyone. Literature on HR flexibility has generated a large number of definitions of this term and is sometimes difficult to have a full picture of these contributions. The first purpose of this chapter is to provide a review of the relevant definitions and classifications of HR flexibility that have appeared during recent years. In this vein, researchers have emphasized two distinct strategies of HR flexibility: external and internal HR flexibility. On the one hand, external HR flexibility refers to alterations in the volume of labor employed, and it coincides with the use of "flexible employment contracts" (Storey et al., 2002) such as short-term temporary workers, temporary help agency or part-time employees (Kalleberg, 2001). On the other hand, internal HR flexibility refers to the flexibility manifested by the pool of human resources in the organization at a certain point of time. From this point of view, higher flexibility can be attained when managers develop and encourage employees to adopt permeable and expandable work roles (Tsui et al., 1997).

Of these two strategies, recent studies have recommended to increasingly rely on internal HR flexibility strategies, which are assumed to result in mutual gains for both the organizations and their employees (Valverde et al., 2000: 650; Kalleberg, 2001: 482). The relevance of internal HR flexibility in modern organizations demands a clear definition of the concept. In this regard, some recent studies have attempted to advance the conceptualization of internal HR flexibility, by delimiting employee features that may contribute to develop organizational responses to external challenges (e.g. Pulakos et al., 2000, Shafer et al., 2001). However, a review of these studies shows a lack of congruence among the dimensions and variables that the previous literature has considered as components of this concept, mainly for two reasons: first, because existing studies belong to a variety of different research fields, and second, because some of these studies lack a clear theoretical basis to guide the definition of the internal HR flexibility components (Breu et al., 2001). Therefore, it is important to address the study of internal HR flexibility with a well-delimited theoretical approach that provides a structured logic for the conceptualization of this concept.

In line with several scholars (Dyer and Shafer, 1999; Chadwick and Cappelli, 1999, 2002), I believe that the Resource-Based View (RBV) provides interesting insights for developing a definition of internal HR flexibility that complements prior conceptualizations of the term. The RBV, critical for the development of Strategic Human Resource Management (SHRM) research, seems appropriate to address the conceptualization of the internal side of HR flexibility due to its organization-focused orientation that emphasizes the contribution of the internal resources of the firm to the organization's success and the consideration of human resources as one of the key resources of a firm (Wright et al., 1994). In particular, some authors have taken the RBV approach to identify a set of criteria that define flexible resources (Sanchez, 1995, 1997), which provides a systematic framework to delimit the features that define a flexible workforce. Thus, the second purpose of our research is to apply these criteria to the analysis of internal HR flexibility, thus providing a definition of this concept from the RBV premises.

3. BACKGROUND

The purpose of this section is to provide an overview of the HR flexibility concept, given the relevance of employees' flexibility for modern organizations and the lack of agreement in the literature regarding its meaning. In doing so, I first introduce different modes of conceptualizing HR flexibility starting with the model of the flexible firm and following with the dimensional interpretation of HR flexibility, based on the distinction between internal and external HR flexibility. Following this, in the second section I propose a general framework for the definition of HR flexibility based on the RBV premises and in the third section, I apply this framework to the definition of HR flexibility.

3.1. Modes of Conceptualizing Human Resource Flexibility

HR flexibility emerges as a key aspect in organizations operating in uncertain environments and whose efforts are oriented towards the development of rapid and diverse responses. It is believed that neglecting social issues in favor of technological and cost variables can lead to the failure of flexibility initiatives in organizations (Upton, 1995; Karuppan, 2004). The *Model of the Flexible Firm*, formulated by Atkinson (1984), Atkinson and Gregory (1986) and Atkinson and Meager (1986) is the most well known framework on HR flexibility and one of the first studies to deal with the multidimensionality of HR flexibility. Other recent studies coincide with the idea that human resource flexibility is a multidimensional concept and complement the Atkinson et al. framework.

3.1.1. The Model of the Flexible Firm

One of the most popular models of HR flexibility is the *Model of the Flexible Firm*, developed at the Institute of Manpower Studies (Atkinson, 1984; Atkinson and Gregory, 1986; Atkinson and Meager, 1986). According to this model, different groups of employees provide the organization with different sources of flexibility, as depicted in Figure 1.

First, there is a core group of employees ("primary job market") who are full-time, manifest high levels of commitment towards the organization and possess a broad pool of skills. Employees within this group are usually managers, designers or technicians and due to their attributes, they can easily move between functions and roles, accept-

Figure 1. The model of the flexible firm

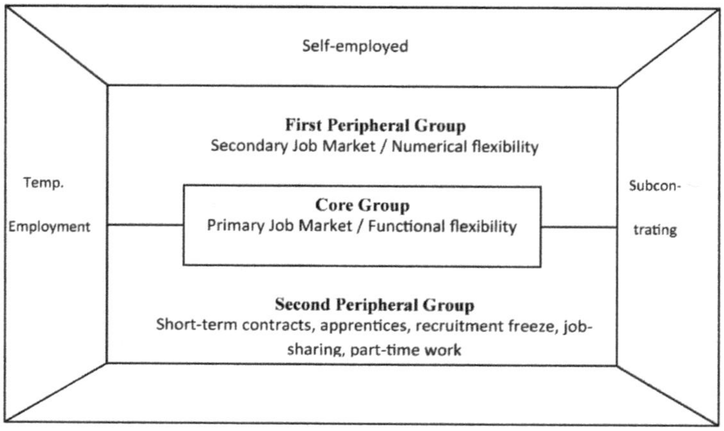

ing responsibilities that go beyond their job descriptions (Valverde et al., 2000: 651). In other words, they manifest "functional flexibility".

Second, organizations can also obtain flexibility from peripheral groups of employees, by altering the size of the workforce according to organizational needs (i.e. "numerical flexibility"). In this regard, Atkinson distinguishes between a first and a second peripheral group of employees. The former group – secondary job market – is made up of full-time employees who do not perform activities others than the duties specified in their job descriptions. The second peripheral group comprises fixed-term or part-time employees, among others. Organizations can reduce or increase the number of peripheral employees without incurring high costs because these employees do not belong to the central activities of the organization (Mayne et al., 1996).

Despite the popularization of the *Model of the Flexible Firm* in studies on HR flexibility, several authors concur that the work of Atkinson and his colleagues suffers from a lack of solid theoretical underpinnings (Pinch et al., 1991; Blyton and Morris, 1992; Kalleberg, 2001). Kalleberg (2001) considers that this model does not provide a clear definition of what constitutes the "core" and the "peripheral" workforce. Kalleberg (2001) also points out that the *Model of the Flexible Firm* assumes that core and peripheral employees belong to completely separate parts of the organization, thus ignoring the potential interrelationships and collaborative processes that can emerge between core and peripheral groups.

These weaknesses do not prevent Atkinson's model from representing an important advance in disentangling the meaning of HR flexibility. The distinction between functional and numerical flexibility has been adopted by a number of studies in the literature since the publication of the model (e.g. Gooderham and Nordhaug, 1997; Creagh and Brewster, 1998; Friedrich et al., 1998) and it constituted a critical step in the multidimensional interpretation of HR flexibility. In line with this framework, a further set of studies has contributed to the multidimensional definition of human resource flexibility. In the next section, I provide a more detailed review of the dimensions of HR flexibility.

3.1.2. A Multidimensional Interpretation of HR Flexibility

The *Model of the Flexible Firm*, together with other studies within the SHRM field, introduced classifications of HR flexibility types based on two interrelated dimensions: a) scope of the analysis and b) nature of variations in the workforce.

In terms of the *scope of the analysis*, HR flexibility can be interpreted as the flexibility that can be attained from the external labor market, beyond organizational boundaries and current workforce (external HR flexibility) or as the flexibility manifested by the pool of human resources in the organization at a certain point of time (internal HR flexibility) (Looise et al., 1998; Arulampalam and Booth, 1998; Kalleberg, 2001). The *nature of the variations in the workforce* can involve qualitative modifications in the tasks performed by the workforce or quantitative modifications in the number of employees or in the number of hours worked (Figure 2).

External HR Flexibility

External HR flexibility focuses on the set of potential employees available to the organization and refers to the improvements in organizational flexibility as a result of alteration in the volume of labor employed. It usually coincides with the use of the external labor market when the organization needs it (e.g. with temporary employees). External HR flexibility has been the subject of numerous studies, which associate HR flexibility with atypical working practices, broadly called flexible employment contracts (Storey et al., 2002) or flexible working patterns (Volberda, 1998). This type of flexibility resembles the "quasi spot contract" in Tsui et al.'s (1997) model, which rests on a pure economic exchange model and attempts to create a market-like flexibility so that the employer is free to hire and fire employees. According to this model, the employer offers short-term, purely economic inducements in exchange for well-specified contributions by the employee.

Much of the literature considers that external HR flexibility is related to *quantitative* modifications in the amount of work available to organizations, providing the organization with a higher or lower number of employees according to its needs (e.g. with seasonal employment to cover "peaks" of work demands) (Peiró et al., 2002). It corresponds with the notion of numerical flexibility introduced in Atkinson's model. Numerical flexibility[4] entails flexibility "on the extensive margin" (Beatson, 1995; Michie and Sheehan-Quinn, 2001) and refers to the ability to adjust the volume of work to changes in environmental forces.

From this quantitative interpretation, organizations can attain higher numerical flexibility through a set of non-standard employment relations or atypical working patterns, whose aims are to limit either the duration of the employment relation or the number of hours worked (Mayne et al., 1996). First, the organization has greater freedom to adjust the volume of work to changing circumstances when contracts have a predetermined duration. This can be achieved through

Figure 2. A multidimensional view of HR flexibility

	External	Internal
Quantitative	Modifications in the number of employees. Contract flexibility (e.g. short-term contracts)	Modifications in the number of hours worked by current employees. Time flexibility (e.g.
Qualitative	Modifications in the content of work carried out by external employees. Contract flexibility (e.g.	Modifications in the content of work of the existing workforce. Functional flexibility (e.g. job rotation)

temporary workers hired for finite periods, or through temporary help agency employees. These employees may be highly skilled (e.g. consultants, independent professionals) or low skilled (e.g. food service) (Kalleberg, 2001). Second, organizations can alter the number of hours worked in the organization by recruiting part-time workers.

Qualitative external HR flexibility only appears in specific cases, such as when the organization uses independent consultants that assume responsibility for more than one task in the organization (Looise et al., 1998).

Internal HR Flexibility

When referring to *internal HR flexibility*, attention centers on the current workforce of the organization and it is defined as the extent to which existing employees contribute to the organization's flexibility. Several authors have interpreted HR flexibility from this point of view (Table 1), such as Molleman and Slomp (1999), who associate labor flexibility with the responsiveness of a fixed team of employees (referred to as the "labor system" in their study). Molleman and Slomp's (1999) study is formulated under the constraint of the "autonomy of the labor system". This refers to the extent to which the organization can deal with external challenges through its current set of employees, without employing outer-system capacity (i.e., without recruiting new employees). Similarly, Chow (1998) and Karuppan (2004) adopt an internal vision of HR flexibility, as both scholars propose that flexibility depends on employees' adjustments in the development of their tasks according to the organization's needs (Table 1).

Internal flexibility is similar to the "mutual investment" approach in Tsui et al.'s (1997) framework[5]. According to these authors, higher flexibility can be attained when managers develop and encourage employees to adopt permeable and expandable work roles. Managers adopting this approach care about employee well-being and professional careers and, in exchange, the employee's obligations and contributions include working on job assignments that fall outside prior agreements or expertise. Employees managed through this approach easily accept job transfers when requested by the employer to do so.

The meaning of internal HR flexibility can also be examined by analyzing its nature. To this end, I include the concepts of quantitative and qualitative HR flexibility in the analysis of internal HR flexibility (Figure 2).

On the one hand, *quantitative* internal HR flexibility is related to "time flexibility" (Peiró et al., 2002), and implies that the workforce of the organization is willing to enhance the number of hours worked (e.g. overtime work or variable working time).

On the other hand, *qualitative* internal HR flexibility entails modifications in the content of work in accordance with employee qualifications, and is usually associated with the ease with which

Table 1. Definitions of internal human resource flexibility

Study	Definition
Molleman and Slomp (1999)	Responsiveness of a *labor system* (fixed team of employees) to variation in the supply and demand of labor, without resorting to outer-system capacity
Chow (1998)	Organizational abilities to constantly motivate employees who possess multiple and firm-specific skills so as to adjust their tasks to the needs of the organization
Karuppan (2004)	Job enlargement, i.e. the number of tasks or operations performed, the number of machines operated, intradivisional jobs that a worker can perform, and intradivisional workstations at which an operator can work

employees move between tasks, for instance through job rotation or multi-tasking (Looise et al., 1998). This is the most commonly adopted definition of HR internal flexibility, and it corresponds with the idea of functional flexibility in the *Model of the Flexible Firm*. That is, qualitative internal HR flexibility refers to the adaptability of the internal workforce to face non-routine circumstances and events that demand creativity and initiative (Huang and Cullen, 2001: 34). It is usually associated with the employee's ability to undertake a range of tasks and/or employ a variety of skills (Blyton and Morris, 1992). For this reason, functional flexibility has also been termed "flexibility on the intensive margin" (Michie and Sheehan-Quinn, 2001).

Functional flexibility involves the processes of increasing employees' range of skills in such a way that they acquire the capacity to work across traditional occupational boundaries (Cordery, 1989; Cordery et al., 1993). In this regard, functional flexibility is linked to the human capital of the organization (i.e. employee collective skills). This strategy allows organizations to quickly respond to changes in products, production methods and technology since it enables the firm to allocate employees where and when needed easily and without extra costs (Atkinson and Meager, 1986). Consequently, the operational concept that underlies functional flexibility is *labor substitutability* or the extent to which one employee in a job can be substituted by another. This substitutability can be complete, when employees carry out other jobs as required (e.g. a waiter is transferred for the whole day from the restaurant to the bar), or incomplete, which comprise "boundary loosening approaches" oriented towards the combination of certain aspects of different jobs (e.g. the restaurant hostess helps in the bar in the early evening before she gets busy or goes into the kitchen) (Riley and Lockwood, 1997).

The conceptualizations of HR flexibility presented in this section on the whole adopted a practice-oriented perspective. That is to say, previous studies on HR flexibility included a set of recommendations for managers that strived to attain high levels of flexibility in their organizations. While this *managerial perspective* of HR flexibility offers helpful recommendations for daily practice in organizations pursuing flexibility, from a theoretical perspective they are to some extent incomplete. According to Looise et al. (1998), the analysis of HR flexibility has ignored relevant frameworks in the human resource management field. For this reason, the connection between the HR flexibility field and SHRM research has been weak, thus preventing researchers from analyzing either the antecedents of HR flexibility (including the role of HRM strategies) or the consequences that HR flexibility strategies have for employees (e.g. satisfaction, commitment, etc.) and organizational outcomes. The model of the flexible firm formulated by Atkinson and the subsequent studies on the concept of HR flexibility can be considered as suitable starting points for developing a more comprehensive definition of HR flexibility. I coincide with Looise et al. (1998) in considering that greater efforts should be made to define HR flexibility from a more solid theoretical framework. In the following sections, I focus on the conceptualization of HR flexibility based on the RBV.

3.2. A Framework of HR Flexibility from a RBV Approach

The RBV approach is a prominent framework in the SHRM literature. The RBV, with its emphasis on the internal workings of organizations, conceives employees as sources of sustainable competitive advantage, and it has enriched theoretical and empirical research on the influence of human resource-related issues on organizational performance (Lado and Wilson, 1994; Wright et al., 1994; Boxall, 1996).

In line with several scholars (Dyer and Shafer, 1999; Chadwick and Cappelli, 1999, 2002), I believe that the RBV provides interesting insights for developing a definition of HR flexibility that complements prior conceptualizations of the term. Others scholars, such as Wright and Snell (1998), Lepak et al. (2003), Bhattacharya et al. (2005) and Beltrán et al. (2009) have adopted the RBV perspective to analyze HR flexibility.

In this section I present a general framework that can serve as a guide to adapt the RBV to the definition of HR flexibility and to identify the features that characterize a flexible workforce. In doing so, I try to clarify two questions that should be taken into account when applying the RBV premises in the field of HR flexibility. First, I discuss the internal focus that the RBV posits on the study of HR flexibility and that differentiates it from previous definitions of the term. Secondly, I introduce the concept of flexible resources from a RBV approach.

3.2.1. The Internal Nature of HR Flexibility: A Review of Prior Studies

The RBV has been proved to be a valid framework in the field of SHRM, and it provides a theoretical basis for moving knowledge about HR flexibility forward. The RBV has an organization-focused orientation, as it considers that sources of competitive advantage come from the internal resources of the firm. Resources are stocks of available factors that are owned or controlled by the organization (Peteraf, 1993). Thus, the application of the RBV in the conceptualization of HR flexibility means that the level of analysis lies in *internal human resource flexibility*. By focusing on internal HR flexibility, a longer-term perspective on organizational competitiveness is adopted. As Murphy (1999) states, organizations that rely too heavily on contingent workers (i.e. external HR flexibility) run the risk of depleting the pool of available high-quality workers and are likely to fail to develop the core workforce they need. Although short-term gains might be realized by replacing permanent employees with contingent workers, the long-run implication of an over-reliance on contingent workers can be fatal to organizations.

Thus, from the perspective of the RBV, it is important to examine which features of the organization's current workforce are relevant in achieving higher levels of organizational flexibility. Specifically, the conceptualization of human resources from the RBV highlights the consideration of employees as "cognitive and emotional beings". Thus, from this theoretical model, the definition of HR flexibility should be based on the employee attributes that can best contribute to generating flexibility in organizations. By focusing on the skills, attitudes and behaviors of employees, the RBV assumes that individual members are the important resource in organizations, rather than the practices and procedures used by the firm (e.g. use of temporary work contracts) (Wright et al., 1994). Consequently, a further benefit of the RBV as the guiding framework is that it examines HR flexibility from the point of view of the variables that may be influenced by organizational human resource decisions and practices (Wright et al., 1994; Dyer and Shafer, 1999). Defining the dimensions of HR flexibility in terms of employee characteristics is important in order to make advances in the activities that best promote them. For example, it is important to know what kind of skills are necessary to promote flexibility in order to design training activities that encourage them or staffing practices that select employees with these skills (Pulakos et al., 2000: 614).

Some recent studies have attempted to advance the conceptualization of internal HR flexibility by delimiting a set of employee attributes (skills, attitudes, behaviors, etc.) that best contribute to the generation of flexibility in organizations. For example, Wright et al. (1994), by adopting a logical-incremental perspective (Quinn, 1980), suggest three ways by which employees can contribute to the organization's flexibility. According to these authors, to be flexible employees should manifest a proclivity to detect external changes and appropriate responses, together with the abilities to implement the required activities.

Pulakos et al. (2000) develop a taxonomy of "adaptive job performance" from an organizational behavior point of view. These authors, through a review of previous literature on adaptability at the individual, team and organizational levels[6], offer a conceptualization of flexible-related aspects of employee performance that differentiates between eight dimensions (handling emergencies or crisis situations, handling work stress, solving problems creatively, dealing with uncertain work situations, etc.).

Breu et al. (2001) review several conceptualizations of HR flexibility and introduce an operationalization of workforce agility based on twelve attributes, such as responsiveness to external change, speed of skill development, knowledge sharing, virtual team working, implementation of collaborative technologies, etc.

Shafer et al. (2001) carried out an exploratory study in a healthcare organization that allowed them to extract a set of agile attributes, in terms of people's flexible competencies and behaviors. Dyer and Shafer (2002), following the conclusions of Shafer et al.'s (2001) exploratory study and focusing exclusively on employee behaviors, suggested their own conception of HR flexibility, which (as the same authors affirm, pp.15) is a *tentative list* of requisite flexible attributes. These authors group flexible behaviors into three categories: 1) proactivity (includes initiative and improvisation), 2) adaptiveness (includes assumption of multiple roles, rapid redeployments and spontaneous collaboration), and 3) generativeness (referring to willingness to learn and share knowledge and information). Similarly, Kara et al. (2002) carried out an exhaustive review of previous literature on flexibility and proposed four human resource flexibility indicators, namely, keenness to improve skills, functional flexibility, work group and willingness to change.

In Table 2, I offer a summary of the abovementioned flexible variables. In elaborating this table, I have attempted to establish correspondences between the concepts considered by these studies in order to identify common themes and differences among researches.

A review of these studies shows that some limitations still exist regarding the conceptualization and operationalization of internal HR flexibility. On the one hand, as regards the conceptualization of this term, there is still much debate over what constitutes internal HR flexibility. A review of previous studies reveals that definitions of this concept are dispersed across various literatures (Breu et al., 2001: 23), such as strategic management (Wright et al., 1994; Shafer et al., 2001; Bhattacharya et al., 2005), quality management (Verdú, 2002), psychology (Pulakos et al., 2000), information technologies (Breu et al., 2001) or manufacturing flexibility (Kara et al.., 2002), which makes it difficult to draw unequivocal conclusions about which aspects to include as components of internal HR flexibility. In addition, some of these studies employ an exploratory methodology (Breu et al., 2001; Shafer et al., 2001; Kara et al., 2002) to elaborate tentative lists of flexible employee characteristics, rather than relying on a theory that delimits the dimensions of the concept and their interrelationships. Finally, studies differ in their focus when defining these dimensions, with some studies emphasizing employee attributes and characteristics (Wright et al., 1994; Pulakos et al., 2000; Breu et al., 2001; Shafer et al., 2001; Kara et al., 2002; Bhattacharya et al., 2005) and others combining this employee-focused definition with organizational activities associated with the improvement of internal HR flexibility, such investments in training (Volberda, 1998; Verdú, 2002).

For these reasons, there is no agreement on the dimensions that integrate the internal HR flexibility concept, as can be seen in Table 2. For example, while employee willingness to share information and knowledge is considered a di-

Table 2. A review of internal human resource flexibility characteristics

Wright et al. (1994)	Pulakos et al. (2000)	Breu et al. (2001)	Shafer et al. (2001)	Dyer and Shafer (2002)	Kara et al. (2002)
Detection of environmental changes			Capacity to pursue new business opportunities and threats	Initiative to spot threats and opportunities in the marketplace	
Capabilities to identify strategies in response to the detected changes					
Implementation of strategies according to environmental changes		Responsiveness to external change	Take action to minimize potential effects of threats		
		Benchmarks for skill assessment			
	Learning work tasks, technologies, and procedures	Speed of skill development		Abilities to learn	Craft (increasing labor skills)
		Speed of adaptation to new work environments		Assumption of multiple roles Rapid redeployments	Functional flexibility
		Speed of information access			
		Speed of information systems change			
		Use of mobile technologies			
		Workplace independence			
	Demonstrating interpersonal adaptability	Collaborative technologies		Spontaneous collaboration	Group work
		Virtual team working			
		Knowledge sharing		Knowledge and information sharing	
		Employee empowerment			
			Commitment		
				Improvisation	
			Accountable		
			Efficacy		
					Willingness to change
	Handling emergencies or crisis situations				
	Handling work stress				
	Solving problems creatively		Focused		

continued on following page

Table 1. Continued

Wright et al. (1994)	Pulakos et al. (2000)	Breu et al. (2001)	Shafer et al. (2001)	Dyer and Shafer (2002)	Kara et al. (2002)
	Dealing with uncertain and unpredictable work situations		Comfortable with ambiguity		
	Demonstrating cultural adaptability				
	Demonstrating physically oriented adaptability				
			Business-driven		

mension of internal HR flexibility by Breu et al. (2001), this question is ignored by other scholars. A similar discrepancy appears with regards to employee keenness to change or organizational commitment, among others. The lack of consensus over the internal HR flexibility concept presents a hurdle to the understanding of the meaning of this term as well as to the analysis of its antecedents and consequences in the organization (Breu et al., 2001). Therefore, a clear theoretical perspective must be introduced to define this concept that will help to integrate previous approaches and clarify the meaning of the construct (Looise et al., 1998).

I seek to bridge these gaps by applying the RBV premises in order to offer a systematic theoretical analysis of internal HR flexibility. In doing so, I apply the features of flexible resources to the definition of internal HR flexibility to extract the dimensions that integrate this concept from the RBV point of view. The next section provides an overview of the features of flexible resources from the RBV that will guide the rest of the chapter.

3.2.2. Resource Flexibility Dimensions

The RBV has formed an integrating ground for most of the work in SHRM in the past decade. As Wright et al. (2001: 703) affirm, "the RBV has become by far, the theory most often used within SHRM, both in the development of theory and the rational for empirical research". The RBV provides a rich and integrated approach that helps focus and organize research efforts in the strategic human resource management field and enables the practice of this field to become a truly strategic discipline (Ulrich, 1997; Colbert, 2004). The RBV has provided a rationale for how a firm's human resources could constitute a potential source of sustainable competitive advantage, but its premises have been also considered central to address flexibility issues (Sanchez, 1995, 1997). The novelty and speed of external challenges that characterize dynamic environments recommend the analysis of flexibility from the point of view of the firm's strategic choices and actions that are appropriated to those challenges, with researchers considering flexibility as an organizational property (Evans, 1991; Volberda, 1998). Firms improve flexibility and therefore, their chances for being successful in an uncertain environment, by having access to flexible resources that allow them to create a range of strategic options (Sanchez, 1997; Wright and Snell, 1998). Therefore, the RBV has been applied to study the flexibility of different resources -flexible production technologies, information systems, etc.- and how they provide the organization with strategic alternatives to counteract external pressures. (Parthasarthy and Sethi, 1993; Garud and Kotha, 1994; MacDuffie, 1995). If we consider that people are critical resources in today's organizations, the RBV provides as well an interesting framework to address the study of their flexibility

(Wright and Snell, 1998; Dyer and Shafer, 1999; Bhattacharya et al., 2005).

Similarly to the role that value, rareness, durability and inimitability play in the generation of sustainable competitive advantages (Barney, 1991), the RBV suggests a set of criteria that define flexible resources, which can be applied to the analysis of HR flexibility. The RBV approach considers that flexible resources, when managed through appropriate organizational activities and in accordance with information about external events, provide the organization with a "strategic option bundle" (i.e., with access to various viable paths for both exploration and exploitation of the environment) (Helfat and Raubitschek, 2000; Mathews, 2002; Sanchez, 2004).

According to the RBV, organizational resources are flexible when they are suppliers of *novel services*. This is possible when resources 1) can be used in several ways or 2) can be easily transformed so as to be valid under novel circumstances (Penrose, 1959; Galunic and Rodan, 1998; Sanchez, 2004). Drawing from these studies, it is possible to extract the dimensions that define resource flexibility. The first question (use of resources in various ways) corresponds to the intrinsic flexibility of resources. Concerning the second question (transformation of resources), the organization can obtain new resources from existing ones so long as they are easily modifiable or combinable with other resources.

Therefore, from the RBV approach, resource flexibility is a multidimensional concept, whose components are: a) *intrinsic flexibility*, that is, resource applicability to different situations, b) *modification flexibility* or the extent to which a resource can be easily transformed (with low costs and time) in order to be used in new circumstances, and c) *relational flexibility*, which facilitates the combinability of one resource with others.

Resource Intrinsic Flexibility

The intrinsic flexibility of a resource refers to its applicability in multiple situations. Intrinsic flexibility corresponds to resource versatility, which refers to the number of alternative uses to which the resource can be applied without being modified. Examples of versatile (or multiple-use) resources are: cash, general-purpose machines, mail order and internet-based marketing channels, information technologies, flexible production systems, a group of engineers with capabilities valid for a range of products, etc. (Sanchez, 1995, 2004; Ghemawat and Del Sol, 1998). All these resources are valuable in situations characterized by high resource obsolescence, rapid technological changes or uncertainty about future technology, among others factors (Wernerfelt and Karnani, 1987; Sanchez, 1997).

In contrast, non-versatile resources are called specialized or specific resources (Sanchez, 1997; Ghemawat and Del Sol, 1998) and are only highly efficient in limited uses. This kind of resource constrains the options available for the organization to increase or decrease production volume or to change from one product to another. For instance, Friedli et al. (2004), when exposing the obstacles to becoming flexible faced by a manufacturing organization (*Fashion Inc.*), detected that specificity of implicit and explicit knowledge required for the smooth working of production sites imposed constraints on the relocation of such facilities. As a result, when the need to expand the product range arose, costs of switching production facilities forced the firm to outsource production or to adjust its internal structure.

The relevance of resource intrinsic flexibility derives from the fact that the firm can apply these resources to a variety of situations (e.g. Sanchez, 1995, 1997). When the organization detects specific environmental contingencies that require rapid responses and when it activates the appropriate resource allocation routines (Burgelman, 1994; Malone and Crowston, 1994), intrinsic flexibility guarantees that internal resources are applied in different situations according to organizational needs, and at the right moment.

In summary, resources that manifest high intrinsic flexibility guarantee their application by the organization to a wide range of alternative uses. Consequently, the firm can face new situations with its existing resources, without the need to acquire, develop or modify them. In other words, the organization has a great number of paths or degrees of freedom to adjust its strategy (Evans, 1991; Teece et al., 1997).

Resource Modification Flexibility

Apart from their general applicability, "resources also differ in their flexibility to be incrementally improved" (Sanchez, 2004: 526). Thus, the second dimension of resource flexibility corresponds to modification flexibility. This refers to resource malleability, or the amenability of resources to change and their capacity to accommodate multiple transformations (Johnson et al., 2003). Modification flexibility involves changes in the nature of the resource that make it valid for application to new uses (i.e. resource extendability), without incurring high costs that offset the benefits obtained from the new resource (Sanchez, 2004).

For instance, machines equipped with electronic controls are highly malleable, since the organization can easily add new functionalities by, for example, adding a fax card to the computer. Other examples of malleable resources are certain organizational techniques, such as management by objective systems, whose evaluation criteria can be easily adapted to each employee and position within the organization (Wright and Snell, 1998). Nonetheless, the degree of malleability of a resource is dependent upon its nature, and is higher in knowledge-based resources (Galunic and Rodan, 1998). Thus, employee know-how, organizational learning or commitment are highly malleable resources that are easily modified and improved with suitable organizational activities.

The possession of malleable resources enables the organization to easily transform (with low costs and time) available resources in order to guarantee their appropriateness in novel situations. Thus, the impact of modification flexibility on organizational responses is similar to the benefits provided by intrinsic flexibility, but in this case the focus is on the longer term, and organizations need to identify potential improvements in current resources that can make them suitable for different uses. In this regard, firms should analyze internal knowledge, extend existing resources with the accumulated organizational knowledge and leverage knowledge into new applications (Mendelson, 2000; Pavlou and El Sawy, 2004).

To continue with the example used previously, when analyzing the transformation process of *Fashion Inc.* to become a more flexible organization, Friedli et al. (2004) observe that the success of this process was greatly due to the transformation of internal production facilities. Specifically, this organization divided its production sites into three sub-units, each of them providing flexible responses to the specific demands of a group of customers.

Resource Relational Flexibility

The final dimension in my analysis of resource flexibility is relational flexibility, which facilitates the emergence of a new combination of resources. Combinable resources can easily be reconfigured, that is to say, linked to each other so that they can jointly achieve a broader objective (Galunic and Rodan, 1998). While modification flexibility implies resource transformations, relational flexibility only requires altering linkages between resources. These new combinations do not necessarily involve changes in the nature of the resources (Mathews, 2002).

Resource relational flexibility has received attention from a variety of fields. In the innovation literature, Henderson and Clark (1990: 12) state that the essence of architectural innovation is the reconfiguration of an established system to link together existing components in a new way. In the field of production management, modular product designs constitute an alternative to traditional design systems, which can increase a range of product model variations, the creation of new families of derived product models and the development of new products through the substitution of upgraded components of previous products (Sanchez, 1995). Resource combinability (or "relatedness") has also received attention in studies analyzing relations between industries, organizations or organizational segments (Fan and Lang, 2000). For instance, in territorial agglomerations of organizations, firms share some knowledge-based resources that are combined with each organization's specific knowledge. This can result in the generation of new resources that represent a source of flexibility. On the other hand, cooperation agreements favor resource combinability and enhance the likelihood of a wider pool of knowledge being generated. In such cases, intangible resources are more easily combined with others such as organizational cultures, information and communication technologies, managerial processes or customer base.

In summary, the range of strategic options that organizations can develop in order to respond to environmental changes is dependent upon the availability of flexible resources. Resource flexibility is a multidimensional term, including intrinsic, modification and relational flexibility (Table 3).

The above resource characteristics play an important role in organizations operating in changing environments, allowing them to develop appropriate responses in light of external challenges. These dimensions have appeared in studies on the flexibility of different resources (e.g. flexible production technologies, information systems, etc.), but their application in the field of HR flexibility is still scarce. I believe that the reasoning employed by the RBV to study the features of flexible resources can also be used to advance the conceptualization of HR flexibility.

In conclusion, the definition of HR flexibility from the point of view of the RBV should take into account two questions:

Table 3. Summary of resource flexibility dimensions

Dimension	Definition	Examples
Intrinsic flexibility	Extent to which resources are versatile, i.e. they can be employed in a number of alternative uses without high costs and time	• Cash • General-purpose machines • Mail-order channels • Information technologies • Polyvalent employees
Modification flexibility	Extent to which characteristics of resources can be easily altered (malleable resources)	• Machines equipped with electronic controls • Employee know-how • Organizational learning • Employee commitment • Organizational reputation • Organizational culture
Relational flexibility	Extent to which resources can be easily combined, i.e. linked to each other so that they can jointly contribute to common goals	• Modular production systems • Organizational culture • Managerial processes • Customer base • Information technologies

1. HR flexibility is an internal characteristic of the organization, and it is defined in terms of employee features.
2. The flexibility of organizational resources can be assessed by attending to three dimensions, namely intrinsic, modification and relational flexibility.

Since human resources are one of the most important resources in current organizations, they can be considered flexible when they fulfill the criteria of flexible resources. To date, practically no study has clearly delimited these requirements (Peiró et al., 2002). I aim to fill this gap in next section by defining the characteristics of human resources that meet the requisites of flexible resources.

3.3. Definition of the HR Flexibility Dimensions from the RBV

In this section, I apply the concepts of internal, modification and relational flexibility to the analysis of HR flexibility in order to provide a more systematic definition of HR flexibility from the employee perspective. Following the premises of the RBV approach, I argue that HR flexibility is a multidimensional construct, as depicted in Figure 3.

3.3.1. Human Resource Intrinsic Flexibility

The first dimension dealt with in my discussion of resource flexibility in the previous section was the adaptability of resources to different uses; in other words, their versatility or intrinsic flexibility. HR intrinsic flexibility implies that employees are able to work on different tasks and under diverse circumstances and that the costs and time needed to mobilize employees into new duties or jobs are low (van den Berg and van der Velde, 2005). This kind of employee usually possesses a wide repertoire of competences[7] (MacDuffie, 1995; Riley and Lockwood, 1997) or a set of general abilities such as leadership skills, problem-solving abilities, etc. that allow them to carry out a high number of tasks easily and without high costs.

The conceptualization of HR intrinsic flexibility under the RBV is similar to that of "functional flexibility" in the *Model of the Flexible Firm* (Atkinson, 1984; Atkinson and Gregory, 1986; Atkinson and Meager, 1986), as it refers to changes in the nature of work that allow the organization to respond to novel contingencies. Human resource intrinsic flexibility is manifest in the organization either through employees accepting responsibilities with a similar level of complexity to those in their everyday job (i.e. cross-functional movements), or when they perform duties belonging to different organizational levels such as administrative or leadership activities (i.e. vertical movements). For instance, in research by Rosenblatt

Figure 3. Dimensions of the HR flexibility concept

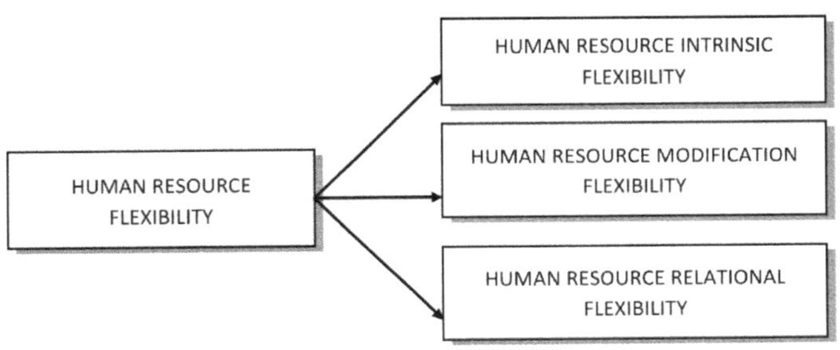

and Inbal (1999) among schoolteachers, employee intrinsic flexibility involves teaching more than one subject and performing school roles other than their lessons, such as implementation of databases. Furthermore, as discussed above, the labor substitutability implicit in functional flexibility can be complete (when employees move between jobs) or incomplete (when they assume certain tasks belonging to other jobs).

In summary, organizations with a versatile workforce can easily assign employees to the appropriate tasks, in accordance with their knowledge of environmental challenges and demands and their consequences for organizational processes (Tüselmann, 1996; Sparrow, 1998; Storey et al., 2002). This type of employee provides the organization with the freedom to modify its strategies, thus reducing the need for the alternative forms of extended labor supply, such as casual or short-term labor (Riley and Lockwood, 1997: 418).

3.3.2. Human Resource Modification Flexibility

The second dimension considered in the analysis of resource flexibility from the RBV was resource malleability. A resource is considered malleable when its characteristics can be modified without difficulty. In terms of HR flexibility, and following the same reasoning, employee malleability indicates the extent to which the workforce's characteristics are easily modifiable. Since the RBV conceives human resources as the pool of skills possessed by employees and the behaviors they exhibit, malleability should be analyzed in terms of both skills and behaviors. That is, employees will manifest modification flexibility when they innovate their skills or modify their behaviors ahead of need (Boudreau and Ramstad, 1997; Wright and Snell, 1998; Breu et al., 2001). While HR intrinsic flexibility corresponds to immediate employee features (i.e. people perform tasks in accordance with their possibilities), modification flexibility entails changes in individual traits. In order to advance the conceptualization of HR modification flexibility, I propose the differentiation between "skill malleability" and "behavioral malleability".

Skill malleability refers to how easily and quickly employees obtain the abilities needed to carry out new tasks (Maurer et al., 2003). In this sense, employees manifest skill malleability when they train and retrain as necessary, anticipate future skill requirements, demonstrate enthusiasm for learning new approaches to conduct their jobs and perceive every event in the organization as an opportunity to learn something significant for the future (Arulampalam and Booth, 1998; Pulakos et al., 2000; Dyer and Ericksen, 2005). For instance, in environments characterized by rapid technological innovations, employees must continuously learn new ways to perform their job and try to keep their skills as up-to-date as possible in order to make valuable contributions to the organization's goals. According to Ilgen and Pulakos (1999), employee abilities or willingness to engage in continuous learning and actually learn effectively should be a formal aspect of performance definition and assessment.

Indicators of skill malleability should reflect the processes of development planning, learning, and applying new knowledge and skills to changing organizational conditions. People who are committed to continuous learning search for new information about themselves and about emerging performance requirements that suggest a learning gap. Moreover, they are willing to devote the time, energy, and financial resources to gain the education needed to close the gap and improve their performance.

The concept and definition of *behavioral malleability* has received relatively little research attention in the human resource management field. However, useful contributions stemming from the organizational behavior literature can help researchers to understand this term[8].

As the starting point for my reasoning, I adopt the analysis of employee behavior in organizations provided by Wright and Snell (1998). These authors conceive employee behaviors as *scripts,* which are sequences of routines adopted by employees in their jobs. Behaviors are considered rigid (or non-malleable) when employees who have applied a particular script in the repetitive handling of situations, select the same script to deal with a novel situation. In contrast, if employees look for new sequences of actions with which to perform their daily activities or to face new circumstances, their behaviors are malleable. Thus, behavioral malleability refers to adjustments in the routines usually followed by employees. An employee with malleable behaviors adapts his or her responses to new circumstances, based on improvisation and not on fixed patterns of action.

Several concepts in the organizational behavior literature bear a similarity to the idea of behavioral malleability, such as proactivity, improvisation, initiative or role innovation, among others. These terms broadly refer to change-oriented employees who are willing to face novel problems through innovative solutions (Dyer and Ericksen, 2005). In my opinion, one of the concepts that best represents the notion of behavioral malleability is that of *taking charge* (Morrison and Phelps, 1999), close to the idea of individuals' *initiative* at work (Frese et al., 1996). These concepts refer to employee efforts to alter the way they perform their duties in the context of their jobs, work units or organizations.

In their study of employee initiative, Frese et al. (1996) used an interesting example that may clarify the meaning of behavioral malleability. These authors introduced the example of an accountant who does not receive the information required to calculate the salaries of the organization's employees on time. This accountant decides to work overtime in order to finish the salary calculation in time to pay the staff salaries. Furthermore, the accountant takes steps to ensure that this situation does not happen again in the future, by introducing changes in his or her job. In this case, for example, he or she may decide to install new computer software that captures the required information immediately. This example shows that one of the features of behavioral malleability is employee ability to manage changing conditions by providing immediate solutions to job problems. The decisions adopted by this accountant (i.e. working overtime and introducing new software) involve alterations to his or her work routines. Employees show behavioral malleability when they take decisions to avoid further mistakes or introduce changes in the way daily activities are performed.

In short, HR modification flexibility indicates the ease with which employees demonstrate change from previous states (Ilgen and Pulakos, 1999). Two concepts make up this second dimension of HR flexibility. First, skill malleability indicates improvements in employees' abilities and knowledge, and is linked to the keenness of employees to continuously master their skills and to avoid specialization in only one area of knowledge. Second, behavioral malleability refers to changes in the way of acting when faced with novel situations.

3.3.3. Human Resource Relational Flexibility

The final dimension in the analysis of flexible resources considers that available resources provide the organization with a higher number of strategic options when they can be easily combined. In terms of HR flexibility, this latter dimension indicates how easy it is to coordinate individuals to work together, and resembles the idea of *collective mind* or *social capital*[9] (Weick and Roberts, 1993; Youndt and Snell, 2004). According to Haiyang and Yan (2002) some employees have better team capabilities than others, as they can easily understand their role within the group and possess suitable abilities to engage in mutual interaction with other employees. When this happens, members of the organization or an organizational unit can easily be integrated into some form of higher-order systems or clusters of resources, such as in technological areas (e.g. printed circuit board assembly) (Teece et al. 1997; Galunic and Rodan, 1998), resulting in the combination of different sources of expertise. Thus, relational flexibility refers to employee mobilization into cooperative tasks (Forsythe, 1997; Breu et al., 2001).

In this regard, Shafer et al. (2001) demonstrated that flexible organizations follow the principle of "reliance on each other", based on the respect for the ideas and work of others, on expressing appreciation for other's contributions, and on giving support and cooperation, making each other stronger. This dimension of HR flexibility can also include cooperative activities with external employees, such as employee collaboration with customers or suppliers. The existence of HR relational flexibility is manifest in the organization through specific employee actions, which can be used to operationalize this concept. Employees with higher levels of relational flexibility may a) converge with their colleagues and share the same opinions, b) assist others when needed, and c) supplement the actions carried out by others (Weick and Roberts, 1993). Broadly speaking, these indicators of HR relational flexibility are linked to employees' organizational citizenship behaviors (Organ, 1988; Shore and Wayne, 1993) and include their efforts to spontaneously collaborate with and help colleagues, to communicate and to affiliate with others.

In summary, human resource relational flexibility includes collectivistic actions among employees in order to jointly work towards common goals. Collectivist employees emphasize the link between themselves and the group to which they belong and as such, they pursue the well-being of the whole group, even if this sometimes requires that their personal interests be disregarded (Van Dyne et al., 2000; Man and Lam, 2003).

3.3.4. Interrelationships between Human Resource Flexibility Dimensions

A review of the literature on RBV and HR flexibility provides several theoretical arguments that sustain the existence of interdependencies between the dimensions of HR flexibility. In this section, I suggest a broader interpretation of the HR flexibility concept, by discussing how HR intrinsic, modification and relational flexibility are interrelated.

For instance, some authors hold that versatile employees (i.e. manifesting intrinsic flexibility) are likely to be better prepared to engage in collaborative activities (relational flexibility) (Motowidlo and Schmit, 1999). Collaboration between employees entails an extension of employee job responsibilities, since team members need to understand others' responsibilities so as to effectively contribute to the group's goals. Since employees showing intrinsic flexibility can perform several activities beyond their job descriptions, they will probably also be prepared to engage in collaborative activities, which entail combination of tasks within the same unit (Motowidlo and Schmit, 1999). In this vein, scholars such as Horwitz and Townshend (1993: 919) or Molleman and Slomp (1999: 1838) assert that that the attainment of high and diverse qualifications by employees increases their performance in team activities. Similarly, Zammuto and O'Connor (1992: 708), when studying a specific type of compartmentalization (incorporation of manufacturing and staff functions associated with producing a product or component in the same unit) highlights the fact that employees in such work units needs to be multi-skilled because of the variety of tasks performed.

However, at the same time, employees manifesting higher relational flexibility may also show a greater tendency to alter their skills and behaviors (modification flexibility). The participation of employees in teams or groups and, in general, the collaboration with other members of the organization extends the employee vision of their own role in the organization. Employees manifesting higher relational flexibility develop a "team-oriented mindset" (LeBlanc and McInerney, 1994), oriented towards the attainment of common goals and the coordination of individual efforts. That is to say, employees with higher relational flexibility will be likely to assume as part of their jobs a broader set of activities and competencies than those that form part of their everyday job, leading to a flexible role orientation (Parker et al., 1997). As a result of this flexible role orientation, employees assume that in order to be efficient in these activities, they need to continuously develop their pool of competences, which favors the malleability of employee skills. Furthermore, when collaborating with others, behavioral scripts often change in order to face diverse and unpredictable situations, which are the result of the social interaction within the group.

Still other authors support the existence of interrelations between the two components of HR modification flexibility (skill malleability and behavioral malleability). According to Simon (1993), continuous learning processes associated with employee skill malleability demand the adoption of new cultural values and structural mechanisms, which eventually change the viewpoint and the role of the individual in the organization. Organizational learning values create an environment in which people widen the focus from the immediate outcomes of their performance to continuous learning by the organization as a whole. Hence, organizational learning values might foster extra-role behaviors, because by promoting strategic thinking, individuals develop an organizational system approach, which expands their perspectives beyond their formal role (Senge, 1990, 1993). This approach may thereby lead them to invest extra efforts in the organization as a whole by making innovative suggestions to improve the organization and volunteering for roles and tasks that are not obligatory, i.e., by showing higher behavioral malleability (Somech and Drach-Zahavy, 2004).

In conclusion, HR flexibility dimensions are interrelated and the interpretation of HR flexibility should be made from a wider point of view. Employees with a higher level of flexibility in one dimension are also likely to manifest better flexibility levels in other dimensions.

4. CONCLUSION

In today's organizations employee performance should no longer be assessed in absolute terms, but rather according to people's responsiveness to changing work demands. The rapid pace of change in job requirements places employees in a situation where they constantly need to demonstrate a capacity to engage in new learning and cope with change. Hesketh and Neal (1999), when referring to this broader conception of employee performance, introduce the term "adaptive performance", also adopted by Murphy and Jackson (1999:345) and Pulakos et al. (2000). Other authors such as Lau (1999) refer to this term as "flexible performance". Regardless of its designation, it is clear that more attention should be paid to the flexibility of the workforce, by considering not only the staff's efficacy levels attained in the daily tasks that their jobs comprise but also employee's ability to cope with change. Flexibility is thus a critical component of employee performance at work, and is separate from task performance.

The purpose of this chapter was to develop a conceptualization of human resource flexibility based on the RBV premises. In doing so, I followed the same logic used in the SHRM field to justify the competitive relevance of the organization's workforce. Research from the SHRM literature (e.g. Wright et al., 1994) assumes that human resources contribute to the organization's sustainable competitive advantages when they satisfy the requirements of strategic resources.

HR Flexibility as a Multidimensional Concept

I considered that human resources are flexible when they fulfill the features of flexible resources. Drawing from this reasoning, I concluded that HR flexibility is a multidimensional concept. Specifically, I assumed that employees are flexible when they show intrinsic flexibility (i.e. they can easily move between tasks and roles), modification flexibility (i.e. they alter their skills and/or behaviors to adapt to new circumstances), and relational flexibility (i.e. they participate in collaborative activities). Furthermore, I defended the existence of interrelations between these dimensions.

The proposed definition complements previous studies on the internal side of labor flexibility. On the one hand, by considering questions such as skill malleability or behavioral malleability, the RBV perspective introduces a proactive element into the HR flexibility concept, focused on the change-initiating role of employees that allows organizations to anticipate rather than simply respond to external changes. That is, from the RBV, labor flexibility depends not only on what employees can do now, but also on what they will be able to do in the future (for example, by improving their skills). This definition differs from initial conceptualizations of HR flexibility, which emphasized the relevance of functional flexibility (e.g. Molleman and Slomp, 1999; Karuppan, 2004) and therefore, of employees' abilities to move or reallocate once a triggering episode has occurred.

On the other hand, the RBV provides a systematic approach for the definition of HR flexibility that complements the absence of clear theoretical underpinnings in the field of labor flexibility, therefore allowing the dimensions of this concept and their interrelationships to be identified. The RBV pays greater attention to the underlying dimensions of HR flexibility, therefore presenting a framework for integrating previous conceptualizations of the term. For instance, questions such as learning of work tasks, technologies and procedures (Pulakos et al., 2000), benchmarks for skill assessment (Breu et al., 2001) or willingness to change (Kara et al., 2002) can be considered as integrants of a broader concept corresponding to modification labor flexibility.

Several studies have developed measurement scales based on this conceptualization of HR flexibility (Bhattacharya et al., 2005; Beltrán et al., 2009) which may be useful in empirical researchers on HR flexibility. Furthermore, these scales provide value to managers, given that these measurement instruments can be used in the HR audit processes aimed at evaluating the efficacy of the human resource function when promoting workforce adaptability. In addition, managers could use the HR flexibility scale to separate the highly flexible groups of employees from their less flexible colleagues, which can help to detect strengths and weaknesses according to the nature of the environmental challenges. This can provide valuable information for firms, enabling them to more easily identify where improvements may be needed in the workforce when novel circumstances emerge.

HR Flexibility in the Knowledge Society

Organizational learning and knowledge management are gaining increasing significance in the knowledge society. Knowledge Management (KM) helps to create learning organizations that are able to cope with the continuous changes in the environment. Defining KM is not easy, and multiple definitions and KM schools have provided diverse dimensions and meanings of this term. The notion of Organizational Knowledge Management Systems (OKMS) helps to integrate the wide-ranging perspectives surrounding KM. OKMS is the system that enhances organizational learning in the firm through facilitation of knowledge exchange and sharing (Yahya and Goh, 2002). The technical aspect of OKMS comprises an advanced assembly of software and hardware in the firm that supports knowledge and organizational learning through free access to and increased sharing of knowledge. However, authors such as Meso and Smith (2000) have proposed a socio-technical view of OKMS, by assuming that these systems are made up of a combination of technologies, organizational infrastructure, culture, knowledge and people. This perspective considers the core element of OKMS to be people, and in particular employees. While much of the early KM literature focused on technological issues, the importance of human and social factors has since gained recognition. Thus, if HRM refers to the management of people, and if people are the most valuable resource in KM, then HRM and KM are closely interrelated (Hislop, 2003; Svetlik and Stavrou-Costea, 2007). Consequently, HR flexibility can be also linked to KM.

On the one hand, external HR flexibility (i.e. short-term temporary workers, temporary help agency or part-time employees) has been posited to affect KM because employee turnover affects the balance and location of experience in the organization and, therefore, the organization's ability to learn (Carley, 1992). When a firm relies on external HR flexibility (i.e. when members of the organization leave and are replaced by new personnel) the firm loses the expertise and experience of the departing employee and gains the experience of the person joining the organization. Firms can benefit from external HR flexibility by obtaining gains from specialization in knowledge acquisition and storage, as long as the organization is able to coordinate the efforts of many specialists (e.g. through rules, routines, group problem solving, etc.) (Grant, 1996). However, external HR flexibility may imply certain risks to KM, because without proper mechanisms for transferring experience among old and new employees, knowledge may be lost. An empirical study by Carley (1992) demonstrated that organizations learn less and more slowly the higher their employee turnover rate, since portions of the organization's memory leave with the departing personnel. Other authors have posited that organizations with a high usage of temporary employees risk losing valuable knowledge (Alvesson, 2000), especially when explicit and tacit knowledge is created within the firm and is firm specific (Grant, 1996). Furthermore, the use of external HR flexibility based on short-term contracts makes it difficult to build stable relationships among employees and with the firm (Díaz-Mayans and Sánchez, 2004) and negatively affects employee involvement and commitment in the firm (Amuedo-Dorantes, 2001). According to the norm of reciprocity, if employees perceive that the firm is not concerned about their well-being or labor conditions, they will not make the effort to benefit the firm (Davis-Blake et al., 2003). This lack of commitment on the part of temporary employees will threaten the success of KM in the firm, given that for KM initiatives to be effective, employees must be *willing* to share their knowledge and expertise (Hislop, D., 2003).

On the other hand, internal HR flexibility (i.e. intrinsic, modification and relational flexibility) provides many advantages for the success of KM in the firm. Internal HR flexibility helps to enhance employees' organizational commitment. Committed employees are more likely to be highly motivated, and will probably be more willing to provide extra discretionary effort and share their knowledge within the organization (Hislop, 2003). Furthermore, Grant (1996) suggested that a necessary condition for the success of KM in the organization is the existence of common knowledge, that is, those elements of knowledge that are common to all organizational members or the intersection of their individual knowledge sets. Intrinsic flexibility facilitates the creation of this common knowledge, by assuring that individuals are able to invade one another's functional boundaries. Thus, intrinsic flexibility sustains the idea of redundancy posited by Nonaka and Takeuchi (1995) in that it builds links between individuals. Other benefits of internal HR flexibility for the success of KM stem from the fact that modification flexibility facilitates knowledge accumulation. In

Grant's (1996) framework, knowledge absorption depends upon the recipient's ability to add new knowledge to existing knowledge (Grant, 1996). Finally, relational flexibility assures the cooperative behavior required to increase the organization's ability to learn by increasing the ambiguity inherent in the organization (Carley, 1992).

REFERENCES

Ahmed, P., Hardaker, G., & Carpenter, M. (1996). Integrated flexibility: A key to competition in a turbulent environment. *Long Range Planning, 29*(4), 562–571. doi:10.1016/0024-6301(96)00048-9

Alvesson, M. (2000). Social identity in knowledge-intensive companies. *Journal of Management Studies, 37*(8), 1101–1123. doi:10.1111/1467-6486.00218

Amit, R., & Schoemaker, P. (1993). Strategic assets and organizational rent. *Strategic Management Journal, 14*, 33–46. doi:10.1002/smj.4250140105

Amuedo-Dorantes, C. (2001). From "temp-to-perm": Promoting permanent employment in Spain. *International Journal of Manpower, 22*, 625–647. doi:10.1108/EUM0000000006234

Arulampalam, W., & Booth, A. L. (1998). Training and labour market flexibility: Is there a trade-off? *British Journal of Industrial Relations, 36*(4), 521–536. doi:10.1111/1467-8543.00106

Atkinson, J. (1984). Manpower strategies for flexible organisations. *Personnel Management*, August, 28-31.

Atkinson, J., & Gregory, D. (1986). Is flexibility just a flash in the pan? *Personnel Management*, September, 26-29.

Atkinson, J., & Meager, N. (1986). *Changing working patterns. How companies achieve flexibility to meet new needs*. London, UK: National Economic Development Office (NEDO).

Barney, J. B. (1991). Firm resources and sustained competitive advantage. *Journal of Management, 17*(1), 99–120. doi:10.1177/014920639101700108

Beatson, M. (1995). *Labour market flexibility. Research Series, 48*. Sheffield: Employment Department.

Beltrán-Martín, I., Roca-Puig, V., Escrig-Tena, A., & Bou-Llusar, J. C. (2009). Internal labour flexibility from a resource-based view approach: Definition and proposal of a measurement scale. *International Journal of Human Resource Management, 20*(7), 1576–1598. doi:10.1080/09585190902985194

Bhattacharya, M., Gibson, D. E., & Doty, D. H. (2005). The effects of flexibility in employee skills, employee behaviors, and HR practices on firm performance. *Journal of Management, 31*, 622–640. doi:10.1177/0149206304272347

Blyton, P., & Morris, J. (1992). HRM and the limits of flexibility. In Blyton, B., & Turnbull, P. (Eds.), *Reassessing human resource management*. London, UK: Sage Publications.

Boudreau, J., & Ramstad, P. (1997). Measuring intellectual capital: Learning from financial history. *Human Resource Management, 36*(3), 343–356. doi:10.1002/(SICI)1099-050X(199723)36:3<343::AID-HRM6>3.0.CO;2-W

Boxall, P. F. (1996). The strategic HRM debate and the resource-based view of the firm. *Human Resource Management Journal, 6*(3), 59–75. doi:10.1111/j.1748-8583.1996.tb00412.x

Breu, K., Hemingway, C., & Strathern, M. (2001). Workforce agility: The new employee strategy for the knowledge economy. *Journal of Information Technology, 17*, 21–31. doi:10.1080/02683960110132070

Brewster, C., Gunnigle, P., & Morley, M. (1994). Continuity and change in European industrial relations: Evidence from a 14 country survey. *Personnel Review*, *23*(3), 4–20. doi:10.1108/00483489410064531

Bueno, E. (1996). *Organización de empresas. Estructura, procesos y modelos*. Madrid, Spain: Pirámide.

Burgelman, R. (1994). Fading memories: A process theory of strategic business exit in dynamic business environments. *Administrative Science Quarterly*, *39*, 24–56. doi:10.2307/2393493

Campbell, D. J. (2000). The proactive employee: Managing workplace initiative. *The Academy of Management Executive*, *14*(3), 52–66. doi:10.5465/AME.2000.4468066

Carley, K. (1992). Organizational learning and personnel turnover. *Organization Science*, *3*(1), 20–40. doi:10.1287/orsc.3.1.20

Chadwick, C., & Cappelli, P. (1999). Alternatives to generic strategy typologies in strategic human resource management. In Wright, P., Dyer, L., Boudreau, J., & Milkovich, G. (Eds.), *Research in personnel and human resources management, supplement 4*. Greenwich, CT: JAI Press, Inc.

Chadwick, C., & Cappelli, P. (2002). *Functional or numerical flexibility? Which pays off for organizations?* Discussion Paper, Institute for Labor and Industrial Relations, University of Illinois.

Chow, I. H. S. (1998). The impact of rules and regulations on workforce flexibility in Hong Kong. *International Journal of Human Resource Management*, *9*(3), 494–505. doi:10.1080/095851998341035

Colbert, B. A. (2004). The complex resource-based view: Implications for theory and practice in strategic human resource management. *Academy of Management Review*, *29*(3), 341–358.

Cordery, J. (1989). Multi-skilling: A discussion of proposed benefits of new approaches to labour flexibility within enterprises. *Personnel Review*, *18*(3), 13–22. doi:10.1108/EUM0000000000772

Cordery, J., Sevastos, P., Mueller, W., & Parker, S. (1993). Correlates of employee attitudes toward functional flexibility. *Human Relations*, *46*(6), 705–723. doi:10.1177/001872679304600602

Creagh, M., & Brewster, C. (1998). Identifying good practice in flexible working. *Employee Relations*, *20*(5), 490–503. doi:10.1108/01425459810238792

Davis-Blake, A., Broschak, J. P., & George, E. (2003). Happy together? How using non-standard workers affects exit, voice, and loyalty among standard employees. *Academy of Management Journal*, *46*, 475–485. doi:10.2307/30040639

Diaz-Mayans, M. A., & Sánchez, R. (2004). Temporary employment and technical efficiency in Spain. *International Journal of Manpower*, *25*, 181–194. doi:10.1108/01437720410535981

Dyer, L., & Ericksen, J. (2005). In pursuit of marketplace agility: Applying precepts of self-organizing systems to optimize human resource scalability. *Human Resource Management*, *44*(2), 183–188. doi:10.1002/hrm.20062

Dyer, L., & Shafer, R. A. (1999). From human resource strategy to organizational effectiveness: Lessons from research on organizational agility. In Wright, P., Dyer, L., Boudreau, J., & Milkovich, G. (Eds.), *Research in personnel and human resources management, supplement 4*. Greenwich, CT: JAI Press, Inc.

Dyer, L., & Shafer, R. A. (2002). Dynamic organizations: Achieving marketplace and organizational agility with people. In Peterson, R., & Mannix, E. (Eds.), *Understanding the dynamic organization*. Mahwah, NJ: Lawrence Erlbaum Associates.

Evans, J. S. (1991). Strategic flexibility for high technology manoeuvres: A conceptual framework. *Journal of Management Studies*, *28*(1), 69–89. doi:10.1111/j.1467-6486.1991.tb00271.x

Fan, J. P. H., & Lang, L. H. P. (2000). The measurement of relatedness: An application to corporate diversification. *The Journal of Business*, *73*, 629–660. doi:10.1086/209657

Forsythe, S. (1997). Human factors in agile manufacturing: A brief overview with emphasis on communication and information infrastructure. *Human Factors and Ergonomics in Manufacturing*, *7*(1), 3–10. doi:10.1002/(SICI)1520-6564(199724)7:1<3::AID-HFM1>3.0.CO;2-7

Frese, M., Kring, W., Soose, A., & Zempel, J. (1996). Personal initiative at work: Differences between East and West Germany. *Academy of Management Journal*, *39*(1), 37–63. doi:10.2307/256630

Friedli, T., Billiger, S., Kickuth, M., & Fleisch, E. (2004). Managing flexibility strategically: A case study on repositioning. *Proceedings of the Production and Operation Management Society, World Conference 2004,* Cancún (Mexico).

Friedrich, A., Kabst, R., Weber, W., & Rodehuth, M. (1998). Functional flexibility: Merely reacting or acting strategically. *Employee Relations*, *20*(5), 504–523. doi:10.1108/01425459810238800

Galunic, D., & Rodan, S. (1998). Resource recombinations in the firm: Knowledge structures and the potential for Schumpeterian innovation. *Strategic Management Journal*, *19*, 1193–1201. doi:10.1002/(SICI)1097-0266(1998120)19:12<1193::AID-SMJ5>3.0.CO;2-F

Garud, R., & Kotha, S. (1994). Using the brain as a metaphor to model flexible production systems. *Academy of Management Review*, *19*, 671–698.

Ghemawat, P., & Del Sol, P. (1998). Commitment versus flexibility. *California Management Review*, *40*(4), 26–42.

Gooderham, P. N., & Nordhaug, O. (1997). Flexibility in Norwegian and UK firms: Competitive pressure and institutional embeddedness. *Employee Relations*, *19*(6), 568–580. doi:10.1108/01425459710367916

Gouswaard, A., Kraan, K. O., & Dhondt, S. (2001). *Flexibility in balance: Flexibility of labour and its consequences for employers and employees.* TNO-Report, R990440.

Haiyang, L., & Yan, Z. (2002). *Founding team comprehension and behavioral integration: Evidence from new technology ventures in china.* Academy of Management Proceedings.

Helfat, C. E., & Raubitschek, R. S. (2000). Product sequencing: Co-evolution of knowledge, capabilities and products. *Strategic Management Journal*, *21*, 961–979. doi:10.1002/1097-0266(200010/11)21:10/11<961::AID-SMJ132>3.0.CO;2-E

Henderson, R. M., & Clark, K. B. (1990). Architectural innovation: The reconfiguration of existing product technologies and the failure of established firms. *Administrative Science Quarterly*, *35*, 9–30. doi:10.2307/2393549

Hesketh, B., & Neal, A. (1999). Technology and performance. In Ilgen, D., & Pulakos, E. (Eds.), *The changing nature of performance. Implications for staffing, motivation, and development* (pp. 21–55). San Francisco, CA: Jossey-Bass.

Hislop, D. (2003). Linking human resource management and knowledge management via commitment. A review and research agenda. *Employee Relations*, *25*(2), 182–202. doi:10.1108/01425450310456479

Horwitz, F. M., & Townshend, M. (1993). Elements in participation teamwork and flexibility in South Africa. *International Journal of Human Resource Management, 4*(4), 917–932. doi:10.1080/09585199300000064

Huang, H. J., & Cullen, J. B. (2001). Labour flexibility and related HRM practices: A study of large Taiwanese manufacturers. *Canadian Journal of Administrative Sciences, 18*(1), 33–39. doi:10.1111/j.1936-4490.2001.tb00242.x

Johnson, J. L., Lee, R. P. W., Saini, A., & Grohmann, B. (2003). Market-focused strategic flexibility: Conceptual advances and an integrative model. *Journal of the Academy of Marketing Science, 31*(1), 74–89. doi:10.1177/0092070302238603

Kalleberg, A. L. (2001). Organizing flexibility: The flexible firm in a new century. *British Journal of Industrial Relations, 39*(4), 479–504. doi:10.1111/1467-8543.00211

Kara, S., Kayis, B., & O'Kane, S. (2002). The role of human factors in flexibility management: A survey. *Human Factors and Ergonomics in Manufacturing, 12*(1), 75–119. doi:10.1002/hfm.10004

Karuppan, C. M. (2004). Strategies to foster labor flexibility. *International Journal of Productivity and Performance Management, 53*(6), 532–547. doi:10.1108/17410400410556192

Lado, A. A., & Wilson, M. C. (1994). Human resource systems and sustained competitive advantage: A competency-based perspective. *Academy of Management Review, 19*(4), 699–727.

Lau, R. S. M. (1999). Critical factors for achieving manufacturing flexibility. *International Journal of Operations & Production Management, 19*(3), 328–341. doi:10.1108/01443579910249741

LeBlanc, P., & McInerney, P. (1994). Need a change? Jump on the banding wagon. *The Personnel Journal, 73*(1), 72–76.

Lepak, D., Takeuchi, R., & Snell, S. A. (2003). Employment flexibility and firm performance: Examining the interaction effects of employment mode, environmental dynamism, and technological intensity. *Journal of Management, 29*(5), 681–703.

Looise, J. C., van Riemsdijk, M., & de Lange, F. (1998). Company labour flexibility strategies in The Netherlands: An institutional perspective. *Employee Relations, 20*(5), 461–482. doi:10.1108/01425459810238747

MacDuffie, J. (1995). Human resource bundles and manufacturing performance: Organizational logic and flexible production systems in the world auto industry. *Industrial & Labor Relations Review, 48*, 197–221. doi:10.2307/2524483

Malone, T. W., & Crowston, K. (1994). The interdisciplinary study of coordination. *ACM Computing Surveys, 26*(1). doi:10.1145/174666.174668

Man, D. C., & Lam, S. K. S. (2003). The effects of job complexity and autonomy on cohesiveness in collectivistic and individualistic work groups: A cross-cultural analysis. *Journal of Organizational Behavior, 24*, 979–1001. doi:10.1002/job.227

Mathews, J. A. (2002). A resource-based view of Schumpeterian economic dynamics. *Journal of Evolutionary Economics, 12*, 29–54. doi:10.1007/s00191-002-0106-z

Maurer, T. J., Wrenn, K. A., Pierce, H. R., Tross, S. A., & Collins, W. C. (2003). Beliefs about "improvability" of career-relevant skills: Relevance to job/task analysis, competency modelling, and learning orientation. *Journal of Organizational Behavior, 24*, 107–131. doi:10.1002/job.182

Mayne, L., Tregaskis, O., & Brewster, C. (1996). A comparative analysis of the link between flexibility and HRM strategy. *Employee Relations, 18*(3), 5–24. doi:10.1108/01425459610116447

Mendelson, H. (2000). Organizational architecture and success in the IT industry. *Management Science*, *46*, 513–529. doi:10.1287/mnsc.46.4.513.12060

Meso, P., & Smith, R. (2000). A resource-based view of organisational knowledge management systems. *Journal of Knowledge Management*, *4*(3), 224–234. doi:10.1108/13673270010350020

Michie, J., & Sheehan-Quinn, M. (2001). Labour market flexibility, human resource management and corporate performance. *British Journal of Management*, *12*, 287–306. doi:10.1111/1467-8551.00211

Molleman, E., & Slomp, J. (1999). Functional flexibility and team performance. *International Journal of Production Research*, *37*(8), 1837–1858. doi:10.1080/002075499191021

Morrison, E. W., & Phelps, C. C. (1999). Taking charge at work: Extra-role efforts to initiate workplace change. *Academy of Management Journal*, *42*(4), 403–419. doi:10.2307/257011

Motowidlo, S. J., & Schmit, M. J. (1999). Performance assessment in unique jobs. In Ilgen, D., & Pulakos, E. (Eds.), *The changing nature of performance. Implications for staffing, motivation, and development* (pp. 56–86). San Francisco, CA: Jossey-Bass.

Murphy, K. R. (1999). The challenge of staffing a post-industrial workplace. In Howard, A. (Ed.), *The changing nature of work* (pp. 295–324). San Francisco, CA: Jossey-Bass.

Murphy, P. E., & Jackson, S. E. (1999). Managing work role performance: Challenges for twenty-first century organizations and their employees. In Ilgen, D., & Pulakos, E. (Eds.), *The changing nature of performance. implications for staffing, motivation, and development*. San Francisco, CA: Jossey-Bass.

Nonaka, I., & Takeuchi, H. (1995). *The knowledge-creating company*. New York, NY: Oxford University Press.

Organ, D. W. (1988). *O.C.B: The good soldier syndrome*. Lexington, MA: Lexington Books.

Parker, S. (2000). From passive to proactive motivation: The importance of flexible role orientations and role breath self-efficacy. *Applied Psychology: An International Review*, *49*(3), 447–469. doi:10.1111/1464-0597.00025

Parker, S. K., Wall, T. D., & Jackson, P. R. (1997). "That's not my job": Developing flexible employee work orientations. *Academy of Management Journal*, *40*(4), 899–929. doi:10.2307/256952

Parthasarthy, R., & Sethi, S. P. (1993). Relating strategy and structure to flexible automation: A test of fit and performance implications. *Strategic Management Journal*, *14*, 529–549. doi:10.1002/smj.4250140704

Pavlou, P. A., & El Sawy, O. (2004). *From IT competence to competitive advantage in turbulent environments: A dynamic capabilities model*. Working Paper, University of California.

Peiró, J. M., García-Montalvo, J., & Gracia, F. (2002). How do young people cope with job flexibility? Demographic and psychological antecedents of the resistance to accept a job with non-preferred flexibility features. *Applied Psychology: An International Review*, *51*(1), 43–66. doi:10.1111/1464-0597.0078z

Penrose, E. T. (1959). *The theory of growth of the firm*. London, UK: Basil Blackwell.

Peteraf, M. A. (1993). The cornerstones of competitive advantage: A resource-based view. *Strategic Management Journal*, *14*, 179–191. doi:10.1002/smj.4250140303

Pinch, S., Mason, C., & Witt, S. (1991). Flexible employment strategies in British industry: Evidence from the UK Sunbelt. *Regional Studies*, *25*(3), 207–218. doi:10.1080/00343409112331346427

Pulakos, E. D., Arad, S., Donovan, M. A., & Plondon, K. E. (2000). Adaptability in the workplace: Development of a taxonomy of adaptive performance. *The Journal of Applied Psychology*, *85*, 612–624. doi:10.1037/0021-9010.85.4.612

Quinn, J. B. (1980). *Strategies for change: Logical incrementalism*. Homewood, IL: Richard D. Irwin.

Riley, M., & Lockwood, A. (1997). Strategies and measurement for workforce flexibility: An application of functional flexibility in a service setting. *International Journal of Operations & Production Management*, *17*(4), 413–419. doi:10.1108/01443579710159996

Rosenblatt, Z., & Inbal, B. (1999). Skill flexibility among school teachers: Operationalization and organizational implications. *Journal of Educational Administration*, *37*, 345–366. doi:10.1108/09578239910285589

Sanchez, R. (1995). Strategic flexibility in product competition. *Strategic Management Journal*, *16*, 135–159. doi:10.1002/smj.4250160921

Sanchez, R. (1997). Preparing for an uncertain future. Managing organizations for strategic flexibility. *International Studies of Management and Organizations*, *27*, 71–94.

Sanchez, R. (2004). Understanding competence-based management. Identifying and managing five modes of competence. *Journal of Business Research*, *57*, 518–532. doi:10.1016/S0148-2963(02)00318-1

Senge, P. M. (1990). *The fifth discipline: The art and practice of the learning organization*. New York, NY: Doubleday.

Senge, P. M. (1993). Transforming the practice of management. *Human Resource Development Quarterly*, *4*, 5–32. doi:10.1002/hrdq.3920040103

Shafer, R. A., Dyer, L., Kitty, J., Amos, J., & Ericksen, G. A. (2001). Crafting a human resource strategy to foster organizational agility: A case study. *Human Resource Management*, *40*(3), 197–211. doi:10.1002/hrm.1011

Shore, L. M., & Wayne, S. J. (1993). Commitment and employee behavior: Comparison of affective commitment and continuance commitment with perceived organizational support. *The Journal of Applied Psychology*, *78*, 774–780. doi:10.1037/0021-9010.78.5.774

Simon, H. A. (1993). Altruism and economics. *The American Economic Review*, *83*, 156–161.

Somech, A., & Drach-Zahavy, A. (2004). Exploring organizational citizenship behaviour from an organizational perspective: The relationships between organizational learning and organizational citizenship behaviour. *Journal of Occupational and Organizational Psychology*, *77*, 281–298. doi:10.1348/0963179041752709

Sparrow, P. (1998). The pursuit of multiple and parallel organizational flexibilities: Reconstituting jobs. *European Journal of Work and Organizational Psychology*, *7*(1), 79–95. doi:10.1080/135943298398970

Storey, J., Quintas, P., Taylor, P., & Fowle, W. (2002). Flexible employment contracts and their implications for product and process innovation. *International Journal of Human Resource Management*, *13*(1), 1–18. doi:10.1080/09585190110092758

Svetlik, I., & Stavrou-Costea, E. (2007). Connecting human resources management and knowledge management. *International Journal of Manpower*, *28*(3/4), 197–206. doi:10.1108/01437720710755209

Teece, D. J., Pisano, G., & Shuen, A. (1997). Dynamic capabilities and strategic management. *Strategic Management Journal*, *18*, 509–533. doi:10.1002/(SICI)1097-0266(199708)18:7<509::AID-SMJ882>3.0.CO;2-Z

Tsui, A. S., Pearce, J. L., Porter, L. W., & Tripoli, A. M. (1997). Alternative approaches to the employee-organization relationship: Does investment in employees pay off? *Academy of Management Journal*, *40*(5), 1089–1121. doi:10.2307/256928

Tüselmann, H. J. (1996). Progress towards greater labour flexibility in Germany. The impact of recent reforms. *Employee Relations, 18*(1), 50–67. doi:10.1108/01425459610110236

Ulrich, D. (1997). Measuring human resources: An overview of practices and a prescription for results. *Human Resource Management, 36*(3), 303–320. doi:10.1002/(SICI)1099-050X(199723)36:3<303::AID-HRM3>3.0.CO;2-#

Upton, D. M. (1995). Flexibility as process mobility: The management of plant capabilities for quick response manufacturing. *Journal of Operations Management, 12*(3-4), 205–224. doi:10.1016/0272-6963(95)00004-C

Valverde, M., Tregaskis, O., & Brewster, C. (2000). Labor flexibility and firm performance. *International Advances in Economic Research, 6*, 649–661. doi:10.1007/BF02295375

van den Berg, P. T., & van der Velde, M. E. G. (2005). Relationships of functional flexibility with individual and work factors. *Journal of Business and Psychology, 20*(1), 111–129. doi:10.1007/s10869-005-6994-9

Van Dyne, L., Vandewalle, D., Kostova, T., Latham, M. E., & Cummings, L. L. (2000). Collectivism, propensity to trust and self-esteem as predictors of organizational citizenship in a non-work setting. *Journal of Organizational Behavior, 21*(3), 3–23. doi:10.1002/(SICI)1099-1379(200002)21:1<3::AID-JOB47>3.0.CO;2-6

Verdú, A. J. (2002). *Relación entre flexibilidad y desempeño organizativo. Una aproximación desde la perspectiva de la gestión de la calidad total. Elche (Alicante)* (Hernández, U. M., Ed.).

Volberda, H. W. (1996). Toward the flexible form: How to remain vital in hypercompetitive environments. *Organization Science, 17*(4), 359–374. doi:10.1287/orsc.7.4.359

Volberda, H. W. (1998). *Building the flexible firm. How to remain competitive.* New York, NY: Oxford University Press.

Weick, K. E., & Roberts, K. H. (1993). Collective mind in organizations: Heedful interrelating on flight decks. *Administrative Science Quarterly, 38*, 357–381. doi:10.2307/2393372

Wernerfelt, B. (1984). A resource-based view of the firm. *Strategic Management Journal, 5*, 171–180. doi:10.1002/smj.4250050207

Wernerfelt, B., & Karnani, A. (1987). Competitive strategy under uncertainty. *Strategic Management Journal, 8*, 187–195. doi:10.1002/smj.4250080209

Wright, P. M., Dunford, B. B., & Snell, S. A. (2001). Human resources and the resource based view of the firm. *Journal of Management, 27*(6), 701–721. doi:10.1177/014920630102700607

Wright, P. M., McMahan, G. C., & McWilliams, A. (1994). Human resources and sustained competitive advantage: A resource-based perspective. *International Journal of Human Resource Management, 5*, 301–326. doi:10.1080/09585199400000020

Wright, P. M., & Snell, S. A. (1998). Toward a unifying framework for exploring fit and flexibility in strategic human resource management. *Academy of Management Review, 23*, 756–772.

Yahya, S., & Goh, W. K. (2002). Managing human resources toward achieving knowledge management. *Journal of Knowledge Management, 6*(5), 457–468. doi:10.1108/13673270210450414

Youndt, M. A., & Snell, S. A. (2004). Human resource configurations, intellectual capital, and organizational performance. *Journal of Managerial Issues, 16*(3), 337–360.

Zammuto, R. F., & O'Connor, E. J. (1992). Gaining advanced manufacturing technologies' benefits: The roles of organization design and culture. *Academy of Management Review, 17*(4), 701–728.

KEY TERMS AND DEFINITIONS

External HR Flexibility: Flexibility that can be attained from the external labor market, beyond organizational boundaries and current workforce.

HR Intrinsic Flexibility: Employees are able to work on different tasks and under diverse circumstances and that the costs and time needed to mobilize employees into new duties or jobs are low.

Internal HR Flexibility: Flexibility manifested by the pool of human resources in the organization at a certain point of time.

Modification Flexibility: Employees innovate their skills or modify their behaviors ahead of need.

Relational Flexibility: How easy it is to coordinate individuals to work together.

ENDNOTES

1 Work Supported by the Bancaixa Foundation (Grant ref. P1.1A2008-06) and by the Generalitat Valenciana (Ref. GVEPRE/2008/047).

2 Other authors (e.g. Arulampalam and Booth, 1998) use the term *labor market flexibility*.

3 See, for example, the *Canadian Journal of Administrative Sciences*, vol. 18, No.1, or the *European Journal of Work and Organizational Psychology*, vol.7, No.1.

4 As Kalleberg (2001) states, numerical flexibility has been referred to as "flexible staffing arrangements" (Houseman, 2001), "market-mediated work arrangements" (Abraham and Taylor, 1996), "contingent work" (Polivka and Nardone, 1989) and "non-standard" work arrangements (Kalleberg et al., 1997).

5 Tsui et al. (1997) in a further development of a previous study (Tsui et al., 1995) suggest four approaches for the employee-organisation relationship (i.e. employer's expectations of employee contributions). The first two approaches are associated with the creation of distinct types of flexibility for organisations.

6 The development of this taxonomy is based on the job performance model developed by Campbell et al. (1993), which treats alternative models of content and structure of job performance. According to Campbell et al.'s (1993) model, job performance is synonymous with behaviour and can be classified into the following components: 1) job-specific task proficiency, 2) non-job-specific-task proficiency, 3) written and oral communication, 4) demonstrating effort, 5) maintaining personal discipline, 6) maintaining peer performance, 7) supervision/leadership, 8) management/administration. The concept of adaptive performance is considered by Pulakos et al. (2000) as a performance component not included in the previous eight dimensions.

7 Human resources versatility has been given different terms, such as, "multi-skilling", "ability extension" and "polyvalence" (Cordery, 1989).

8 Crant (2000) offers a review of this literature.

9 Social capital is broadly defined as an asset that inheres in social relations at different levels (individuals, communities, networks, societies, etc.) (Coleman, 1988; Shaw et al., 2005). At the organizational level, Leana and Van Buren (1999) proposed a definition of *organizational social capital*, as a resource reflecting the character of social relations within the firm. It refers to individual's levels of collective goal orientation and is comprised of associability (i.e. employee willingness and abilities to engage in collective actions) and trust.

Chapter 2 Outsourcing the HR Function in the New Economy: Literature, Facts, and Research Agenda

Mireia Valverde
Universitat Rovira i Virgili, Spain

Sergi Romeu
Universitat Rovira i Virgili, Spain

José Luís Gascó
University of Alicante, Spain

ABSTRACT

In the current changing economic environment, organisations are searching for ever more flexibility in their quest towards competitiveness. One of the avenues towards flexibility is the outsourcing of some business processes, whether it is to achieve cost cutting or acquire expertise from external organisations. The HR function or some of its components have also been subjected to outsourcing for some time. However, the decision of what to outsource, how, when, and to whom in HR is not exempt from risks. This chapter reviews the existing literature on outsourcing the HR function that has addressed these questions and generates and agenda for research in order to orient researchers' efforts to reach more conclusive evidence about the practice of HR outsourcing.

INTRODUCTION

The new economy immerses businesses in an ever more competitive world. New and changing factors, such as the unstoppable development of new technologies, the de-regulation of markets, more volatile demand or the global "liquid" business environment (Bauman, 2007) brings organisations to find new ways of coping and surviving this environment. In this highly competitive, information overloaded environment, organisations find it difficult to be specialists in all of their processes, at the same time that they deal with the challenges of retaining talent (skills and commitment) in all their activities. In the quest to achieve efficiency and competitiveness, organisations may consider

DOI: 10.4018/978-1-61350-207-5.ch002

to externalise or outsource some of their processes as a means to optimise the use of their resources.

Outsourcing organisational activities is not a new phenomenon. However, some of the characteristics of the current business environment mentioned above, not only enable its growth but also allow for whole new sectors to develop under the premises of outsourcing, like in the case of externalising customer services in call centers. Furthermore, outsourcing is now not only more possible but also may make more sense in the new economy, where organisations are conceptualised more as networks than as self-contained units of operation. The benefits of outsourcing have been well reported in the literature. However, outsourcing is a choice not devoid of certain risks and flaws.

In this context, outsourcing Human Resource Management (HRM) activities (otherwise known as HRO – Human Resource Outsourcing) is a possibility that organisations need to consider. But, what activities should be outsourced (some, such as recruitment and selection, have been long carried out by external companies)? Should companies consider outsourcing HR activities on a one to one basis, or part away with whole blocks of the function? Should they do it on a once-off or ongoing basis? Should they outsource for specific purposes when the need arises or make it part of their business strategy? Is outsourcing of the HR function for every organisation? And more consequentially, what are the implications of outsourcing HRM activities for the internal HR departments and their so contested and intended strategic role in the organisation? How may it affect the HR department's competency and skill set?

To date, in order to answer these questions, the study of outsourcing has been mainly supply lead, that is, consultants tending to write on why outsourcing is good and providing data on the extent to which it is used (e.g. Rippin & Dawson, 2001) and it may be used (e.g. Syedain, 2009). However, few theoretical reflections and empirical contributions have been carried out in this regard.

Therefore, this chapter aims to review the literature on HRM outsourcing and put forward a research and practice agenda for this topic. The chapter draws from both practitioner and academic literatures, and it covers the questions that have been posed with regards to why companies should outsource some or many of their HR function, what HR activities are most commonly or are most likely to be outsourced, and how companies undertake and manage this decision.

The chapter is arranged along the following structure: First, we will deal with the concept of outsourcing as a general business process and its rationale. Then, we will move on to the definition of HRO (Human Resource Outsourcing). This will be followed by an analysis of the triggers that make companies take the decision to outsource some of their HR activities, and the related advantages that this decision may bring about. A parallel section deals with the problems and limitations to HRO. The following section deals with how much of the HR function gets outsourced, detailing the activities that are more likely to be outsourced, and making special emphasis on available international comparative data. Next, just before concluding, there is a key section of the chapter, devoted to identifying and outlining the research lines that future investigations should take in the topic of HRO.

OUTSOURCING BUSINESS PROCESSES, OUTSOURCING HRM

The concept of outsourcing has been very loosely defined in the literature, many times to the point that almost any form of procurement of goods and services by an organisation could be included in this concept (Gilley & Rasheed, 2000). In general, and from an intuitive point of view, however, it is widely understood that outsourcing consists of contracting out some activities that may otherwise be carried out by the company. In this sense, Gilley & Rasheed (2000) focus on this idea that in order to have outsourcing, companies must be able to choose to carry out those activities inside or outside the company. Otherwise, if the company could not undertake the outsourced activity, it would be a case of simple procurement. Therefore, the element of choice is key in the concept of outsourcing.

Over time, there has been an evolution in the use and concept of outsourcing. Duffy (2001) situates its origins in tax collection outsourcing in the Roman Empire, and points out that it regained impetus in eighteenth-century England, establishing itself in different forms and with diverse emphasis to date. In modern times, the concept of core business (Prahalad & Hamel, 1990), and the widely accepted prescription that non-core activities are those that should be considered for sourcing out (Mullin, 1996), has not only popularised, but also intensified the amount and diversity in which outsourcing takes place nowadays.

Within these developments, the focus of outsourcing activities has changed, originating a variety of outsourcing forms that have developed in an increased implication between client and provider companies. For example, in some cases, companies simply subcontract part of their activities. In subcontracting, a part of the business is carried out by an external company, but the process is done under the strict indications of the client, who determines what they want and how, with little room for the provider to specify conditions. Another materialisation of outsourcing is in the form of externalisation, where there is a higher level of implication and communication between client and provider companies, and whereby the provider is beginning to be responsible for the results of outsourced activities. In a further level of outsourcing involvement, the provider company brings added value to the client, and in this sense, it becomes a business partner as the client company can avail of their knowledge, experience and creativity (Altés, 2003). The currently in use development in outsourcing is the idea of BPO (Business Process Outsourcing), where outsourcing goes one step ahead by externalising most or all of the activities that make up a complete business process or function, as for example, when a company uses a call centre provider in order to manage all their customer services. Other whole functions can be subjected to business process outsourcing, such as procurement, finance and accounting, human resources, or supply chain (e.g. Accenture BPO services, 2010).

In sum, outsourcing can be considered "the contracted use and leverage of third-party resources, assets and skills, with guaranteed levels of quality, resilience and value to cost criteria and measurement, to deliver services previously provided in-house" (Bravard & Morgan, 2006).

The main reasons why companies decide to outsource some of their activities or full processes are:

- Cost control or reduction: given the constant pressure to lower prices, organisations are in turn hard pressed to find alternatives in their functioning in order to reduce their cost or, at least, to transform some structural or permanent costs into variable.
- Improvement in the time to deliver certain processes: particularly in the case of large organisations, some internal processes may become stagnant or bureaucratised, with difficulties to shed some inefficient dynamics. In these cases, outsourcing may be a way to make some processes more agile.
- The search for a higher degree of specialisation and functional quality, in order to be fully up to date in certain techniques, legislation or business processes. Indeed, outsourcing may be a form of obtaining expertise in areas where one individual organisation may not be able to invest sufficient internal resources. Thus, some companies outsource in order to access new technologies and skills.

- Apart from these, Kakabadse & Kakabadse's (2005) empirical research shows a number of related reasons why companies decide to outsource, such as the aim to achieve best practice, improve service quality, focus on core competencies, reduce the number of employees and/or capital costs and/or production costs, grow in-house expertise, enhance capability for change, and enhance the organisation's position in the value chain.

Finally, it is important to consider the fact that more companies may consider outsourcing some of their processes simply because it is now not only possible, but also convenient. Indeed, Click & Duening (2005) remark that a series of factors in the current business context that make outsourcing more useful than ever: the availability of broadband Internet (and the security systems associated with Internet transactions, a higher level of business specialisation and educational attainments of the workforce, inexpensive data storage, and the possibility of online analytical processing.

Having stated the concept and logic behind outsourcing business activities in general, we will now focus on the specific concept of outsourcing the human resource management function, normally referred to as HRO (Human Resource Outsourcing). On the same line as with other activities and functional areas, there has been little effort to date to agree on a single definition of HRO, with many authors implicitly assuming what HRO means (Woodall et al., 2009), or explicitly stating different degrees of inclusion of activities.

In this sense, we can observe a main division in defining HRO according to whether a specific outsourcing activity occurs on an ad-hoc or a continuous basis. Thus, the classic text by Mary Cook (1999) defines human resource management outsourcing as "having a third- party service provider or vendor furnish, on an ongoing basis, the administration of an HRM activity that would normally be performed in-house" (Cook, 1999, p. 4). This definition emphasises the fact that outsourcing involves a continuation of the outsourced service, as opposed to other authors that also include the more occasional outsourcing of specific tasks at specific moments in time, as in the case of Turnbull (2002), who considers that HRO consists of placing responsibility for various elements of the HR function with a third-party provider, regardless of the duration of this shift in the provider of the activity or service.

Other differences in definitions of human resource management outsourcing are based on the emphasis on the type of activities subject to being outsourced (for example, emphasising the fact that non-core activities should be the ones to consideration for outsourcing); or the differentiation between outsourcing single tasks as opposed to whole processes, in what has recently been presented as a progression towards business process outsourcing (Houtzagers & Janssen, 2008). In any case, the fact that few studies highlight where they place the limits on the concept of outsourcing HRM makes it difficult to carry out comparative analyses, as well as posing some questions to the validity of such studies. Thus, there is a need to reach either a consensus on the definition of HRO or at least to produce a classification of HRO activities and approaches.

WHY DO COMPANIES OUTSOURCE HR? PERCEIVED AND ACTUAL BENEFITS AND THE DECISION TO OUTSOURCE HR ACTIVITIES

The study of the reasons for outsourcing HR activities (e.g. Woodall et al., 2009) does not neatly fit into a single conceptual set of studies. Instead, it is intertwined in many occasions with studies about perceived benefits of outsourcing (and many of these, in turn, fail to clearly separate between benefits at pre- and post-outsourcing experience), as well as with the literature about the "make or buy decision" (e.g. Greer et al., 1999; Nuñez, 2009) which, again, has many points of clash and differentiation with that about the "triggers" to outsource (e.g. Alewell et al., 2009) and even with the studies that consider the determinants of outsourcing (e.g. Klaas et al., 2001) and that will be dealt with separately in this chapter.

In any case, there is an emergent body of knowledge about the general area of why companies outsource. Even though this body of knowledge is somehow disparate and presents some difficulties in discerning between scientific evidence and provider-led prescription, there are a limited number of aspects where there is consensus (Sisson & Storey, 2000). The reasons most traditionally acknowledged for outsourcing HR activities are:

1. **Cost reduction:** this is the main and most cited reason for outsourcing, whether for any business activity (Kakabadse & Kakabadse, 2002), or specifically for HRO (Lever, 1997); and whether from a practitioner (Accenture, 2008), empirical (Woodall et al., 2009), or theoretical (Klaas, 2008) point of view. By outsourcing the most transactional human resource practices, organisations expect to obtain cost advantages, as a result of achieving the same level of service as they would if they were to carry out those tasks internally.
2. **To obtain specialised services:** It may be difficult for any given HR department to be a specialist in all of its areas, particularly in small and medium size firms (Klaas et al., 2005). But even in larger organisations, the characteristics of the new economy, where knowledge evolves rapidly and legislative frameworks are affected by constant changes, keeping up to date would require HR departments with an important number of personnel. Instead, HR providers, by specialising in a smaller number of services and offering them to many clients, may develop and update their knowledge about their specialised topics.
3. **Allowing the HR department to undertake a more strategic role** (Kakabase & Kakabase, 2005). This rationale tends to appear in conjunction with the idea that, by outsourcing some more transactional activities, the HR department will be able to focus on activities that are considered core to the function, and those that add value. In this sense, in a climate where talent management is one of the main responsibilities with which HR departments are charged, outsourcing may allow them to concentrate on developing strategies and policies to acquire, develop and retain talent, while representing a business partner role, in terms of Ulrich (1998).

Recent studies have added new items to the reasons and benefits that HRO may bring about, as summarised in Table 1. In this context, it is important to highlight two aspects: First, the fact that the main reason to trigger the outsourcing of some (or indeed most) of HR activities may be that such decision is taken by someone else other than the HR director, usually the board of directors or the very CEO of the organisation (Woodall et al., 2009). In this sense, the rationale for HRO may be unclear for the HR department, representing in fact a non-reason to outsource, and allegiance to this process may be difficult to establish. Second, as it has been pointed out, there may be an important gap between the initial rational to undertake HRO and the results obtained by it. In this sense, Shen (2005) presents a number of examples and counter-examples where the objectives and advantages that have been considered for undertaking HRO have turned into the opposite intended outcomes (for example, companies that aimed to achieve cost reductions but ended up spending more money than expected, including the price of the service provided, the person who would monitor the process in-house, etc.).

This moves us to the consideration of disadvantages of HRO, as shown in the next section, and points to the need to carry out more research

Table 1. Summary of reasons to outsource HR activities

Reason for HRO	Example references
Cost reduction	Woodall et al., 2009 ; CIPD, 2009; Caiyun et al., 2008 ; Shen, 2005 ; Pin & Sáenz-Díez, 2002; Greer et al., 1999.
Using specialised services / Improving service quality	CIPD, 2009; Caiyun et al., 2008; Shen, 2005; Cooke et al., 2005 ; Pin & Sáenz-Díez, 2002
Allowing the HR department to undertake a more strategic role / moving HR up the value chain	Woodall et al., 2009; Houtzagers & Janssen, 2008; Cooke et al., 2005
Improving or focusing on HR department's core competencies	Caiyun et al., 2008; Shen, 2005; Cooke et al., 2005; Pin & Sáenz-Díez, 2002
Organisational learning from specialist provider	CIPD, 2009; Cooke et al., 2005; Pin & Sáenz-Díez, 2002
Minimising risks	Woodall et al., 2009; Shen, 2005;
Restructuring the organisation and/or the HR function	Alewell et al., 2009; Caiyun et al., 2008
Fitting HR department manpower capacities to respond to organisational contraction and growth	Woodall et al., 2009; Alewell et al., 2009; Cooke et al., 2005
Decision made elsewhere in the organisation	Woodall et al., 2009

about the intended and actual results of undertaking HRO initiatives.

THE OUTSOURCING EXPERIENCE BACKLASH: PROBLEMS AND LIMITATIONS

In the same vein as in the case of the reasons for outsourcing, there is some degree of overlap between studies on reasons why not to outsource HR activities and those identifying the disadvantages of such initiatives. As above, both will be treated in conjunction.

Even though it does not feature highly in the literature, whether practitioner or academic based, the decision to outsource HR activities does not always result in the benefits that were originally expected. Some problems and limitations to HRO initiatives are (Woodall et al., 2009; Kosnik et al., 2006; Cooke et al., 2005; Shen, 2005; Lawler & Mohrman, 2003; Cook, 1999):

- Internal resistance from HR department's members. This has been a common topic since the expansion of outsourcing HR activities in the 1990's, as HR department members would see service providers as "usurpers" of their jobs and thus seen with reticence. This problem is particularly notorious when the decision to outsource is made by top management instead of the managers of the HR department.
- Lack of skills of the HR department, both regarding to setting up the outsourcing process, and to monitoring its ongoing practice. This may pose special difficulties in reassigning existing HR personnel from carrying out transactional tasks to performing added-value, strategic activities.
- Reduction in staff in the HR department and the consequently possible low morale among these employees.
- Loss of skills and knowledge in the HR department, as there is little or no incen-

tive to keep investing in learning initiatives as they trust that the provider company constitutes their repository of knowledge about the outsourced activity at hand.

- Perception of loss of control and/or information about the outsourced activities in favour of the provider, with the subsequent sensation of dependence toward them. In relation to this aspect is also the fear that the provider may use this situation of dependence in order to exert pressure on their conditions (such as price).
- Lack of skills and knowledge on the part of the provider company, resulting in an insufficient quality of the services provided. Sometimes, a related problem is the provider's poor knowledge about the client company. This is a particularly problem for small and medium size client organisations, which do not constitute a large account of the provider company, and thus may get a limited amount of their time and attention.
- Failure to establish clear and satisfactory relationships between the internal HR department and the service provider.
- Failure to achieve cost savings, particularly when one of the initial motivations for outsourcing has been reducing costs of the HR function. This problem is particularly worrying when the outsourcing initiative is aimed at reducing time of the HR department members to the tasks outsourced (or even reduce headcount in the HR department) and it ends up consuming as much time or even more.
- The danger of the provider company using the know-how of the client company in order to generate standardised products for other clients.

Thus, the problems and limitations related to outsourcing the HR function may arise from the agents involved in the process, both in terms of capacities (lack of skills from the provider, loss of know-how, etc.) and of attitudes (loss of motivation, uncertainty, low morale, etc.). These problems are not to be underrated, particularly in a business context subjected to constant changes, with few solid bases where to consolidate lasting relationships, both between the organisation and its employees, and between the organisation and its providers (Rifkin, 2004). Even though there is limited published evidence about the low or non-use of outsourcing (Woodall et al., 2009; Kosnik et al., 2006; Valverde et al., 2003), such problems may result in a process of involution and prompt companies to terminate outsourcing contracts or stay off outsourcing initiatives altogether.

HOW MUCH HRM IS OUTSOURCED? A PRUDENT LOOK AT HRO PRACTICES

As with other aspects of Human Resource Outsourcing, the amount of provider produced literature is much larger than the evidences gathered by scientific research, leading to a situation whereby very basic but important questions about HRO remain unanswered, such as: How much of the HR function is outsourced? What types of activities tend to be outsourced? What areas of HR practice are outsourced? Even though there is inconclusive evidence about these matters, this section provides a cautious approximation to the answers that have been advanced.

To the question of how much outsourcing does take place in HRM, the starting point is the classic forecast by Adams (1991), who considered that there was no evidence that outsourcing HR was not and would not become an overwhelming trend of the function. Yet, throughout the decade of the 1990's, most evidences seemed to point out towards a considerable expansion of outsourced HR activities (Harkins et al., 1995; Maurer & Mobley, 1998; Siegel, 2000). Other sources based on a European wide survey did not consider the progression of the trend of outsourcing as spectacular, and consider it as an "evolution rather than a revolution" (Vernon et al., 2000). And yet another group of authors were even more cautious about the growth of HRO, both in the United States (e.g. Mise, 2001) and in Europe (Valverde et al., 2003; 2006). Simultaneously, most outsourcing providers have kept making predictions about the future growth of HRO (Shen, 2005) and its current favourable positioning in the context of the current crisis and companies' efforts to cut costs (Capital Humano, various 2008 & 2009; Roberts, 2009). However, most current evidence from the UK (CIPD, 2009) states that only 29% of UK organisations use HRO. In sum, and given the inconclusive nature of the existing evidence, outsourcing HR can be considered a common, but not an overwhelming trend within the HR function, and current trends are unclear about whether they will move towards outsourcing or in-sourcing (Allen, 2009).

Regarding the types of activities that tend to be outsourced, there is an agreement that the most transactional, routine and administrative aspects of the function (such as payroll, legal aspects, training, contracting, etc.) are the most likely to be outsourced, while the more strategic activities tend to be kept in-house (such as establishing HR policies and strategies, identifying and developing top talent, organisational change and development, etc.) (Woodall et al., 2009), as following the logic of outsourcing non-core business (Prahalad & Hamel, 1990). Other studies, however, have warned that in some organisations, HR department is very reluctant to outsource even transactional activities, and keeps these tasks among their responsibilities (Valverde et al., 2003; 2006).

At this point it is also interesting to highlight that some areas of HR tend to be wholly outsourced (legal activity, payroll and pensions), while others are only partially outsourced (training, recruitment and selection) (CIPD, 2009). This is an approximation to the need proposed by Gilley & Rasheed (2000) that it is necessary to study not only the breadth of outsourced activities, but also their depth.

In relation to the specific activities that are outsourced, European wide evidence based on the Cranet/Cranfield Project (http://www.cranet.org/home/Pages/Default.aspx) (Vernon et al., 2000) suggests that the most frequently outsourced HR activities are as shown in Table 2.

As shown in Table 2, the most common activity to be outsourced is training and development. Other sources in the United States point to the same widespread use of training providers (Bassi et al., 1997). This may be seen as a possible contradiction in terms if we consider the context of the knowledge economy, where it is essential that organisations avail of an intellectual capital that is quickly updated, flexible, innovative and creative. In this sense, training and development would be retained in-house as part of the core competencies of the organisation. However, the figure of 77% of companies that use training does not account for how much of this is standardised, packaged training activities that are outsourced on account of price or convenience, or highly sophisticated and specific training and development activities (Vázquez-Palacios,

Table 2. Most frequently outsourced activities in Europe. Source: Vernon et al., 2000.

HR activity	% of organisations outsourcing some of the activity
Training and development	77
Recruitment and selection	59
Compensation and benefits	30
Outplacement	29

2010). Gainey & Klaas (2005) have attempted to redress this lack of accuracy by proposing to distinguish between generic or "off-the-shelf" training (which can be standardised, having little to do with core-competencies of the client firm) and more specific or "tailor-made" programs (which may involve a lot more time investment on the part of both). Thus, the decision to outsource training will depend on what type of training activities are being considered, as well as on the availability of providers and the capabilities that the organisation already has in-house, among other factors (Galanaki et al., 2008).

Another common group of activities to be outsourced are those related to recruitment and selection, and have been so for a long time (Capital Humano, 2007). It is in this area where smaller companies rely more on outside providers (Klaas et al., 1999; 2005), as they can benefit from the economies of scale and the visibility that specialist recruitment companies may afford. Also, recruitment and selection service providers can keep up to date with the new pools of potential candidates as the search for talent is every time placed in different sources of recruitment, including social networks, job sites, and recruitment fairs where an individual organisation may not be able to have a continuous presence (Solís & Corral, 2010).

The next area where outsourcing is common is compensation and benefits. In this case, it is again difficult to evaluate whether only the most transactional aspects of payroll are outsourced or also other aspects. However, this is a more technical area, where the availability of potent software (which are every time moving more to server-based platforms) and the capacity to update criteria and formats to keep up to date with current norms and legislations, makes it convenient for many companies to consider outsourcing (Ang, 2010). The complexities of managing the day to day of compensation and benefits are even greater when we take into account international companies with expatriate employees (with the consequent need to adjust the fiscal implications of their wages) (Gómez & Martí, 2010), and companies that offer flexible compensation plans, and thus compensation is a lot more than simply paying a salary at the end of each month. It is important to highlight that compensation is sometimes not outsourced, but moved onto other functions of the organisation, such as accounting or finance.

Finally, outplacement services are also used by less than one third of organisations with a greater difference between different countries, as outplacement agencies have widely different regulations across European countries. For example, in France, Belgium or the Netherlands, the use of outplacement providers is higher as the legislation obliges companies undertaking staff reduction processes to take measures to help employees cope with the loss of a job and find a new occupation. On the other extreme, Spain not only does not have this requirement, but has a legal gap about the status of outplacement agencies, which are supposed to be only non-profit organisations (Wensels, 2010). With the current recessionary situation of the economy, the market for outplacement services should be well placed to grow, particularly as HR departments search to project an image of "employee champions" (Ulrich, 1998) who care for the employees that must leave the organisation while not having to be involved in the potentially more conflictive aspects of laying off employees.

In relation to the variability between HRO practices between countries, some studies have collected data on the use of HR providers, making a similar emphasis to the Cranfield Survey, but each one of them adding more refined information about outsourcing the HR function in their particular country. In this regard, Valverde et al. (2003) evaluated the distribution of HR tasks made by different agents (top and line management, the HR department, and HR service providers) and found that only about 5% of all HR activities and responsibilities were outsourced in Spanish companies of more than 200 employees. Furthermore, they identified different patterns of use of HR providers

(Valverde et al., 2006): of the few companies that were more prone to outsource, half of them would partially outsource activities in a wide range of HR areas, while the other half outsourced specialist services (about 20% of task delivery such as acting as trainers or interviewing candidates, about 17% of high level specialist HR responsibilities such as advising on labour relations, aligning HR with corporate strategy, etc., and about 12% of all technical and administrative tasks, such as administering contracts, storing and handling employee information or organising means and resources for training activities), while they kept generalist tasks in-house. Regarding the general functions were HRO is more prevalent in Spain, Pin & Sáenz-Díez (2002) identified similar trends to the European average in the areas of recruitment and selection, training, where 60% of organisations used HR service providers; while more companies used HR providers for legal advice, and less for outplacement (20%) and compensation and benefits (less than 10%). In Greece, Galanaki & Papalexandris (2005) found that the use of HR providers for recruitment and selection, training and outplacement was similar to the European average, while compensation and benefits services represented only 1,75% of the whole HR services providers market. Instead, in the United States, the available information points to the fact that the area most likely to be outsourced is precisely payroll and benefits management, followed by training (American Management Association, 1997; cited in Pin & Sáenz-Díez, 2002, p.12). Australia has more points of contact with European countries, and the most commonly outsource HR activity is training and development (48% of companies do so), followed by recruitment and selection (34%), HR information systems (29%), change management (22%) and performance appraisal (21%) (Sheehan, 2009).

Other European countries offer not dissimilar trends: In Germany, Alewell et al. (2009) recently found that the highest use of HR providers was in the areas of training and legal advice (around 60% of companies used them), followed by temporary agency work (about 50%), consulting, headhunting, payroll accounting and placement services (between 20 and 40%). Only a minimal fraction of their sample outsourced the whole HR function, while about 10% had considered it but rejected carrying out, and most had never considered it. Recent evidence in the UK has attempted to convey not only the breadth of HR activities outsourced, but also the depth, by measuring what activities are outsourced, but also whether they are so on a partial of full basis. Thus, a CIPD (2009) survey has shown that legal advice, payroll and pensions are the most outsourced function, with about a third of companies outsourcing them partially, and another third fully. Training and recruitment and selection are, on the other hand the most partially outsourced functions (with half the sample outsourcing parts of these activities), but only less than 10% of companies outsource them fully. Conversely, some activities are more likely to be outsource fully in the UK, in particular employee assistance and outplacement, where about a third of the surveyed companies outsource the whole activity. Other areas follow similar patterns to the European average. Another interesting attempt to capture the depth of outsourcing initiatives was carried out by Tremblay et al. (2008) in Canada, by asking companies to rate which proportion of different activities they outsourced to external organisations. Their results are difficult to compare with those of other countries, as they express depth but not breadth, and thus there is no information about how many companies used or not used each service, but the extend to which they did. In this sense, benefits and labour relations were the activities where a greater proportion of the activity was outsourced, while recruitment and selection was the area where companies outsourced to a lesser extent, which shows that in this activity, the client company avails of a provider to initiate the recruitment and selection process, but keeps an important part of it in-house (notably, the final decision making).

WHAT'S NEXT IN OUTSOURCING: AN AGENDA FOR PRACTICE AND RESEARCH

There is no doubt that the conditions of the new economy present a scenario that is highly conducive to the practice of outsourcing human resource management activities: on the demand side, the rapidly changing business environment makes the constant updating of knowledge in a wide diversity of areas a necessity, and organisations may not be able to keep up unless they outsource some of their activities. On the supply side, the development of new technologies allows HR service providers to serve their client companies with an increasing ease, particularly in terms of location and easily adaptable service platforms that allow every time more customisation and thus a closer adaptation to the needs of their clients. It is within this context that some trends in Human Resource Outsourcing have been identified to the extent that we can talk about the consolidation of the HR providers market is indeed occurring in the twenty-first century.

However, the practice of HRO has produced numerous examples of failures from small to large scale, and many client organisations have not been satisfied with their HRO experiences (CIPD, 2009). These failures may be due to a number of factors, both on the client and the provider side, such as unclear objectives for the service outsourced, poor planning, lack of effective monitoring, etc., by the client organisation; or lack of knowledge and skills about the activity or the client company, excessive standardisation, etc. by the service provider.

But beyond specific actions of specific companies, there remain a number of tensions that are affecting and will continue to affect the practice of HRO. First, the tension between the exponential growth of technological developments that place computer networks at the centre of outsourcing processes and the not corresponding development of legislative and normative labour and fiscal frameworks between different countries. Even though the markets for products, services and employees have less and less barriers and become ever more globalised, national laws, particularly in Europe, are resisting to abandon their idiosyncrasies in favour of a more homogeneous treatment of contractual and payment relationships. Similarly, in the United States, the complexity of certain laws (for example, the federal Family and Medical Leave Act, see Cadrain, 2010) is so great that companies need special expertise to monitor a relatively menial, non-value adding activity. A second tension is that created between the objectives of cutting costs (which the rhetoric continues to establish as the initial impetus for the decision to outsource) and maintaining or improving the quality of the service provided. And finally, a related third debate affects directly the members of the HR departments, and consists of the tension between the possible outcomes of outsourcing some or many HR activities: will the process really free them up from transactional chores in order to give them time to become strategic partners to the business, or will it just decrease the HR department's resources, power, influence and control over the HR function? In practice, then, even though outsourcing HR activities and processes seems like a most useful option for HR departments nowadays, its practical application may pose certain difficulties and reticence.

While the agenda for practice in HRO is linked to contextual technological and legal processes and to tuning the relationships between clients and suppliers, the need for scientific research on the topic is so great that myriad topics are still part of the agenda for research.

The research to date in the topic of outsourcing the HR function shares a number of characteristics: First, the most basic questions about what, why, and how outsourcing HR happens have been answered inconclusively. This may be due to two main factors: On one hand, the service provider literature is pervasive and often mixed (even though with different results) with rigorous scientific research. On the other hand, results are difficult to compare and contrast because of the lack of objective and shared indicators about different aspects of outsourcing. Second, very few attempts to endow the study of HRO with a theoretical underpinning have been made. It was not until 1999 that Klaas et al. proposed to observe the outsourcing process and its impact under a transaction cost lens. It has taken almost a decade for other researchers to put forward existing theories to frame HRO research: For example, Klaas (2008) presents how transaction cost economics, the resource based view, the HR architecture literature and institutional theory may explain the growth of HRO; Nuñez (2009) uses transaction cost theory to explain the make or buy decision for health and safety services; Gospel & Sako (2010) use the theoretical perspectives of the administrative structure of the firm, transaction costs, and resources and capabilities to explain the evolution of shared services versus outsourcing; Klaas et al. (2010) use institutional theory to explain how SME's (small and medium sized companies) may innovate through outsourcing. Even though these are very few cases, it points to a promising future in the study of HRO. In sum, we agree with Cooke et al. (2005) that there is a paucity of academic research on the topic of outsourcing the HR function in general, not only regarding theoretical models, but also basic descriptive studies, while in the meantime, a great deal of practitioner lead, prescriptive literature keeps being produced to encourage the idea that all companies should outsource their HR activities.

Given this situation, we now propose a number of areas where research on HRO should make further research insights, presented in the following focuses of study areas:

1. Aggregate Descriptive Studies about HRO Practice

As shown in previous sections, the basic question of how much HRO takes place in companies is yet to be systematically addressed. In this sense, there is not even agreement on whether the HRO is a growing trend or not. Even though this may not seem a highly interesting area of scientific enquiry, if serious advances are to be made about studying the area, it is necessary to first describe the phenomenon before we can produce other theoretical, empirical or prescriptive insights. On the same line, there is a need to engage in studies to find out what gets outsourced (what activities or processes, which parts of them, etc.), who is involved in the outsourcing decision making and other processes, and how, and how outsourcing HR takes place. The reasons for the lack of information about these aspects may lie in the difficulties to finance large scale representative studies, or in the lack of definition about the measurements to be used. The latter topic will be addressed in the next section.

Following the needs to describe HRO, some studies have pointed out to the need to address the determinants of outsourcing HR by organisations. The work by Klaas et al. (2001) opened up the inroads into the topic and identified idiosyncratic HR practices, strategic HR involvement, positive HR outcomes, promotional opportunities, demand uncertainty, and pay level as determinants of specific types of HRO activities. However, few other attempts to follow this avenue have been made, with the notable attempts by Tremblay et al. (2008) who studied how various strategic and organisational characteristics, transactional variables and risk factors influenced the adoption of HRO, and Galanaki & Papalexandris (2007), who looked at the influence of internationalisation as a determinant of HRO. However, these are not enough studies to reach conclusive evidence: for example, out of these three studies, two (carried out in USA and Greece) found no influence of firm size on the adoption of HRO, but one (carried out in Canada) found strong support to say that there is a lot more reliance on HRO of payroll, recruitment and labour relations in smaller and larger companies, as opposed to medium sized ones. This is only a small example, but one that reminds us again of the incoherence between scientific fact and popular visibility, as most non-scientific literature portrays an image of larger companies being more engaged in outsourcing HR. Whether this is purely an assumption, or an interest of HRO providers in acquiring larger contracts with larger organisations, it clearly shows the fact that further research is needed in the topic of HRO determinants.

Other avenues of inquiry within the general area of descriptive studies of HRO should also tap into the examination of the HRO practices carried out in the public sector, both because it may be a more controversial issue, and also because it may have specific nuances in the decision-making and ongoing implementation of such practices (Coggburn, 2007). Finally, another group of companies that should be studied are those who decide to undertake an extreme process of HRO (Bradbury & Waechter, 2009), or as it is known in the practitioner arena as end-to-end HR business process outsourcing, that is, the few organisations that have decided to fully part with their HR department.

2. Conceptual and Theoretical Contributions

A first objective within the realm of conceptual and theoretical contributions needed in the study of HRO is to agree on a definition. As seen in the corresponding section, the use of the term "human resource outsourcing" or "outsourcing human resource management" is used quite loosely. In this sense, some contributions need to be made to provide a definition that can reach consensus among authors, or at least to generate some criteria that help researchers and practitioners alike to mark the limits to their use of the term HRO. This first conceptual contribution would also aid future studies to be compared and contrasted.

A second necessary contribution, related to the first one, consists of providing some standards for the measurement of HRO. As stated before, a decade ago Gilley & Rasheed (2000) already posed the need to separate the measurement of the breadth from that of the depth of outsourcing in general. Regarding breadth, even though different studies have used different categorisations of HR activities being outsourced, these may be easily identified and collapsed in order to make comparisons between studies (for example, some studies measure "training and development" (Vernon et al., 2000), while others break it down into more specific categories, such as "training needs analysis", "training programme design", "development of training materials", etc. (Woodall et al., 2009)). However, the issue of depth has not yet been resolved, and even though some studies have made attempts to incorporate measures of depth in their data collection, their results are difficult to compare: for example, CIPD (2009) asked their respondents to differentiate between "part of the activity outsourced" and "full activity outsourced", while Galanaki & Papalexandris (2007) asked about the "percentage of a HRM function" that the company had outsourced"; and more similarly, Tremblay et al. (2008) asked to ascribe a proportion of an activity outsourced to external organisations in four discrete categories (0%, 1-49%, 50-99%, 100%).

A third area of theoretical contributions where more future research should focus is on the use of existing theories, in order to support the argumentation and proposals of new studies. Some important insights have already been made to initiate this avenue, as explained above in the paragraphs about the characteristics of the research to date, but few later empirical studies have actually used the frameworks of transaction cost theory, the resource-based view or institutional theory to frame their questions or their findings. Also, other theories and concepts should be considered in the study of HRO. Here, we suggest the study of the outsourcing relationship as part of the value chain (Gereffi & Kaplinsky, 2001), and the conceptualisation of HRO as networks (Coe & Hes, 2006). Both views may help in pursuing studies that focus in the relationships among agents involved in the outsourcing process or the governance of the process (or the chain), and may evaluate different levels of analysis, from individual to organisational (Rainnie et al., 2008).

Finally, a much needed conceptual contribution to the study of outsourcing the HR function would be the provision of a holistic *conceptual framework* that incorporated both the process of outsourcing and the different explanatory aspects that affect it.

3. Human Resource Outsourcing Topics at the Organisational Level

A number of topics in HRO need to be further studied, particularly with a strong theoretical background, extensive empirical data and above all, with the scientific canons of rigour and independence. Among these topics are:

The triggers for outsourcing, otherwise known as the "make or buy decision". There is agreement in some obvious triggers, but most research has been carried out quantitatively with the possible triggers being pre-defined by researchers. In this field, more qualitative in-depth research is needed, as recent results from a qualitative study (Woodall et al., 2009) have shown that there may be other triggers that researchers have previously not considered, but may determining in the decision to outsource HR activities. Also, it is worth considering non-triggers, that is, the lack of reason for outsourcing other than succumbing to the pressure of HR service providers, or deciding to outsource simply out of mimicking what other companies do. The latter has recently been considered by Galanaki & Papalexandris (2007) and Klaas et al. (2010), but more evidence needs to be gathered in this respect. Thus, not only the topic of triggers needs to be further studied, but also the methods need to be re-evaluated.

On a related note, there needs to be a stricter differentiation between the related topics of triggers, determinants, perceived benefits or advantages, and actual benefits or advantages. Thus, research needs to separate these topics both in concept and in its chronological ordering (previous to undertaking an outsourcing initiative, or after having implemented it, that is, pre- and post-outsourcing).

A topic that has mostly been covered in books but has received little attention in the scientific literature is that of the choice of the service provider. This is a theme of important relevance and practical implications, particularly in a context of growing number and variety of HR service providers. Many questions remain to be answered in this area: who makes the choice of provider, what criteria are used in order to make this choice, how much time and effort is devoted to making a choice, how many offers of service are considered, how do the corporate philosophy and culture of the provider align with those of the client company, etc.

Next, a wide area that needs to be addressed is that of the effects of HRO on different stakeholders. For example, in the case of the HR department, the practitioner literature insists on how outsourcing "can help boost HR's value contribution to the organisation" (Jonkeren, 2009). Yet, there is hardly any evidence showing how outsourcing the HR function may affect the role and contribution of the HR department to the organisation. Instead, Woodall et al. (2009) have identified some potential problems for HR staff when faced with outsourcing HR activities, such as staff reductions, difficulties in reassigning personnel to higher valued added activities, and actual deskilling. They point out to unclear trends in this matter, so the topic deserves further research. Many other topics related to the effects of HRO on HR departments need to be addressed: how the actual structure of HR departments will be affected by outsourcing; what new skills HR department members will need in order to deal with their serviced providers; how committed the HR departments will be to the ongoing HRO relationship (Lievens & De Corte (2008) have started research in this regard); how emotions and sensations of lost control appear and are dealt with; how the confidentiality of employee information is handled, etc.

Along the same lines, the impact on other agents needs to be established, particularly on the clients of the HR function: middle managers, and employees. Delmotte & Sels (2008) have started by looking at their perceptions and satisfaction with the outsourcing process, but many more insights are needed in this regard.

Finally, the study of the governance of the outsourcing relationship needs to be addressed. The relevance of the topic is fundamental, as the governance of the process, even from its inception, may determine the value, effectiveness, etc. of the HRO relationships that are eventually developed. Again, the practitioner literature offers a bright picture on the topic (e.g., Marchetti & Wilson, 2006), but scientific evidence needs to be provided to understand the topic.

4. Bridging Research and Practice in HRO

A final key area where more insights need to be made in HRO represents the acid test for both theory and practice, and it has to do with the study of the success factors of HRO initiatives. This is a wide area of inquiry, and little research has been carried out about it even though most HRO books devote many pages to how to make HRO work. In this case, then, it is not a matter of disagreement between the practitioner and academic literature, but the profusion of the former and the dearth of the latter.

In the study of the characteristics of successful HRO, many aspects need to be evaluated: how to actually measure HRO success (in terms of its impact on organisational performance, of savings for the organisation, of satisfaction on the HR department members or other stakeholders such as employees receiving the HRO service or middle managers), how the relationship between client and provider evolves, what problems arise in this relationship, etc. Because many HRO activities are deployed for a long period of time, longitudinal studies should aid to make contributions on this matter.

In sum, HRO is here to stay, and even if it is not as pervasive as some commentators suggest, it is a common and varied practice that deserves to be investigated in order to aid managers to make informed decisions about outsourcing HR activities and researcher to explore the production of services within value chains that are not in the same organisation.

CONCLUSION

This chapter on outsourcing the HR function in the new economy has highlighted what is known and what is inferred about the topic, what is conclusive scientific research and what is an effort from HR providers to paint a brighter picture. Whether the objective is cost cutting or upgrading quality and knowledge, HRO is a well developed tool at the service of organisations, and they must consider carefully whether it will be useful and fitting to their own characteristics. In this sense, outsourcing HRM activities in this chapter have been analysed not only as a "threat" to non-efficient HRM departments in terms of cost (Torrington & Mackay, 1986; Fitz-Enz, 2009), but also as a conscious choice to focus on developing competencies for this department that are value-adding to the organisation.

The scientific community however, has not yet provided enough insights to clearly guide such choice or decision. Because of this lack of knowledge about the topic, this chapter has presented a literature review summarising some of the most up-to-date research, and has made a contribution by setting a research agenda that identifies the topics that have yet to be addressed and those that have not sufficiently been demonstrated, both in general and in particular, given the characteristics of the new economy that pose opportunities and challenges to the outsourcing of HRM activities.

REFERENCES

Accenture. (2008). *Beyond cost reduction*. Retrieved July 30, 2010, from http://www.accenture.com/NR/rdonlyres/6FBF58AC-7A26-43F9-BCED-90F6E232B70C/0/177090038HRBPO-BeyondCostReductionandInnovationv3.pdf

Accenture. (2010). *Business process outsourcing*. Retrieved from https://microsite.accenture.com/explore_outsourcing/Business_Process_Outsourcing/Pages/Business_Process_Outsourcing.aspx

Adams, K. (1991). Externalisation versus specialisation: What is happening to personnel? *Human Resource Management Journal*, *1*(4), 40–54. doi:10.1111/j.1748-8583.1991.tb00236.x

Alewell, D., Hauff, S., Thommes, K., & Weiland, K. (2009). Triggers of HR outsourcing decisions-An empirical analysis of German firms. *International Journal of Human Resource Management*, *20*(7), 1599–1617. doi:10.1080/09585190902985228

Allen, A. (2009). Shaping up or shipping out? *People Management*, November, 4-9.

Altés, J. (2003). El outsourcing en el departamento de recursos humanos. *Capital Humano*, July-August, 23-27.

Ang, J. L. (2010). Nómina y seguros sociales. In *M. J. Pérez López (Coord.), Externalización de funciones de recursos humanos*. Madrid, Spain: Pearson.

Bassi, L. J., Cheney, S., & Van Buren, M. (1997). Training industry trends 1997. *Training & Development*, (November): 46–59.

Bauman, Z. (2007). *Liquid times: Living in an age of uncertainty*. Cambridge, MA: Polity.

Bauman, Z. (2007). *Tiempos líquidos*. Barcelona, Spain: Tusquets.

Bradbury, M. D., & Waechter, G. D. (2009). Extreme outsourcing in local government: At the top and all but the top. *Review of Public Personnel Administration*, *29*(3), 230–248. doi:10.1177/0734371X09332572

Bravard, J. L., & Morgan, R. (2006). *Smarter outsourcing: An executive guide to understanding, planning and exploiting successful outsourcing relationships*. Dorchester, MA: Pearson Education.

Cadrain, D. (2010). A leave law that just won't go away. *HRMagazine*, (July): 49–52.

Caiyun, G., Zhiqiang, L., & Bing, L. (2008). *Research on human resource management outsourcing*. Wireless Communications, Networking and Mobile Computing, 2008. WiCOM '08. 4th International Conference. IEEE.

Capital Humano. (2007). *RPO o como rentabilizar los procesos de selección* (p. 215).

Capital Humano. (2008). *Outsourcing de RR.HH. Para reducir costes en tiempos de crisis*, (p. 226).

Capital Humano. (2009). *El outsourcing ayuda a afrontar la crisis al 33,1 por ciento de las empresas*, (p. 235).

Capital Humano. (2009). *Los servicios de outsourcing mantienen su crecimiento a pesar de la crisis* (p. 238).

CIPD. (2009). *HR outsourcing and the HR function: Threat or opportunity?* London, UK: Chartered Institute of Personnel and Development.

Click, R. L., & Duening, T. N. (2005). *Business process outsourcing: The competitive advantage*. London, UK: Wiley.

Coe, N., & Hess, M. (2006). *Global production networks: Debates and challenges*. Presented at GPERG Workshop, University of Manchester.

Coggburn, J. D. (2007). Outsourcing human resources: The case of the Texas Health and Human Services Commission. *Review of Public Personnel Administration, 27*(4), 315–335. doi:10.1177/0734371X07301976

Cook, M. F. (1999). *Outsourcing human resources functions. Strategies for providing enhanced HR services at lower cost*. New York, NY: AMACOM.

Cooke, F. L., Shen, J., & Mcbride, A. (2005). Outsourcing HR as a competitive strategy? A literature review and an assessment of implications. *Human Resource Management, 44*(4), 413–432. doi:10.1002/hrm.20082

Delmotte, J., & Sels, L. (2008). HR outsourcing: threat or opportunity? *Personnel Review, 37*(5), 543–563. doi:10.1108/00483480810891673

Duffy, M. N. (2001). Outsourcing a 401(K) plan. *Journal of Accountancy, 191*(5), 30–35.

Fitz-enz, J. (2009). Disruptive technology for human resources. *Employment Relations Today*, (Winter): 1–10. doi:10.1002/ert.20219

Gainey, T. W., & Klaas, B. S. (2005). Outsourcing relationships between firms and their training providers: The role of trust. *Human Resource Development Quarterly, 16*(1), 7–25. doi:10.1002/hrdq.1121

Galanaki, E., Bourantas, D., & Papalexandris, N. (2008). A decision model for outsourcing training functions: Distinguishing between generic and firm-job-specific training content. *International Journal of Human Resource Management, 19*(12), 2332–2351. doi:10.1080/09585190802479579

Galanaki, E., & Papalexandris, N. (2005). Outsourcing of human resource management services in Greece. *International Journal of Manpower, 26*(4), 382–396. doi:10.1108/01437720510609564

Galanaki, E., & Papalexandris, N. (2007). Internationalization as a determining factor of HRM outsourcing. *International Journal of Human Resource Management, 18*(8), 1557–1567. doi:10.1080/09585190701502687

Gereffi, G., & Kaplinsky, R. (2001). The value of value chains. [Special Issue]. *IDS Bulletin, 32*(3).

Gilley, K. M., & Rasheed, A. (2000). Making more by doing less: an analysis of outsourcing and its effects on firm performance. *Journal of Management, 26*(4), 763–790.

Gómez, S., & Martí, C. (2010). Externalización y expatriados. Proceso de expatriación. In *M. J. Pérez López (Coord.), Externalización de funciones de recursos humanos*. Madrid, Spain: Pearson.

Gospel, H., & Sako, M. (2010). The unbundling of corporate functions: The evolution of shared services and outsourcing in human resource management. *Industrial and Corporate Change, 19*(5). doi:10.1093/icc/dtq002

Greer, C., Youngblood, S., & Gray, D. (1999). Human resource management outsourcing: The make or buy decision. *The Academy of Management Executive, 13*(3), 85–96. doi:10.5465/AME.1999.2210317

Hall, L., & Torrington, D. (1998). *The human resource function. The dynamics of change and development*. London, UK: FT-Pitman Publishing.

Harkins, P., Brown, S., & Sullivan, R. (1995). Shining new light on a growing trend. *HRMagazine, 40*(12), 75–79.

Hervás, J. (1999). El outsourcing se debe contratar sólo por razones estratégicas. *AEDIPE*, March, 23-25.

Houtzagers, G., & Janssen, S. (2008). Shifting from HR outsourcing to business process sourcing. *IHRIM Journal, 7*(1), 30–42.

Jonkeren, S. (2009). HR's time to shine: Challenging economy spells opportunity to demonstrate value. *IHRIM Journal*, *13*(1), 3–7.

Kakabadse, A., & Kakabadse, N. (2005). Outsourcing: Current and future trends. *Thunderbird International Business Review*, *47*(2), 183–204. doi:10.1002/tie.20048

Kakabadse, A. P., & Kakabadse, N. (2000). Outsourcing: A paradigm shift. *Journal of Management Development*, *19*(8), 668–778. doi:10.1108/02621710010377508

Kakabadse, A. P., & Kakabadse, N. (2002). Trends in outsourcing: Contrasting USA and Europe. *European Management Journal*, *20*(2), 189–198. doi:10.1016/S0263-2373(02)00029-4

Klaas, B. S. (2008). Outsourcing and the HR function: An examination of trends and developments within North American firms. *International Journal of Human Resource Management*, *19*(8), 1500–1514. doi:10.1080/09585190802200280

Klaas, B. S., Klimchak, M., Semadeni, M., & Holmes, J. J. (2010). The adoption of human capital services by small and medium enterprises: A diffusion of innovation perspective. *Journal of Business Venturing*, *25*, 349–360. doi:10.1016/j.jbusvent.2008.12.002

Klaas, B. S., Mcclendon, J., & Gainey, T. (1999). HR outsourcing and its impact: The role of transaction costs. *Personnel Psychology*, *52*(1), 113–136. doi:10.1111/j.1744-6570.1999.tb01816.x

Klaas, B. S., Mcclendon, J., & Gainey, T. (2001). Outsourcing HR: The impact of organizational characteristics. *Human Resource Management*, *40*(2), 125–138. doi:10.1002/hrm.1003

Klaas, B. S., Yang, H., Gainey, T., & McClendon, J. A. (2005). HR in the small business enterprise: Assessing the impact of PEO utilization. *Human Resource Management*, *44*(4), 433–448. doi:10.1002/hrm.20083

Kosnik, T., Wong-MingJi, J. D., & Hoover, K. (2006). Outsourcing vs insourcing in the human resource supply chain: A comparison of five generic models. *Personnel Review*, *35*(6), 671–683. doi:10.1108/00483480610702728

Lawler, E. E., & Mohrman, S. A. (2003). *Creating a strategic human resources organisation*. Palo Alto, CA: Stanford University Press.

Lever, S. (1997). An analysis of managerial motivations behind outsourcing practices in human resources. *Human Resource Planning*, *20*(2), 37–47.

Lievens, F., & De Corte, W. (2008). Development and test of a model of external organizational commitment in human resources outsourcing. *Human Resource Management*, *47*(3), 559–579. doi:10.1002/hrm.20232

Limam, A. (2009). Enabling HR transformation in the European public sector: e-HR and outsourcing. *IHRIM Journal*, *13*(1), 41–48.

Marchetti, C., & Wilson, A. (2006). Outsourcing: Letting go to gain control. *CriticalEYE*, *15*, 1–6.

Maurer, R., & Mobley, N. (1998). Outsourcing: Is it the HR department of the future? *HRFocus*, (November): 9–10.

Mise, I. J. (2001). Payroll and HR outsourcing. *CMA Management*, *75*(1), 30–32.

Mullin, R. (1996). Managing the outsourced enterprise. *The Journal of Business Strategy*, *17*(4), 28–32. doi:10.1108/eb039792

Nuñez, I. (2009). Outsourcing occupational safety and health: An analysis of the make or buy decision. *Human Resource Management*, *48*(6), 941–958. doi:10.1002/hrm.20323

Pin, J. R., & Sáenz-Díez, I. (2002). *Outsourcing de recursos humanos*. Barcelona, Spain: IESE Business School. IRCO (International Research Center on Organizations), OP 03/4.

Prahalad, C., & Hamel, G. (1990). The core competence of the corporation. *Harvard Business Review*, *68*(3), 79–91.

Rainnie, A., Barrett, R., Burgess, J., & Connell, J. (2008). Introduction: Call centers, the networked economy and the value chain. *The Journal of Industrial Relations*, *50*(2), 195–208. doi:10.1177/0022185607087897

Rifkin, J. (2004). *The end of work*. New York, NY: Penguin.

Rippin, S., & Dawson, G. (2001, September 27). How to outsource the HR function. *People Management*, 42-43.

Roberts, B. (2009). Outsourcing in turbulent times. *HRMagazine*, *54*(11), 42–47.

Sheehan, C. (2009). Outsourcing HRM activities in Australian organisations. *Asia Pacific Journal of Human Resources*, *47*(2), 236–253. doi:10.1177/1038411109105444

Shen, J. (2005). Human resource outsourcing: 1990-2004. *Journal of Organisational Transformation and Social Change*, *2*(3), 275–296. doi:10.1386/jots.2.3.275/1

Siegel, G. B. (2000). Outsourcing personnel functions. *Public Personnel Management*, *29*(2), 225–236.

Sisson, K., & Storey, J. (2000). *The reality of human resource management*. Oxford, UK: Oxford University Press.

Solís, E., & Corral, A. (2010). Reclutamiento y selección de personal. Externalización. In *M. J. Pérez López (Coord.), Externalización de funciones de recursos humanos*. Madrid, Spain: Pearson.

Syedain, H. (2009). Out of the ordinary. *People Management*, *15*(23), 20–22.

Torrington, D., & Mackay, L. (1986). Will consultants take over the personnel function? *Personnel Management*, February, 34-37.

Tremblay, M., Patry, M., & Lanoie, P. (2008). Human resources outsourcing in Canadian organizations: An empirical analysis of the role of organizational characteristics, transaction costs and risks. *International Journal of Human Resource Management*, *19*(4), 683–715. doi:10.1080/09585190801953699

Turnbull, J. (2002). Inside outsourcing. *People Management*, 10–11.

Ulrich, D. (1998). A new mandate for human resources. *Harvard Business Review*, (January-February): 124–134.

Valverde, M., Quijano, S., & Soler, C. (2003). *HRM as a shared function: An exploratory study of responsibilities and agents in the managing of people*. Paper presented at the IV International Workshop on Human Resources, Cadiz, May.

Valverde, M., Ryan, G., & Soler, C. (2006). Distributing HRM responsibilities: A classification of organisations. *Personnel Review*, *35*(6), 618–636. doi:10.1108/00483480610702692

Vázquez-Palacios, I. (2010). Externalización del área de formación. In *M. J. Pérez López (Coord.), Externalización de funciones de recursos humanos*. Madrid, Spain: Pearson.

Vernon, P., Philips, J., Brewster, C., & Ommeren, J. (2000). *European trends in HR outsourcing*. Cranfield, UK: Cranfield School of Management & William M. Mercer.

Wensels, O. (2010). Recolocación. In *M. J. Pérez López (Coord.), Externalización de funciones de recursos humanos*. Madrid, Spain: Pearson.

Woodall, J., Scott-Jackson, W., Newham, T., & Gurney, M. (2009). Making the decision to outsource human resources. *Personnel Review*, *38*(3), 236–252. doi:10.1108/00483480910943313

KEY TERMS AND DEFINITIONS

Business Process Outsourcing: Contracting out some activities that may otherwise be carried out by the company.

Human Resource Outsourcing (HRO) / Outsourcing HRM: "Having a third- party service provider or vendor furnish, on an ongoing basis, the administration of an HRM activity that would normally be performed in-house" (Cook, 1999, p. 4). As explained in the chapter, many streams of research consider also HR outsourcing any use of a third-party provider regardless of the duration of this activity or service.

Research Agenda: Evaluation of research lines that researchers of a topic should pursue in order to improve science's understanding of that topic.

Chapter 3
Talent Management: A New Perspective in Human Resource Management

Roberto Luna-Arocas
University of Valencia, Spain

ABSTRACT

Talent management is a new concept in the scientific literature, but not in the practitioner arena. In this sense, this chapter underlines the powerful perspective of talent management integrating practices from organizations and providing a scientific approach. Some previous results from different research are expressed in relation to strategy and organizational performance. Talent Management is clearly a concept close to high performance work system, with the difference of a more strategic approach. More than practices are goals for the organization to achieve excellence at work. So Talent Management is clearly linked with a strategic approach to the organization in order to attract, develop, evaluate, and retain professionals with high capacities.

INTRODUCTION

Today's businesses have to cope with hard competition and continuous change (Hamel & Prahalad, 1994) and Human Resources Management (HRM) and Talent Management (TM) is clearly affected by this turbulent environment. As Farndale et al. (2009) confirm, competition between employers has shifted from the country level to the regional and global levels (Ashton & Morton, 2005). Indeed, as Ulrich (1997, p.304) emphasises "under environment conditions of low change, attention to HR practices had little impact on business results, but under environmental conditions of high change, executive attention to HR practices had a large impact on business results". Moreover, Lawler (2005) defend that HR "can and should add more value to corporations. The best way to do this is by being a business partner- by directly improving the performance of the business. This can be accomplished by effective talent management" (p. 165).

In response to this situation, it is argued that organizations are flattening, relying on self-managed teams, becoming highly matrixed, and reconfiguring the structure of work (Shippman et al., 2000). Lepak and Snell (2002) argue that given pressures for both efficiency and flexibil-

DOI: 10.4018/978-1-61350-207-5.ch003

ity firms are exploring the use of differentiated employment modes to engage workers and allocate work. Indeed, HRM "is under pressure to reduce costs, to improve its services, to increase its impact, and to provide a more satisfying work experience for its own employees" (Kochanski & Ruse, 1996, p. 19). Also, Calvo and Garcia (2010) in the analysis of talent management in professional services firm observed that a lack of flexibility within the organization tended to lead to a reduction in potential intellectual capital, and increased excess workload, together with a decline in the company's competitiveness within the marketplace, as seen in a drop in its level of corporate attractiveness.

In this sense, Farndale et al. (2009) according with Bhatnagar (2007) argue that "while companies are facing significant talent management challenges in several regions of the world, such as Europe and North America, the challenges are most acute for young professionals and new managers in the emerging markets such as the BRIC economies of Brazil, Russia, India, China and the economies of Central and Eastern Europe" (p.161). Indeed, there is a shortage of leadership talent (Farndale et al., 2009) and assuming talent management as a source of competitive advantage has been key in HRM (Frank & Taylor, 2004; Lewis & Heckman, 2006).

In spite of the great deal of research on HRM in the last decades, there is still a significant gap between what happens in theory and practice (Pfeffer & Sutton, 1999). This is especially the case in the area of "talent management" (TM). There is much talk about 'talent' and 'talent management', driven largely by the belief that while traditional approaches to HRM have served MNCs well in the 20th century, the contemporary business environment requires new and innovative approaches in the development and deployment of human resources (Caligiuri, 2006; Lengnick-Hall & Andrade, 2008). However, while much is claimed for it, much confusion also surrounds it. A cursory trawl of the literature will reveal several variations on the meaning of TM: informal individually-focused talent management (Tansley, Turner & Foster, 2007); TM as a relabeling of human resource planning (Lewis & Heckman, 2006); TM as succession management (Hirsch, 2000) and strategic TM (Cappelli, 2008; Zuboff, 1998; Boudreau & Ramstad, 2005).

Talent management comes up as a new perspective from strategic human resource management (SHRM) linked more than ever with business strategy. Therefore, TM offers SHRM goals covered with different HRM practices in a configurational way. This systemic perspective have attracted practitioners more than other scientific HRM approaches.

Regardless of how it is conceived, of critical importance in this regard is the debate on TM and firm performance. If a focused talent strategy is capable of delivering a return of investment of up to twenty two points over other firms in the same sector, why it is not widespread in firms? Indeed, practices should warranty the use of talents in the organizations and this can be done not only with practices but also "cultures and systems in which these great people can actually use their talents, and even better, management practices that produce extraordinary results from almost every-body" (O'Reilly & Pfeffer, 2000, p. 63).

MAIN PURPOSE

HRM is growing as a discipline that takes information from many other fields. This multidisciplinary approach makes more complex but also richer the research in the next future. In this sense, HRM and High Performance Work System (HPWS) literature has concentrated during the last years to analyse configurations in organizations to fit strategy and environmental factors. Our main purpose in this chapter is to analise the theoretical construction of TM in the line of HPWS perspective as a system. Also we present different results obtained and the evolution of HRM during the last few years from this perspective. Finally, we present the results in the research of TM in its relationship with organizational satisfaction and job performance.

Wright and McMahan (1992) indicated the need for the creation of a horizontal fit among HR practices, so that they complement each other rather than compete against one another. This horizontal fit has been called internal alignment (Becker & Gerhart, 1996; Huselid, 1995) and it is completely coherent with a talent management systemic perspective.

TALENT MANAGEMENT AND STRATEGIC COMPETENCIES FOR COMPETITIVENESS

HRM needs to advance scientifically and professionally in order to assume a proactive role in these new sceneries. One new perspective is assuming organizations as a bank of talents, that is, focusing on TM. A talent strategy may be conceived as the attraction, the development and retention of people with excellent competencies appropriate to the work context. Generally, TM is associated with Competency Based Management (CBM) where competencies are aligned to organizational values and goals (Hayton & McEvoy, 2006; Kochanski & Ruse, 1996; Sharma & Bhatnagar, 2009; Ulrich, 1998; Ulrich et al., 1995). Also competencies activate actual performance and employee potential. Indeed, competencies refer to "an individual´s demonstrated knowledge, skills or abilities" (Ulrich et al., 1995, p. 474). Also Ulrich (1998) defined intellectual capital as the function of competence by commitment. Therefore, firms with high competence, but low commitment, will have talented but unengaged employees. Conversely, in the context of having high commitment with low competence, the organisations talent pool is more limited, but commitment is high. Intellectual capital requires both aspects. And this equation can be analysed at different levels: firm, units or individual. Arguably there are at least two ways to increase competence in a firm: aligning competencies with business strategy and generating competences through more than one mechanism (buy, build, borrow, bounce and bind). Similarly, commitment can be generated thorough different ways: (a) reducing demands, (b) increasing resources so employee can find new resources to accomplish work and (c) turning demand into resources. But most importantly, by way of consequence, the intellectual capital index will be positively associated with productivity.

There is a clear consensus about the importance of employee behaviours more than the HRM practices in the value creation of an organization (e.g. MacDuffie, 1995; Schuler & Jackson, 1987; Wright & Snell, 1998; Colvin & Boswell, 2007). Discretionary employee behaviours that are difficult to specify in advance are particularly valuable to organizations because, unlike routine job performance, they are difficult for competitors to imitate or replicate and thus may provide a source of sustainable competitive advantage (Colvin & Boswell, 2007). TM offers a strategic perspective where behaviours are aligned with performance goals in the organization. Therefore, micro and macro perspectives are connected in a configurational model of HPWS.

This focus in HRM research has involved a shift in the outcomes of interest from individual level impacts to outcomes at the level of the organization (Becker & Gerhart, 1996; Delery & Shaw, 2001). However, the shift has not simply been towards using organizational performance measures as the dependent variables of interest, but also a growing interest in how employees relate to the organization and how human resources contribute to the business strategy of the organization. Driving much of this interest is the underlying idea that HRM can play a more central role in organizational success by enhancing the contribution of employees to organizational value creation (Wright & McMahan, 1992) thorough TM.

Thus, competencies are dealt with as values that must be developed and are interpreted as provoking benefits both for employees and organization (Van der Heijde & Van der heijden, 2006). In this sense, CBM is the base for a more integrated model dedicated to further addressing how organisations which are suffering a shortage of talent might respond to the challenge (Chiavenato, 2002; Michaels et al., 2003; Peters, 2005; Schlemenson, 2003). Thus, all efforts are concentrated on TM (Casado, 2011; Cubeiro, 2004; Chiavenato, 2002; Gubman, 2000; Jerico, 2000; Michaels et al., 2001; Peters, 2005; Schlemenson, 2003). And TM is based on strategies of talent banks trying to focus on four main actions: (1) staffing and atraction of talents, (2) detecting and identifying talent in the organization, (3) developing talents and (4) retention and loyalty of talents. These SHRM goals guide HPWS in the organization.

TM, through HPWS literature, has been linked with positive results including higher efficiency, less turnover and higher commitment, and a growth in what Sternberg (1999) has termed organisational intelligence. Also, TM as a HPWS have been linked with improvements in the market value of organizations and a higher ROA. Indeed since the last recession, the TM proposition is more strategic now than ever.

One of the streams of research that seems to hold "the promise of creating a truly strategic approach to talent management" (Lewis & Heckman, 2006, p. 145) is the Resource-Based View (RBV) perspective. The HR arquitecture is a value-creating system that raises the question of the appropriate locus of strategic value creation (Becker & Huselid, 2006). These authors follow "going forward, however, we need more theoretical work on the "black box" between the HR architecture and firm performance, and less emphasis on the "black box" within the HR architecture (p. 900). Following Lee and Miller (1999), a dedicated and talented workforce may serve as a valuable, scarce, non-imitable resource that can help firms execute an appropriate positioning strategy. Indeed, competences are one category of possible resources that enable firms to achieve performance and (sustained) competitiveness. A prerequisite for sustained competitive advantage consists of a unique combination of acquiring and retaining competent workers, and adequate HR policies and practices which promote the importance of investing in them. In the RBV literature, terms like resources and capabilities "are used interchangeably and refer to the tangible and intangible assets firms use to develop and implement their strategies" (Ray et al., 2004, p. 24). Strategy experts argue that an organization´s internal resources-including the capabilities of its workforce- could give rise to competitive advantage (Barney, 1991; Wernerfelt, 1984). Competencies benefit firms when they provide customer value, resist imitation, and enable a firm to create new business through product and service-extensions and through innovation (Hagan et al., 2006). Therefore, RBV literature offers clearly a theoretical approach to TM and HPWS literature.

DEVELOPING A TALENT MANAGEMENT MINDSET

A growing number of organizations see competence as the key to enduring performance and to making HR most effective (Kochanski & Ruse, 1996, p.19). Prahalad and Hamel (1990) argue that key employees are the ones who have unique and very valuable competencies that developed in a right organizational context generate key organizational competencies for sustained competitive advantage. Ulrich et al. (1995), have situated the competencies in a framework referred to as the HR Value Proposition "an integrated blueprint for the future of HR" (Ulrich & Brockbank, 2005, ix).

Against the backdrop of this value proposition, a question that arises is how can we develop TM as a talent strategy in firms? Michaels, Handfield-Jones and Axelrod (2001) argue that talent mindset is the key concept, something which they refer to as a passionate belief that to achieve excellence in business we need talents and these have to be underscored by the company values and goals. This cannot be done by the HRM Department, it cannot depend on HR processes, but rather demands a unique and different mental predisposition. Michaels et al., (2001) in their treatise on the future of HRM point to the existence of a shortage of talent which they suggested would result in a "war of talent". This war is predicated on the necessity to increase the skill level of up to 60% of the workforce in order to compete in the age of the smart economy. The firms surveyed by these authors were all considered High Performance Companies and results demonstrated that only 20% considered that they have enough leaders with talent and only one out four establish talent banks as a priority. But results also indicated that 48% of leaders were likely to leave the organization in the next two years. Therefore, "we are seeing a shortage of skilled management talents" (Gubman, 1996, p. 34).

But is there a shortage of talent? Results of several studies indicate that talent mindset is not often present in firms. Right Management Consultants (Philadelphia, a subsidiary of Manpower; Credit Union Management, 2004) found that four out of ten managers and executives are considered excellent leaders, exhibiting management talents their employers value most. Moreover, from a practitioner perspective, TM is predicted to be the highest challenge for Human Resources up to 2015. Indeed, a study with 1350 managers in 27 countries done by The Boston Consulting Group (BCG) and the European Association of Personnel Management (EAPM) identified five main challenges for the future, TM being the number one challenge. Also, HR Focus, a prestige practitioner journal found that 75% organisations they surveyed were worried about retaining and developing key employees into the future.

Gupta (2001, p. 2), noted that "If I am puzzled as you are with the crisis of TM, I encourage you to engage in some serious study of this issue. We need to learn more before it´s too late". More recently Boudreau and Ramstad (2005) in an article published in *Human Resource Management* indicate that a talent science is needed given its incremental importance to stimulate talent decisions, while Tolich (2005) in *Administrative Science Quarterly* notes the dearth of TM and the consequences of our inability to deliver on the promise and potential of the knowledge economy. In 2007 the *Harvard Business Review* gives a special emphasis on TM (HBR Spotlight, *How to manage the most talented*) with two articles: (1) "Leading clever people" (Goffee & Jones, 2007) and (2) "Crisis at the Summit" (Parsons & Pascale, 2007). And finally, The Economist published in 2006 an article entitled *The battle for brainpower* claiming that organizations are much concerned about TM.

But what composes a talent mindset? Some research in Benchmarking has been done from the *American Productivity and Quality Centre* and *the Centre for Creative Leadership* (McCauley & Wakefield, 2006) in order to analyse TM. They include eight main practices, several of them focused on organizational culture (integration, involvement, key employees), others linked with the measurement from the competency model, and others associated with the organisational context.

HPWS TO ENHANCE TALENT MANAGEMENT AND PERFORMANCE

Hiltrop (1999) also supported the importance of HPWS for strategies that strive to attract and retain talented personnel, arguing that high performance organizations beat their competitors in a number of HR factors such as team work, healthy relationship between collaborators, training and the development of opportunities that are offered to workers.

Luna-Arocas and Camps (2008) mention that in short, scholars conceptualize HPWSs as a set of distinct but interrelated HR practices that, taken together, select, develop, retain, and motivate a workforce (Becker & Huselid, 1998; Guthrie, 2001; Huselid, 1995). However, Becker and Gerhart (1996) have suggested that among empirical studies which have examined the link between HR practices and firm performance, there is little consistency in regards to the individual HR practices that have been included within these studies. Similarly, Way (2002) analyzed six exemplary empirical studies (Arthur, 1994; Huselid, 1995; MacDuffie, 1995; Ichniowski et al., 1997; Becker & Huselid, 1998; Guthrie, 2001).

The differences in the practices used by researchers come from the variety of perspectives in HR (Becker and Gerhart, 1996). Three main traditional approaches have been distinguished (Delery and Doty, 1996): universalistic, contingency and configurational. The universalistic approach assumes a direct relationship between HPWS and organizational performance. The contingent approach contemplates the value of strategy, among other aspects of the organization, as moderating the relationship between HPWS and organizational performance. And finally, the configurational approach considers that some variables in HPWS can act jointly showing consistent patterns related to organizational performance. In general terms, as is defined by Becker and Gerhart (1996, p. 786) "there may be a best HR system architecture, but whatever the bundles or configurations of policies implemented in a particular firm, the individual practices must be aligned with one another and be consistent with the HR architecture if they are ultimately to have an effect on firm performance. In this sense, the best practice and contingency hypotheses are not necessarily in conflict-they simply operate at different levels of an HR system". Also, some authors have contemplated universalistic and configurational perspectives together to explain HPWS impact over organizational performance (e.g., Shih, Chiang and Hsu, 2006).

The main trend in research on the HRM-firm performance linkage comes from a systems view of HRM by considering the overall configuration of HR management practices (Ferris et al., 1998). Moreover, some practices are argued to support better organizational performances (Jackson and Schuler, 1995; Huselid, 1995) and these HRM practices, as a system, can contribute to firm performance helping the organization to achieve strategic goals (Bowen and Ostroff, 2004). Following these two authors, we can distinguish two features of an HRMS: content (based on the HRM practices needed to achieve organizational goals) and process (how the HRM system can be designed and administered).

In this sense, HRM research has focused on the practices, as a system, directly related with organizational performance. This is called HPWS and is concerned as a means to maximize firms´competitive advantage (e.g., Guthrie, 2001; Huselid, 1995; Sun, Aryee and Law, 2007; Takeuchi et al., 2007) and a way to improve firm performance (Arthur, 1994; Combs, Liu, Hall, and Ketchen, 2006; Datta, Guthrie, and Wright, 2005; Guthrie, 2001). Indeed, HPWS understand HR management as a system concerning the management practices that provide employees with skills, information, motivation and attitude that results in a work force that becomes a source of competitive advantage (Huselid, 1995; Collins and Smith, 2006). Thus, HPWS refer to a group of separate but interconnected HR management practices designed to enhance employee and firm performance outcomes through improving workforce competence, attitudes, and motivation (Huselid, 1995). TM goes beyond putting practices together to achieve strategic goals. Therefore, different organizations can use different practices to achieve the same organizational goals.

Also, a main premise of this research stream is that a system of internally coherent HR practices aligned with an organizational strategy, rather than individual HR practices used in isolation, may enhance organizational performance (Lepak, et al., 2006). Literature has much empirical evidence in support of these HPWS or systems of HR practices (Lepak, et al., 2006). It is the combination of practices in a bundle, rather than individual practices, that shapes the pattern of interactions between and among managers and employees (Cutcher-Gershenfeld 1991). Thus, research that focuses on the impact of individual HR practices on performance may produce misleading results, with a single practice capturing the effect of the entire HR system (Ichniowski, Shaw, and Prennushi 1993).

Folllowing MacDuffie (1995), implicit in the notion of a "bundle" is the idea that practices within bundles are interrelated and internally consistent, and that "more is better" with respect to the impact on performance, because of the overlapping and mutually reinforcing effect of multiple practices.

From the HRM literature, the concept of HPWS has emerged as a core construct encompassing the extent to which firms invest in the attraction, selection, management, and retention of the best possible human capital (Huselid, 1995; Lepak et al., 2006; Lepak & Snell, 2002) linking HPWS with the new TM approach. That is, HPWS are indicative of the value firms place on their human capital as a source for competitive advantage, with higher HPWS levels indicative of greater value firms place in their human capital.

Summing up, from the literature about TM, we distinguish five main components: alignment with the organization, that is, the adequacy of value that involves commitment and loyalty; managers that manage with talent, that is, a model about how employees can use and relate with others concerning talent; talent effectiveness, that is, the way people can use their talents; the freedom and autonomy to use their talents, that is, a basic characteristic of empowerment; and last, the development of talent, that is, the update of feedback about competencies and the development of potential in organizations. Therefore we propose a model of TM to analyse its relationship with other organizational variables.

EMPIRICAL EVIDENCE

In this chapter we present two main empirical results. First, the role of strategy and organizational structure with HPWS on firm performance. Second, the role of job satisfaction as a mediator variable in the impact of TM over individual performance. Results from the first study of Camps and Luna-Arocas (2009) showed that the originality of this work lies in conveying the fact that in order for these companies competing in quality to implement HR commitment practices it is important that they have a strong company culture based on shared values. Once the image one has of the company is perceived homogenously, making important investments and efforts in achieving optimized paying systems, training workers or selecting them properly becomes more coherent (See Figure 1, Figure 2).

TM is associated with a high performance system in organizations. But from a behavioural perspective, one of the main indicators of a high performer employee is their own perception of the work and effort they do (Bailey, 1993; Kuvaas, 2008; Peccei & Rosenthal, 2001; Tam et al., 2002). In this sense, the main model proposed postulates that the application of a TM strategy will have a direct effect on individual performance. Although this is quite useful for building a model of TM as was expressed by Lewis & Heckman (2006), as earlier literature has indicated, other variables related with involvement/commitment and satisfaction are also important in explaining the effectiveness and impact of a TM strategy. In this sense the literature review supports the idea of a mediating role of job satisfaction in the TM→ Individual performance relationship.

Figure 1. Structural model results. Standard regression weights

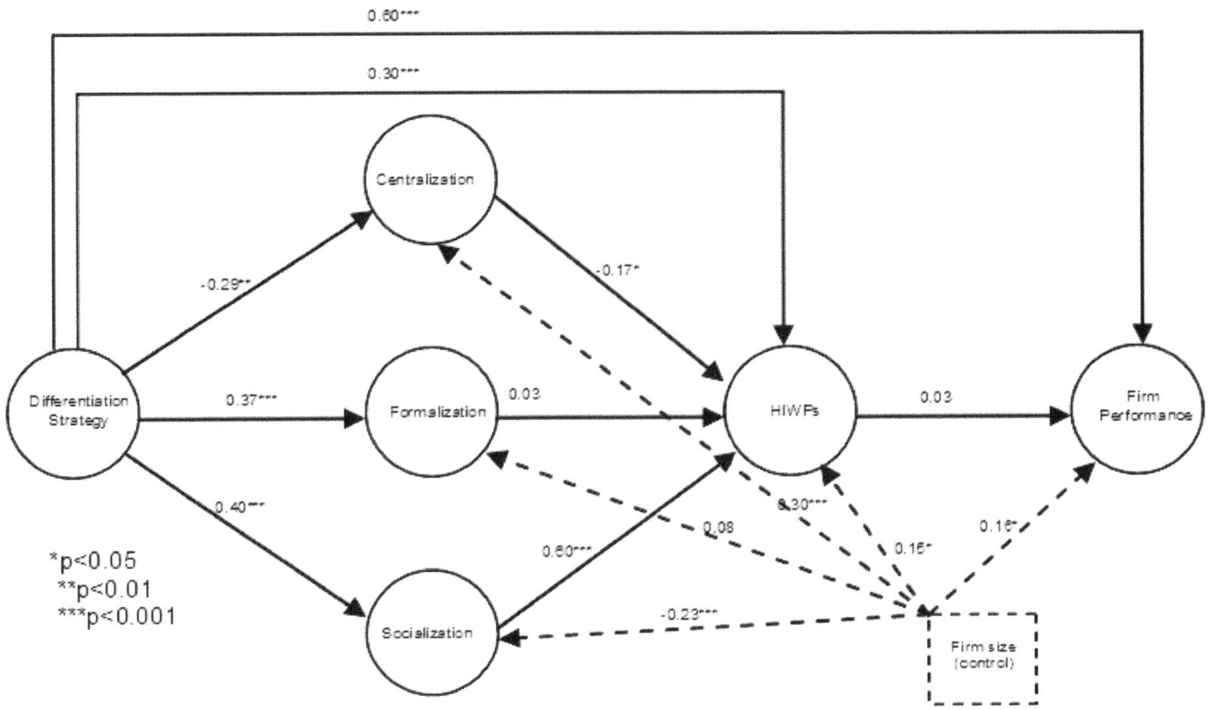

Figure 2. Model comparisons

Model	c^2 (d.f.)	c^2 difference (d.f.)	RMSA	CFI
Model 1: theoretical model	451.76 (379)		0.032	0.971
Model 2: centrali. → perform.		0.69 (1)		
Model 3: formali. → perform.		5.12 (1)*	0.032	0.973
Model 4: sociali. → perform.		0.18 (1)		

*p<0.025; each model compared to the theoretical model

Studies suggest that HR practices affect organizational outcomes by shaping employee behaviors and attitudes (Huselid, 1995). More specifically, HPWS increase organizational effectiveness by creating conditions where employees become highly involved in the organization and work hard to accomplish its goals, in other words, by increasing their employees' commitment to the organization and job satisfaction (Eisenberger et al., 1997). When Varma et al. (1999) carried out research into 39 organizations in order to evaluate the effectiveness of HPWS initiatives, they discovered that organizations that concentrated their changes on the internal work-culture of the company as well as on their HR practices as part of HPWS design achieved important improvements in operational and financial performance.

Several studies have linked HR practices with individual work performance (Kuvaas, 2008; Edgard & Geare, 2005). The basic assumption is that the HR practices from the best practices perspective improve motivation, skills, attitudes and behaviours of their employees (Kuvaas, 2008). Although there are some controversial results Wright and Boswell (2002) indicate that HR practices impact on individuals constitute an opportunity to research in the future. Lee and Bruvold (2003) argue that investment in employee development facilitate more obligation by employees toward the organizations, and this elicitate employees´ motivation to effort and work hard supporting organizational effectiveness. Also MacDuffie (1995) noted that to get a high performance organization linked to HR practices employees should be motivated to apply their skills and knowledge through discretionary effort. In this sense, TM is analysed as HPWS encompassing all the literature that links these practices with organizational performance. The added value from our perspective is a new model of HPWS called TM focused on a different general HR strategy. Also, the relationship between empowerment (development, one of the dimensions in TM) and performance has been contemplated in the literature (Logan & Ganster, 2007; Peccei & Rosenthal, 2001) (See Figure 3, Figure 4).

These results (Figure 3) reflect on the study that was presented at the International Human Resource Management Conference in Birminghan (Luna and Morley, 2010). Moreover, some new interesting research has been done by these authors relating also TM with Organizational Performance (growth and return of investment, perceived ROI) with the innovation strategy as mediator variable (Luna and Morley, 2011). Promising research is growing in the literature that sure will consolidate the talent management strategy concept and its powerful implications in the relation between micro and macro perspectives.

DISCUSSION OF RESULTS

Concerning the first research results from Camps and Luna (2009), the mediator role of organizational structure between organizational strategy and HR practices (in the sense used by Ferris et

Figure 3. Path analysis 3 TPs -> POP mediated by job satisfaction. MADE-C talent model.

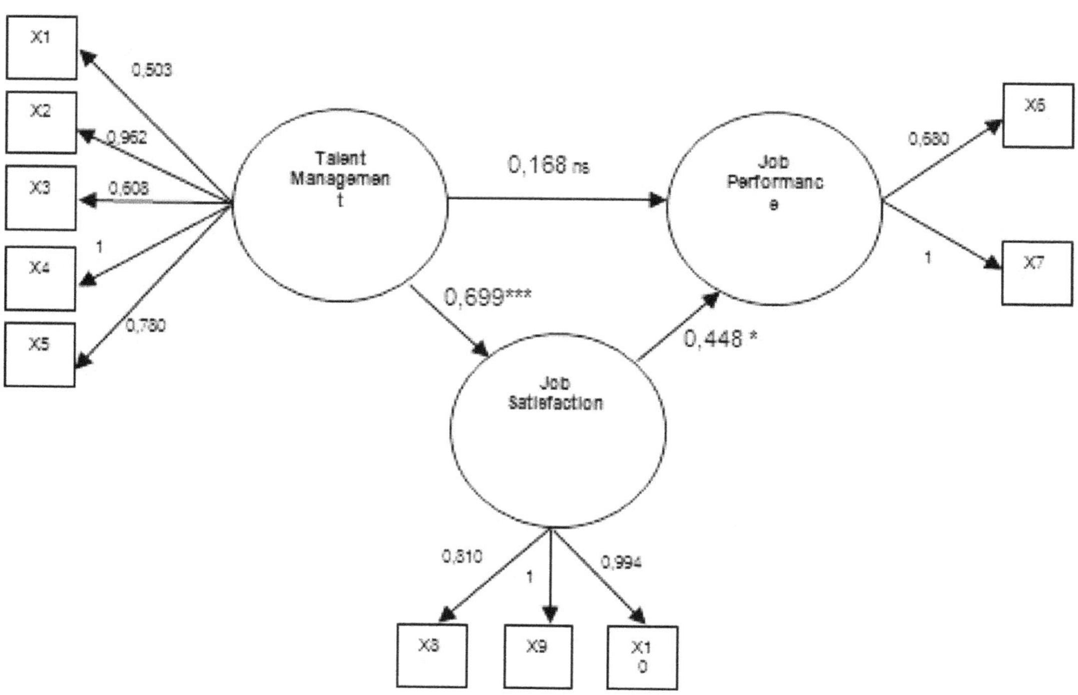

Figure 4. Path analysis TPs -> POP mediated by job satisfaction

Model	χ^2 (d.f.)	RMR	RMSA	GFI	NFI	CFI
Model 5: model	839,204 (45)	0,351	0,299	0,385	0,000	0,000
Model 6 TPs→ POP mediated by Job Satisfaction	89,159 (32)	0,052	0,095	0,916	0,894	0,928

al., 1999 and Bowen & Ostroff, 2004) needs to be clarified in an empirical way. Much work has to be done in that direction. The intention in this first study is to advance in this effort by selecting an organizational strategy (differentiation), a construct of HR practices configuring a HPWS, and a component of the organizational structure (control mechanisms), and by trying to show the mediator role of these control mechanisms between the other two constructs.

Results obtained in this first study provide support for most of the hypothesized relationships presented in the model. First, this research provides empirical evidence that differentiation strategy enhances HPWS. The causal link between differentiation strategy and HPWS has been not completely validated in the literature. To date the evidence in favour of this connexion has been mixed at best (Becker & Gerhart, 1996; Delery & Doty, 1996; Huselid, 1995). This first study supports, in line with much of this literature, the positive direct relationship between these two constructs.

Also, the mediator role of control mechanisms between differentiation strategy and HPWS has been partially supported. This study supports the literature that established the negative relationship between centralization and differentiation strategy, and between centralization and HPWS (Govindarajan & Fisher, 1990; Ashmos et al., 1990; Wooldridge & Floyd, 1990). Therefore, centralization mediates, in a negative way, the relationship between differentiation strategy and HPWS implementation. Moreover, the results of this study support the causal positive relationship between differentiation strategy and formalization. In this sense, the results of the study support the causal positive relationship between differentiation strategy and formalization, in contradiction with other studies (Govindarajan y Fisher, 1990). However, the hypothetic relationship between formalization and HPWS is not significant. Therefore, we can conclude that formalization level in an organization which follows a differentiation strategy does not affect the implementation of HPWS in a significant way. Also in this study a strong organizational culture clearly mediates the relationship between differentiation strategy and HPWS.

Finally, in the same line that some works such as Ostroff (2000) and Lee and Miller (1999), this study found a positive direct relationship between organizational strategy and firm performance.

These findings can be interpreted as an indication that within the differentiation strategy context, HPWS implementation is easier when strong values and low levels of centralization are found in the organization. An eminently current HR literature is therefore corroborated, which favours the intensification of a "humanizing" work practice, especially if the aim is to achieve a correct implementation of competitive strategies which require the workers' implication in the process.

Nowadays, there is general agreement that human capital can be a source of competitive advantage, and that HR practices have a direct influence on the human capital of a firm (Delery & Shaw, 2001).

The results of the second study (Luna and Morley, 2010) are especially in line with extent contributions to the literature that focus on how a HPWS, in this case TM, influences individual perceptions of the psychological contract (e.g. Rousseau & Greller, 1994), in this case Job Performance. Therefore, we conclude that strategies centred on TM affect directly Job Performance, but thorough Job Satisfaction and this can be done if we attract people close to the values of the organization (Alignment). Moreover, this staff should be measured and developed (Developing TM) in competencies, and empowered (Empowerment) as much as the position can extend. But mainly, employees should have the talent mindset, that is, the competence of TM (Talent Competence). This Talent Model indicates the importance of job satisfaction (JS) as a source of positive perceptions in organizations and points to how perceptions in this regard can influence the perceived individual performance of employees. The model does not postulate that we have to pursue JS as a main key to job performances (JP). On the contrary, it suggests that if we develop and institutionalise a comprehensive system of TM, this can affect both JS (directly) and JP (indirectly). Only in this sense we can understand both dependent variables.

More research should be done in order to integrate HPWS and TM literature. This will bring about practical implications in HRM. Also, recent studies are linking TM strategies with organizational performance. In this sense, TM literature has to demonstrate its contribution to the competitive advantage of organizations.

REFERENCES

Arthur, J. B. (1994). Effects of human resource systems on manufacturing performance and turnover. *Academy of Management Journal*, *37*(3), 670–687. doi:10.2307/256705

Ashmos, D. P., McDaniel, R. R., & Duchon, D. (1990). Differences in perception of strategic decision-making processes: The case of physicians and administrators. *The Journal of Applied Behavioral Science*, *26*, 201–218. doi:10.1177/0021886390262010

Ashton, C., & Morton, L. (2005). Managing talent for competitive advantage. *Strategic HR Review*, *4*(5), 28–31. doi:10.1108/14754390580000819

Bailey, T. (1993). *Discretionary effort and the organization of work, employee participation and work reform since Hawthorne*. Discussion Paper, Teacher College, University of Columbia, New York.

Barney, J. B. (1991). Firm resources and sustained competitive advantage. *Journal of Management*, *17*, 99–120. doi:10.1177/014920639101700108

Becker, B., & Gerhart, B. (1996). The impact of human resource management on organizational performance: Progress and prospects. *Academy of Management Journal*, *39*, 779–801. doi:10.2307/256712

Becker, B. E., & Huselid, M. A. (1998). High performance work systems and firm performance: a synthesis of research and managerial implications. In Rowland, K. M., & Ferris, G. R. (Eds.), *Research in personnel and human resource management* (pp. 53–101). Greenwich, CT: JAI Press.

Becker, B. E., & Huselid, M. A. (2006). Strategic human resource management: Where do we go from here? *Journal of Management*, *32*(6), 898–925. doi:10.1177/0149206306293668

Bhatnagar, J. (2007). Talent management strategy of employee engagement in Indian ITES employees: Key to retention. *Employee Relations*, *29*, 640–663. doi:10.1108/01425450710826122

Boudreau, J. W., & Ramstad, P. M. (2005). Talentship and the new paradigm for human resource management: From professional practises to strategic talent decision science. *Human Resource Planning, 28*(2), 17–26.

Bowen, D. E., & Ostroff, C. (2004). Understanding HRM-firm performance linkages: The role of the "strength" of the HRM system. *Academy of Management Review, 29*, 203–221.

Caligiuri, P. M. (2006). Developing global leaders. *Human Resource Management Review, 16*, 219–228. doi:10.1016/j.hrmr.2006.03.009

Calvo Babío, N., & García Rodríguez, R. (2010). Talent management in professional services firms: A HR issue? *The International Journal of Organizational Analysis, 18*(4), 392–411. doi:10.1108/19348831011081877

Camps, J., & Luna-Arocas, R. (2009). High involvement work practices and firm performance. *International Journal of Human Resource Management, 20*(5), 1056–1077. doi:10.1080/09585190902850273

Cappellli, P. (2008). Talent management for the twenty-first century. *Harvard Business Review, 86*(3), 74, 76–81.

Casado, J. M. (2011). *El mantra del talento.* Deusto.

Chiavenato, I. (2002). *Gestión del talento humano.* McGraw-Hill.

Collins, C. J., & Smith, K. G. (2006). Knowledge exchange and combination: The role of human resource practices in the performance of high-technology firms. *Academy of Management Journal, 49*, 544–560. doi:10.5465/AMJ.2006.21794671

Colvin, A., & Boswell, W. (2007). The problem of action and interest alignment: Beyond job requirements and incentive compensation. *Human Resource Management Review, 17*(1), 38–51. doi:10.1016/j.hrmr.2006.11.003

Combs, J., Liu, Y., Hall, A., & Ketchen, D. (2006). How much do high-performance work practices matter? A meta-analysis of their effects on organizational performance. *Personnel Psychology, 59*(3), 501–528. doi:10.1111/j.1744-6570.2006.00045.x

Credit Union Management. (2004, November 7). Many managers lack management skills. www.Cumanagement.org.

Cubeiro, J. C. (2004). *En un lugar del talento.* Pearson Education.

Cutcher-Gershenfeld, J. (1991). The impact on economic performance of a transformation in industrial relations. *Industrial & Labor Relations Review, 44*, 241–260. doi:10.2307/2524806

Datta, D., Guthrie, J., & Wright, P. (2005). Human resource management and labor productivity: does Industry matter? *Academy of Management Journal, 48*(1), 135–145. doi:10.5465/AMJ.2005.15993158

Delery, J. E., & Doty, D. H. (1996). Modes of theorizing in strategic human resource management: Tests of universalistic, contingency, and configurational performance predictions. *Academy of Management Journal, 39*(4), 802–836. doi:10.2307/256713

Delery, J. E., & Shaw, J. D. (2001). The strategic management of people in work organizations: Review, synthesis and extension. *Research in Personnel and Human Resources Management, 20*, 165–197. doi:10.1016/S0742-7301(01)20003-6

Edgard, F., & Geare, A. (2005). HRM practice and employee attitudes: Different measures – Different results. *Personnel Review, 34*, 534–549. doi:10.1108/00483480510612503

Eisenberger, R., Cummings, J., Armeli, S., & Lynch, P. (1997). Perceived organizational support, discretionary treatment, and job satisfaction. *The Journal of Applied Psychology, 82*(5), 812–820. doi:10.1037/0021-9010.82.5.812

Farndale, E., Scullion, H., & Sparrow, P. (2010). The role of the corporate HR function in global talent management. *Journal of World Business, 45*(2), 161–168. doi:10.1016/j.jwb.2009.09.012

Ferris, G. R., Arthur, M. M., Berkson, H. M., Kaplan, D. M., Harrell-Cook, G., & Frink, D. W. (1998). Toward a social context theory of the human resource management–organization effectiveness relationship. *Human Resource Management Review, 8*, 235–264. doi:10.1016/S1053-4822(98)90004-3

Frank, F. D., & Taylor, C. R. (2004). Talent management: Trends that will shape the future. *Human Resource Planning, 27*(1), 33–41.

Goffee, R. E., & Jones, G. (2007). Leading clever people. *Harvard Business Review, 85*(3), 72–79.

Govindarajan, V., & Fisher, F. (1990). Strategy, control systems, and resource sharing: Effects on business unit performance. *Academy of Management Journal, 33*(2), 259–285. doi:10.2307/256325

Gubman, E. (1996). The gauntlet is down. *The Journal of Business Strategy*, (Nov/Dec): 33–35.

Gupta, A. (2001). Talent management in crisis. *Mid-American Journal of Business, 16*(2), 2.

Guthrie, J. (2001). High involvement work practices, turnover and productivity: Evidence from New Zealand. *Academy of Management Journal, 44*(1), 180–190. doi:10.2307/3069345

Hagan, C. M., Konopaske, R., Bernardin, H. J., & Tyler, C. L. (2006). Predicting assessment center performance with 360-degree, top-down, and customer-based competency assessments. *Human Resource Management, 45*(3), 357–390. doi:10.1002/hrm.20117

Hamel, G., & Prahalad, C. K. (1994). *Competing for the future*. Boston, MA: Harvard Business School Press.

Hayton, J. C., & McEvoy, G. M. (2006). Competencies in practice: An interview with Hanneke C. Frese. *Human Resource Management, 45*(3), 495–500. doi:10.1002/hrm.20126

Hiltrop, J. M. (1999). The quest for the best: Human resource practices to attract and retain talent. *European Management Journal, 17*(4), 422–430. doi:10.1016/S0263-2373(99)00022-5

Hirsch, W. (2000). *Succession planning demystified, report 372*. Brighton: Institute of Employment Studies.

Huselid, M. A. (1995). The impact of human resource management practices on turnover, productivity and corporate financial performance. *Academy of Management Journal, 38*(3), 635–672. doi:10.2307/256741

Ichniowski, C., Shaw, K., & Prennushi, G. (1993). *The effects of human resource management practices on productivity.* Working paper, Graduate School of Industrial Administration, Carnegie Mellon University.

Ichniowski, C., Shaw, K., & Prennushi, G. (1997). The effects of human resources management practices on productivity: A study of steel finishing lines. *The American Economic Review, 87*(3), 291–313.

Jackson, S. E., & Schuler, R. S. (1995). Understanding human resource management in the context of organizations and their environments. *Annual Review of Psychology, 46*, 237–264. doi:10.1146/annurev.ps.46.020195.001321

Jerico, P. (2000). *La gestión del talento: Del profesional con talento al talento organizativo*. Pearson Education.

Kochanski, J., & Ruse, D. (1996). Designing a competence-based human resources organization. *Human Resource Management, 35*, 19–34. doi:10.1002/(SICI)1099-050X(199621)35:1<19::AID-HRM2>3.0.CO;2-#

Kuvaas, B. (2008). An exploration of how the employee-organization relationship affects the linkage between perception of developmental human resource practices and employee outcomes. *Journal of Management Studies*, *45*(1), 1–25.

Lawler, E. E. III. (2005). From human resource management to organizational effectiveness. *Human Resource Management*, *44*(2), 165–169. doi:10.1002/hrm.20059

Lee, C., & Bruvold, N. T. (2003). Creating value for employees: Investment in employee development. *International Journal of Human Resource Management*, *14*, 981–1000. doi:10.1080/0958519032000106173

Lee, J., & Miller, D. (1999). People matter: Commitment to employees, strategy and performance in Korean firms. *Strategic Management Journal*, *20*(6), 579–593. doi:10.1002/(SICI)1097-0266(199906)20:6<579::AID-SMJ37>3.0.CO;2-C

Lengnick-Hall, M. L., & Andrade, L. (2008). Talent staffing systems for effective knowledge management. In Vlaiman, V., & Vance, C. (Eds.), *Smart talent management: Building knowledge capital for competitive advantage*. Edward Elgar Publishing.

Lepak, D. P., Liao, H., Chung, Y., & Harden, E. H. (2006). A conceptual review of human resource management systems in strategic human resource management research. In Martocchio, J. J. (Ed.), *Research in personnel and human resources management* (*Vol. 25*, pp. 217–271). Oxford, UK: Elsevier. doi:10.1016/S0742-7301(06)25006-0

Lepak, D. P., & Snell, S. A. (2002). Examining the human resource architecture: The relationships among human capital, employment, and human resource configurations. *Journal of Management*, *28*, 517–543.

Lewis, R. E., & Heckman, R. J. (2006). Talent management: A critical review. *Human Resource Management Review*, *16*, 139–154. doi:10.1016/j.hrmr.2006.03.001

Logan, M. S., & Ganster, D. C. (2007). The effects of empowerment on attitudes and performance: The role of social support and empowerment beliefs. *Journal of Management Studies*, *44*(8), 1523–1550.

Luna, R., & Morley, M. (2011). *Talent management and organizational performance: The mediator role of innovation*. Communication at the VIII International Workshop on Human Resource Management, 12-13th May, Sevilla (Spain).

Luna-Arocas, R., & Camps, J. (2008). A model of high performance work practices and turnover intentions. *Personnel Review*, *37*(1), 26–46. doi:10.1108/00483480810839950

Luna-Arocas, R., & Morley, M. (2010). *Strategic management and job performance: The mediating role of job satisfaction*. Communication at the 11th International Human Resource Management Conference, 9-12th June, Birmingham (UK).

MacDuffie, J. P. (1995). Human resource bundles and manufacturing performance: Organizational logic and flexible production systems in the world auto industry. *Industrial & Labor Relations Review*, *48*, 197–221. doi:10.2307/2524483

McCauley, C., & Wakefield, M. (2006). Talent management in the 21st century. *Journal for Quality and Participation*, *29*(4).

Michaels, E., Handfield-Jones, H., & Axelrod, B. (2001). *The war for talent*. Harvard Business School Press.

O'Reilly, C. A., & Pfeffer, J. (2000). Unlocking the hidden valued in organizations. *Employment Relations Today*, *27*(2), 63–80. doi:10.1002/ert.3910270208

Ostroff, C. (2000). *Human resource management and firm performance*. Working paper, Arizona State University.

Parsons, G. D., & Pascale, R. T. (2007). Crisis at the summit. *Harvard Business Review*, *85*(3), 80–89.

Peccei, R., & Rosenthal, P. (2001). Delivering customer-oriented behaviour through empowerment: An empirical test of HRM assumptions. *Journal of Management Studies*, *38*, 831–857. doi:10.1111/1467-6486.00261

Peters, T. (2005). *Tom Peters essentials: Talent*. Dorling Kindersley.

Pfeffer, J., & Sutton, R. I. (1999). Knowing "what" to do is not enough: Turning knowledge into action. *California Management Review*, *42*(1), 83–108.

Prahalad, C. K., & Hamel, G. (1990). The core competence of the corporation. *Harvard Business Review*, *68*(3), 79–91.

Ray, G., Barney, J. B., & Muhanna, W. A. (2004). Capabilities, business processes, and competitive advantage: Choosing the dependent variable in empirical tests of the resource-based view. *Strategic Management Journal*, *25*, 23–37. doi:10.1002/smj.366

Rousseau, D. M., & Greller, M. M. (1994). Human resource practices: Administrative contract makers. *Human Resource Management*, *33*(3), 385–401. doi:10.1002/hrm.3930330308

Schippmann, J. S., Ash, R. A., Battista, M., Carr, L., Eyde, L. D., & Hesketh, B. (2000). The practice of competency modeling. *Personnel Psychology*, *53*, 703–740. doi:10.1111/j.1744-6570.2000.tb00220.x

Schlemenson, A. (2003). *The strategy of the talent*. Paidos.

Schuler, R. S., & Jackson, S. E. (1987). Linking competitive strategies with human resource management practices. *The Academy of Management Executive*, *1*, 207–219. doi:10.5465/AME.1987.4275740

Sharma, R., & Bhatnagar, J. (2009). Talent management - Competency development: Key to global leadership. *Industrial and Commercial Training*, *41*(3), 118–132. doi:10.1108/00197850910950907

Shih, H., Chiang, Y., & Hsu, C. (2006). Can high performance work systems really lead to better performance? *International Journal of Manpower*, *27*(8), 741–763. doi:10.1108/01437720610713530

Sternberg, R. (1999). The theory of successful intelligence. *Review of General Psychology*, *3*(4), 292–316. doi:10.1037/1089-2680.3.4.292

Sun, L. Y., Aryee, S., & Law, K. S. (2007). High performance human resources practices, citizenship behavior and organizational performance: A relational perspective. *Academy of Management Journal*, *50*, 558–577. doi:10.5465/AMJ.2007.25525821

Takeuchi, R., Lepak, D. P., Wang, H., & Takeuchi, K. (2007). An empirical examination of the mechanisms mediating between high performance work systems and the performance of Japanese organizations. *The Journal of Applied Psychology*, *92*(4), 1069–1083. doi:10.1037/0021-9010.92.4.1069

Tam, Y. M., Korczynski, M., & Frenkel, S. J. (2002). Organisational and occupational commitment: Knowledge workers in large corporations. *Journal of Management Studies*, *39*(6), 775–801. doi:10.1111/1467-6486.00311

Tansley, C., Turner, P., & Foster, C. (2007). *Talent: Strategy, management, measurement. Research into practice*. London, UK: Chartered Institute of Personnel and Development.

Tolich. (2005). Phillip Brown and Anthony Hesketh: The mismanagement of talent: Employability and jobs in the knowledge economy. *Administrative Science Quarterly, 50*(2), 306-307

Ulrich, D. (1997). Measuring human resources: An overview of practice and a prescription for results. *Human Resource Management, 36*(3), 303–320. doi:10.1002/(SICI)1099-050X(199723)36:3<303::AID-HRM3>3.0.CO;2-#

Ulrich, D., Brockbank, W., Yeung, A., & Lake, D. G. (1995). Human resource competencies: An empirical assessment. *Human Resource Management, 34*(4), 473–495. doi:10.1002/hrm.3930340402

Ulrich, D., Brockbank, W., Yeung, A. K., & Lake, D. G. (1995). Human resource competencies: An empirical assessment. *Human Resource Management, 34*(4), 473–495. doi:10.1002/hrm.3930340402

Van der Heijde, C., & van der Heijden, B. (2006). A competence-based and multidimensional operationalization and measurement of employability. *Human Resource Management, 45*(3), 449–476. doi:10.1002/hrm.20119

Varma, A., Beatty, R. W., Schneier, C. E., & Ulrich, D. O. (1999). High performance work systems: Exciting discovery or passing fad? *Human Resource Planning, 22*(1), 26–38.

Way, S. A. (2002). High performance work systems and intermediate indicators of firm performance within the US small business sector. *Journal of Management, 28*(6), 765.

Wernerfelt, B. (1984). A resource-based view of the firm. *Strategic Management Journal, 5*, 171–180. doi:10.1002/smj.4250050207

Wooldridge, B., & Floyd, S. W. (1990). The strategy process, middle management involvement, and organizational performance. *Strategic Management Journal, 11*, 231–241. doi:10.1002/smj.4250110305

Wright, P., & McMahan, G. (1992). Theoretical perspectives for strategic resource management. *Journal of Management*, 18.

Wright, P. M., & Snell, S. A. (1998). Toward a unifying framework for exploring fit and flexibility in strategic human resource management. *Academy of Management Review, 23*, 756–772.

Zuboff, S. (1988). *In the age of the smart machine: The future of work and power*. New York, NY: Basic Books.

KEY TERMS AND DEFINITIONS

Competency Based Management (CBM): Individual´s demonstrated knowledge, skills or abilities (competencies) aligned to organizational values and goals.

High Performance Work System (HPWS): Practices, as a system, directly related with organizational performance.

Human Resources System: Perspective of HR that considers all practices interrelated and influencing each other.

Talent Management (TM): Strategies developed in the firm to get the most for each employee in coherence with managerial goals.

Talent Mindset: Employee´s mentality focused on quality and excellence performance.

Chapter 4
Ethics and Learning Organizations in the New Economy

Alexis Bañón
Polytechnic University of Valencia, Spain

Manuel Guillén
University of Valencia, Spain

Ignacio Gil
Polytechnic University of Valencia, Spain

ABSTRACT

In the context of the new economy, business organizations need to learn faster, and to maintain and to improve knowledge, producing creative solutions based on their knowledge, skills, and new technologies to develop a customer responsive culture in a more economic and efficient way.

In order to achieve this, CEOs and Human Resource (HR) policies should potentially contribute to knowledge development by creating authentic learning organizations. The authors propose in this study that learning improvements in organizations are not just a matter of techniques or aptitudes, but also a matter of feelings, attitudes, and, above all, of the moral habits of their members. The authors strongly suggest complementing currently established conceptions of knowledge management and organizational learning through an explicit inclusion of ethics and ethical learning in organizations.

INTRODUCTION

In the context of the New Economy, but especially in a moment of general economic crisis, organizations require the ability to understand what is happening outside their environment in order to create a competitive advantage. And what is probably more important, new business organizations need to learn faster, and to maintain and to improve knowledge, producing creative solutions based on their knowledge, skills and new technologies to develop a customer responsive culture in a more economic and efficient way.

DOI: 10.4018/978-1-61350-207-5.ch004

In order to achieve this, CEOs and Human Resource (HR) policies should potentially contribute to knowledge development by creating authentic learning organizations. These organizations enable a learning environment for all members to consciously transform organizations and their contexts into situations "where people continually expand their capacity to create the results they truly desire, where new and expansive patterns of thinking are nurtured, where collective aspiration is set free, and where people are continually learning to see the whole together" (Senge, 1990, p. 3).

Moreover, we propose in this study that learning improvements in organizations are not just a matter of techniques or aptitudes but, also, a matter of feelings, attitudes and, above all, of the moral habits of their members. In this chapter, we strongly suggest complementing currently established conceptions of knowledge management and organizational learning through an explicit inclusion of ethics and ethical learning in organizations, to build up ethically healthy organizations where the process of learning is easier than in ethically unhealthy organizations.

Our purpose is to show that organizational ethical healthiness is an essential facilitator of the processes of learning by learning organizations in the context of the New Economy and global crisis.

This chapter is divided into five sections. The first section analyzes organizational learning and the concept of learning organizations. The second describes the explicit need to consider ethics and ethical learning competence among agents in a learning organization context. The third section distinguishes between ethically healthy organizations from ethically unhealthy organizations, as concepts that help to better understand the ethical quality of an organization. The fourth section lays down the argument that the ethical healthiness of the organization is an essential contextual facilitator for learning processes and for constructing learning organizations. Finally, some conclusions and future lines of research are put forward.

1. LEARNING ORGANIZATIONS

For many years, we have been hearing that we live in a "knowledge society" (Toffier, 1990; Bell, 1973; Drucker, 1968). As is widely accepted today, the concept of knowledge is based on two primary elements: information (explicit knowledge) and know-how (tacit knowledge) (Simmonds et al., 2001; Nonaka, 1991).

Information is considered to be the knowledge that can be transmitted without loss of meaning and truth, once the syntactical rules required for interpreting it are known. Thus, knowledge as information implies knowing what something means, and that it can be written down (Grant, 1996; Nonaka, 1994). In this sense, defining knowledge as information whose validity has been established through evidence allows distinguishing it from opinion, speculation, beliefs, or other types of unproven information (Liebeskind, 1996).

On the other hand, know-how, as tacit knowledge, is a much more complex concept than information. It can be defined as the accumulated practical skill or experience through time that allows one to do something efficiently. Therefore it is a personal quality -that involves both cognitive and technical elements- difficult to formalize and to communicate because it is not easy to write down (Grant, 1996; Nonaka, 1994).

Knowledge appears then as the key element to define the learning process (individual and organizational) because it can be understood as the result of transforming information into knowledge (Nonaka, 1994), although other understandings conceive learning as an outcome (Nicolini and Meznar, 1995; Dodgson, 1993).

But, if we understand learning as a process, can we distinguish individual learning from organizational learning? Is it possible to distinguish different types of learning according to the subject of the learning involved, individual or organizational? According to Weick (1991) "individual learning occurs when people give a different response to the same stimulus" while "organizational learning occurs when groups of people give the same response to different stimuli" (p. 121). Therefore, it seems reasonable to talk about organizational learning as something different from individual learning (Kim, 1993). The question now is if we can associate organizational learning to a well-managed knowledge organization or if having organizational learning is not necessarily the same as managing knowledge correctly. In other words, is organizational learning of an organization the same as being at the head of a learning organization?

According to the literature, learning organizations appeared as a result of the adjustments and pressures modern organizations had to face to remain competitive in the business environment (O'Keeffe, T. 2002). These types of organizations facilitate the learning of their members to be able to continuously transform themselves (Pedler *et al.*, 1997). They accomplish their goals based on a culture of trust that generates free and reliable communication, allowing co-operation between individuals and groups (Argyris, C. 1999), mainly focusing on five features: systems thinking, personal mastery, mental models, shared vision and team learning (Senge, P.M. 1990). All the above mentioned contextualizes Senge's (1990) definitions of Learning Organization quoted by Malhotra (1996) as the organization as "a group of people continually enhancing their capacity to create what they want to create".

Academic literature does not always agree as to the answer to the question of whether the Learning Organization and Organizational Learning are different concepts (McGill et al., 1992) Nevertheless, we understand that Organizational Learning and Learning Organization are distinguishable based on Ang & Joseph's (1996) idea that differentiates process -Organizational Learning- from structure -Learning Organization-. In this sense, and according to Malhotra (1996), a Learning Organization is an "Organization with an ingrained philosophy for anticipating, reacting and responding to change, complexity and uncertainty" (p. 2).

In synthesis, all these authors suggest that the Learning Organization creates the conditions for a constant Learning Organizational Environment. But, what are those conditions, or what should those conditions be?

2. EXPLICIT CONSIDERATION OF ETHICS AND ETHICAL LEARNING

Organizations where knowledge is managed have in common the fact that knowledge (tacit and explicit) is shared between human beings. This premise is not only true for creating and processing knowledge, but also for its management. But, what kind of knowledge should people and organizations acquire? Is it enough with information and know how? Knowledge has different effects and, broadly understood, its concept not only develops theoretical or scientific dimensions (*knowing*) and practical or technical-artistic ones (*knowing how*) but, also, ethical conduct (*knowing how to live*) (Guillén, Fontrodona and Rodriguez, 2007). Therefore, we strongly suggest that humans must learn by acquiring not only knowledge and skills (technical habits), but also moral virtues (ethical habits).

Unfortunately, moral dimension has received scant attention in knowledge management literature, but also in general management. As Sumantra Ghoshal (2005) plainly described, "many of the worst excesses of recent management practices have their roots in a set of ideas that have emerged from business school academics over the last 30 years", the question is that "by propagating ideologically inspired amoral theories, business schools have actively freed their students from any sense of moral responsibility" (2005, p. 75).

Morality or ethics has to do with the right and proper completion of a human person with human excellence (through rational judgement and practical action). The moral dimension demands a kind of moral or ethical learning understood as the acquisition of theoretical moral principles of conduct and of practical moral virtues (Guillén, 2006). And in this sense, ethical knowledge or competence could be described as the possession of certain principles or behavioural norms for achieving good ethics, combined with practical habits for personal improvement. Some of these relevant virtues described in business ethics literature are fairness, responsibility, loyalty, honesty, integrity, humility, tolerance, determination, enthusiasm and courage, among others (Solomon, 1992).

By broadening the object of management to the study of moral human conduct, a new line of research opens up that complements the current findings in social sciences and allows the study of aspects that have not been discussed in previous research. We maintain that an assumed evaluative ethical "asepsis" ends up being extremely limited in understanding human behaviour in organizations. The contribution of ethics complements the study of organizations and the behaviour of its members by addressing it in terms of good or bad in relation to how it affects the improvement of the person or people involved and of society as a whole, but also explaining some requisites that are indispensable to understand organizations as groups of human beings.

In this way, together with the climate of desirable and demandable *competencies* and *skills –technical habits–* for a correct management of knowledge, we consider it important to complement the current scope by also considering *ethical or moral habits*. This allows us to take a deeper look at aspects such as fairness, trust, reciprocity or commitment, essential for explaining learning organizations.

Ethical knowledge is both theoretical and practical, and implies a way of growing in the human sense, of becoming a better human being, or what the ancient Greeks called being "virtuous".

Ethical or moral virtue (human excellence in Greek) is a stable character habit in people that enables the achievement of ethical goodness. The habits of the will are those denominated by the Greek philosopher as moral or ethical virtues. As we have just stated, to act correctly does not only require a sound choice of methods, but also a correct inclination towards the goals. In reference to the acquisition of these habits, Aristotle wrote: "we become just by doing just acts, temperate by doing temperate acts, brave by doing brave acts" (Nicomachean Ethics, II, 1, 1103b). The moral virtues are those operative habits which make human beings behave in a way which contributes to making their personalities flourish.

For example, the virtue of sincerity is the habit required as a result of repeatedly telling the truth. In the same way as occurs in other areas of theoretical and practical learning, such as in sports or in handling vehicles, repeating acts generates habits. In a moral sense, moral habits are generated, and are called virtues when they contribute to enriching the person as a human being. Someone who habitually tells the truth acquires the habit of sincerity, while a person who is normally fair acquires the moral virtue of fairness, and the same thing happens with any number of other virtues such as prudence, strength, hard work, order or patience.

Both, moral virtue and competence are kinds of human knowledge. While competence embodies notions of learned skills and technical efficiency, *virtues* offer perspectives to shape the way we live bounded up in ideas of morality (Macaulay and Lawton 2006). Virtue and competence are distinguishable concepts although they are interrelated because they require one another: "Competencies embody certain virtues, whereas virtues require competence in order to successfully implement them through virtuous actions" (Macaulay and Lawton 2006, p. 702).

Having arrived at this point, the question should be asked: What exactly is the relevance of ethical knowledge and learning, and how does it affect learning organizations? To be able to answer to this question, we should first consider how the ethical personal level of analysis affects the organizational level.

3. THE ETHICAL HEALTHINESS OF THE ORGANIZATION

In this section, we examine in more detail the role played by the organization in terms of *means* or context, and in terms of a climate capable of enabling or impeding the process of transmission, reception and creation of knowledge on the part of its members. The interesting question to be addressed here lies in analyzing to what extent the ethical quality of the organization affects this process.

If we look at the concept of a "virtuous person," by analogy, we can probably discover elements that could be present in the concept of a "virtuous organization". Collier (1995) described "virtuous organizations" based on the idea that the organization is a "moral person" and that it is meaningful to speak about organizations as "virtuous". She then lists four features of virtuous organizations: appropriate purpose, discernment of purpose in relation with human flourishing, action to fulfil the purpose and qualities to attain goods. The first two characteristics are related to principles and judgement, and the second two have to do with practices.

In a similar way, avoiding the debate of considering organizations as moral individuals, Guillén (2006) describes the ethical quality of an organization by using the clinical analogy and by distinguishing between ethically healthy and unhealthy organizations. This study adopts the use of such terminology.

3.1. Ethically Unhealthy Organizations

An ethically unhealthy organization "is that which contributes, to a lesser or greater extent, to the human *impoverishment* of its members, of the people involved in achieving its mission and those affected by its activity" (Guillén, 2006, p.36). In biological terms, we say that an organism is sick when its members and the functions they perform suffer some kind of pathology or dysfunction. In ethical terms, the "pathology" is equivalent to actions that lack a human quality, without moral competence or virtues such as honesty, sincerity or veracity, to give just a few examples.

If a member of the organization ends up being selfish, bigheaded or a liar as a result of belonging to that organization, we can state that this is the case of an unhealthy organization; an organization that destroys or impoverishes the human quality of the people in the organization and those it serves.

Does the organization I work for allow me to grow in human virtues? Am I better or worse as a person since I began working at that organization? If the answer is no, if instead of being more sincere, more hardworking, more generous or more honest we are the complete opposite, then it is likely that the organization in which we work suffers some kind of ethical illness and in such cases, it might be a good idea to ponder to what extent we are one of the causes of that situation.

The ethical sickness of an organization can be explained by the poor moral quality of its aims or its means, or of the people that make it up (its moral competence). This would be the case of organizations in which the objectives or reward systems might lead to dehumanizing their workers. Generally speaking, any management mechanism an organization uses and the people that manage it have an influence on the behaviour of its members, in the same way as they affect it by giving a good or bad example. When these aims, means or people lead to a deterioration of the moral virtues of members, then we are dealing with an ethically unhealthy firm.

The ethically unhealthy firm destroys the human element, and thus does the same to human relationships in the deepest sense, as it does with the trust these relationships are based upon. Although it may be efficient in economic terms, this organization is destructive in ethical terms and is doomed to a lack of union, to disintegration.

The same happens to ethically unhealthy firms as it does to organisms under the attack of some type of virus; it tends to spread and propagate to the other cells. This would be the case of negative behaviour or reactions guided by jealousy, envy or suspicion. To deny this fact would be to deny the reality of the ethical condition of human beings, elements that constitute the organization.

Just as we can refer to the ethical sickness of an organization, we can also talk of the opposite; the ethical health of an organization. When ethically positive behaviour is accepted, praised, and promoted within an organization, we have what can be classed as a "healthy" organization in the ethical sense.

3.2. Ethically Healthy Organizations

An ethically healthy organization "is one that contributes, to a greater or lesser extent, to the human enrichment of its members, of the people involved in achieving its mission and those affected by its activity" (Guillén, 2006, p. 37). The ethical health of an organization can be explained by the good moral quality of its aims, of its means or the people that constitute it. An organism is healthy when its members and the functions they carry out are correct. In ethical terms, correctness is equivalent to acting with human quality in line with moral standards.

The ethical health of an organization can be explained by the possibilities it provides to its members for their personal human enrichment, for their growth in virtues, as well as those of the people it serves. All of this can be attained precisely via the development and practices of the work of each of its members.

Some examples might help to further clarify this concept. "In an ethically healthy organization, constructive and not destructive criticism is habitually provided; people are able to work in teams, members are aware that this enriches people as individuals and the organization as a whole, both at an intellectual level and in terms of habits and emotions; collaboration with others is common, members do not try to "trip each other up"; knowledge is shared experience and does not remain with isolated members for selfish reasons or due to a fear of opportunism; people cooperate to work towards the common good of the organization and of society in general, not exclusively for self gain: in short, people attempt to work using their human qualities, thereby contributing to "building" trust and hence building the organization and stable human relationships" (Guillén, 2006, p. 37).

The concept of "ethical health" is a gradual one, i.e. health is understood as the absence of sickness, but this sickness can vary in size. Indeed, and talking in the strictest sense, even in healthy organisms, there are always small imperceptible pathologies. Precisely because it is people that make up organizations, moral defects are always present and can lead to small anomalies, misunderstandings, etc. Therein lays the possibility of making mistakes, of doing wrong. This is a fact of universal experience, all human beings do things wrong in moral terms. It is for this very reason that the concept of ethical health, in the same way as biological health, allows us to apply suitable remedies, to make rectifications after making mistakes. We are thus faced with a concept of moral health that is dynamic, that can improve or worsen, that grows or decreases, that is always gradual, precisely because human beings are free and we are responsible for our own actions, and are capable of recognizing mistakes and of finding the means to change and rectify. In this sense, it is possible to talk of degrees in the ethical quality of organizations. Perfect health would not exist, but it is possible to have a permanent trend towards that goal.

Following this logic, the healthy organism is not only free of sickness, but its health gets better the more it moves towards a complete state of wellbeing, towards total health. The concept of excellence can thus be used to qualify a permanent search for ethical health. Moral excellence presupposes the existence of ethical health along with a *permanent and complete* effort towards improvement. One can thus define an ethically excellent organization as "one that permanently makes an effort to contribute to the *full* human development of *all* its members, of *all* the people involved in achieving its mission and *all* those affected by its activity" (Guillén, 2006, p. 38).

Evidently, the concepts of ethical *sickness, health* and *excellence* are not exact terms, as they refer to human groups in a permanent state of dynamism. In an excellent organization, there may be, and indeed there is, occasional behaviour of a lower ethical quality. However, we can talk of excellence when this dimension forms a habitual part of the means, aims and behaviour of the people in the organization. When the ethical quality of the behaviour of members of the organization ceases to be a common goal for all, we may still be talking about organizations that are more or less ethically healthy, but we can no longer state that they are excellent.

In the same way that the relationship between health and excellence is a question of degree, in a positive sense, so is the relationship between health and sickness. When we talk of ethical sickness, this emphasizes the existence of more or less stable behaviour that encourages the destruction of the human qualities in people, their debasement, and their lack of union, and in the long-term, if measures are not taken, the *dysfunction* of the organization.

The term ethical health therefore allows us to describe organizations that, in general terms, do not encourage dishonest, unfair, deceitful, damaging behaviour. They are thus organizations that allow or even contribute to human development. To the extent that this contribution becomes a mission and permanent task in the organization, with the stable purpose of the continuous improvement of all those affected by the activity of the organization, we can then talk of a total ethical bill of health, of complete ethical quality or ethical excellence.

Having reached this point, and returning to the question that we proposed in previous section: To what extent can we say that the ethical quality of the organization influences or may influence the process of the transmission, reception and creation of knowledge, the process of learning in organizations?

4. THE ROLE OF ORGANIZATIONAL ETHICAL HEALTHINESS IN ORGANIZATIONAL LEARNING

Learning organizations find themselves in a permanent process of knowledge communication. We suggest that this communication requires members in the organization that are capable of being good transmitters, receivers, and also good message creators. In order to explain the role of organizational ethical healthiness this section firstly focuses on an individual level of knowledge transmission and creation understood as a process of constant dialogue, and then moves on to examine the organizational plane.

4.1. Understanding Learning Processes as a Permanent Dialogue

Particular people acquire knowledge in organizations and this knowledge is communicated, by in large, through dialogue to the other members. Each individual can play different roles or go through different stages of the process of communicating knowledge.

The first stage involves a *knowledge receiver*, understood as the person that receives knowledge that allows him/her to carry out tasks, but also to develop skills, competencies and habits that he/she can then maintain, which do not necessarily have an impact on the rest of the organization. The analogy here could be that of a "pool" or store of knowledge. Logically, people can be either good or bad pools of knowledge.

A second stage of the process of the communication of knowledge is that which acts as a knowledge transmitter, i.e. the person that transfers acquired knowledge to other members of the organization. Closing the circle would mean a person who is both, a receiver and transmitter of knowledge, the stage where they do not only receive but also impart knowledge either by informing or by training (teaching). Returning once more to the hydraulic analogy, this person would represent an excellent channel for knowledge, as this channel not only receives, but also gives.

But the transmission and reception of knowledge alone (explicit and tacit) are insufficient. Today, it is not enough to have good pools and channels of knowledge. Firms in the New Economy are complex organizations facing complex environments. Their reality does not allow them to make do with faithfully receiving and transmitting what they have learnt. They need *all* their members to vindicate their condition as free people with understanding, free will and feeling for creating value in their tasks via a level of involvement that generates new knowledge, ideas, principles, values, procedures, attitudes, behaviour etc.

Learning organizations in the New Economy require the creation of knowledge and for this reason cannot afford to waste opportunities wherever they may stem from, either from within the organization or outside it. Therefore, firms must not only promote downward learning but also upward and horizontal learning. The organizational structure must promote and absorb learning and/or knowledge creation situations from boss to subordinate, subordinate to boss, among peers and not only within the firm but also outside it.

Firms today must focus their attention on all aspects that are necessary not only for learning, but also for teaching, teaching how to learn and learning how to create knowledge. Within the framework of the communication of knowledge, it is not enough to be good receivers and communicators of messages, it is also essential to be able to create new messages. It is necessary for each member of the organization to be a good "pool", a good "channel" as well as an authentic "source" of knowledge or knowledge creator. The "source" not only gives and receives, but also creates. Learning, creating and teaching are the obligations of the members of an organization that wishes to be called a learning organization.

In response to our initial question on the relevance of moral habits or ethical virtues in knowledge management, an answer is already beginning to surface. In the communication of knowledge in organizations, it is not only knowledge and technical skills that are involved, but also attitudes and moral habits or virtues.

Those that are sceptical about recognizing the place of ethics in business might argue that knowledge management, in so far as it can be parameterized, can also be managed using protocol and measurements that allow for its creation, processing and retention. But, can ethics and ethical knowledge be managed?

If we distinguish between explicit and tacit knowledge, the former is liable to be formalized, and can hence be measured and managed. Conversely, tacit knowledge is not a formalized type of knowledge. On the contrary, at times it is neither understood nor perceived; it is *silent knowledge.* Due to its nature, this type of knowledge can be neither measured nor managed in a controllable way, and the same can be said about tacit ethical knowledge. Nevertheless, tacit knowledge and the most ambiguous and non-delimitable explicit knowledge will only have an effect on the organization in so far as enough conditions of trust exist for these effects to occur and remain over time. As we go on to suggest, organizational ethical healthiness is essential in learning organizations.

4.2. Organizational Ethical Healthiness: A Contextual Learning Facilitator

In this study, we maintain that; *the better the ethical health of an organization the easier it should be for its members to act as good pools, channels and sources of knowledge*. We are aware that empirical support will be necessary to validate this proposition, but the purpose of this study is to present just such a thesis. We maintain that moral competence, ethical virtues, or their absence as well as the greater or lesser presence of ethical aims and means, have a direct repercussion on the way people work, teach and create knowledge around them. Put differently, theoretical and practical knowledge, the acquisition of intellectual, psycho-affective and technical competencies is inseparable from moral knowledge and indeed has a direct repercussion on the way in which this knowledge develops.

We agree with Zamagni (2010)[1] when he points out that to "make the most of the tacit knowledge of our employees, we have to know how to establish reciprocal relationships with them, because I can force someone to arrive at work at 8 in the morning or to be in the office for 8 hours, but I can't force them to contribute their best ideas, their intellectual capital to the organization. (…) The only way of taking advantage of people's tacit knowledge (….) is through reciprocity" (p.2). In other words, the way of getting the best out of each member of the firm is through the reciprocal, mutual and parallel growth between the firm and its members in an atmosphere of reciprocal moral trust, which is the best way of managing tacit knowledge.

When it is someone's duty to transmit information in an organization, that person will perform the task better from an ethical point of view and hence also from a professional one, in so far as the knowledge acquired will be transmitted with *high fidelity* to the rest of the organization. From the organizational perspective, this attitude or moral competence, of being a right "channel" of knowledge, represents an element of union, a unity in a *win - win* relationship based on a job well done. Furthermore, this moral competence has repercussions for the personal development of the agent and his/her contribution to the development of the organizational fabric.

Being an element of union and unity becomes a behavioural norm or ethical principle of action for whoever wants to carry out their work correctly in the organization. "Behaving intelligently, using information correctly, analyzing it and making sure it reaches the people that have the right to access that information generates trust and builds organizations. Work provides the chance for the development of personal and organizational welfare." (Guillén, 2006).

In addition to the moral virtues of fidelity and unity, members of the organization also require *humility*. Without humility among superiors, subordinates and peers, it is not possible to have a climate that is capable of recognizing limitations and of learning from those that can teach. The firm understood in this sense must be capable of detecting the generators and builders of knowledge regardless of their hierarchical level. When the intermediate managers are capable of recognizing their limitations and learn from those that can teach them, be they superiors or subordinates, they are in a condition to become generators or builders of trust, the course along which tacit knowledge truly flows.

Another essential moral competence for knowledge creation is *prudence* or practical wisdom for determining the degree to which it is possible to cooperate in the development of knowledge. The nature of the virtue of prudence allows organizations to discover which ethical principles are applicable in each particular case and to what extent it is necessary to evaluate the degree of commitment of the organizations with its members, and vice versa.

Together with the virtues of humility and prudence, and in line with a concept in the field of knowledge management understood as communication, a good transmitter of knowledge requires another moral competence, that of having a conversational nature. In this sense, *dialogue* is a key component that demands reciprocity and the willingness to listen. Dialogue enables the establishment of the dual teaching-learning flow in both directions and senses: horizontal – from left to right and right to left- and vertical –up-down and bottom-up–.

Regardless of whether or not the firm is looking to manage that *knowing how to live*, going further than *knowing* and the *know-how* of its members, ethical health is an essential condition for the transmission and generation of both explicit and tacit knowledge. Without trust on an ethical plane, it is extremely difficult to get out of the plane of obligation, whilst conversely, when we can trust that the organization in which we work has aims that are in harmony with our wellbeing and that of society, we are dealing with the most suitable breeding ground for a fluid communication of knowledge, so that all the members of the organization can truly be pools, channels and sources of knowledge. It is for this reason that learning organizations require permanent communication and articulation via continuous dialogue between the organization's members: people.

5. CONCLUSION

The New Economy and current worldwide economic crisis demands that firms should be capable of reinventing themselves if necessary. Today's problems do not only demand internal free and reliable communication, but a higher level of cooperation and trust between individuals and firms. In this study, we propose that organizational ethical healthiness constitutes an essential facilitator for building authentic learning organizations. Today, technical learning is essential but it should be complemented with the ethical one.

State of the art equipment and technology are no longer a key differentiating element as the cost of acquiring the latest technology is within the reach of practically anyone. In any case, both information and technology can be bought. The same thing cannot be said, however, for aspects such as innovation or knowledge creation, and even less so in the case of behavioural dimensions such as personal involvement or initiative resulting from a high level of commitment on the part of workers towards the organization.

Current organizational reality implies making a major leap towards placing their value added in those elements that are inherent to each firm and are non-transferable. If we understand the organization as a group of people arranged in such a way as to achieve an objective or aim, are these people, their actions, relationships, contributions, originality and singularity not the DNA of each firm?

Despite the fact that it may seem obvious to highlight the importance of people in the organization, there is no doubt that 'people' concentrate and represent knowledge and organizational knowledge. Both of these aspects are manifested in human relationships and are built up in atmospheres of trust, respect, desire, commitment etc. It is precisely for this reason that within such a context, it appears logical to assume that businesses firmly committed to ethics, in addition to knowledge management and lifelong learning, can easier develop to their full potential.

Ethical virtues of the members of the organization, such as humility or loyalty, are essential for a learning organizations to attain their goals, but also other virtues such as the audacity to conquer new ideas, goals and high standards; the courage to avoid difficulties; the constancy to overcome failures in the learning process and the benevolence, trustworthiness and reliability to share ideas.

In order to be able to build ethically healthy organizations, and in order to achieve ethical excellence, HR policies are required that consider this third type of knowledge, knowing how to live. However, having an interest in achieving stable, good behaviour among the members of an organization is insufficient. The ethical quality of organizational behaviour does not solely depend upon human resource policies. Top management must ensure that not only behaviour, but also the organization's aims and means are really designed to contribute to the common good, to the ethical welfare of the members of the organization and the people it serves. This constitutes a truly ethically healthy organization, and we propose that this is an essential facilitator of learning processes.

There is no question that the proposals made here call for future research. Empirical verification of the theoretical propositions made here is necessary. Furthermore, much deeper studies should be carried out to better understand unexplored fields such as the relationship between ethical and technical learning in organizations, its influence on other organizational outputs such as organizational trust and organizational commitment, or its relationship with economic results.

Logically, some practical implications follow from the proposals made here. Top management should be the first to set an example in terms of the moral quality of their behaviour, intentions and the means they use to facilitate learning in their organizations. Nonetheless, this does not reduce in the slightest the personal responsibility of each member of the organization in making it a place where everyone reaches personal fulfilment through the exercise of professional work.

Professionally, technical learning cannot be separated from ethical learning, precisely because we are free, responsible human beings. The final outcome of work in organizations combines the external or explicit (measurable) results with internal or tacit ones, such as the capacity to generate and construct trust between agents. The most suitable organizational climate for lifelong learning requires trust in its technical and affective dimension, but above all in an ethical one. It is a fact that humans learn whilst carrying out their work, and in that learning the most technical aspects live alongside the most human ones. We either become enriched by our work or debased by it. This is a personal decision, but one which the management of the firm can do much to influence.

REFERENCES

Ang, S. A., & Joseph, A. D. (1996). *Organizational learning and the learning organization: Trigger events, processes, and structures*. Paper for Academy of Management Meetings, Cincinnati, OH, August. Retrieved August 1, 2011, from www.ntu.edu.sg/home/adjoseph/webpages/publications/aom10.pdf

Argyris, C. (1999). *On organizational learning* (2nd ed.). Oxford, UK: Blackwell Publishing.

Aristotle,. (1980). *The Nicomachean ethics* (Ross, D., Trans.). Oxford, UK: Oxford University Press. Also published in World's Classics collection. (Original work published 1925)

Bell, D. (1973). *The coming of post-industrial society: A venture in social forecasting*. New York, NY: Basic Book.

Collier, J. (1995). The virtuous organization. *Business Ethics (Oxford, England)*, *4*(3), 143–149. doi:10.1111/j.1467-8608.1995.tb00245.x

Dodgson, M. (1993). Organizational learning: A review of some literature. *Organization Studies*, *14*(3), 375–394. doi:10.1177/017084069301400303

Drucker, P. (1968). *The age of discontinuity: Guidelines to our changing society*. New York, NY: Harper & Row.

Easterby-Smith, M., Crossan, M., & Nicolini, D. (2000). Organizational learning: Debates past, present and future. *Journal of Management Studies*, *37*(6), 783–796. doi:10.1111/1467-6486.00203

Ghoshal, S. (2005). Bad management theories are destroying good management practices. *Academy of Management Learning & Education*, *4*(1), 75–91. doi:10.5465/AMLE.2005.16132558

Grant, R. M. (1996). Toward a knowledge-based theory of the firm. *Strategic Management Journal*, *17*, 109–122.

Guillén, M. (2006). *Ética en las organizaciones. Construyendo confianza.* Madrid, Spain: Prentice-Hall (Pearson).

Kim, D. H. (1993). The link between individual and organizational learning. *Sloan Management Review*, *34*(1), 37–50.

Liebeskind, J. P. (1996). Knowledge, strategy, and the theory of the firm. *Strategic Management Journal*, *17*, 93–107.

Macaulay, M., & Lawton, A. (2006). From virtue to competence: Changing the principles of public service. *Public Administration Review*, (September- October): 702–710. doi:10.1111/j.1540-6210.2006.00635.x

Malhotra, Y. (1996). *Organizational learning and learning organizations: An overview* [WWW document]. Retrieved May 1, 2011, from http://www.brint.com/papers/orglrng.htm

McGill, M. E., Slocum, J. W., & Lei, D. (1992). Management practices in learning organizations. *Organizational Dynamics*, *21*(Summer), 5–17. doi:10.1016/0090-2616(92)90082-X

McHugh, D., Groves, D., & Alker, A. (1998). Managing learning: What do we learn from a learning organization? *The Learning Organization*, *5*(5), 209–220. doi:10.1108/09696479810238215

Naquin, S. S., & Elwood, F. H. (2003). Redefining state leadership and management development: A process for competence-based development. *Public Personnel Management*, *32*(1), 23–46.

Nicolini, D., & Meznar, M. B. (1995). The social construction of organizational learning: Concepts and practical issues in the field. *Human Relations*, *48*(7), 727–746. doi:10.1177/001872679504800701

Nonaka, I. (1991). The knowledge-creating company. *Harvard Business Review*, *69*(6), 96–104.

Nonaka, I. (1994). A dynamic theory of organizational knowledge. *Organization Science*, *5*(1), 14–37. doi:10.1287/orsc.5.1.14

O'Keeffe, T. (2002). Organizational learning: A new perspective. *Journal of European Industrial Training*, *26*(2), 130–141. doi:10.1108/03090590210422012

Oakley, J., & Dean, C. (2001). *Virtue ethics and professional roles*. Cambridge, UK: Cambridge University Press. doi:10.1017/CBO9780511487118

Pedler, M., Burgogyne, J., & Boydell, T. (1997). *The learning company: A strategy for sustainable development* (2nd ed.). London, UK: McGraw-Hill.

Senge, P. M. (1990). *The fifth discipline*. London, UK: Century Business.

Serenko, A., Bontis, N., & Hardie, T. (2007). Organizational size and knowledge flow: A proposed theoretical link. *Journal of Intellectual Capital*, *8*(4), 610–627. doi:10.1108/14691930710830783

Simmonds, P. G., Dawley, D. D., Ritchie, W. J., & Anthony, W. P. (2001). An exploratory examination of the knowledge transfer of strategic management concepts from the academic environment to practicing managers. *Journal of Managerial Issues*, *13*(3), 360–375.

Solomon, R. C. (1992). *Ethics and excellence: Cooperation and integrity in business*. New York, NY: Oxford University Press.

Solomon, R. C. (2004). Aristotle, ethics and business organizations. *Organization Studies*, *25*(6), 1021–1043. doi:10.1177/0170840604042409

Toffier, A. (1990). *Powershift: Knowledge, wealth and violence at the edge of 21st century*. New York, NY: Bantam Books.

Wang, C. L., & Ahmed, P. K. (2003). Organizational learning: A critical review. *The Learning Organization*, *10*(1), 8–17. doi:10.1108/09696470310457469

Weick, K. E. (1991). The nontraditional quality of organizational learning. *Organization Science*, *2*(1), 116–124. doi:10.1287/orsc.2.1.116

Zamagni, S. (2010). *Speech at the 1st Symposium Society, Economy and Values*, IESE Business School, Barcelona, Spain.

KEY TERMS AND DEFINITIONS

Business Ethics: Form of applied ethics to the business environment.

Learning Organizations: Organization that facilitates the learning of its members and, therefore, continuously transforms itself.

Organizational Learning: Area of knowledge that studies models and theories in relation to the way an organization learns and adapts.

ENDNOTE

1 Extract from the speech made by Stefano Zamagni at the I Symposium "Society, Economy and Values" held at the campus of the IESE and organized by the University of Navarra. Taken from: Lucas, A., 2010: 'Humanismo y reciprocidad: El regreso al mundo vital', *Aceprensa*, 26 May, 2010.

Section 2
HRM Policy Implementation in the New Economy

Chapter 5

E-Recruitment: The Move towards a Virtually Organized Recruitment Process[1]

Anna B. Holm
Aarhus University, Denmark

ABSTRACT

E-recruitment, also known as online or Web-based recruitment, is little discussed in research from an organizational perspective. The aim of this chapter, therefore, is to analyze and discuss the process of e-recruitment, its key constituents, and organizing principles, based on the results of a qualitative study conducted in 2008-2009, together with data from industrial reports, articles from practitioner magazines, and in-depth interviews. The chapter provides a summary of e-recruitment properties and a composite matrix of the overall elements of e-recruitment organizing. E-recruitment is viewed as a case of virtual organizing, i.e. the organization of processes and activities, which, via technology and human agents, facilitate time- and space-independent interaction and collaboration. The chapter closes with a brief discussion of implications for HR managers and professionals and avenues for future research.

INTRODUCTION

The Internet and the availability of supporting information and communication technologies (ICTs) have radically changed recruiting (Cappelli, 2001), e.g. by significantly changing the traditional process of job advertising, CV screening, short-listing, and communication with candidates (Coleman, 2006). Reports from practice suggest that e-recruitment generates a higher application turnover (Minton-Eversole, 2007), considerably shortens the time it takes for a vacancy to be filled (Taylor, 2001), and is highly cost-efficient (Coleman, 2006). Surveys indicate that job seekers increasingly expect to find work through the Internet, rather than through other sources (Bartram, 2000).

E-recruitment is frequently discussed in practitioner magazines and industry reports, and is a widespread managerial practice in many parts of the world. A report from the US-based Society for Human Resource Management (SHRM, 2008b) indicates that, in the USA, national online job boards and the career sections of company websites are the primary recruiting sources for 47% and 35% of companies respectively. In 2001, Taleo

DOI: 10.4018/978-1-61350-207-5.ch005

Research (formerly iLogos Research) found that 76% of European 500 companies had a career section on their websites, 55% posted job positions there, and 55% accepted website submission of applications (Lermusiaux & Snell, 2001). In a following survey, Taleo Research concluded that, in 2003, 93% of Global 500 companies were using e-recruitment: 96% of those in Pacific Asia, 94% in Europe and 96% in North America (Lermusiaux & Snell, 2003).

Despite being a widespread managerial practice and the second largest e-business application (Lee, 2005), e-recruitment has received rather limited attention in academic research from an organizational perspective (Parry & Tyson, 2008). Both e-recruitment and its parental term, electronic HRM (e-HRM), are in the process of academic conceptualization, and require further empirical investigation (Bondarouk & Ruël, 2009). This might explain why it is difficult to find a clear and universally accepted definition of e-recruitment. For example, The Dictionary of Human Resource Management (Heery & Noon, 2001) defines e-recruitment as the recruitment of employees on the Internet. According to the dictionary, e-recruitment assumes a number of different forms, including advertising vacancies on websites, Internet job boards that act as labour market intermediaries, and internet directories which list job vacancies and connect directly to company websites (p.112). Lee (2005) further broadens the definition by incorporating a more general notion of electronic means which organizations employ to "fill open positions effectively and efficiently". Put another way, it still remains unclear exactly which organizational processes and principles constitute efficient e-recruiting practice, and how they are embedded in an organization.

One of the main aims of this chapter is to discuss the scope and key constituents of e-recruitment, and, in so doing, identify which organizational processes the term includes. The overall research question guiding the study is to determine whether e-recruitment should be viewed as a technological means of supporting recruiting services, or as a broader organizational phenomenon largely under managerial discretion.

The paper is based on two assumptions. Firstly, e-recruitment is viewed as subordinate to e-HRM, also called virtual HRM (Lepak & Snell, 1998). Broadly defined, e-HRM is a way of executing HRM to acquire, develop and retain human capital with the use of ICTs. Secondly, as e-recruiting goes beyond physical organizational boundaries, it enters the realm of ICTs and cyberspace. My research thus takes a starting point in virtual organizing, which is understood as the organization of processes and activities, using technology and human agents, to facilitate time- and space-independent interaction and collaboration (Holm & Ulhøi, 2010).

The empirical data used in the research stems from a variety of sources. Data collection and analysis started with a detailed review of professional magazines and a study of industry reports. Key search concepts included: e-recruitment, online recruitment, web-based recruitment, and e-HRM. Subsequently, a number of qualitative interviews were conducted with recruiters and Internet job intermediaries and technology providers from Denmark in 2008 and 2009. The interviews were fully transcribed and analyzed.

The findings show that e-recruiting is not just about using technology for administrating recruiting tasks, but that it is also a way to support and integrate multiple tasks and processes in an organization. Furthermore, the research confirms that e-recruitment as an organizational concept is not limited to the formal boundaries of an organization.

The rest of the chapter is structured as follows. The next section explains the theoretical background of the study, which is rooted in academic research on e-HRM, e-recruitment, and organization theory, and argues for studying the concept of e-recruitment from the conceptual perspective of virtual organizing. This is followed by a description of the research design, including the choice of qualitative research methods and a list of the interviewed organizations. The next section presents the findings, a summary of key e-recruitment properties, and the organizing elements of e-recruitment. This is followed by a discussion of the general and specific findings, and the conclusions of the paper. The last section discusses the limitations of the paper and implications for HR professionals, and outlines avenues for future research in the field of e-recruitment and e-HRM.

RESEARCH FRAMEWORK

The recruitment of personnel is a key strategic HR function, the main purpose of which is to develop internal and external resources of qualified applicants, the success of which is crucial to HRM (McKenna & Beech, 2008, p. 179). The primary aim of an organization's recruiting practices and activities is to identify, attract, and influence the job choices of competent candidates (Barber, 1998, p. 5; Ployhart, 2006). The most common sources for recruitment outside the organization, also referred to as external recruitment, include newspaper advertisements, private and public employment agencies, Internet job boards, corporate websites, employee referrals, colleges and universities, search firms, job fairs, etc. (Ployhart, Schneider & Schmitt, 2006, p. 280). Increasingly, Internet-based sources are becoming the main recruitment source for many, especially large, organizations (Parry & Tyson, 2008).

According to various definitions, e-recruitment comprises activities and tools, such as organizational website recruitment (Thompson, Braddy, & Wuensch, 2008), advertising on Internet job boards (Borstorff, Marker, & Bennett, 2007), e-recruitment systems for the collection of applications, data management and coordination, also called hiring management systems (Cappelli, 2001), search engines and online résumé databases for identifying and attracting potential applicants (Lee, 2005), electronic communication with applicants, etc. In other words, the term e-recruitment is rather broad and relates to different sets of specific practices, activities and tools. Since e-recruitment ranges from advertising vacancies on a corporate website to the use of back-office e-recruitment systems, in this paper it is treated as an umbrella term covering practically all recruitment-related activities which use, or are supported by, the Internet and related information and communications technologies (ICTs).

The Internet and ICTs were used by recruiters even before the introduction of the term e-recruitment, but were often limited to the automation of the internal process or advertising on a corporate website (Lee, 2005). HR departments use various information systems (IS) for streamlining operations and alleviating much of the administrative burden, and a good deal of research has already been carried out on HR Information Systems (HRIS) (Ruël, Bondarouk, & Looise, 2004). However, this operational use of ICTs does not automatically turn HR recruiting tasks into e-recruiting, i.e. the process of recruiting by electronic means. The study of e-recruitment as a process also has to include the processes initiated by organizations on the Web: attracting qualified candidates often takes place through Internet-based intermediaries, such as external online job databases and job boards. In many, especially large, organizations, this way of recruiting is supported by an e-recruitment system – a back-office system for administrating the recruitment process, which is normally designed to allow applicants to submit their data electronically. Contemporary e-recruitment systems allow practically instant communication with applicants, and the submission of applications and résumés normally takes place in real-time, most often directly to the organization's internal e-recruitment system. Increasingly, e-recruitment is used as a means of marketing the organization, and is even integrated in overall marketing campaigns

(Cappelli, 2001). Furthermore, companies using e-recruitment are moving towards integrated HR systems of staffing, career development, training, performance and other key activities, which often require a reorganization of the HR process and better alignment of HR information and database architectures (Cullen, 2001). Using e-recruiting thus goes beyond the goal of improving efficiency by easing the administrative burden, and might even lead to wider consequences for the organization in general and its environment.

In the case of e-recruitment, the Internet and other ICTs also have a big influence on the way organizations handle recruitment (Lievens & Harris, 2003). At the same time, recruitment is a function embedded in a wider organizational HRM system, and in many organizations recruitment is managed and performed by HR personnel and departments. Viewing e-recruitment and related activities from a broader e-HRM perspective thus appears to be well in line with existing definitions of e-HRM. For example, Lepak and Snell (1998) define e-HRM, which they also call virtual HRM, as a network-based structure built on partnerships, and typically mediated by information technologies to help the organization acquire, develop and deploy intellectual capital. Ruël, Bondarouk & Looise (2004) see e-HRM as a way of implementing HR strategies, policies and practices in organizations through the conscious and direct support, and/or with the full use of web-technology-based channels.

Lepak and Snell (1998) identify three areas of HRM where organizations can offer HR services by electronic means. The first, operational HRM, concerns basic HR activities in the administrative area. The second, relational HRM, deals with more advanced HRM activities which support the basic business process. The third area, transformational HRM, concerns HRM activities of a strategic character, and is also called strategic HRM (SHRM). According to Ruël, Bondarouk and Looise (2004), three types of e-HRM can therefore be distinguished: operational e-HRM, relational e-HRM and transformational e-HRM. Following this line of thought, Figure 1 classifies areas of HRM with examples of each type from traditional and electronic HRM practices, in which e-recruitment is distinguished as being part of the relationship type of e-HRM.

Since recruitment is influenced by labor market variables, various partners and intermediaries, technological infrastructure, and society in general, it is considered adequate to investigate organizations' e-recruitment from an open-system perspective on organizations. Open-system theories treat organizations as complex sets of dynamically intertwined and interconnected elements, including inputs, processes, outputs, feedback loops, and the environment in which it operates (Shafritz, Ott & Jang, 2005, p. 477). Most importantly, open-system perspectives recognize the dynamic interaction of the system with its environment (Robbins, 1983, p. 10). Thus, from an open-system perspective, the recruitment pro-

Figure 1. HRM and e-HRM types

HRM Type	Traditional HRM	Electronic HRM
Operational	Salary administration Personnel data administration	HRIS
Relational	Recruiting and selection of personnel Training of personnel	e-Recruitment e-Learning
Transformational	Competence management Performance appraisal	HR internal portals Talent management systems

Source: Adapted and elaborated from Lepak and Snell (1998)

cess represents an organizational effort to attract and retain human resources from the external environment to ensure the input of labor needed by the organization to produce output.

The idea of researching e-recruitment from an open-system perspective, especially at a higher, strategic HRM level, is not new. For example, Wright and Snell (1991) used an open-system model for generating HRM strategies, where the inputs in HR systems consist of the competencies, skills, abilities, etc., of individuals, imported from external environments. The throughput process of the system is then characterized by individuals' behavior, and the outputs consist of both performance variables (e.g. productivity, efficiency) and affective outcomes (e.g. job satisfaction). Here, the recruitment function has a strategic importance, since it must ensure that the individuals found and hired for an organization will perform well enough for the organization to achieve its goals.

From an organizational perspective, e-recruitment relies on an organizational design which involves formal organizational arrangements, including formal structures, processes and systems (Nadler, Tushman & Nadler, 1997). E-recruitment can also be considered as a case of virtual organizing (Holm & Ulhøi, 2010), which 1) is understood as the organization of activities facilitating time- and space-independent interaction and collaboration, 2) requires the appropriate organization, technology and human agents, and cannot exist without all three components, and 3) is a strategic choice characterized by directionality and different levels of organizing (ibid.). Thus, the challenge in studying e-recruitment from an organizational perspective is to include technology, its users, organizations and their environment in a cohesive research framework. Moreover, e-recruitment takes place and involves activities at different levels of an organization – i.e. recruiter, recruitment team, and even recruiting organization – and relies on resources provided by third parties, e.g. online job boards and e-recruitment systems. The recruitment effort is directed to either internal or external human resources from the wider society in which an organization operates. In this way, the e-recruitment process is organized and performed at multiple levels; therefore, investigating e-recruitment in a holistic way also requires a multi-level research design (Holm, 2009).

Research Design

Since the main aim of the study was to investigate e-recruitment from an organizational perspective, the collection of primary data was based on the conceptual framework for virtual organizing (Holm & Ulhøi, 2010). This framework helped both to make the research operational and give it a common direction, gain initial insights, and develop concepts (Corbin & Strauss, 2008, p. 40). The choice of the research design is based on the presumption that social science exploration would permit a discovery of generalizations, leading to descriptions and an understanding of the researched phenomenon (Stebbins, 2001, p. 3). Given the research questions asked, the deductive approach was considered the most suitable, since the aim of deductive research is typically to deduce particular explanations from general theories and to compare the researched phenomena with these (Gibbs, 2007, p.5).

The explorative study employed data collection techniques from qualitative research methods (Corbin & Strauss, 2008). The data sources include industry reports, professional magazines, newspaper articles, websites, online forums and discussion boards in English, and a number of qualitative semi-structured interviews conducted in Denmark in Danish. The secondary data from practitioner-oriented research provided a good deal of insight into e-recruiting, its principles, practice, benefits and disadvantages, major trends, and challenges for the future. A variety of reports and data were available from professional organizations and the intelligence units of established consultancies, e.g. the US-based Society for Human Resource Management (SHRM), the UK- and Ireland-based

Figure 2. Organizations sampled for interviewing

Fictional Name	Primary Industry	No. of employees	Nationality
Dairy Giant	Dairy products	16,000-	Danish MNC
Personnel Partner	Employment agency and personnel search	200-	Danish branch of an MNC
Big Drinks Company	Soft drinks and breweries	45,000-	Danish MNC
e-Recruitment Hosting	Recruitment software and hosting	10	Danish
Jobs on Internet	Job portal and personnel search	120-	Danish
Banking Monster	Investment and banking	34,000-	Pan-Scandinavian
e-HRM Partner	HR Software developer	50-	Danish
Enterprise System Solutions	Enterprise Systems developer	100-	Danish branch of an MNC
Clean Energy Production	Energy systems	20,000-	Danish MNC

Chartered Institute of Personnel and Development (CIPD), the UK Institute for Employment Studies (IES), Taleo Research (formerly known as iLogos Research), and others. Key search concepts included: e-recruitment, online recruitment, web-based recruitment, and e-HRM.

Based on the framework, I developed a set of research themes for conducting semi-structured interviews. During the course of the iterative data collection and analysis, some initial questions were reviewed, and some were either reformulated or removed, and new questions added. All the interviews were conducted in Denmark in 2008 and 2009. Except for a few exploratory in-depth interviews, all the interviews were semi-structured and lasted between one and two hours.

The organizations chosen for interviews were: 1) employers or recruitment intermediaries who used e-recruiting in their daily activities, 2) e-recruitment technology providers, and 3) an Internet job portal. This variety of respondents was considered necessary to achieve a higher validity of results and to cover most of the aspects of e-recruitment organizing. The selected organizations varied in the size and nature of their activities, and more than half of them were either Danish Multinational Corporations (MNC) or Danish branches of foreign MNCs. Almost all respondents were either in high-level managerial positions, or were executive recruitment partners or senior consultants. The full list of organizations interviewed is provided in Figure 2. In addition, a small number of professional presentations, face-to-face conversations with practitioners, and online professional forums and discussion boards were used as data sources.

FINDINGS

E-Recruitment Properties

Before discussing e-recruitment as a way of organizing recruitment, it was considered necessary to determine exactly what is meant by the term. E-recruitment, as suggested earlier, is more of an "umbrella", or composite, term characterized by diverse properties. The analysis of a substantial number of practitioner articles, professional websites and industry reports revealed a broad range of practices, tools and processes related to what the professional community calls online-, web- or e-recruiting. A summary of the e-recruitment properties as identified from the data analysis is presented in Figure 3. Many of the recruitment tools, techniques and sources evolved together with changes and innovations in online technologies. For example, social networks, newsgroups and other web resources facilitating user-input and the exchange of information emerged with Web 2.0 technology.

Recruiters might choose to use only some of the many e-recruitment techniques, or prefer one e-recruitment source over another, or they might use the entire arsenal of possibilities associated with e-recruiting. For the interviewed recruiting professionals, online job boards, corporate career websites, and e-recruitment systems are reported to be of particular significance.

Job Databases and Career Portals

Job databases and career portals, also called employment websites, normally refer to an online platform where organizations advertise their employment opportunities locally, regionally, nationally and internationally. The fact that successful employment websites can attract a substantial amount of online traffic, which can then be directed to the organization's own website, makes them one of the main recruitment sources for employers. According to a report by SHRM (2008a), 48% of the US-based companies in the survey used national online job boards as their primary recruitment source. For job seekers, employment websites provide access to employment opportunities and other resources for successful job searches.

The Danish organizations in the study used job portals and career websites for all external recruitment, and were often in stable cooperation partnerships. Some sourced up to 70% of employees through advertising on job portals. According to a respondent from Jobs on Internet, most of their revenue was generated by regular customers. He explained their business concept in the following way: "We provide certain services, depending very much on customer size. Our key account management department deals with our biggest customers... " (my translation). Some organizations in the study had even automated their job posting on employment websites, so that job announcements were fed electronically into the repository of internet job databases when ready. In addition, job databases sometimes exchange job postings electronically with other job databases and various online resources.

For both employers and job seekers, job databases and career portals also serve other purposes than pure job advertising and job searches. For example, they often contain a résumé repository or database, which employers can access to find, screen and contact candidates. Interested candidates can subscribe to automated notifications of job openings that match their employment objective. Online tools can help them write and build a résumé, test their personality and cognitive abilities, and put them in touch with other job–seekers, or even employers, through discussion forums. Several of the analyzed employment websites provide information on labour market statistics, analyses and forecasts, information on education, training and certification requirements and possibilities, individual career fields, links to additional job search resources located at other sites, and many others.

Corporate Websites

Corporate career websites, organizations' own online platform for placing job vacancies, provide information about career opportunities and company employment policies, allow applicants to apply for jobs electronically, and facilitate communication with prospective candidates. According to many of the interviewed recruitment professionals, they are often the main recruitment source for external candidates.

Recruiters regard corporate career websites as the main medium for communication with prospective candidates, and they are often used as shop windows to present the organization as an attractive employer and an exciting place to work. They are often designed and maintained by

Figure 3. E-recruitment properties

Property	Short description	Purpose	Comments
Internet job boards and career portals	National, regional, local and international job-posting and search websites.	Job advertising, and a way of directing interested applicants to the corporate career website.	Offered by third parties, which serve as Internet job intermediaries, e.g. careerbuilder.com, hotjobs.com.
E-recruitment system	Stand-alone information system or a module of a larger ERP system.	Applicant registration, receiving and registering applications, monitoring the recruitment process, screening and evaluating candidates' résumés.	Can be deployed onsite or implemented as a Software-as-a-Service (SaaS) solution.
Corporate career website	Organization's own web page, containing information about the organization and its vacancies.	Applicant attraction, submission of résumés, and communication with candidates.	Reported as one of the most efficient e-recruitment sources.
Résumé databases and applicant pools	Electronic databases with résumé repositories.	Locating suitable candidates through screening and evaluating their résumés.	Can be internally or externally operated, and / or part of an e-recruitment system.
Various (third-party) websites	Websites of the mass media, professional organizations, educational institutions, and others.	Job advertising and informing about the organization.	Any website which can generate good exposure to the target audience.
Corporate intranet	Intranet portal pages listing current vacancies in the organization.	Advertising vacancies inside organizations.	Related to internal recruitment.
Social networking sites	Websites based on Web 2.0 technology, allowing users to open and update their own pages, share content, and communicate with each other.	Advertising jobs, promoting the organization, searching, screening and contacting qualified candidates.	E.g. Facebook, LinkedIn, MySpace.
Industry-specific web resources	Web-based professional newsgroups, portals, forums, blogs, facilitating communication and information-sharing with users of a specific profession.	Advertising jobs, searching, and approaching potential candidates.	E.g. www.pharmiweb.com, www.nuclearcareersonline.com.
News- and discussion groups, forums and blogs	Various interactive websites, where users can share news, opinions, have discussions and blog.	Advertising jobs and promoting the organization, searching and approaching potential candidates.	E.g. twitter.com.
Search engines	Interactive search engines for both recruiters and job seekers.	Specialized candidate and job searches, and matching jobs with résumés .	Can be part of an e–recruitment system or a service provided by a third party
Electronic mail and electronic mailing lists	Client and server e-mail lists of subscribers registered for receiving job alerts.	To inform interested applicants about job openings and other relevant information.	Can be part of an e-recruitment system
Online pre-screening tools	Web-based pre-screening, self-assessment and testing tools.	Automated screening of résumés and (self-) testing of candidates online.	Can be part of an e-recruitment system. Enables recruiters to filter out suitable applications.

communications professionals, and offer visitors a rich multimedia content, e.g. video presentations and virtual tours of company premises, recorded video testimonials of existing employees, live news feeds, and sometimes self-assessment test and résumé-building tools. In the three large organizations interviewed for this paper, the task of running the corporate career website was carried out jointly by the HR, IT and communications departments.

A communications manager from Clean Energy Production explained why his company's management was so serious about maintaining the career website: "It is extremely important. Because much of our communication is not personal, and the website expresses who we are. That is also why we've tried to make it live with some video interviews, "Ask an employee" and other various clips. Anything we can do to show what Clean Energy Production stands for. Not because we want to show off with the IT, not at all, but it's about giving as natural an impression of Clean Energy Production as a workplace as possible…" (my translation).

E-Recruitment System

An e-recruitment system is an online job application and processing system which enables employers to advertise job openings, store candidate data, assess incoming applications, schedule interviews and monitor the recruitment process. For job seekers, it serves as an interface which allows them to submit their applications online.

The interviewed organizations with e-recruitment systems had no great difficulties in acquiring and deploying such systems, and expressed overall satisfaction with the software. Typically, staff members received little or no initial training in using e-recruitment systems, and described most system interfaces as being well-designed and user-friendly. In cases where application submission was automated, the recruitment personnel were particularly satisfied with not having to manually process the paper-based résumés. The large organizations included in the sample, which typically receive a substantial number of applications, had dropped traditional paper-based applications altogether, requiring candidates to register and submit their data exclusively in electronic form. A respondent from Clean Energy Production gave an example from his own organization: "We use software N. We get 1,400 applications from around the world each week… so we must have an electronic system to administrate them. And we use N throughout the entire recruitment process to register and administrate the process" (my translation).

In organizations with an e-recruitment system, many recruitment activities were automated and standardized. A recruiter from Personnel Partner explained how it works in their organization: "… For us it is entirely electronic. We use the system from vendor N… So we put our [job] ads in there, and it connects to those [internet job databases] with whom we have cooperation agreements… Within 24 hours, job ads are kicked out to all the other [job databases], with everything going electronically. There are few places which we have to look at manually – to put data correctly in the appropriate categories… It works great!" (my translation) .

In organizations which did not use an e-recruitment system, recruiters compensated for this by using other electronic means. For example, a recruitment partner from Dairy Giant explained how their organization coped without a dedicated e-recruitment system: "If you look at the administrative part of it [i.e. recruitment], we have tried to make it as electronic as possible, but currently we do not have an electronic recruitment system ... But everything we possibly can run electronically we do, in the form of emails, databases, Excel sheets and things like that. So it runs as electronically as possible… without an e-recruitment system…" (my translation).

The many examples of using an e-recruitment system for administrating and monitoring the recruitment process suggest that it should be seen as an operational HR tool. However, e-recruitment software offers a lot more functionality than just the possibility of administrating individual recruitment cycles and tracking candidates. It is often used as a tool for time- and space-independent collaboration and communication between recruitment stakeholders, who may come from any part of the organization. A recruitment partner from Big Drinks Company explained how their e-recruitment system enabled instant communication with line managers, and allowed quick and efficient processing of applications for the next phases of the recruitment process. "I cooperate… with managers in Asia, and it can be quite difficult to run a recruitment process if you work from Copenhagen and Kuala Lumpur. And here we use the Internet, where we can go on and look at the applications online. He [i.e. the Asian manager] has access to the same things as I have, we can rate the candidates, discuss them, etc. So it's really much easier with the system!" (my translation).

A modern e-recruitment system is also an efficient tool for communicating with applicants, and in some cases it even permits candidates to track their application process online in real-time. The system's automatic response function removes the administrative burden of preparing and mailing the traditional standardized reply, since applicants are informed immediately about the result of the initial screening of their résumés by email or SMS. A consultant from e-Recruitment Hosting explained how it worked: "…In the application phase, it is most important that all candidates receive an auto reply. You do not need to confirm that the application is received, the system does it fully automatically" (my translation). The same procedure applied after the pre-screening phase at Big Drinks Company, where the system was used to inform rejected candidates. The same respondent said: "The system is also used to send a rejection letter. It happens through an automatic emailing, where all [rejected] applicants who we did not have face-to-face contact with receive an email. We click on the persons who should receive the rejection mail, and then it is sent off" (my translation).

The e-recruitment solutions available in the market are generally perceived as being somewhat similar in functionality regarding requisitioning, candidate self-service, job board posting, and applicant tracking (Otter & Holincheck, 2009). However, recent developments in e-recruitment software include expanding the functionality of candidate management and self-management. And, according to Gartner Inc. (2009), many e-recruitment vendors are adding social software capabilities and onboarding solutions. Another new functionality is workforce planning, and some vendors are even pursuing the concept of talent acquisition in their software development - a holistic view of all the sources of talent.

Social Networking Sites

An important e-recruitment-related trend is the growing use of social and professional networks sites, e.g. Facebook, LinkedIn and MySpace, which give recruiters the possibility to source, contact and screen both active and passive job candidates. The survey conducted by SHRM (2008a) concluded that an increasing number of companies used social networking sites as an HR tool. In the two years prior to the study, the proportion of such companies had grown from 21% to 44%.

However, some of the recruiters I interviewed avoided using these channels for ethical reasons, while the rest considered them as a perfectly legitimate communication medium for approaching qualified individuals.

General Findings

One of the main questions asked in the interviews was how e-recruitment had changed the process of recruiting. The traditional (paper-based) recruitment process was characterized by a repetitive

Figure 4. E-recruitment sources, techniques, and tools at different recruitment phases

Recruitment phase	E-recruitment techniques, tools and sources
Identification of hiring needs	None
Preparation of job description and specifications	E-recruitment system
Attracting applicants	Corporate intranet, Internet job boards and career portals, corporate career website, social networking sites, résumé databases and applicant pools, various (third-party) websites, industry-specific web resources, news- and discussion groups, forums and blogs, search engines electronic mail and electronic mailing lists
Processing and pre-screening applications	E-recruitment system, automatic pre-screening tools
Communicating with candidates	E-recruitment system, corporate career website, electronic mail and electronic mailing lists

cycle of phases, i.e. job analysis and identification of hiring needs, preparation of job description and specifications, attracting applicants, processing and pre-screening applications, and communicating with candidates. According to the respondents, with e-recruitment the phases were the same, but the cycle lead time and task sequence differed. One of the reported changes was that attracting, sorting, and contacting candidates could now take place simultaneously. A recruitment partner from Banking Monster explained this in the following way "…The moment a candidate logs in and registers his application, we can see it here in the HR department… For each [recruitment] project there is a dedicated HR partner… There are also one or two line managers assigned to the project, and regardless of where they are located in the company, they can see simultaneously that there is an application submitted for the job opening. They can sit and follow it from hour to hour if they have the time. And when they can, and have time …they decide whether to proceed further with the candidate …" (my translation).

As pointed out by many respondents, e-recruitment techniques and technologies were mostly utilized at the stages of candidate attraction and communication. Although the e-recruitment system would be used for administrative and data-sharing purposes from the very start of a new hiring process, most of the qualified decisions are made prior to attracting candidates. In this way, attracting qualified applicants is characterized by the use of diverse online media. This is illustrated in Figure 4.

According to a number of industry reports and practitioner articles, many organizations with a large pool of applicants make at least some use of automatic résumé screening tools and assessment software. None of the interviewed organizations in this study did, however. The screening and identification of potential candidates was carried out manually by the recruitment personnel, often in cooperation with line managers responsible for the final hiring decision. The main reason for not using software for pre-screening was that many vacancies required diverse qualifications, and creating evaluation questionnaires or tweaking the software for each new opening would have been a daunting task.

Most of the interviewed recruiting professionals were convinced that attracting the best candidates requires a good organizational reputation and brand, and even a consistent promotional effort in conjunction with other marketing activities. Corporate career websites play a distinct role here, and are widely perceived as a powerful tool for promoting and reinforcing the corporate image, brand and reputation.

For most of the companies, communication with candidates is turning into a major recruitment challenge, e.g. to ensure that candidates do not regard the recruitment process as a bad experience, since this can have a negative effect on the employer brand. A respondent from Clean Energy Production expressed his reservations in the following way: "I think the candidates often feel like numbers in a row, especially if they are not chosen for the first interview ... So I think there is a close link between employer-branding and e-recruitment, and no matter how you look at it, candidates will always assess how they have been treated... It is very important to me that people get a reason why they were not recruited and not just a rejection in the e-recruitment system. So I would say that the weakness of such a system is that candidates may feel impersonalized" (my translation).

Many recruitment professionals are beginning to recognize candidate management as a critical new component of the recruitment process. In consequence, some e-recruitment software vendors have already begun to offer improved functionality for integrated candidate relationship management and sourcing. HR software provider e-HRM Partner gives a simple example: "When we have to send a rejection... the system may contain a variety of rejection texts. So when you evaluate the candidates, you can choose that 8 [applicants] must receive rejection text 1, 120 should receive rejection text 2, and so on. This makes them feel that they have received a rejection which is not a standard machine rejection..." (my translation).

Organizing for E-Recruitment

As can be seen from the findings, e-recruitment activities vary in their complexity and significance, and often range from intra- to inter-organizational levels of organizing. I therefore asked the organizations in the study how their organizational formal and informal structures were geared to an e-recruitment process. As a result of my analyses, I have developed a matrix containing the key organizational design elements for organizing e-recruitment. This is presented in Figure 5. The elements in the matrix constitute an aggregate composition of research findings, and were not all present in each of the visited organizations.

CONCLUSION

The present study has shown that e-recruitment as an organizational concept and social phenomenon is not limited to the internal boundaries of an organization, and should not be perceived simply as some technological solution to a recruitment need. The research presented here has also confirmed that e-recruitment can be viewed and studied from the perspective of virtual organizing. E-recruitment can thus be defined as *the organization of recruitment processes and activities, which, by means of technology and human agents, facilitate time- and space-independent collaboration and interaction in order to identify, attract, and influence competent candidates*.

The analyses of the various data sources confirm the suggestion that e-recruitment is a highly composite phenomenon which relies on diverse organizational arrangements. Organizing for e-recruitment involves employers, various online intermediaries and technology providers, and even job seekers. Therefore, research into e-recruitment from an organizational perspective requires a theoretical approach or framework which takes account of the environment in which the organization operates. E-recruitment is directed towards, and carried out together with, external partners and customers, and is therefore affected by the broader norms and traditions of the society in which the organization operates. As a result of its emphasis on the external environment,

Figure 5. The organizing of e-recruitment

Level \ Elements	Organizational	Technological	Socio-cognitive
Employee	Recruitment partner, manager, assistant, line manager.	Client software, PC, internet / intranet connection, email.	Basic technology familiarity and computer literacy, self-moderation.
Groups / Teams	Recruitment group, part of HR department, well-established recruitment routines.	E-recruitment system, résumé database, local network and group servers.	Online group moderation.
Organization	Collaboration between HR, IT, communications departments and employer-brand manager.	Corporate website, Web integration of e-recruitment system, connectivity with wider ERP system.	Corporate ethics. Confidentiality.
External partners and customers	Stable collaboration with online job and résumé databases, cooperation with e-recruitment software vendors.	Digital connectivity and data exchange with online partner resources.	Trust. Formalization.
Society	Candidate management and online communication through corporate website, general and industry- specific web resources. Employer-branding.	Internet connectivity. Web 2.0	Confidentiality. Reputation. Employer brand.

e-recruitment thus gains a strategic dimension, which in turn increases the significance of how well the recruitment function performs.

While traditional recruitment may not require any particular technology, e-recruiting is entirely dependent on a wide range of modern ICTs. By virtue of the Internet and supporting ICTs, communication and information exchange between employers and job seekers has no geographical limits, and takes place on a 24/7 basis. Therefore, when researching e-recruitment, technology cannot be seen merely as a contingency, since this would limit the research to establishing the rationality of automating recruitment tasks and changes in routines; here, technology plays a wider societal role, affecting the livelihoods of individuals and organizations.

This explorative research also suggests that recruiting with the help of electronic means does not eliminate the phases of a traditional recruitment cycle; rather, it modernises and advances the whole process, with some tasks and activities being performed concurrently, or being fully automated. On the other hand, the immediate availability of diverse communication media and channels takes the process to another level of organizational complexity, which might require both changes in organizational structures and management and additional qualifications and skills from recruitment professionals. Clearly, recruiters are devoting increasing resources, time and effort to communicating with candidates, particularly during the attraction phase. However, since the Internet gives job seekers endless possibilities to find employment elsewhere, there is also a clear trend among staffing professionals towards dedicating more resources to candidate management in an attempt to keep them loyal and interested in their organization.

LIMITATIONS AND IMPLICATIONS

There are several limitations to the research presented here. First, this is an explorative study which provides initial insights into the research subjects. Second, the primary data used in the analysis stem from 9 organizations and 11 interviews. Although they use various e-recruitment techniques and technologies, all the organizations in the study come from one country – Denmark. The Danish context might therefore have had an effect on how the recruitment practices were shaped and executed in those organizations.

As mentioned earlier, research on e-recruitment lags somewhat behind practice, which means that there are many interesting avenues for future research. If approached from an organizational perspective, research on e-recruitment can address the issue of organizational change and identify the effect of e-recruitment on the recruitment process. Such changes may potentially occur at different organizational levels, and be related to tasks, people, structures and management.

More studies are needed to show how different elements of e-recruitment contribute to the efficiency and effectiveness of the recruitment process. Other dependent variables than time and cost of recruitment need to be looked at. Of these, the effect of e-recruiting activities on employer brands, employee turnover and job satisfaction are of importance for both private and public employers.

A better understanding of the organizational concept of e-recruitment might also be achieved by studying how the external environment and society in general affect the organization's recruitment practices, and which factors lead to the adoption of e-recruitment. In the same way, research on how potential candidates expect to be hired, and their criteria for choosing one employer rather than another, can give good insights into which recruitment methods to adopt. The research should also try to identify which combination of recruitment sources will result in attracting candidates who can best help organizations to achieve their performance objectives.

ACKNOWLEDGMENT

I would like to thank John P. Ulhøi, Alistair Anderson, and reviewers and participants of the 3rd International Workshop on Human Resource Information Systems, Milan, 2009, for their valuable comments and suggestions for developing and improving this manuscript.

REFERENCES

Barber, A. E. (1998). *Recruiting employees*. Thousand Oaks, CA: SAGE.

Bartram, D. (2000). Internet recruitment and selection: Kissing frogs to find princes. *International Journal of Selection and Assessment*, *8*(4), 261–274. doi:10.1111/1468-2389.00155

Bondarouk, T. V., & Ruël, H. (2009). Electronic human resource management: Challenges in the digital era. *International Journal of Human Resource Management*, *20*(3), 505–514. doi:10.1080/09585190802707235

Borstorff, P. C., Marker, M. B., & Bennett, D. S. (2007). Online recruitment: Attitudes and behaviours of job seekers. *Journal of Strategic E-commerce*, *5*(1 & 2), 1–23.

Cappelli, P. (2001). Making the most of on-line recruiting. *Harvard Business Review*, *79*(3), 139–146.

Coleman, A. (2006). Does e-recruitment work? *Director (Cincinnati, Ohio)*, *59*(12), 27.

Corbin, J., & Strauss, A. (2008). *Basics of qualitative research: Techniques and procedures for developing grounded theory* (3rd ed.). Thousand Oaks, CA: SAGE.

Cullen, B. (2001). e-Recruiting is driving HR systems integration. *Strategic Finance*, *83*(1), 22–26.

Gibbs, G. (2007). *Analyzing qualitative data*. London, UK: SAGE.

Heery, E., & Noon, M. (Eds.). (2001). *Dictionary of human resource management*. Oxford, UK: Oxford University Press.

Holm, A. B. (2009, May). *Virtual HRM: A case of e-recruitment*. Paper presented at the HRIS 2009 – 3rd International Workshop on Human Resource Information Systems, Milan, Italy.

Holm, A. B., & Ulhøi, J. P. (2007, December). *From physical towards virtual organization of business activities*. Paper presented at the 2007 Annual Conference of The Danish Academy of Management, Aarhus, Denmark.

Holm, A. B., & Ulhøi, J. P. (2010). A framework for conceptualising virtual organising. *International Journal of Business and Systems Research*, *4*(3), 293–311. doi:10.1504/IJBSR.2010.032952

Katz, D., & Kahn, R. L. (1966). *The social psychology of organizations*. New York, NY: John Wiley & Sons.

Lee, I. (2005). The evolution of e-recruiting: A content analysis of fortune 100 career websites. *Journal of Electronic Commerce in Organizations*, *3*(3), 57–68. doi:10.4018/jeco.2005070104

Lepak, D. P., & Snell, S. A. (1998). Virtual HR: Strategic human resource management in the 21st century. *Human Resource Management Review*, *8*(3), 215–234. doi:10.1016/S1053-4822(98)90003-1

Lermusiaux, Y., & Snell, A. (2001). *Best practices for European 500 career web site recruiting*. Taleo Research.

Lievens, F., & Harris, M. (2003). Research on internet recruiting and testing: Current status and future directions. In Cooper, C. L., & Robertson, I. T. (Eds.), *International review of industrial and organizational psychology* (*Vol. 16*, pp. 131–165). Chichester, UK: John Wiley & Sons, Ltd.

Miles, M. B., & Huberman, M. A. (1994). *Qualitative data analysis* (2nd ed.). Thousand Oaks, CA: SAGE.

Minton-Eversole, T. (2007). E-recruitment comes of age, survey says. *HRMagazine*, *52*(8), 34–34.

Otter, T., & Holincheck, J. (2009). *Magic quadrant for e-recruitment software*. Gartner, Inc.

Parry, E., & Tyson, S. (2008). An analysis of the use and success of online recruitment methods in the UK. *Human Resource Management Journal*, *18*(3), 257–257. doi:10.1111/j.1748-8583.2008.00070.x

Ployhart, R. E. (2006). Staffing in the 21st century: New challenges and strategic opportunities. *Journal of Management*, *32*, 868–897. doi:10.1177/0149206306293625

Ployhart, R. E., Schneider, B., & Schmitt, N. (2006). *Staffing organizations: Contemporary practice and theory* (3rd ed.). Mahwah, NJ: Lawrence Erlbaum Associates, Publishers.

Robbins, S. (1983). *Organization theory: The structure and design of organizations*. Englewood Cliffs, NJ: Prentice Hall Inc.

Ruël, H., Bondarouk, T., & Looise, J. K. (2004). E-HRM: Innovation or irritation. An explorative empirical study in five large companies on web-based HRM. *Management Review*, *15*(3), 364–380.

Shafritz, J. M., Ott, J. S., & Jang, Y. S. (2005). *Classics of organization theory* (6th ed.). Belmont, CA: Thomson Wadsworth.

SHRM. (2008a). *Online technologies and their impact on recruitment strategies: Using social networking websites to attract talent*. Society for Human Resource Management.

SHRM. (2008b). *SHRM staffing research: Online technologies and their impact on recruitment strategies*. Society for Human Resource Management.

Stebbins, R. A. (2001). *Exploratory research in the social sciences*. Thousand Oaks, CA: SAGE.

Taylor, C. (2001, May). E-recruitment is powerful weapon in war for talent. *People Management, 10*.

Thompson, L. F., Braddy, P. W., & Wuensch, K. L. (2008). E-recruitment and the benefits of organizational Web appeal. *Computers in Human Behavior*, *24*(5), 2384–2398. doi:10.1016/j.chb.2008.02.014

Wright, P. M., & Snell, S. A. (1991). Toward an integrative view of strategic human resource management. *Human Resource Management Review*, *1*(3), 203–225. doi:10.1016/1053-4822(91)90015-5

ADDITIONAL READING

Barber, A. E. (1998). *Recruiting employees*. Thousand Oaks, CA: SAGE.

Cappelli, P. (2001). Making the most of on-line recruiting. *Harvard Business Review*, *79*(3), 139–146.

Ployhart, R. E. (2006). Staffing in the 21st century: New challenges and strategic opportunities. *Journal of Management*, *32*, 868–897. doi:10.1177/0149206306293625

Ployhart, R. E., Schneider, B., & Schmitt, N. (2006). *Staffing organizations: Contemporary practice and theory* (3rd ed.). Mahwah, NJ: Lawrence Erlbaum Associates, Publishers.

KEY TERMS AND DEFINITIONS

E-HRM: A way of executing HRM to acquire, develop and retain human capital with the use of ICTs.

E-Recruitment: Organization of recruitment processes and activities, which, by means of technology and human agents, facilitate time- and space-independent collaboration and interaction in order to identify, attract, and influence competent candidates.

Recruitment: Practices and activities carried out by an organization for the primary purpose of identifying, attracting and influencing the job choices of competent candidates.

Virtual Organizing: Organization of processes and activities using technology and human agents to facilitate time- and space-independent interaction and collaboration.

ENDNOTE

1 This research was supported and financed by the Department of Business Administration, Business and Social Sciences, Aarhus University, Denmark. The usual disclaimer applies.

Chapter 6
The Importance of Psychological Contracts in Human Resource Management within the New Global Economy

Inocencia M. Martínez-León
Technical University of Cartagena, Spain

ABSTRACT

Human resources are considered the company's most valuable resource. For that reason, organizations require a Human Resource Management (HRM) that provides the adequate work force within the new global economy. The psychological contract includes solid agreements between company and its employees, beyond the written contract, specifying their contributions, expectations, beliefs, promises, and obligations between both parties. Their management requires the definition of the concept of psychological contract, the analysis of its main characteristics and contents; and the identification of its stages of development. Electronic Human Resource Management (e-HRM) introduces a way of implementing HRM strategies through Web-based tools, improving the psychological contract management. Information phase, Intranet, and internal electronic mail have an important role in socialization stage. The consolidation of psychological contract (maintenance phase) is favoured by Intranet, business-to-employees, internal electronic mail, database, videoconference, and groupware. Finally, in breach phase, intranet, electronic mail, database, videoconference, and groupware are important E-HRM.

1. INTRODUCTION

In the global economy, human resources are considered to be the most important asset of an organization, but very few organizations are able to fully harness its potential. Human resources are also considered the most important component of a corporation's competitive advantage in global markets. As human resources are considered the company's most valuable resource, it is necessary to improve their management.

Human Resource Management (HRM) refers to all of the dedicated activities that an organization uses to affect the behaviours of all the people who work for it (Jackson & Schuler, 2003). HRM is defined as "the policies, practices and systems

DOI: 10.4018/978-1-61350-207-5.ch006

that influence employees' behaviour, attitudes and performance" (De Cieri and Kramar, 2005). Ideally, these comprise "activities undertaken to attract, develop, and maintain an effective workforce within an organization" (Daft 2000), who ensure the effective functioning and survival of the organization and its members. Organizations need to hire the right numbers of people, with the right qualifications, for the right jobs, in the right locations. Line managers' expectations are to attract and retain excellent employees from the workforce pool. The impact of HRM policies and practices on firm performance is widely accepted and become an important topic in the fields of HRM, industrial relations, and industrial and organizational psychology (Kleiner, 1990; Boudreau, 1991; Jones & Wright, 1992). For these reasons, organizations try to optimize their HRM.

2. MAIN PURPOSE

The classical employment relationship, open-ended as to both its task content and its duration, was tying stable rates of pay to indefinite duration jobs. Employers offered job security and internal career opportunity with little risk of exposure to their workers (Kluytmans and Ott, 1999; Rousseau & Schalk, 2000), because workers did not leave, and certainly not to the competitor (Kluytmans and Ott, 1999). Employees were obedient and disciplined when carrying out their function and when requesting reallocation; and provided maximum performance at present function (Kluytmans y Ott, 1999).

In return, employees expected the reward of loyalty through continuation of the labour relationships and job security (Hiltrop, 1995; Kluytmans and Ott, 1999), irrespective of economic circumstances; the control of career by employer (Kluytmans and Ott, 1999) and payment of a guaranteed wage (Garud & Shapira, 1997; Rousseau and Shperling, 2003).

In the new global economy, increasing international competition, deregulation and globalization of markets have demanded greater flexibility and productivity of organizations. Firms have made the organizational structure change from a hierarchical to a flat form, accompanied by rapid reduction of the permanent core workforce, supplemented by a larger group of temporary contract and contingent employees, which has become a common employment staffing practice in today's modern organizations (Byrne, 1993; Allred *et al.*, 1996; Nicholson, 1996). New forms of labour contract has therefore emerged, such as fixed-term contracts, temporary contracts, etc (atypical contract). They are more short-term contracts and provide fewer career development activities (Pieters, 2009). Companies introduce new strategies, methods and techniques to attract and retain talented workers (Hiltrop, 1995, 1999; Ermel and Bohl, 1997; Rousseau and Shperling, 2003). Large companies had a core group of professional employees who were dependable, who would work hard and who would sacrifice the present for the future (Hiltrop, 1995).

For *organization*, new contractual forms provide greater flexibility and a good deal of enforceability (Marsden, 2004; Silla *et al.*, 2005). Instead of career paths and job security, a new kind of security, called "employability security" is established (Hiltrop, 1995; Schein, 1996). This is the promise that the employee's skills will be enhanced, and access to other tasks and assignments will be facilitated. Moreover, a new pay system is applied, based on flexibility bands to encourage the members to develop new skills (Hiltrop, 1995; Kinnie *et al.*, 1999; Silla *et al.*, 2005).

On the other hand, new contractual forms are getting *employees* to broaden their skills, by organizational learning and training programs, and identify the development of additional skills as a prerequisite for advancement (Schuler and Jackson, 1987). Workers receive higher salaries linked to performance and skills, rewards linked to performance, absorbing risk and exposure to market volatility (Rousseau and Shperling, 2003; Silla *et al.*, 2005). They consequently assume more responsibilities, and take risks (Schuler and Jackson, 1987). Workers may also manage the organizational intangible resources (knowledge, culture and business reputation, etc.).

This important change in employment relationship has generated a radical change in the relationship established between employer and employee, which must be adequately managed. Today's organizations need, among other actions, (a) to reach solid agreements with their employees, beyond the written contract, specifying the contributions and inducement that each party will receive from the other; (b) careful consideration of expectations in order to overcome ambiguity with their employees (Sims, 1994); and (c) examining and exploring the changing employment relationship. This new approach affects the employees, because the new psychological contract assumes that employees are not automatically loyal to their company as a whole, but that, as professionals, they will be moved by a new kind of loyalty to their own disciplines and skills. It also assumes that people will care more about their own development and that their skills are highly transferable between companies and industries (Hiltrop, 1995).

For those reasons, the psychological contract has reached a great importance nowadays. It is necessary to define the concept of psychological contract, analyze its main characteristics, contents, typologies; and identify its stages of development.

Electronic Human Resource Management (e-HRM) has both a negative and positive impact on employment relationship. Certain jobs are becoming obsolete on the one hand, and new job opportunities are created. IT tools improves information management, sharing information and knowledge, communication and coordination among different organizational members. Thus, these tools facilitate HRM, organizational memory, business management and a corporations' ability to compete in the global economy. These utilities have some important effects in psychological contract management.

As a strong importance of the psychological contract has been reported in the literature, the aim of this study is to examine the role of psychological contract in the actual HRM, and identify the facilitating role of e-HRM on its adequate management. This chapter has been divided into four parts. The first part deals with psychological contract term, analyzing its different subjects, conceptualizations, contents and typologies. Next, the state of psychological contract is examined, considering three different phases. Third, e-HRM and their positive influence on psychological contract are studied. As a conclusion, section 5 presents a discussion on the results of the study, and suggests fields in which further study would be fruitful.

3. DEVELOPMENT: PSYCHOLOGICAL CONTRACT

In reality, a written contract which is explicitly agreed between employee and employer may not exist. When a worker is asked "Is that (task, obligation or benefit) specified in your contract? The reply is more often that it is not. This situation suggest that employees rarely have detailed knowledge about what is included in their contract of employment, or they don't remember what was agreed at the outset of employment. In 2000, Cooke *et al.* demonstrated work and employment-related agreements are often made on an *ad-hoc* and on-going basis between the employer and the employee as and when it becomes necessary to do so.

Under these circumstances, a new concept has emerged: the "psychological contract", which refers to the relationship between employee and employer in terms of the unwritten expectations (Rousseau, 1989), beliefs, obligations, promises, and perceived perceptions that exist, they are not in a formal written contract, but in implicit expectations, often covertly held and infrequently discussed (D'Annunzio-Green and Francis, 2005), which determine the employment relationship. This broad conceptual term is increasingly used as a framework to study the employment relationship. This concept also has various meanings and interpretations among different subjects and constituencies. It is important, therefore, to mention that the conceptual model of psychological contract that is emerging from the literature review requires us to analyze its subjects, main characteristics, content, state and the e-HRM tools' influence on it.

3.1. Written Contract and Psychological Contract: Main Differences

The labour and capital relation includes written contract relation and unwritten psychological contract relation. Gradually, the psychological contract becomes the core content of HRM (Guest, 2004), in detriment to written contract. The psychological contract emphasizes that apart from formal contents regulated by the written contract, there are hidden, unwritten and largely unspoken sets of congruent expectations (Herath and Kishore, 2007), beliefs, perceptions, promise and obligations in the mutual relation between the organization and employees, which (a) mediates the unwritten employment bargain between the individual employee and the organizational agent to satisfy stakeholder requirements (Bathmaker, 1999), and (b) constitutes the important factors to decide employees' attitudes and behaviours (Rousseau, 1995).

The written contract is generally understood and usually includes four key factors: worktime required and available knowledge, skills and psychosocial abilities, and a wage-effort bargain (Huiskamp, 1995). However, the psychological contract fills the perceptual gaps in the employment relationship and shapes day-to-day employee behaviour in ways that cannot necessarily be discerned from a written contract (O'Donnell and Shields, 2002). Hence, employees weigh their obligations, expectations and norms towards the organization against the obligations, expectations and norms of the organization towards them, because they are not created by only an explicit agreement, and adjust their behaviour on the basis of critical outcomes. Although psychological contract is an unwritten contract it also has a central role in work behaviour by better specifying the dynamics of the employment relationship, and can be a powerful determinant of workplace behaviour and attitudes (van de Ven, 2004).

In this sense, Ring and Van de Ven (1994) saw this "informal and psychological contract increasingly compensating or substituting formal contractual safeguards as reliance on trust among parties that increases over time".

3.2. Development of the Concept of Psychological Contract

A considerable amount of definitions has been published on psychological contract. The development of this concept can be divided into two parts. The first part deals with the work of early theorists from 1930 to 1980, which is known as the so called "classical approach" to psychological contract. The second part deals with the "modern approach" to psychological contract, and begins with the publication of Rousseau's work in 1989.

The *classical approach*, arguably, the idea of psychological contract appeared in Bernard (1938), continuing with the seminal work of Argyris (1960), Levinson et al. (1962) and Schein (1965) (Robinson, 1996; Morrison and Robinson, 1997; Zhao *et al.*, 2007).

Bernard (1938) revealed the psychological relations between employees and employers developed as result of an explicit process of co-operative exchanges between them, which covered employee's implicit expectations of both material and non-material rewards (March and Simon, 1958).

In connection to the previous theories of Bernard (1938) and March and Simon (1958), Argyris (1960) defined the psychological contract as "the expectations of employer and employee which operate over and above the formal contract of employment". He introduced two different aspects. First, he only considered the material or transactional inducement and ignored nonmaterial or relational aspect of inducement (introduced by Barnard, 1938). Second, his approach was somewhat more close to the relationship between groups of employees and their supervisors, rather than a two-way relationship between an individual employee and supervisor (Taylor and Tekleab, 2004).

Shortly after Argyris (1960), Levinson *et al.* (1962) focused on employee-organization relationship as a dynamic contracting process involving an individual employee and organization.

Then, Schein (1965) focused on the need to recognize that the contents of psychological contract between employee and employer could incorporate not only material and economic dimensions but also non-material benefits that would satisfy both parties (Taylor and Tekleab, 2004).

In summary, the concept focused on mutual and reciprocal exchange relations between employee and employer (Argyris, 1960, Schein, 1965). In this step, more exploratory and less empirical research on the concept of psychological contract was done.

The beginning of *Modern Approach to Psychological Contract* was Rousseau's (1989) article titled "Psychological and implied contracts in organizations" (Rousseau, 1989). Rousseau (1989) defined this term as "an individual's beliefs in mutual obligations that a person and another party such as an Employer have¨, (either a firm or another person). This belief is predicted in the perception that a promise has been made (e.g. of employment or career opportunities) and a consideration offered in exchange for it (e.g. accepting a position, foregoing other job offers), binding the parties to some set of reciprocal obligations". This proposal introduced the cognitive-perceptual or individual's mental model to psychological contract, describing it as cognitive-perceptual process or mental model (Rousseau & McLean Parks, 1993; Rousseau, 1995; Rousseau & Tijoriwala, 1998; Rousseau and Schalk, 2000). Hence, the psychological contract represents a cognitive model existing at the individual level rather than an exchange model incorporating different parties.

Beside the significant contribution of Rousseau and her colleagues, some critics challenged the cognitive-perceptual framework of psychological contract of an individual's mental model of relationship with his or her organization and viewed it as a narrower approach to psychological contract that focuses only on an individual employee but doesn't cover the organizational side (Morrison & Robinson, 1997; Guest, 1998; Coyle- Shapiro & Kessler, 2000).

However, some researchers emphasized the need to include both parties in psychological contract (employer and employee) (Tekleab & Taylor, 2003; Guest 2004). Moreover, many authors agree that the psychological contract is based on expectations (Argyris, 1960), perceived promises (Conway & Briner, 2005), beliefs (Morrison & Robinson, 1997) or obligations (Rousseau, 1990).

Despite these ideas, scholars have reached a consensus about the nature of the psychological contract.

The Modern Approach of Psychological Contract is dominated by more empirical studies and less exploratory studies.

3.3. Subject of Psychological Contract

Traditionally, psychological contract is held by employees alone (Shore & Tetrick, 1994; Morrison & Robinson, 1997). However, the inclusion of the employer's perspective is already a reality.

In this sense, Guest (1998) argues that neglecting the employer's perspective may be misrepresenting the core of the psychological contract, and the reciprocal obligations between two parties. In this study, the employment relationship is considered as the exchange process between two parties. For both reasons, employee and employer should be considered as *subjects* of the psychological contract.

Usually, "organizations become party to the psychological contracts as principals who directly express their own terms or through agents who represent them" (Rousseau, 1995). However, they can not "feel" a psychological contract with employees and respond accordingly (Rousseau, 1989). As Rousseau (1995) and Morrison and Robinson (1997) acknowledge, the organization is comprised of multiple agents with whom the psychological contract is made, but the authors do not consider these as proxies for the organization as there may be differences in what each agent is offering or expecting. Indeed, the individual may have concurrent psychological contracts: one that is 'personal' with an organizational agent and one that is 'impersonal' with the organization. The discretionary actions of organizational agents are viewed as actions of organization by the workers (Rousseau, 1989) and may contribute to employees' perceptions of procedural fairness (Moorman *et al.,* 1998).

Hence, we set out to examine the characteristics, content and state of the psychological contract from the employer and employee perspective.

3.4. Main Characteristics of Psychological Contract

The psychological contract is the unwritten *implicit contract* (Argyris, 1960; Schein, 1965; Rousseau, 1989). Its content extends beyond the written contract and determines the bilateral and mutual relationship as a crucial factor in understanding the behaviour of both employee and employer in psychological contract in the work place (March and Simon, 1958; Schein, 1965). It forms in the minds of both parties and its content is highly *personal* for them. Consequently, the creation of a psychological contract may result from implicit means relying on an individual's interpretation of actions and events within an organization (Coyle-Shapiro and Kessler, 2000).

Proper development of psychological contract requires a *social process* (Herriot and Pemberton, 1997), because the employment relationship is founded on social exchange[1]. The psychological contract includes an exchange of terms and conditions of the agreement, *reciprocal* (Rousseau, 1989; Morrison y Robinson, 1997; Tabernero *et al.*, 2005) and *balance* between both parties. Consequently, it contains reciprocal exchange agreement and needs to be well-balanced, to survive at the time.

The psychological contract is thus *subjective* (Schein, 1965; Robinson and Rousseau, 1994; Rousseau, 1995), because its content is open to interpretation. It comprises subjective beliefs regarding an exchange agreement between an individual and, in organizations, typically the employing firm and its agents (Rousseau, 1995). Employees and organizational agents may hold different views on the content of the psychological contract and the degree to which each party has fulfilled the mutual obligations of exchange (Coyle-Shapiro and Kessler, 2000). Psychological contract is consequently open to misunderstandings (Schein, 1965; Robinson and Rousseau, 1994).

All the previous characteristics justifying each subject possess a unique psychological contract based upon his/her own understanding of the reciprocal obligations in the employment relationship between employee and employer (Turnley and Feldman, 1999). As a result, psychological contract can reach an unlimited number of formats.

Thus, the psychological contract is *dynamic* (Rousseau, 1989; Thomas and Anderson, 1998; Reyes and Martinez, 2007), which means it changes over time during the employment relationship (Anderson and Shalk, 1998), as do the organization and the individual. The psychological contract develops within a dynamic environment in which the individual is often interacting with multiple organizational agents who may each be sending a variety of messages, both verbal and non-verbal (Shore and Tetrick, 1994). The organizations operate in a similar environment, changing their organizational basis (structure, strategy, and culture), and their needs and goals. Subsequently, both reasons justify the changing relationships between constituent or configured parts.

Furthermore, the psychological contract is itself a *flexible* and still evolving concept (Guest, 2004), requiring it to undergo continual revision.

A psychological contract must constantly be *renegotiated and revaluated* (Rousseau & Parks, 1993; Conway & Briner, 2005), to be capable of adapting to the changing conditions. This is a fact that differentiates it from the formal contracts. Therefore, the nature of the perceived content could change.

Finally, the psychological contract is *obligatory* as expectations of perceived obligations must be fulfilled in order to prevent a breaching of the psychological contract which might result in adverse consequences.

3.5. Content of Psychological Contract

The content of the psychological contract includes different combinations of terms like perceptions, expectations, beliefs, promises and obligations (Kotter, 1973; Schein, 1978; Rousseau, 1989; Herriot and Pemberton, 1995). Schein (1965) highlighted the need to recognize that the contents of psychological contract between employee and employer could incorporate not only material and economic dimensions but also non-material benefits that satisfy both the parties (Taylor and Tekleab, 2004). This content can be analyzed from the worker's and employer's perspective.

The psychological contract is the *perception* of both parties of their relationship and the things they offer each other in this relationship (Herriot and Pemberton, 1995). Later, Herriot and Pemberton (1997) define this term as "the *perception* of both parties to the employment relationship, organization and individual, of the reciprocal promises and obligations implied in that relationship". Slightly adapting a previous definition by Herriot and Pemberton (1997), Guest and Conway (2002) define the psychological contract as "the *perceptions* of both parties to the employment relationship, organizational and individual, of the reciprocal obligations implied in that relationship".

Levinson *et al.* (1962) develop the psychological contract as "a series of mutual *expectations* of which the parties to the relationship may not themselves be dimly aware but which nonetheless govern their relationship to each other". Building upon this work, McLean Parks *et al.* (1998) states that a psychological contract is the idiosyncratic set of reciprocal *expectations* held by employees concerning their obligations (i.e., what they will do for the employer) and their entitlements (i.e., what they expect to receive in return).

Rousseau (1989) defines the psychological contract as "an individual's *beliefs* regarding the terms of conditions of a reciprocal exchange agreement between the focal person and another party". This definition is apparently focused exclusively on the employee perspective. However, researchers tend to downplay the aspect of mutuality and introduce the employer perspective (organizational agent).

Turnley *et al.* (2003) stated that "specific, psychological contracts are comprised of the *obligations* that employees believe their organization owes them and the *obligations* the employees believe they owe their organization in return".

Building upon these previous works and conceptual background, the content of the psychological contract includes perceptions, expectations, beliefs, promises, and obligations (Anderson and Schalk, 1998). The core of it concerns the exchange of promises and commitments (Guest & Conway, 2002) and mutual obligations and expectations.

After this brief analysis, we can summarized that the psychological contract consists of a set of perceptions, expectations, beliefs and obligations, based on explicit or implicit promises exchanged, which should cover the viewpoints of both parties (employer and employee) to the employment relationship. Psychological contract only exists within the context of an exchange relationship, mutuality is an underlying feature of it. In addition, psychological contract is implicit and subjective because its content is open to interpretation. It is "obligatory" since its content must be fulfilled in order to avoid contract "breaching". Furthermore, the psychological contract is dynamic due to mutual bargaining and negotiation mechanisms (Conway & Briner, 2005). Finally, the psychological contract is of a highly personal, idiosyncratic and self-constructed nature (Rousseau, 1989) and, consequently, the psychological contract of each individual is unique, and can reach an unlimited number of formats.

The content of psychological contract is adjusting and changing along with an organizations' development in different periods and employees' extending apperception of psychological contract.

3.5.1. Main Content of Psychological Contract

The content of the psychological contract includes perceived terms or "clauses" of the employment relationship. However, both subjects currently have no consensus on what psychological contracts contain or what the psychological contract is or what it actually encompasses. Hence, Anderson and Schalk (1998) contended that most employees are able to describe the content of their contract, and every worker has one. On the contrary, the employer has a general psychological contract with all employees of the same professional level, with some specifications to each worker.

Psychological contract has varied in recent years due to economic, social and existing business changes, and organizational flexibility required in terms of temporal relations, downsizing or part-time labour, imposed by the current financial crisis (Kluytmans and Ott, 1999). Traditionally, loyalty, commitment and confidence of employees are rewarded with job security, promotion, training and professional development, and protection against any problems. Currently, employers require workers with high commitment to business objectives, shared responsibility of success, quality performance, flexibility, judgement, strategic skills and continuous improvement. Employer in turn provides employability, learning capacity, flexibility, a performance-based compensation, greater participation and involvement, wide tolerance of change (challenging work) and a job but not a professional career (Csoka, 1995; Silla *et al.*, 2005).

Many previous works indicate distinct content and dimensions of employee's and employer's psychological contract (Rousseau, 1990; Herriot *et al.*, 1997; Hutton and Cummins, 1997; Guest and Conway, 1999, 2002; De Vos *et al.*, 2003; Jannsens *et al.*, 2003; Cable, 2008). A thorough preliminary investigation of existing research, the most important employee and employer psychological obligations are included in Table 1.

For employee obligations, Table 2 shows several variables of the psychological contract content, based upon previous work. They are related to the main term of contract, loyalty, flexibility, other employee factors' and organizational support.

For employer obligations, Table 3 exhibits the different variables of the psychological contract content. They should be grouped in extrinsic and intrinsic rewards, employer obligations' to employee, arrangement, and employee obligation's.

The heterogeneous nature of the employee obligations and employer obligations warrant

Table 1. Summary of employee and employer psychological obligations

EMPLOYEE OBLIGATIONS		EMPLOYER OBLIGATIONS	
Loyalty	An employee's perceived socio-emotional link and attachment toward their employing organization (for example, not accepting every job offer that comes along, working for the organization for at least several years).	Offering possibilities for development and/or promotion within the organization (such as possibilities for development, chances of promotion).	Career development
Flexibility	Willingness to be flexible in carrying out the work that needs to be done (for example, working overtime, taking work home).	Offering challenging, interesting job content (such as work in which employees can use their capacities, challenging tasks).	Job content
Ethical conduct	Willingness to conduct oneself ethically towards the organization (for example, not making confidential information public, dealing honestly with resources and budgets).	Offering respect and understanding for the personal situation of the employee (for example, flexibility in working hours, understanding of personal circumstances).	Work-private life balance
Honesty	The belief that one's partner stands by its word, fulfils promised role obligations, and is sincere.	Offering appropriate compensation (such as remuneration commensurate with the work, conditions of employment that have favourable tax consequences).	Financial compensation
Fairness	The perception by the individual that a particular activity in which they are a participant is conducted fairly by the organization.	Organization is characterized by high levels of distributive, procedural, and interactional justice.	Fairness
Trust	The mutual expectation that the other party will not behave opportunistically in future transactions of the employment relationship.		Trust
Availability	Willingness to keep one's availability status at an acceptable level (for example, taking training courses that become available, keeping up with trade literature).	systematic activities to develop and improve employees' skills, knowledge and behaviours to enable them to perform job-related duties, accomplish specific tasks and meet the quality requirements of HR for the future.	Training
Employability	The development of professional skills in general (such as e.g. language competence), but also to the acquisition of interpersonal skills, such as learning to cope with change.	Offering a pleasant and cooperative working environment (such as good communication among co-workers, good cooperation within the group).	Social environment
Job involvement	The extent to which people are psychologically attached to their jobs and the degree of importance that work holds in their life.	The perception that the job is more or less secure, whether or not there is objective evidence to support this.	Job security
Job satisfaction	The affective feelings toward the job. The extent to which employees gain enjoyment from their efforts at the workplace.	The belief that other people are generally well meaning and capable of positive interactions.	Humanity
Extra role Behaviour	The discretionary actions of workers that go beyond their existing role expectations and are not mandated by the organization.	The perceptions of justice and equality by individuals in an organization.	Organizational Justice
Personal congruence	The process of creating a balance between authentic cultural makeup and the organization.	Consistency between organizational values and actions.	Perceived consistency

Table 2. Employee obligations in the psychological contract

	Rousseau (1990)	**Herriot, Manning and Kidd (1997)**	**Hutton and Cummins (1997)**	**Guest and Conway (1999)**	**De Vos, Buyens and Schalk (2003)**	**Jannenes, Sels and Van den Brande (2003)**	**Cable (2008)**
EMPLOYEE OBLIGATIONS							
MAIN TERM OF CONTRACT		Work	Getting the Job Done				
				Delivery of the Deal			Intention to Quit
Employee Obligations for Organization							
LOYALTY	Loyalty						
FLEXIBILITY		Flexibility		Flexibility	Flexibility		
	Willing to Accept Transfers		Flexible Citizenship			Open Attitude	
OTHER EMPLOYEE'S FACTORS		Honesty		Fairness Trust	Ethical Behaviour		
					Employability		
	Extra Role Behaviours				Extra Role Behaviours		
						Respect for Authority	
		Self-Presentation					
	Property						
	Overtime Minimum Stay	Labelled Hours					
						Personal Investment	Work Involvement
							Job Involvement
							Job Satisfaction
ORGANIZATIONAL SUPPORT							Perceived Organizational Support

specific mention. Psychological contract differ across a number of factors, including employment level and personal culture of both parties. For this reason, the present research contributes to improve this situation, exhibiting the list of the psychological contract contents' of the most representative research, considering different countries.

Understanding the content of psychological contract is key to understand the factors which lie behind employee turnover.

Table 3. Employer obligations in the psychological contract

	Rousseau (1990)	Herriot, Manning & Kidd (1997)	Hutton and Cummins (1997)	Guest and Conway (2002)	De Vos, Buyens & Schalk (2003)	Jannenes, Sels & Van den Brande (2003)	Cable (2008)
EMPLOYER OBLIGATIONS							
EXTRINSIC AND INTRINSIC REWARDS	Development			Development Opportunities	Career Development		Career Plateau
	Advancement			Promotion			
	Support		Support				
	High Pay	Pay		Fair Payment	Financial Rewards		
		Benefits Recognition		Benefits Package			
EMPLOYER OBLIGATIONS' TO EMPLOYEE				Feedback on Performance			Job - Task Performance
	Training	Training		Training			
	Job Security	Job Security		Reasonable Job Security		Long Term Involvement	
		Fairness	Respect And Fair Practice	Fair Treatment			
		Needs					
		Consult					
		Discretion					
		Humanity				Personal Treatment	
		Environment		Working Environment	Social Atmosphere		
		Justice		Not to Make Unreasonable Demands of Employees		Equal Treatment	
					Work-Life Balance		
ARRANGEMENT				Interesting Work	Job Content	Carefuless Regarding Arrangement	
							Person-Organization Fit
EMPLOYEE OBLIGATION'S							Organizational Citizen Behaviour

3.6. State of Psychological Contract

Compared to the traditional approach of Rousseau and colleagues, Guest (1998) advocates develop theoretical models and research that would allow researchers to go beyond mere description of the content of the psychological contract to an evaluation of its state, introducing consequently the concept of state of the psychological contract.

The dynamism and flexibility of psychological contract explains the existence of several states of its development. The state of the psychological contract refers to "whether the promises and obligations have been fulfilled, whether they are fair, and their implications for trust" (Guest & Conway, 2002). It involves a broader evaluation of the employment relationship, shaped by the content of the psychological contract, but going beyond them.

For Guest and colleagues, evaluation of the state of the psychological contract should represent a qualitative leap in research on the psychological contract, rendering it more explanatory and less descriptive, and making possible the prediction of diverse variables related to attitudes, behaviour and employees' health (Gracia *et al.*, 2007)..

The main phases of psychological contract development are: the process of the psychological contract formation, the development or maintenance, and breaking the psychological contract. A more detailed study is needed to accurately characterize each phase.

1st Phase: The Process of Psychological Contract Formation

Many of the psychological contract contents are explicitly or implicitly addressed during the recruitment and selection phases (Sutton and Griffin, 2004). The recruitment and selection are social processes with important informational qualities for both potential employers and employees. These stages have influence on the development of appropriate expectations and inform the early development of individuals' psychological contract with the organization (Herriot, 1984; 1989). Employers also make implicit and explicit promises during *recruitment stage* which obligate them to give certain things to their workers in exchange for their employees' contributions to the organization. Consequently, many terms of a psychological contract are established during recruitment (Rousseau & Greller, 1994; Shore & Tetrick, 1994).

In the *hiring stage*, the potential employee receives and analyzes an introductory letter spelling out conditions of employment. If both parties agree to the formal aspects of employment relationship (working hours, salary, type of contract, benefits and mutual obligations) (Dunahee and Wangler, 1974), the written contract is signed and established. However, new workers have a limited and incomplete knowledge in terms of arrangement that have been voluntarily involved. People "fill in the blanks" along the way, and they sometimes do so inconsistently (Rousseau, 1995). Therefore, a psychological contract is also created in this stage, beginning the formalization of its content.

When an employee is hired, the human resources department starts his/her *socialization stage*. Socialization is characterized by a process through which newcomers could receive a detailed job description and/or an employee handbook. Hence, employees start to understand, interpret, and respond to their new environment. At this moment, most newcomers have only limited or incomplete information about the terms of their employment relationship (Rousseau, 2001) and, consequently, the psychological contract contents. For that reason, organization facilitates all the learning that took place prior to a person's first day on the job. Worker receives instruction about the correct execution of the job, and the organizational culture and strategy. This motivates newcomers to actively interpret their initial experiences as a basis for predicting future events and for changing their expectations, thereby making their psychological contract schema more complete. Nevertheless, this training is not sufficient to establish the definitive content of psychological contract. This in turn should help them to reduce uncertainty and make their experiences in their new work setting more predictable (Shore & Tetrick, 1994). Thus, the perceptions of promises are adapted based upon newcomers' interpretations of their experiences in the work setting (Rousseau, 2001).

During organizational socialization the psychological contract rapidly develops and becomes more stable. Empirical research demonstrates newcomers change their perceptions of what their employer had promised them based on their perceptions of the employer inducements actually received (De Vos *et al.*, 2003). They also change their perceptions of what they had promised their employer, based on their perceptions of what they actually contributed to their employer (De Vos *et al.*, 2003). Consequently, newcomers use their experiences and knowledge within the work environment as feedback about their initial expectations and they are flexible in adapting their initial expectations based upon this feedback. In the same line, newcomers adapt their promise based upon their interpretations of the employer's actions in the work setting (De Vos *et al.*, 2003). The main conclusion of the foregoing analysis is that the mutuality also operates during the stage of psychological contract formation and that it explains changes in newcomers' psychological contract perceptions (De Vos *et al.*, 2003).

It is important to note that in recruitment, selection and hiring stage, the contract is still an incomplete mindset (Topa and Palací, 2005), for employee and the employer (organizational agent). After the training and socialization stage, the content of the worker's psychological contract changes to introduce his/her feedback (Thomas and Anderson, 1998; Topa *et al.*, 2002) as well as the employer. This adaptation is a result of gradual adjustment and mutuality, which it is more significant in employee psychological contract. Consequently, workers are more flexible in changing their perceptions of their own promises than in changing their perceptions of their employer's promises.

Finally, it is important to note that small and medium-sized companies (SMEs) can build the psychological contract in a different way, as SMEs make less use of formalized recruitment practices, provide less training to their employee, and are less likely to use formal performance based compensation systems (Carison *et al.*, 2006).

2nd Phase: Maintenance of Psychological Contracts

The development of the employment relationship involves a series of ongoing interactions that help to gradually reduce the uncertainty, generating and consolidating the contents of the psychological contract. The dynamism and mutuality of the psychological contract is confirmed in this stage, where the employees' and employers' perceptions, beliefs, expectations, obligations and promises are continuously changing, revaluating and renegotiating. However, our knowledge on the way in which psychological contracts develops, and on the conditions that initiate and influence changes in the psychological contract is still limited.

The constant change of the contract provides increased opportunities for misunderstandings and therefore causing contract breach (Robinson, 1996). The content of the psychological contract are subject to internal and external changes (Rousseau, 1995). The internal changes have their origin in both subjects. The internal changes generated by employee stems from his/her evolution during the employment relationship with the company (employment status, training, wage, work-life balance and promotion for other colleagues). The internal changes generated by the employer are due to economic, productive, organizational and strategic changes. Both kinds of changes can modify the contents of the psychological contract.

When one party changes the content of psychological contract, a mutual feedback process is activated; which involves a continuous revaluation, renegotiation and implementation of newly required obligations to achieve the recent beliefs and expectations set.

Promoting employability is part of a new psychological contract between employer and employee, whereby both are responsible for maintaining the employment situation (Kluytmans and Ott, 1999).

This phase of the psychological contract depends on a balance between employee's and employer's obligations, and the congruence of both subjects' action. Consequently, it depends on the fulfilment of the arrangement. In this sense, Shore and Barksdale (1998) identified four types of psychological contract, on the basis of two dimensions: employee and employer obligations, and the level of obligations. The types of psychological contract are mutual high obligations, mutual low obligations, employer over-obligation and employee under obligations. In the first two situations, the psychological contract is balanced and both parties have high or low obligations, respectively. In this case, the psychological contract is frequently stable. The other types of contracts are not balanced and their breach has a stronger probability to happen, generating the following step.

The effects of the maintenance stage of psychological contract must be analyzed from two perspectives, the employee and the employer. The fundamental premise is that a fulfilled and healthy psychological contract will result in positive individual behaviour, and be associated with positive outcomes for the organization.

The maintenance of a psychological contract by the employee helps his/her professional development, the reduction of social distance between him/her and the organization, and reaches higher performance (Topa and Palací, 2004). Consequently, the worker has more opportunities for promotion, training and improved wage conditions (Topa and Palací, 2004).

For the employer, if employees maintain a satisfied psychological contract, they are more profitable in the performance of tasks, they work more efficient and with higher quality, yield and efficiency; and consequently, the company achieves greater productivity and performance (Topa and Palací, 2004).

The age of the subjects of the psychological is determinant in this phase. Some authors have suggested that employee age affects the breach of psychological contracts differently, and therefore its maintenance phase. Older workers usually provide more stable psychological contracts (Rousseau, 2001), and react to possible breaches of the psychological contract from younger ones.

3rd Phase: Breach of Psychological Contracts

Breach is a "subjective experience, referring to one's perception that another has failed to fulfil adequately the promised obligations of the psychological contract" (Robinson, 1996). Robinson and Morrison (2000) distinguished breach from violation of psychological contract. Breach of psychological contract is a cognitive evaluation of what is received compared to what was promised. Violation would better describe the emotional and affective state that may follow this cognitive evaluation. Breach is less serious and a cognitive appraisal of the event, while violation is more serious and initiating behaviour, attitude, or emotional response beyond the cognitive appraisal. Breach or violation of psychological contract was found to be related to a lot of negative work attitudes and organizational consequences (see below).

However, it is important to note that there are two intermediate stages in the process of breaking the psychological contract. The first, *reneging*, occurs when one party knowingly fails to meet an obligation or recognize the existence of an obligation between two parties, but refuse to comply. The second, *incongruence* which can result when an employee has perceptions of a given promise that differs from those held by the organizational agent or agents responsible for fulfilling that promise

(Morrison and Robinson, 1997). Incongruence is when a promise is established or it can develop as time elapses and perceptions of promises decay or become distorted in memory. Thus, incongruence relates to a gap in the shared understanding between an employee and the organizational agent (e.g., the employer has a different understanding of the promised outcome), as to whether an obligation exists. These different perceptions result because the terms of a psychological contract are inherently perceptual.

This situation comes from three interrelated areas: (a) workers and organizational agents take part in employment relationship with different cognitive schemata, (b) obligation incongruence and (c) insufficient communication (Morrison and Robinson 1997). The direct results of the incongruence are the lack of trust between the parties, reducing performance and prosocial behaviour, and in contrast, intend to stay in business. This is the most common situation in business area (Turnley and Feldman, 2000) and may progress to the next stage, depending on the interpretation of the damaged party and how the other party responds to these circumstances, trying to respect their commitments and therefore, correcting the failure or not.

The intensification of the two previous intermediate stages generates the psychological contract breach as evidence of non-fulfilment arrangement. Seven reasons justify their existence: (a) the "delay" or sense of procrastination on behalf of the party in providing the obligations, (b) the "magnitude of discrepancy" or degree of lack of obligation that the party feels he or she should be receiving, (c) "type/form", where one party's obligations represent different qualifying standards from those expected by the other party; (d) the "inequity" or comparative analysis between what the participants are getting and what others of the same level in the same job type and/or organizational level are receiving; (e) the "reciprocal imbalance" refers to the sense of imbalance between what one party is receiving compared to what it is giving back to the other (adapted from Cassar and Briner, 2005); (f) the absence of "organizational responsibilities" for fairness and consistency; and (g) the incongruence between organizational culture and individual values.

Morrison and Robinson (1997) argued that the cognitive perception of a breach in a psychological contract will not necessarily result in the intense emotional reaction associated with the term psychological contract violation. These authors observe that explanations of the root causes of breach of contract may reduce the intensity of the emotional response.

The results of breach of psychological contract in behaviours, described by Rousseau (1995), are (a) the exit (termination of the relationship), (b) the voice (actions to remedy the violation), (c) the loyalty (silence, willingness to endure), and (d) the destruction (neglect, counterproductive behaviours). Herriot and Pemberton (1995) described those same behaviours as: 'get ahead' (voice), 'get safe' (loyalty), 'get even' (destruction), or 'get out' (exit).

The psychological contract breach has influence on some variables. Guest (2004) categorized the outcomes of non-fulfilment between attitudinal consequences and behavioural consequences. Attitudinal consequences are job satisfaction (Robinson & Rousseau, 1994; Gakovic & Tetrick, 2003; Johnson & O'Leary-Kelly, 2003), organizational commitment (Coyle-Shapiro & Kessler, 2000; Johnson & O'Leary-Kelly, 2003), trust in the organization (Robinson & Rousseau, 1994), emotional exhaustion (Gakovic & Tetrick, 2003) and cynicism (Johnson & O'Leary-Kelly, 2003). Behavioural consequences are job performance (Robinson, 1996; Turnley & Feldman, 2000; Johnson & O'Leary-Kelly, 2003), organizational citizenship behaviour (Robinson, 1996; Coyle-Shapiro & Kessler, 2000; Turnley & Feldman, 2000; Lo & Aryee, 2003), absenteeism (Johnson & O'Leary-Kelly, 2003), turnover intention (Robinson & Rousseau, 1994; Robinson, 1996; Turnley & Feldman, 2000; Lo & Aryee, 2003) and actual turnover (Robinson & Rousseau, 1994; Robinson, 1996).

To reduce these negative effects of psychological contract breach is critical to the management of actual or perceived breaches or violations of the contract.

3.5. E-Human Resource Management: Information Management

Since last decade, technology and Internet development have had a profound effect on the field of HRM, promoting the implementation and application of electronic Human Resource Management (e-HRM). This concept is introduced as "a way of implementing HRM strategies, policies and practices in organizations through the conscious and directed support of and with the full use of web technology" (Rüel *et al.*, 2004). While a variety of definitions of the term e-HRM have been suggested, this paper will use the definition suggested by Strohmeier (2007) who saw it as "the application of information technology for both networking and supporting the interaction of a least two individual or collective actors in their shared performing of HR activities".

According to Strohmeier (2007), Information Technology (IT) has two different implications in e-HRM. First, IT is necessary to connect usually spatially segregated subjects and enable interactions between them irrespective of their working in the same room or on different continents, increasing their connection and integration. Second, IT supports subjects by partially – and sometimes even completely – substituting them in executing HR activities. Consequently, IT are tools for task fulfilment.

Both implications operate in a psychological contract management in several ways: (a) two internal subjects interact in order to perform psychological contract activities (employee and employer), (b) the connection between subjects becomes essential, independent of their location, and (c) the task fulfilment is crucial for both subjects, but especially by the employer. Therefore, e-HRM system provides some benefits, such as decreased transaction times, management of psychological contract content, administer compensation and benefit systems (Strohmeier, 2007). In addition, these systems increase the efficiency of HR processes, reduce administrative costs, give support to training and facilitate the information management between both subjects.

Information management is the management of the processes and systems that create, acquire, organize, store, distribute, and use information (Detlor, 2010). Its goal is to help people and organizations access, process and use information efficiently and effectively. The different state of psychological contract requires using and communicating information, as well as the correct establishment of psychological contract content.

A better study would examine a large group of selected Technology and tools that give support to e-HRM, especially on psychological contract management. Intranet, internal electronic mail, database, videoconference, groupware and business-to-employee are IT that help people carry out personal information management more efficiently and effectively (Jones, 2008). IT tools facilitate personal information management, improving communication, collaboration, knowledge acquisition and management. Now, a new need has arisen, the employees are required to develop and maintain professional networks, where they can exchange knowledge and skills, improving their professional efficiency (Surowiecki, 2004; Cross *et al.*, 2005). IT tools track down and contact co-workers and experts inside and outside the organization, keeping their interests, skills and role profiles updated at all times. Thus, IT tools assist the establishment of organizational memory, its content can be used as a new source of organizational information by the workers. IT tools have consequently an important role in the professional activity and development of the employee and HRM.

Employees should be able to use IT tools to enhance their communication flow with the organization and its members. As it has been inferred previously, the communication process is a fundamental activity for both workers and organization agents, in the different psychological contract's phases. Consequently, if workers improve their skills and capabilities on managing IT resources, a higher degree or performance and job satisfaction will be achieved, leading to an improvement on their psychological contract.

Internet is not an e-HRM, but it is included in this study by its importance in some phases of psychological contract.

Another important factor to consider in the use of IT tools is the age of the staff (personal). It has been broadly demonstrated that younger people use these technologies more frequently, even in their personal environment. Consequently, the age should have a significant impact on the influence of IT tools in the psychological contract management.

In *formation phase*, e-HRM tools are very important to create the new psychological contract.

In *recruitment* stage, Internet is a relevant information source for any newcomer as information about the company's structure, services, target markets, quality certification or customer services can be made accessible through it. Large companies often include charts, news, social responsibility (policies and reports), organizational results, research and development activity reports, newsletters, customers', suppliers' and shareholders' spaces, and online services. Unfortunately, the abundant available information complicates the information search and the detection of interesting information (Martínez-León *et al.*, 2008). Although this tool is not an e-HRM, its great importance in this stage justifies its inclusion in the study.

In another sense, some companies develop an intelligent Web portal to serve as service providers in recruitment tasks. This portal aims at helping people to find a job. When the potential employees find this e-HRM, they start the formation of the psychological contract.

In the *socialization* stage, Intranet and internal electronic mail are e-HRM which help to configure the initial psychological contract. Intranet includes word processing, spreadsheet applications, database management, data analysis, graphics, and communications. In definition, intranet has a wide range of information about organization in general, and its organizational culture, strategy and HRM in particular, which is very important for the newcomer. Intranet's data visualization software, data collection and its ability to provide real-time feedback, makes it far easier to implement training and development initiatives (Denton, 2007), which are essential for any new organizational member.

Moreover, intranet technology allows sharing information, enhancing the organizational communication in general by supporting collaboration across departmental, functional and regional boundaries (Bernard, 1996; Scacchi & Noll, 1997; McNaughton *et al.*, 1999). This tool also supports organizational learning (Martínez-León *et al.*, 2008) and knowledge management (KM) processes (Scott, 1998; Alavi & Leidner, 1999).

The available functionality on the Intranet depends on the level of IT investment, the use of IT tools, its configuration and contents, its updates and internal control, and previous employees' Intranet experience. This is not a "packaged" technology with fixed attributes, but it is moulded and shaped according to the social forces at play in the organization (Hughes, 1987; Williams & Edge, 1996). The computer networks incorporate more significant, private and extensive information than any webpage. Hence, Intranet usually includes community news, message boards, notices of upcoming events and HRM policies and systems (Arnold *et al.*, 2003). When the private network is more developed, it has available collections of documents and relevant information about organizational culture, corporate government, production systems, strategic plans, forthcoming marketing campaign, classified advertising, IT emergency, project management procedures and project evaluation, and newsletters.

If the Intranet radically changes its unidirectional role for communication and information channel, and is configured in a bidirectional sense, a new working space is created (Corso *et al.*, 2009). This new working space is characterized by a creative environment, focused on employees, their needs, specific working conditions and interactions with others. There, workers can find what they need to work, to learn, to meet and to interact with others (Corso *et al.*, 2009).

Intranets, in their actual configuration, emphasize internal information, build important links among organizations and their employees and improve the employee's vision of the organization. Intranets therefore allow the worker to consistently adapt his/her psychological contract content to the real and changing organization's situation.

An Intranet can also facilitate an ad hoc workstation at home, where the employee could connect to the company's network via VPN technologies.

These different possible utilities justify why the same technology can manifest completely differently in different organizational settings and have different effects on psychological contract formation.

Another important point about e-HRM in the socialization stage of the psychological contract formation is internal electronic mail. Electronic mail facilitates the exchange of information between individuals or groups by offline messages, which can contain documents, programmes, graphics and texts (Martínez-León *et al.* 2008). It allows the users (both workers and organizational agents) share employment relationship's information, which improves their understanding of psychological contract obligations and contents. Hence, this tool gives support to their employment relationships, improving the psychological contract management. However, it can create new problems, such as notably "information overload" or transmission of misleading messages, creating undesired confusion. These problems could make difficult the establishment of the initial psychological contract, so face-to-face contact is required to get a better result.

Thus, on the basis of above arguments we can assume that e-HRM tools facilitate the formation of psychological contract.

- **Hypothesis 1:** e-HRM tools have a positive effect on psychological contract formation.
    - **Subhypothesis 1a:** e-HRM tools have a positive effect on socialization stage as intermediate stage of psychological contract formation.
    - **Subhypothesis 1b:** Internet has a positive effect on recruitment stage as intermediate stage of psychological contract formation.

In Maintenance Phase, some e-HRM could help to consolidate the content of psychological contract, such as Intranet, business-to-employee, internal electronic mail, database, videoconference and groupware.

Intranet facilitates sharing, searching, exchanging and storing organizational, departmental and professional information in the organization, which can be exploited by both psychological contract subjects. The available information should be used to establish the definitive psychological contract content and, in the future, renegotiates their contents. In addition, if the computer network is highly advanced, new working space change employees' role (Corso *et al.*, 2009), varying the psychological contract. Consequently, Intranet is then already widely used as a managerial tool in the employment relationship during the maintenance phase of psychological contract.

A business-to-employee (B2E) system is part of a corporate Intranet (Huang *et al.*, 2004). This tool is an e-business initiative which if successfully implemented assists organizations in delivering useful services, information, or products to their disperse employees (Turban *et al.*, 2006). B2E provides mobile employees and organizational agent (managers and HR professionals) the freedom to access essential business services and data irrespective of location and technical infrastructure at hand.

Different typologies of B2E might support different aspects of maintenance of psychological contract: (a) Business-to-Employee (B2E) portal assists organizations in delivering useful information and services to their disperse employees/ organizational agent or with high geographical mobility which in turn creates productive employee relationships needed for psychological contract maintenance; (b) Business-to-employee messaging system is similar to an internal electronic mail, and refers a closed-loop message distribution system is disclosed for accepting documents from a variety or sources among two subjects; and (c) business-to-employee cooperation support proposes online interviews as a tool that allows subjects to gather interview information and assess cultural fit. Online interviews cannot replace the traditional face-to-face interviews, but can definitely provide the means to help disperse employees/ organizational agent or with high geographical mobility with interviewing challenges.

The internal electronic mail offers the promise of rapid communication of essential information (Pazzani, 2000), which should facilitate the adequate development of psychological contract. This tool also creates communication and cooperation channels with organizational people. These utilities improve the subject's vision about his/her employment relationship, favouring the concretion of psychological contract content. Nevertheless, communication is asynchronous, which does not allow both subjects to communicate in real time, reducing the capability of negotiation. For this reason, face-to-face contact is required to get a better result.

The databases are deposits of past data, internal information and knowledge, which permits interchanging information and creates knowledge (learning) (Martínez-León *et al.*, 2008). Both internal information and knowledge should be able to use in the concretion of initial psychological contract content. Thus, this tool creates and maintains an organizational shared intelligence and memory (Ruggles, 1998), which should provide important and useful internal information for organizational members. Consequently, this information should be extremely valuable and important in the creation and maintenance of the psychological contract. Therefore, databases reflect current and changing situations, which should improve the psychological contract arrangements and renegotiation activities among both subjects.

Videoconference permits simultaneous dialogue through a virtual interaction among people (De Geus, 1997; Davenport *et al.*, 1998), and the exchange of documents, files and shows. Consequently, the videoconferencing technologies facilitate the frequent exchange of information between remote subjects, their collaboration and the creation of cooperative settings in a fast and flexible way (Martínez-León *et al.*, 2008). Thus, this tool should be able to create, diffuse and negotiate of psychological contract content through their synchronous interaction with remote employees, colleagues and organizational agents, which enable to overcome geographical and time barriers in extended organization. Therefore, videoconference helps to adequately manage the psychological contract.

The groupware encourages efficient and accurate sharing of information and knowledge across organizational boundaries. Groupware is able to hide the identity of the participant, which makes it easier to share information and improve communication in a much more free-flowing manner than databases or intranet allow (Martínez-León *et al.*, 2008). If both subjects of the psychological contract work in dispersed geographical areas, groupware facilitates the remote communication and transference of useful internal information, making the concretion and renegotiation of psychological contract content easy. Therefore, this tool facilitates the psychological contract management.

Internet is not an e-HRM, but its utilities facilitate the psychological contract maintenance. This technology facilitates the search and exchange of information about customers, suppliers, competitors, target markets and own organization (community manager). Hence, the adequate management of information allows detecting external and organizational changes. These changes can affect the arrangement obligations between the parties. Therefore, actual information allows concreting the definitive psychological contract content and its continue revaluation and renegotiation.

Psychological contract is established at a certain point in time, and it is assumed to be revaluated and renegotiated. The perceptions, expectations, beliefs, promises and obligations of employee and/or organizational agent are able to change; generating misunderstandings. In this situation, all technological tools give an opportunity to get interesting and right information for psychological contract maintenance, and facilitate its exchange and communication; maintaining or facilitating a healthy relationship between two subjects. Hence, IT tools facilitate the necessary revaluation and renegotiation of the psychological contract, getting reciprocal and balanced obligations. Therefore, e-HRM is helpful tools to improve the psychological contract management.

These arguments suggest that:

- **Hypothesis 2:** e-HRM tools have a positive effect on psychological contract maintenance.
    - **Subhypothesis 2a:** Internet has a positive effect on psychological contract maintenance.

In Breach phase, e-HRM helps and improves the psychological contract management, reducing its breaching. These tools diminish the *reneging stage*, because they mitigate the information asymmetry about two parties, disappearance of alienation in favour of a maximum transparency between subjects. e-HRM tools also reduce different incentives for opportunistic behaviour by the contracting parties.

Thus, e-HRM reduces the frequency and intensity of *incongruence stages* between both subjects. Technological tools improve communication, reduce obligation incongruence, and diminish the different perception about contributions and inducements. The main conclusion of the foregoing analysis is that e-HRM generates and provides clear information which solves the main sources of incongruence and conflict. However, these tools do not always reduce the different cognitive schemata of subjects.

In *violation or breach stage*, some e-HRM (intranet, electronic mail, database, videoconference and groupware) could help to solve the differences between both psychological contract subjects'. These tools facilitate the acquisition and exchange of correct information and improve their communication. Consequently, one party can share with the other party any well-grounded non-fulfilment or variations in psychological contract content; reducing the time and negative effect of this situation. However, continuous non-fulfilment of psychological contract obligations from anyone generates many requirements, which could worsen the situation.

Obviously, groupware is the less useful e-HRM in this stage, because it does not always facilitate the direct communication between two parties. Sometimes, groupware is used to anonymously announce any non-fulfilment obligations, which alert all groupware's members about them; giving an opportunity to resolve the controversy. On the other hand, Internet is not an e-HRM, but facilitates acquire external information, which could be useful to avoid breaching the psychological contract.

The previous arguments can be summarized in:

- **Hypothesis 3:** e-HRM tools have a positive effect on psychological contract breach.
    - **Subhypothesis 3a:** e-HRM tools have a positive effect on psychological contract reneging stage.
    - **Subhypothesis 3b:** e-HRM tools have a positive effect on psychological contract incongruence stage.
    - **Subhypothesis 3c:** e-HRM tools have a positive effect on psychological contract breach stage.
    - **Subhypothesis 3d:** e-HRM tools have a positive effect on psychological contract violation stage.
    - **Subhypothesis 3e:** Internet has a positive effect on psychological contract breach.

Therefore, e-HRM tools improve human resources management, with special influence in their psychological contracts, which attract and retain talent in the organization. The breach of psychological contract is minimized, reducing the mentioned losses associated. Hence, HRM might be able to improve the organizational management. Then, the firm creates values and can generate and maintain a competitive advantage.

4. CONCLUSION

The findings of this study contribute substantially to the general view of psychological contract concept and the contribution of e-HRM tools in its adequate management. Recently, a new approach to employment relationship has emerged, where the psychological contract reaches wide importance in organizations as a determinant of the employment relationship. The psychological contract contains a set of perceptions, expectations, beliefs and obligations, based on explicit or implicit promises, exchanged and shared by employee and employer. Its content is reciprocal, subjective, dynamic, and derived from a social process. Hence, the psychological contract is an implicit contract, unique and can reach an unlimited number of formats.

This study also focuses on the different statements of the psychological contract: formation, maintenance and breach. Departing from the psychological contract *formation phase*, all stages and processes to create the psychological contract have been considered. Intranet and internal electronic mail are facilitating tools for the psychological contract formation, during socialization stage. Internet is a useful tool in recruitment stage. In a similar way, the *maintenance* of psychological contract includes some process and can be facilitated by different e-HRM tools (Intranet, business-to-employees, internal electronic mail, databases, videoconference and groupware). Finally, the *breach* of psychological contract has two intermediate stages, and numerous and important consequences. E-HRM tools can apparently reduce the number of incongruence and reneging, and decrease the number and effect of breach or violations of psychological contract. These findings suggest interesting and useful implications for future research and for organizational practice. Three main hypotheses about the role of e-HRM on psychological contract and subhypothesis have been proposed, which should be testing future empirical research carried out in the companies.

Consequently, E-HRM tools can not be viewed as a means of reaching higher levels of productivity and managerial results, without a way of providing basic information to organizational members, which plays a significant role in enhancing employment relationships and, consequently, human resource management.

Some limitations to this study must be noted with respect to future theoretical and empirical research. First, the age of the employees has influence on psychological contract management; whereas older employees have a more stable psychological contract than younger ones. Several reasons justify this reality. First, older employees have more realistic expectations regarding the compensation. Second, they have fewer options to achieve real employment in other organizations (Hedge *et al.*, 2006). As a result, and thirdly, they better regulate their emotions to possible breaching of the psychological contract (Carstensen *et al.*, 2003). Fourth, in a near withdrawal, older workers strive to positively intensify all their experience to keep their job, because it may be the last one before retirement (Carstensen *et al.*, 1999). Fifth, they focus on more positive aspects of relations with the organization, seen as less affected by negative events (Bal *et al.*, 2008). And finally, these workers should receive higher satisfaction of other factors outside work, such as family, hobbies and community (Clark *et al.*, 1996).

However, younger employees should be able to adapt better than older ones in a rapidly changing environment, renegotiating their psychological contract. Thus, younger employees have higher computer literacy than older ones. The benefits of e-HRM tools in the psychological contract management also intensify the actual tendency to generate greater entry for younger employees relative to older ones. Consequently, when organization uses e-HRM tools, organizations are more reluctant to hire older employees and tend to favour younger ones (Aubert *et al.*, 2006).

Secondly, the size of a firm plays an important role in terms of HRM (Kotey and Sheridan, 2004) and the e-HRM tools investments. In fact, larger organizations have more sophisticated recruitment structures and process than smaller one. Barrett and Mayson's (2007) reveal that growth-oriented firms have more formal HRM practices. On the other hand, large organizations invest a lot in technology adoption and use, developing more e-HRM tools' functionalities and getting better HRM.

Thirdly, the efficient usage of e-HRM tools is closely related to a wider comprehension of information management and contribution to organizational performance. IT tools should be understood as a part of organizational information context and its usefulness is influenced by organizational culture, values and principles concerning strategic information management.

Despite these limitations, to our knowledge this research is the first to systematically examine the influence of IT tools on psychological contract management.

REFERENCES

Alavi, M., & Leidner, D. E. (1999). Knowledge management systems: Issues, challenges, and benefits. *Communications of the Association for Information Systems, 1*, article 7, (pp. 2-36).

Allred, B. B., Snow, C. C., & Miles, R. E. (1996). Characteristics of managerial careers in the 21st century. *The Academy of Management Executive, 10*(4), 17–27. doi:10.5465/AME.1996.3145316

Anderson, N., & Schalk, R. (1998). The psychological contract in retrospect and prospect. *Journal of Organizational Behavior, 19*(S1), 637–648. doi:10.1002/(SICI)1099-1379(1998)19:1+<637::AID-JOB986>3.0.CO;2-H

Argyris, C. (1960). *Understanding organizational behavior*. London, UK: Tavistock Publication.

Arnold, M., Gibbs, M., & Wright, P. (2003). Intranets and the creation of local community: 'Yes, an intranet is all very well, but do we still get free beer and a barbeque? In Huysman, M., Wenger, E., & Wulf, V. (Eds.), *Communities and technologies*. Dordrecht, The Netherlands: Kluwer Academic Publishers.

Aubert, P., Caroli, E., & Muriel, R. (2006). New technologies, organisation and age: Firm-level evidence. *The Economic Journal, 116*, F73–F93. doi:10.1111/j.1468-0297.2006.01065.x

Bal, P. M., De Lange, A. H., Cansen, P. G. W., & Van Der Velde, M. E. G. (2008). Psychological contract breach and job attitudes: A meta-analysis of age as a moderator. *Journal of Vocational Behavior*, *72*(1), 143–158. doi:10.1016/j.jvb.2007.10.005

Barrett, R., & Mayson, S. (2007). Human resource management in growing small firms. *Journal of Small Business and Enterprise Development*, *14*(2), 307–320. doi:10.1108/14626000710746727

Bathmaker, S. (1999). So, what's the deal? The state of the psychological contract in a new university. *Journal of Vocational Education and Training*, *51*(2), 265–282. doi:10.1080/13636829900200084

Bernard, R. (1996). *The corporate Intranet*. New York, NY: John Wiley & Sons.

Boundreau, J. W. (1991). Utility analysis in human resource management decisions. In Dunette, M. D., & Hough, L. M. (Eds.), *Handbook of industrial and organizational psychology* (2nd ed., *Vol. 2*, pp. 271–326). Palo Alto, CA: Consulting Psychologists Press.

Byrne, J. A. (1993). The horizontal corporation. *Business Week*, 76-81.

Cable, D. A. J. (2008). *The psychological contract: The development and validation of a managerial measure*. Unpublished Master Doctoral Thesis of University of Waikato.

Carison, D., Upton, N., & Seaman, P. (2006). The impact of human resources practices and compensation design performance. An analysis of family owned SMES. *Journal of Small Business Management*, *44*(4), 531–543. doi:10.1111/j.1540-627X.2006.00188.x

Carstensen, L. L., Fung, H. H., & Charles, S. T. (2003). Socioemotional selectivity theory and the regulation of emotion in the second half of life. *Motivation and Emotion*, *27*(2), 103–123. doi:10.1023/A:1024569803230

Carstensen, L. L., Isaacowitz, D. M., & Charles, S. T. (1999). Taking time seriously: A theory of socioemotional selectivity. *The American Psychologist*, *54*(3), 165–181. doi:10.1037/0003-066X.54.3.165

Cassar, V., & Briner, R. B. (2005). Psychological contract "breach": A multiple component perspective to an over-researched construct? *Revista de Psicología Social*, *25*(1), 125–136. doi:10.1174/0213474052871079

Clark, A., Oswald, A., & Warr, P. (1996). Is job satisfaction u-shaped in age? *Journal of Occupational and Organizational Psychology*, *69*(1), 57–81. doi:10.1111/j.2044-8325.1996.tb00600.x

Conway, N., & Briner, R. B. (2005). *Understanding psychological contracts at work -A critical evaluation of theory and research*. Oxford University Press. doi:10.1093/acprof:oso/9780199280643.001.0001

Corso, M., Giacobbe, A., & Martini, A. (2009). Rethinking knowledge management: The role of ICT and the rise of the virtual workspace. *International Journal of Learning and Intellectual Capital*, *6*(3), 272–292. doi:10.1504/IJLIC.2009.025045

Coyle-Shapiro, J., & Kessler, I. (2000). Consequences of the psychological contract for the employment relationship: A large scale survey. *Journal of Management Studies*, *37*(7), 903–927. doi:10.1111/1467-6486.00210

Csoka, L. S. (1995). A new employer-employee contract? *Employment Relations Today*, *22*(2), 21–31. doi:10.1002/ert.3910220204

D'annunzio-Green, N., & Francis, H. (2005). Human resource development and the psychological contract: Great expectations or false hopes? *Human Resource Development International*, *8*(3), 327–344. doi:10.1080/13678860500199725

Daft, R. (2000). *Management* (5th ed.). Fort Worth, TX: Dryden Press.

Davenport, T. H., Delong, D. W., & Beers, M. C. (1998). Successful knowledge management projects. *Sloan Management Review*, *39*(2), 43–57.

De Cieri, H., & Kramar, R. (2005). *Human resource management in Australia: Strategy, people, performance* (2nd ed.). Macquarie Park, Australia: McGraw-Hill.

De Geus, A. P. (1997). The living company. *Harvard Business Review*, *75*(2), 51–59.

De Vos, A., Buyens, D., & Schalk, R. (2003). Psychological contract development during organizational socialization: Adaptation to reality and the role of reciprocity. *Journal of Organizational Behavior*, *24*(5), 537–559. doi:10.1002/job.205

Denton, K. (2007). Corporate Intranets: How can they give a new meaning to training and development? *Development and Learning in Organizations*, *21*(6), 12–14. doi:10.1108/14777280710828576

Detlor, B. (2010). Information management. *International Journal of Information Management*, *30*(2), 103–108. doi:10.1016/j.ijinfomgt.2009.12.001

Dunahee, M. H., & Wangler, L. A. (1974). The psychological contract: A conceptual structure for management/employee relations. *The Personnel Journal*, *53*(7), 518–530.

Ermel, L., & Bohl, D. (1997). Responding to a tight labor market: Using incentives to attract and retain talented workers. *Compensation and Benefits Review*, *29*(6), 25–29. doi:10.1177/088636879702900605

Gakovic, A., & Tetrick, L. E. (2003). Psychological contract breach as a source of strain for employees. *Journal of Business and Psychology*, *18*(2), 235–246. doi:10.1023/A:1027301232116

Garud, R., & Shapira, Z. (1997). Aligning the residuals: Risk, return, responsibility and authority. In Shapira, Z. (Ed.), *Organizational decision making* (pp. 238–256). Cambridge, UK: Cambridge University Press.

Gracia, F. J., Silla, I., Peiró, J. M., & Fortes-Ferreira, L. (2007). The state of the psychological contract and its relation to employees' psychological health. *Psychology in Spain*, *11*(1), 33–41.

Guest, D. (2004). The psychology of employment relationship: An analysis based on the psychological contract. *Applied Psychology: An International Review*, *53*(4), 541–555. doi:10.1111/j.1464-0597.2004.00187.x

Guest, D. E. (1998). Is the psychological contract worth taking seriously? *Journal of Organizational Behavior*, *19*(S1), 649–664. doi:10.1002/(SICI)1099-1379(1998)19:1+<649::AID-JOB970>3.0.CO;2-T

Guest, D. E., & Conway, N. (1999). Peering into the black hole: The downside of the new employment relations in the UK. *British Journal of Industrial Relations*, *37*(3), 367–389. doi:10.1111/1467-8543.00133

Guest, D. E., & Conway, N. (2002). Communicating the psychological contract: An employer perspective. *Human Resource Management Journal*, *12*(2), 22–38. doi:10.1111/j.1748-8583.2002.tb00062.x

Hedge, J. W., Borman, W. C., & Lammlein, S. E. (2006). *The aging workforce: Realities, myths, and implications for organizations*. Washington, DC: American Psychological Association. doi:10.1037/11325-000

Herath, T., & Kishore, R. (2007). Outsourcing success: Psychological contract perspective. In *AMCIS 2007 Proceedings*, Paper 58. Retrieved from http://aisel.aisnet.org/amcis2007/58

Herriot, P. (1984). *Down from the ivory tower: Graduates and their jobs*. Chichester, UK: Wiley.

Herriot, P. (1989). Selection as a social process. In Smith, M., & Robertson, I. T. (Eds.), *Advances in selection and assessment* (pp. 171–187). Chichester, UK: Wiley.

Herriot, P., Manning, W. E. G., & Kidd, J. M. (1997). The content of the psychological contract. *British Journal of Management*, *8*(2), 151–162. doi:10.1111/1467-8551.0047

Herriot, P., & Pemberton, C. (1995). A new deal for middle managers. *People Management*, *1*(12), 32–34.

Herriot, P., & Pemberton, C. (1997). Facilitating new deals. *Human Resource Management Journal*, *7*(1), 45–56. doi:10.1111/j.1748-8583.1997.tb00273.x

Hiltrop, J. M. (1995). The changing psychological contract: The human resource challenge of the 1990s. *European Management Journal*, *13*(3), 286–294. doi:10.1016/0263-2373(95)00019-H

Hiltrop, J. M. (1999). The quest for the best: Human resource practices to attract and retain talent. *European Management Journal*, *17*(4), 422–430. doi:10.1016/S0263-2373(99)00022-5

Huang, J.-H., Yangb, C., Jinb, B.-H., & Chiua, H. (2004). Measuring satisfaction with business-to-employee systems. *Computers in Human Behavior*, *20*(1), 17–35. doi:10.1016/S0747-5632(03)00047-5

Hughes, T. (1987). The evolution of large technological systems. In Bijker, W., Hughes, T., & Pinch, T. (Eds.), *The social construction of technological systems* (pp. 51–82). Cambridge, MA: MIT Press.

Huiskamp, R. (1995). Regulating the employment relationship: An analytical framework. In Van Ruysseveldt, J., Huiskamp, R., & Van Hoof, J. (Eds.), *Comparative industrial and employment relations*. London, UK: Sage.

Hutton, D., & Cummins, R. (1997). Development of the psychological contract inventory (PsyCon). *Australian Journal of Career Development*, *6*(3), 35–41.

Jackson, S. E., & Schuler, R. S. (2000). *Managing human resources: A partnership perspective 7e*. Cincinnati, OH: South-Western Publishing.

Janssens, M., Sels, L., & Van Den Brande, I. (2003). Multiple types of psychological contracts: A six-cluster solution. *Human Relations*, *56*(11), 1349–1378. doi:10.1177/00187267035611004

Johnson, J. L., & O'Leary-Kelly, A. M. (2003). The effects of psychological contract breach and organizational cynicism: Not all social exchange violations are created equal. *Journal of Organizational Behavior*, *24*(5), 627–647. doi:10.1002/job.207

Jones, G. R., & Wright, P. M. (1992). An economic approach to conceptualizing the utility of human resource management practices. In Rowland, K., & Ferris, G. (Eds.), *Research in personnel and human resources management* (pp. 271–299). Greenwich, CT: JAI Press.

Jones, W. (2008). *Keeping found things found: The study and practice of personal information management*. Burlington, MA: Morgan Kaufman.

Kinnie, N., Purcell, J., Hutchinson, S., Terry, M., Collinson, M., & Scarbrough, H. (1999). Employment relations in SMEs: Market-driven or customer-shaped? *Employee Relations*, *21*(3), 218–236. doi:10.1108/01425459910273071

Kleiner, M. M. (1990). The role of industrial relations in firm performance. In Fossum, J. A., & Mattson, J. (Eds.), *Employee and labour relations* (pp. 4.23–4.43). Washington, DC: BNA Press.

Kluytmans, F., & Ott, M. (1999). Management of employability in The Netherlands. *European Journal of Work and Organizational Psychology*, *8*(2), 261–272. doi:10.1080/135943299398357

Kotey, B., & Sheridan, A. (2004). Changing HRM practices with firm growth. *Journal of Small Business and Enterprise Development*, *11*(4), 474–485. doi:10.1108/14626000410567125

Kotter, J. P. (1973). The psychological contract: Managing the joining-up process. *California Management Review, 15*(3), 91–99.

Levinson, H., Price, C., Munden, K., Mandl, H., & Solley, C. (1962). *Men management and mental health.* Cambridge, MA: Harvard University Press.

Lo, S., & Aryee, S. (2003). Psychological contract breach in a Chinese context: An integrative approach. *Journal of Management Studies, 40*(4), 1005–1020. doi:10.1111/1467-6486.00368

March, J. G., & Simon, H. A. (1958). *Organizations*. New York, NY: Wiley.

Marsden, D. (2004). The 'network economy' and models of the employment contract. *British Journal of Industrial Relations, 42*(4), 659–684. doi:10.1111/j.1467-8543.2004.00335.x

Martínez-León, I. M., Ruiz Mercader, J., & Martínez-León, J. A. (2008). The effect of organisational learning tools on business results. *International Journal of Knowledge and Learning, 4*(6), 539–552. doi:10.1504/IJKL.2008.022887

Mclean Parks, J., Kidder, D. L., & Gallagher, D. G. (1998). Fitting square pegs into round holes: Mapping the domain of contingent work arrangements onto the psychological contract. *Journal of Organizational Behavior, 19*(S1), 697–730. doi:10.1002/(SICI)1099-1379(1998)19:1+<697::AID-JOB974>3.0.CO;2-I

Mcnaughton, R. B., Quickenden, P., Matear, S., & Gray, B. (1999). Intranet adoption and inter-functional co-ordination. *Journal of Marketing Management, 15*(5), 387–403. doi:10.1362/026725799784870270

Moorman, R. H., Blakely, G. L., & Niehoff, B. P. (1998). Does perceived organizational support mediate the relationship between procedural justice and organizational citizenship behaviour? *Academy of Management Journal, 41*(3), 351–357. doi:10.2307/256913

Morrison, E., & Robinson, S. (1997). When the employers feel betrayed: A model of how the psychological contract violation develops. *Academy of Management Review, 22*(1), 226–256.

Nicholson, N. (1996). Career systems in crisis: Change and opportunity in the Information Age. *The Academy of Management Executive, 10*(4), 40–51. doi:10.5465/AME.1996.3145318

O'Donnel, M., & Shields, J. (2002). Performance management and the psychological contract in the australian federal public sector. *The Journal of Industrial Relations, 44*(3), 435–457. doi:10.1111/002218560204400300 8

Pazzani, M. J. (2000). Representation of electronic mail filtering profiles: A user study. In *Proceedings of the 5th International Conference of Intelligent User Interfaces*. New York, NY: ACM Press.

Pieters, Z. (2009). *Expatriation as a career experience*. Master's Thesis. University of Stellenbosch. Retrieved April 15, 2011, from http://hdl.handle.net/10019.1/2343

Reyes Contreras, Y., & Martínez León, I. M. (2007). Los contratos psicológicos: Sus efectos en los resultados de las organizaciones. In *V Encuentro Iberoamericano de Finanzas y Sistemas de Información*. EFSI.

Ring, P. S., & Van De Ven, A. H. (1994). Developmental processes of cooperative interorganisational relationships. *Academy of Management Review, 19*(1), 90–118.

Robinson, S. L. (1996). Trust and breach of the psychological contract. *Administrative Science Quarterly, 41*(4), 574–599. doi:10.2307/2393868

Robinson, S. L., & Morrison, E. W. (2000). The development of psychological contract breach and violation: A longitudinal study. *Journal of Organizational Behavior, 21*(5), 525–546. doi:10.1002/1099-1379(200008)21:5<525::AID-JOB40>3.0.CO;2-T

Robinson, S. L., & Rousseau, D. M. (1994). Violating the psychological contract: Not the exception but the norm. *Journal of Organizational Behavior*, *15*(3), 245–259. doi:10.1002/job.4030150306

Rousseau, D., & Parks, J. (1993). The contract of individuals and organizations. *Research in Organizational Behavior*, *15*(1), 1–43.

Rousseau, D. M. (1989). Psychological and implied contract in organizations. *Employee Responsibilities and Rights Journal*, *2*, 121–139. doi:10.1007/BF01384942

Rousseau, D. M. (1990). New hire perception of their own and their employer's obligations: A study of psychological contracts. *Journal of Organizational Behavior*, *11*(5), 389–400. doi:10.1002/job.4030110506

Rousseau, D. M. (1995). *Psychological contracts in organizations: Understanding written and unwritten agreements*. Thousand Oaks, CA: Sage.

Rousseau, D. M. (2001). Schema, promise and mutuality: The building blocks of the psychological contract. *Journal of Occupational and Organizational Psychology*, *74*(4), 511–541. doi:10.1348/096317901167505

Rousseau, D. M., & Greller, M. M. (1994). Human resource practices: Administrative contract makers. *Human Resource Management*, *33*(3), 385–401. doi:10.1002/hrm.3930330308

Rousseau, D. M., & Mclean Parks, J. (1993). The contracts of individuals and organizations. In Cummings, L. L., & Staw, B. M. (Eds.), *Research in organizational behavior* (pp. 1–47). Greenwich, CT: JAI Press.

Rousseau, D. M., & Schalk, R. (2000). *Psychological contracts in employment: Cross-national perspectives*. Newbury Park, CA: Sage.

Rousseau, D. M., & Shperling, Z. (2003). Pieces of the action: Ownership and the changing employment relationship. *Academy of Management Review*, *28*(4), 553–570.

Rousseau, D. M., & Tijoriwala, S. A. (1998). Assessing psychological contracts: Issues, alternatives and measures. *Journal of Organizational Behavior*, *19*(S1), 679–695. doi:10.1002/(SICI)1099-1379(1998)19:1+<679::AID-JOB971>3.0.CO;2-N

Ruël, H. J. M., Bondarouk, T., & Looise, J. C. (2004). E-HRM: Innovation or irritation: An explorative empirical study in five large companies on Web-based HRM. *Management Review*, *15*(3), 364–381.

Ruggles, R. (1998). The state of the notion: Knowledge management in practice. *California Management Review*, *40*(3), 80–89.

Scacchi, W., & Noll, J. (1997). Process-driven Intranets: Life-cycle support for process reengineering. *IEEE Internet Computing*, *1*(5), 42–49. doi:10.1109/4236.623967

Schein, E. H. (1965). *Organizational psychology*. Englewood Cliffs, NY: Prentice-Hall.

Schein, E. H. (1978). *Career dynamics: Matching individual and organizational needs*. Reading, MA: Addison- Wesley.

Schein, E. H. (1996). Career anchors revisited: Implications for career development in the 21st century. *The Academy of Management Executive*, *10*(4), 80–88. doi:10.5465/AME.1996.3145321

Schuler, R. S., & Jackson, S. E. (1987). Linking competitive strategies with human resource management practices. *The Academy of Management Executive* (1987-1989), *1*(3), 207-219.

Scott, J. E. (1998). Organizational knowledge and the Intranet. *Decision Support Systems*, *23*(1), 3–17. doi:10.1016/S0167-9236(98)00032-3

Shore, L. M., & Barksdale, K. (1998). Examining degree of balance and level of obligation in the employment relationship: A social exchange approach. *Journal of Organizational Behavior*, *19*, 731–744. doi:10.1002/(SICI)1099-1379(1998)19:1+<731::AID-JOB969>3.0.CO;2-P

Shore, L. M., & Tetrick, L. E. (1994). The psychological contract as an explanatory framework in the employment relationship. In Cooper, C. L., & Rousseau, D. M. (Eds.), *Trends in organizational behavior* (*Vol. 1*, pp. 91–109). Chichester, UK: John Wiley & Sons.

Silla, I., Gracia, F. J., & Peiró, J. M. (2005). Diferencias en el contenido del contrato psicológico en función del tipo de contrato y de la gestión empresarial pública o privada. *Revista de Psicología Social*, *20*(1), 61–72. doi:10.1174/021347405287l097

Sims, R. R. (1994). Human resource management's role in clarifying the new psychological contract. *Human Resource Management*, *33*(3), 373–382. doi:10.1002/hrm.3930330306

Strohmeier, S. (2007). Research in e-HRM: Review and implications. *Human Resource Management Review*, *17*(1), 19–37. doi:10.1016/j.hrmr.2006.11.002

Sutton, G., & Griffin, M. A. (2004). Integrating expectations, experiences, and psychological contract violations: A longitudinal study of new professionals. *Journal of Occupational and Organizational Psychology*, *77*(4), 493–514. doi:10.1348/0963179042596487

Tabernero, C., Arenas, A., & Briones, E. (2005). Impacto del feedback negativo sobre los procesos de autorregulación del grupo. La ruptura del contrato grupal relacional. *Revista de Psicología Social*, *20*(1), 93–108. doi:10.1174/021347405287l033

Taylor, S. M., & Tekleab, A. G. (2004). Taking stock of psychological contract research: Assessing progress, addressing troublesome issues, and setting research priorities. In Coyle-Shapiro, J., Taylor, M. S., & Tetrick, L. E. (Eds.), *The employment relationship. Examining psychological and contextual perspectives* (pp. 253–283). Oxford University Press.

Tekleab, A. G., & Taylor, M. S. (2003). Aren't there two parties in an employment relationship? Antecedents and consequences of organization-employee agreement on contract obligations and violations. *Journal of Organizational Behavior*, *24*(5), 585–608. doi:10.1002/job.204

Thomas, H., & Anderson, N. (1998). Changes in newcomers' psychological contracts during organizational socialization: A study of recruits entering the British Army. *Journal of Organizational Behavior*, *19*(S1), 745–767. doi:10.1002/(SICI)1099-1379(1998)19:1+<745::AID-JOB967>3.0.CO;2-I

Topa, G., Alonso, E., & Palací, F. (2002). Una medida de socialización. Aplicación de un estudio longitudinal con Soldados Profesionales Españoles. *Metodología de las Ciencias del Comportamiento*, 544-549.

Topa, G., & Palací, F. (2004). Psychological contract breach or unfulfilment? A meta-analytic review of empiric research. *Acción Psicológica*, *3*, 155–171.

Topa, G., & Palací, F. (2005). La ruptura del contrato psicológico entre los Soldados Profesionales Españoles y su relación con la decisión de permanecer o abandonar. *Revista de Psicología Social*, *20*(1), 45–60. doi:10.1174/021347405287l088

Turban, E., Leidner, D., Mclean, E., & Wetherbe, J. (2006). *Information Technology for management: Transforming organizations in the digital age* (5th ed.). John Wiley & Sons.

Turnley, W. H., Bolino, M. C., Lester, S. W., & Bloodgood, J. M. (2003). The impact of psychological contract fulfillment on the performance of in-role and organizational citizenship behaviors. *Journal of Management*, *29*(2), 187–206. doi:10.1177/014920630302900204

Turnley, W. H., & Feldman, D. C. (1999). A discrepancy model of psychological contract violations. *Human Resource Management Review*, *9*(3), 367–386. doi:10.1016/S1053-4822(99)00025-X

Turnley, W. H., & Feldman, D. C. (2000). Re-examining the effects of psychological contract violations: Unmet expectations and job dissatisfaction as mediators. *Journal of Organizational Behavior*, *21*(1), 25–42. doi:10.1002/(SICI)1099-1379(200002)21:1<25::AID-JOB2>3.0.CO;2-Z

van de Ven, C. (2004). *The psychological contract: A big deal!* Behavioural Sciences Service Centre. Ministry of Defence. The Hague, The Netherlands. Retrieved April 18, 2011, from http://www.internationalmta.org/Documents/2004/2004024P.pdf

Williams, R., & Edge, D. (1996). The social shaping of technology. *Research Policy*, *25*(6), 865–899. doi:10.1016/0048-7333(96)00885-2

Zhao, H., Wayne, S. J., Glibkowski, B. C., & Bravo, J. (2007). The impact of psychological contract breach on work-related outcomes: A meta-analysis. *Personnel Psychology*, *60*(3), 647–680. doi:10.1111/j.1744-6570.2007.00087.x

KEY TERMS AND DEFINITIONS

e-HRM: A way of implementing HRM strategies, policies and practices in organizations through the conscious and directed support of and with the full use of web technology.

Information Technologies: The broad range of technologies involved in information processing and handling, such as recording devices, communication systems, computer Systems (including hardware and software components, and data), and other electronic devices.

Psychological Contract: Includes the implicit and specific terms and conditions of the employment relationship, not regulated nor specified in a formal written contract, which has a central role in work behaviour and determine the relationship between employee and employer (organizational agent).

Psychological Contract Content: Refers to the unwritten expectations, beliefs, obligations, promises, and perceptions that are a powerful determinant of workplace behaviour and attitudes, determining the relationship between employee and employer (organizational agent) and, consequently, the employment relationship.

Psychological Contract State: Refers to a relatively stable base level from which changes in the nature of the employment relationship can be judged.

ENDNOTE

[1] Masterson *et al.* (2000, p. 740) observe, "…an employee is involved in at least two social exchange relationships at work: one with his or her immediate supervisor, and one with his or her organization."

Chapter 7
Strategic and Organizational Considerations Related to an E-Learning Model:
A Case of Study

Nuria Calvo
University of A Coruña, Spain

Paolo Rungo
University of A Coruña, Spain

Ignacio Moreno
ICARINDI S.L., Spain

ABSTRACT

The aim of this chapter is to assess the impact of different organizational factors on the success of e-learning programs, in terms of both self-reported satisfaction and the level of learning. Hence, this study adds to the analysis of the efficacy of e-learning models from an organizational perspective by providing some useful insights, which may help to improve decision-making related to employee's continuing education and satisfaction. This simulation, using a bivariate ordered probit model, shows that economic and indirect economic incentives play a key role in augmenting the level of both satisfaction and learning. This analysis also considers how efficacy of learning programs may thus improve by linking the human resources development policy with results obtained in e-learning courses.

1. INTRODUCTION

The application of latest advancements in Information Technology to business education has contributed to fulfill most wishes of education managers. The demand for continuing education programs with top specialists, the costs of mobility, the fragmentation of learning time due to inopportune meetings, among other difficulties, seem to have come upon a solution with the so called e-learning model (Box, 1999; Kirby, 1999;Bose, 2003).

Technological improvement and the design of high-capacity networks for sharing data have allowed for solving most of the limitations of the traditional learning methodology, by facilitating

DOI: 10.4018/978-1-61350-207-5.ch007

both access to information and the adaptation of programs to individual needs. However, earlier applications of e-learning have shown that those technological tools do not automatically guarantee high levels of learning (Snyder et al., 2000; Ettinger et al., 2006, Fernández Díez de Lastra, R., 2001). Moreover, Ettinger et al. (2006) demonstrate that the mere act of uploading materials of traditional courses to a software platform may reduce motivation and, thus, learning outcomes. Discussion about these problems has generated a second wave of development of e-learning models (Servage, 2005). Due to the different approach to learning under the new model, it has been argued that e-learning needs a different pedagogical system (Roy, 2006). As a consequence, a number of firms, focused on the management of educational contents for e-learning, have born during the last decade. Those firms offer teams of experts in pedagogy, scriptwriters and technicians, who work for creating personalized educational paths and taking advantage of all possibilities of software platforms for enhancing learning. In addition, the development of e-learning educational materials has actually generated an increasing demand for standard rules to facilitate the compatibility of contents and software platforms (Singh and reed, 2002; Orbea, 2008).

Previous research about factors influencing learner satisfaction reveal that learner computer anxiety, instructor attitude toward e-learning, e-learning course flexibility or quality, perceived usefulness, and diversity in assessments are critical factors affecting learners´ perceived satisfaction (Sun, P.; Tsai, R.J.; Chen, Y. & Yeh, D., 2008). Perceived self-efficacy has been also considered as key factor that influences e-learning satisfaction (Liaw, S., 2008). According to existing literature, disposing of both advanced technological instruments and adapted learning programs does not guarantee the optimal management of e-learning within firms (Ettinger et al., 2006; Galagan, 2002; Netteland et al., 2007). In fact, it is necessary to consider where, when and how learning programs take place, to identify the possible difficulties and how the firm should manage learning through a software platform (Zhand and Jasimuddin, 2008; Rahmandad et al., 2009). Learning by using the new training model requires self-motivation and self-management, and it demands both cultural and organizational changes (Redmon and Salopek, 2000; Tynjäla and Häkkinen, 2005; McPherson et al., 2005; Ettinger et al., 2005; Davis and Wong, 2007). Hence, organizations must strategically define what the main objective of e-learning is, and what they should do in order to achieve their goals.

The rest of the paper is organized as follows. Section 2 describes the main objective and the methodology used in the analysis. Section 3 presents data, descriptive statistics, and the model that will be estimated. Section 4 discussed main results, and the last section summarizes conclusions.

2. MAIN GOAL AND METHODOLOGY

The aim of this paper is to assess the impact of different organizational factors on the success of e-learning programs, in terms of both self-reported satisfaction and the level of learning. Hence, this study adds to the analysis of the efficacy of e-learning models from an organizational perspective by providing some useful insights, which may help to improve decision-making related to employee's continuing education and satisfaction. To this end, the case of EVERIS, one of the first firms that introduced e-learning as the standard methodology for its employees, has been studied. From a methodological point of view, the analysis can be divided into two parts. Firstly, both directors and users of e-learning programs at EVERIS have been surveyed. The survey included questions about personal satisfaction, level of learning, and relevant organizational factors such as the existence of incentives, the preference for e-learning compared to the traditional system, and users' capacity to use the technological platform (variables are fully described in the next section).

In order to assess the impact of the explicative variables on satisfaction and learning, we estimated a bivariate ordered probit model, which is specified by seemingly unrelated equations. The main reason for using this methodology is the potential existence of an indirect relationship between satisfaction and learning, which might cause biased results when two different ordered probit models are considered, one for satisfaction and the other for the level of learning. There is a growing empirical literature which makes use of bivariate probit models. Greene and Hanser (2009) cite more than 20 different papers applying this methodology from 1997 to 2007, though first applications are due to Calhoun (1989, 1991, 1994), and more recently, Dawson and Dobson (2010). The methodology used here is the same used in these papers and, explicitly, we have followed Sajaia (2008) for its implementation in Stata©.

3. DEVELOPMENT OF THE ANALYSIS

As starting point of the framework, we analyzed both strategic and organizational factors that have driven a firm such as EVERIS to choose an e-learning model. To this end, we discussed the reasons for the implementation of e-learning with directors of EVERIS, who helped them to better understand relevant variables. As a multinational consulting firm, EVERIS offers solutions for business strategy, development and maintenance of technological appliances. The growth of the number of employees (from 3300 persons in 2004 to 7400 in 2009), as well as the increased volume of projects and its differentiation, justified in 2006 the strategic decision of adopting e-learning for continuing education at EVERIS. On evaluating this choice, the board of directors analyzed the need for aligning with new tendencies of professional education, and they also underlined both the possibility of offering a larger number of courses, and the opportunity for cost-reduction.

In particular, two main goals emerged as reasons for change toward an e-learning model: enlarging the offer of educational programs, and innovating through alternative models of learning. Provided that one value of EVERIS is the building of a learning software platform for continuing education for its employees, these goals were perfectly aligned with their business strategy. In order to carry out the implementation, departments of Marketing, People and Human Resources worked together and, under the direction of their Corporate University, they evaluated open-source applications for collaborative learning. The adoption of e-learning was thus an internal process of the organization, and it took about six months.

Employees at EVERIS are characterized by a higher qualified and technical skills than other professionals, which allowed the firm for not investing in their capacity to use the learning platform. It should be noted, in fact, that this is one of the main investment limitations in other organizations. The management of the e-learning model in this firm involves the development of a supply of courses, which reflects the professional needs of employees and makes use, to this end, of specific personal mapping for professional development. Each personal map consists of mandatory courses, recommended programs, and other optional courses depending on both personal preferences and job skills. This educational system aims to achieve the goal of 40 hours of education per employee and year, by investing about 10 millions of euros yearly.

At present, the learning model of this company takes the form of "blended-learning", combining lectures and online courses available at EVERCAMPUS. Though the firm did not design an explicit incentive scheme, the monetary value of grades achieved by EVERCAMPUS is published and communicated to employees, thus underlining their direct value in terms of employability or market value of professionals.

3.1 Data and Descriptive Statistics

Data have been collected by a survey to 84 employees of EVERIS Madrid, who were attending e-learning courses between march and may 2010, and voluntarily answered our online questionnaire. It should be noted that the questionnaire is independent from both the attendance of any particular course and results obtained (though data about dropout rates for each course is not available, the average at EVERCAMPUS is about 28% of all participants). In addition, given our objective, we have decided to survey employees independently from the specific course they attended, and we focused on specific organizational measures such as incentives, time and space availability for learning, and the skills that permitted to use the learning platform in a profitable way.

Figure 1 presents the functional distribution of the sample of employees surveyed. Although this analysis did not focused on studying satisfaction with a specific e-learning course, most of the contents assessed were related with technical knowledge (programming languages, technical certifications) and management skills.

As it can be observed, about 50% of respondent work in the department of Technology Systems, a characteristic that is likely to affect results. In fact, we expect those employees to be able to handle the learning platform with ease.

Table 1 presents the definitions of variables being analyzed, along with descriptive statistics. Dependent variables, self-assessed satisfaction and self-assessed learning, are categorical variables, which take value from 1 to five.

On considering the impact of organizational factors on Self-Assessed Satisfaction (SAS, hereafter) and Self-Assessed Learning (SAL, herafter), it is important to control for both basic characteristics of the course, such as quality of contents, for example, and the ability of attendants to use the e-learning platform. This is in fact the standard object of study in existing literature, and taking into account these variables allows for a comparison of main results. Figure 2 represents the distribution of SAS when subgroups of respondents that value positively specific aspects of the e-learning course attended are considered. As it can be observed, and as expected, the distribution of SAS suffer a significant change when people have a positive perception of both course material and contents. In other words, satisfaction and the quality of contents are highly correlated, and we will thus take into account this factor in the

Figure 1. Functional distribution of the sample

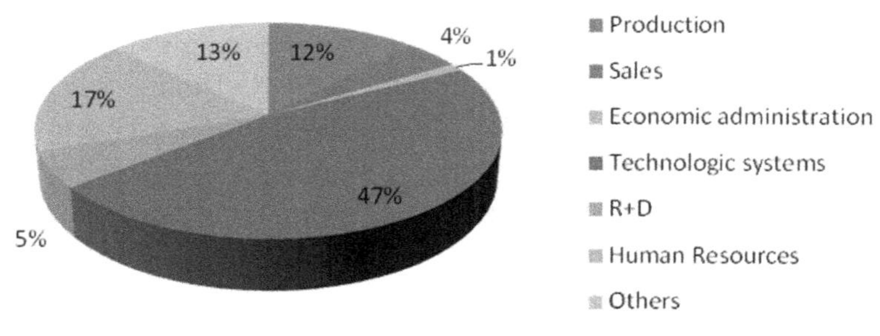

Table 1. Variables and descriptive statistics

Variable	Description	Mean (sd)
SAS	Self-assessed satisfaction, it takes discrete values from 1 (not at all satisfied) to 5 (very satisfied)	4.06 (0.848)
SAL	Self-assessed learning, it takes discrete values from 1 (very low) to 5 (very high)	3.454 (0.851)
TECNO	1: Learning technology is perceived as excellent 0: Otherwise	0.523 (0.502)
INC	1: If there is incentive related to learning (both real and perceived) 0: Otherwise	0.333 (0.474)
TIME	1: Specific time for learning available 0: Otherwise	0.13 (0.339)
EASY	1: If the use of the learning platform is perceived as very easy 0: Otherwise	0.19 (0.395)
MODEL	1: If respondent prefers traditional learning methodology 0: Otherwise	0.623 (0.487)
TECHSUP	1: If respondent works in technical support department 0: Otherwise	0.511 (0.502)
CONTENTS	1: If course contents were perceived as very good 0: Otherwise	0.357 (0.482)
PLACE	1: E-learning course attended out of the workplace 0: Otherwise	0.619 (0.488)

rest of the analysis. In addition, the quality of contents positively affects learning, as Figure 3 points out.

This last point underlines an important feature, which affects our analysis and the specification of the model being proposed: the same set of variables may have an impact on both satisfaction and learning. Figure 4, in fact, demonstrates that SAL and SAS are highly correlated. The rest of this section develops the model being estimated, in which the correlation between SAL and SAS is considered.

3.2 Model Description

The objective of this paper is to assess the impact of different key organizational factors on the levels of both satisfaction and learning related to the e-learning model. This study makes use of categorical dependent variables, which constraint the possibility of specification of the model (see Table 1). It is assumed here that the same set of variables might have an effect on both satisfaction and learning, as in

$$SAL_i = f(x_i) \quad (1)$$

$$SAS_i = g(x_i) \quad (2)$$

Figure 2. Distribution of SAS when all respondents are considered (top left), for the subgroup of people who value postively the contents of the course (top right), for the subgroups that value positively the e-learning platform (bottom right), and for the subgroup of people who value positively the tutoring system (bottom left)

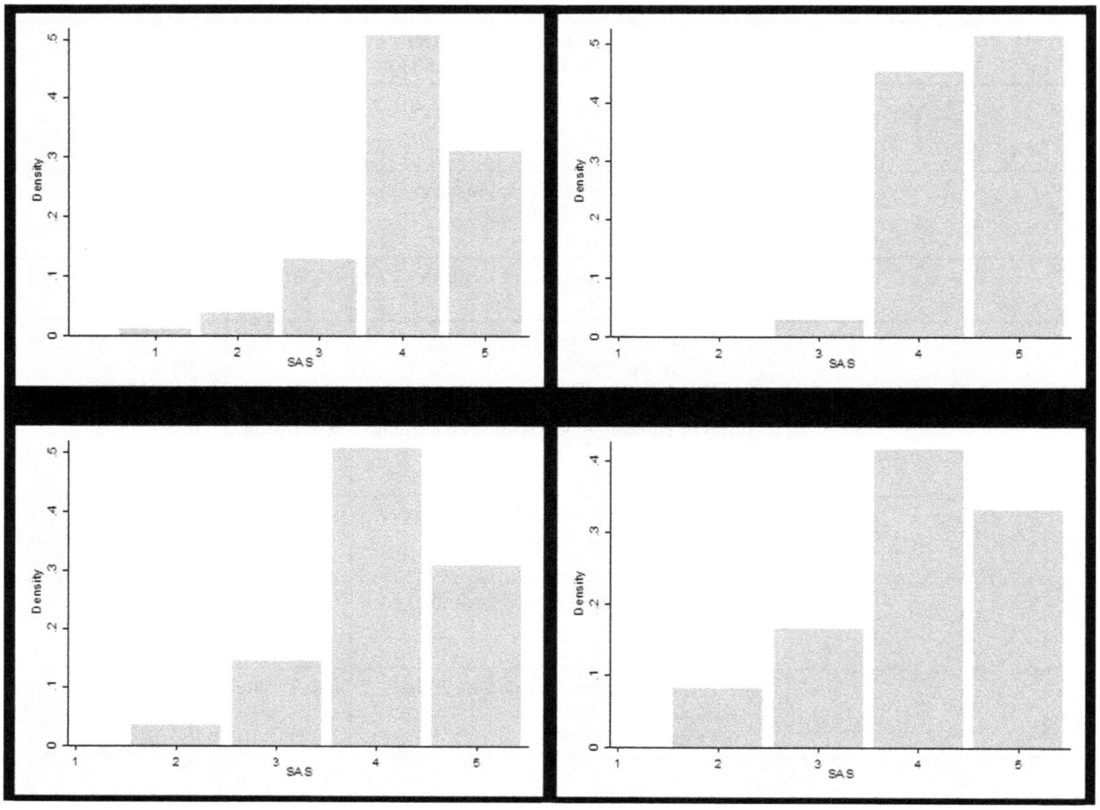

Figure 3. Distribution of SAL when CONTENTS: 0 (left), and when CONTENTS: 1 (right)

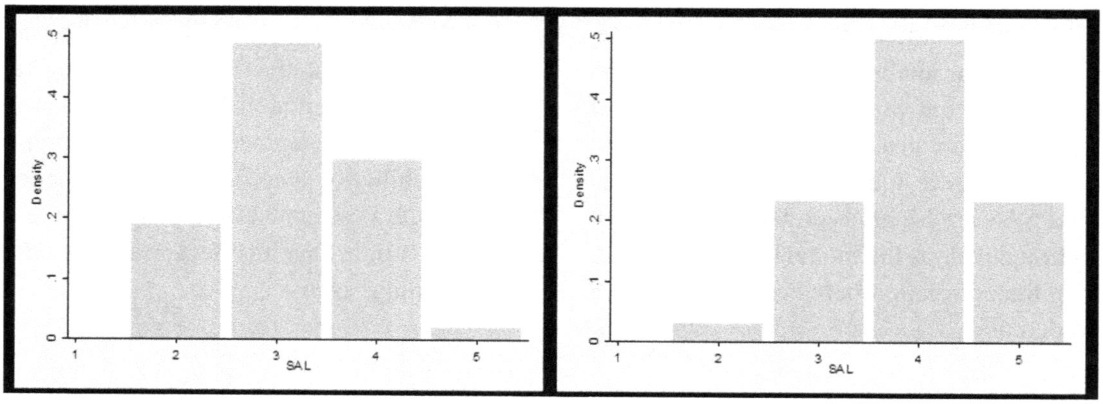

Figure 4. Distribution of SAS when SAL: 2 (top left), SAL: 3 (top right), SAL: 4 (bottom left), SAL: 5 (bottom right)

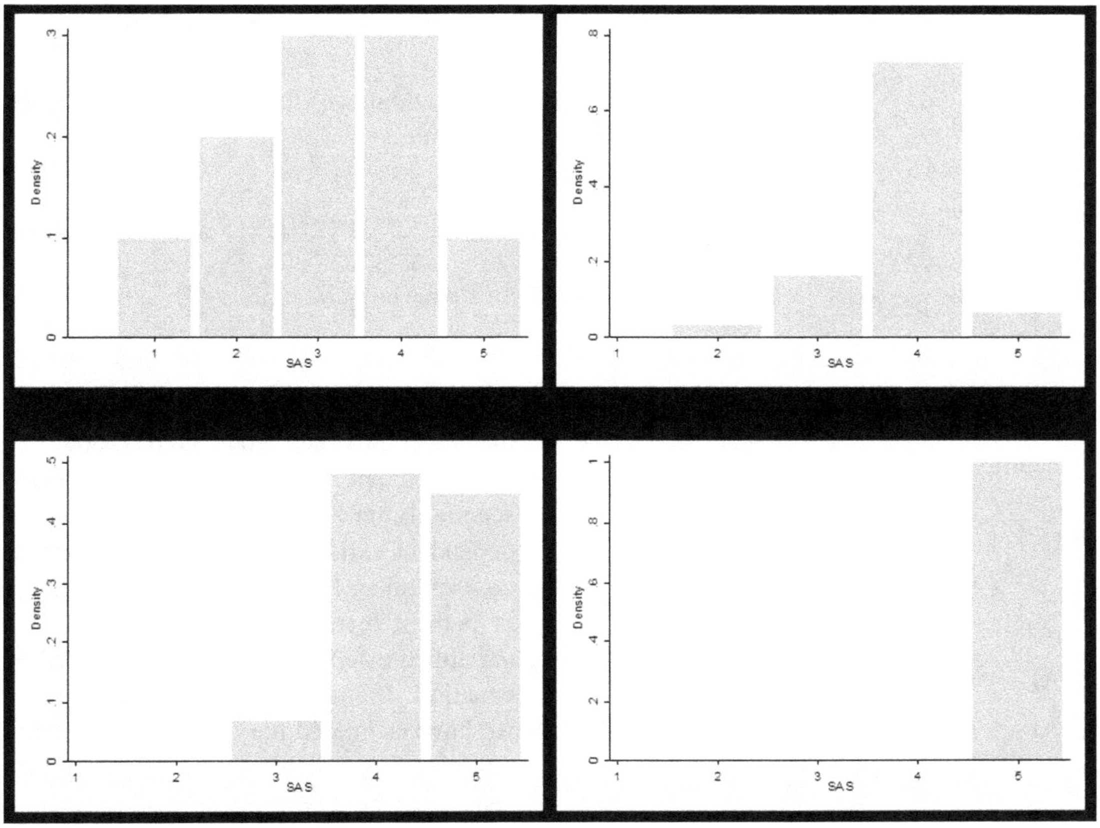

Neither the level of learning nor the level of satisfaction are directly observable from answers to survey questions; hence, following Greene and Hensher (2008), we assume that latent variables SAL_i* and SAS_i* depend on:

$$SAL_i^* = x_1'\beta_1 + \varepsilon_{Ai} \qquad (3)$$

$$SAS_i^* = x_1'\beta_2 + \varepsilon_{Si} \qquad (4)$$

where β_1 and β_2 are vectors of unknown parameters, ε_A and ε_S are error terms, and x_1 is the vector of explicative variables. Explicative variables are, by assumption, exogenous; that is,

$E[x_1' \varepsilon_{Ai}] = 0$, $E[x_1' \varepsilon_{Si}] = 0$.

Provided that the specification proposed in (3) and (4), which represent seemingly unrelated equations, does not imply any identification problem, we can use the same set of explicative variables. In fact, there are not any theoretical reasons to exclude some variable from one particular equation. We can observe in our database the values of the categorical variables SAL_i and SAS_i such that:

$$SAL_1 = \begin{cases} 1 & si SAL_1^* \le c_{A1} \\ 2 & si c_{A1} < SAL_1^* \le c_{A2} \\ 3 & si c_{A2} < SAL_1^* \le c_{A3} ; \\ 4 & si c_{A3} < SAL_1^* \le c_{A4} \\ 5 & si c_{A4} < SAL_1^* \end{cases}$$

$$SAS_1 = \begin{cases} 1 & si SAS_1 \le c_{S1} \\ 2 & si c_{S1} < SAS_1^* \le c_{S2} \\ 3 & si c_{S2} < SAS_1^* \le c_{S3} \\ 4 & si c_{S3} < SAS_1^* \le c_{S4} \\ 5 & si c_{S4} < SAS_1^* \end{cases} \quad (5)$$

where $c_{j1} < c_{j1} < c_{j1} < c_{j1} < c_{j1} < c_{j1}$, j = SAL, SAS. The probability of being $SAL_i = k$ and $SAS_i = 1$ is:

$$\begin{aligned} &\Pr\left(SAL_i = k,\ SAS_i = 1\right) \\ &= \Pr\left(c_{Ak-1} < SAL_i^* \le c_{Ak},\ c_{Sl-1} < SAS_i^* \le c_{Sl}\right) \\ &= \Pr\left(SAL_i^* \le c_{Ak}, SAS_i^* \le c_{Sl}\right) \\ &- \Pr\left(SAL_i^* \le c_{Ak-1}, SAS_i^* \le c_{Sl}\right) \\ &- \Pr\left(SAL_i^* \le c_{Ak}, SAS_i^* \le c_{Sl-1}\right) \\ &+ \Pr\left(SAL_i^* \le c_{Ak-1}, SAS_i^* \le c_{Sl-1}\right) \end{aligned} \quad (6)$$

When error terms are distributed as normal bivariant with correlation ρ, their contribution to the likelihood function can be expressed as:

$$\begin{aligned} &\Pr\left(SAL_i = k, SAS_i = 1\right) \\ &= \Phi\left(c_{Ak} - x_{1i}'\beta_1, c_{Sl} - x_{1i}'\beta_2, \rho\right) \\ &-\Phi\left(c_{Ak-1} - x_{1i}'\beta_1, c_{Sl} - x_{1i}'\beta_2, \rho\right) \\ &-\Phi\left(c_{Ak} - x_{1i}'\beta_1, c_{Sl-1} - x_{1i}'\beta_2, \rho\right) \\ &+\Phi\left(c_{Ak-1} - x_{1i}'\beta_1, c_{Sl-1} - x_{1i}'\beta_2, \rho\right) \end{aligned} \quad (7)$$

The logarithmic likelihood function for observation I is, thus:

$$\ln L_i = \sum_{k=1}^{5}\sum_{l=1}^{5} I\left(SAL_i = k,\ SAS_i = 1\right) \ln \Pr(SAL_i = k,\ SAS_i = 1) \quad (8)$$

By adding (8) for each of the N observation we obtain:

$$\ln \mathcal{L} = \sum_{i=1}^{N}\sum_{k=1}^{5}\sum_{l=1}^{5} I\left(SAL_i = k,\ SAS_i = 1\right) \ln \Pr(SAL_i = k,\ SES_i = 1) \quad (9)$$

Using (9), we can estimate model (3)-(4) by the procedure described in Sajaia (2008). It could be argued that model (3)-(4) takes the form of a seemingly unrelated specification, though both dependent variables might have a direct impact on each other. Hence, either equation (3) or (4), or both of them, should include as explicative variable the dependent variable used in the other equation. Though this might be true in principle, we have estimated the model with a simultaneous bivariate ordered probit specification, and tests carried out permit to exclude a direct effect of either satisfaction or learning on the other variable. Accordingly, the seemingly unrelated specification has been adopted here.

4. RESULTS

We have estimated model (3)-(4) by both a bivariate ordered probit model and two separated ordered probit models for each dependent variable. Table 2 shows results of the bivariate model (column 1) and the two ordered probit models (columns 2 and 3). Firstly, regarding the interdependency between equations (3) and (4), it can be observed that the correlation between error terms is positive and significantly different from zero: the null hypothesis of independency of the two equations does not hold true. In other words, the two ordered probit models produce biased estimations, which may be adjusted by the bivariate model. We will thus consider results of column 1 only. The first relevant result is the positive impact of incentives on both satisfaction and self-assessed learning. INC, in fact, is the variable with the largest effect on both SAL and SAS. In addition, incentives (either monetary incentives or the possibility of advancements in the professional career) have a greater effect on satisfaction compared to learning. The ability to handle the technological platform (variable EASY), in contrast, has not a significant effect on both learning and personal satisfaction. However, a good assessment of the technological aspect of the platform (TECNO) generates an increase in the probability of both being satisfied and awarding a positive evaluation to the level of learning. In other words, results underline that the quality of the learning platform has a significant impact on both learning and satisfaction, independently from the basic ability to handle it.

Assigning a specific amount of time to learning during the working day has interesting consequences. In fact, having time for education boosts the level of learning, whereas its impact on satisfaction is not significant. This variable thus appears to be especially relevant to the goal of optimizing investments in education. On considering the professional group, being a computer technician has a significant impact on both SAS and SAL. Possibly, due to their higher "sensibility" to technological aspects of the e-learning platform, computer technicians report a significantly lower level of satisfaction; in should be observed, in addition, that the impact is much larger for SAS then for SAL.

Preference for an e-learning model compared to classic methodology, based on the meeting of trainer and trainee in a common space and time, is another interesting variable being analyzed. Estimations suggest that people preferring the classic methodology are both less satisfied with e-learning and report to have achieved a lower level of learning. As for TECNO, however, the effect on SAS is larger: people who do not prefer e-learning declare to be unsatisfied, where their level of learning is affected to a considerable lower extent. This might be due to the functional

Table 2. Results of estimations

	(1)		(2)	(3)
	SAL	**SAS**	**SAL**	**SAS**
INC	1.050** (0.288)	1.295** (0.380)	1.044** (0.286)	1.321** (0.387)
EASY	0.514 (0.290)	0.446 (0.340)	0.501 (0.294)	0.461 (0.348)
TIME	0.843** (0.342)	0.141 (0.457)	0.840** (0.346)	0.162 (0.455)
MODEL	0.541** (0.286)	0.718** (0.258)	0.543** (0.285)	0.786** (0.268)
TECNO	0.796** (0.307)	0.759** (0.349)	0.808** (0.316)	0.766** (0.358)
TECHSUPP	0.598* (0.306)	0.976 (0.336)	0.593* (0.306)	1.003 (0.333)
PLACE	0.602** (0.267)	0.598** (0.275)	0.600** (0.267)	0.617** (0.267)
CONTENTS	0.963** (0.302)			
N		77	77	77
ρ	0.377** (0.160)			
Model stat		68.01**	49.22**	49.04**
LR test		5.56		

Note. ** = confidence level 95%; * = confidence level 90%. Robust standard errors in parenthesis.

distribution of respondents, as pointed out in section 3 (Figure 1).

Finally, it can be observed in Table 2 that flexibility (in terms of the possibility of attending the e-learning program out of the workplace), improves both satisfaction and learning.

Estimations carried out allows for predicting the effect of explicative variables on the probability of reporting at the same time both a high level of satisfaction (SAS = 5) and a high level of learning (SAL = 5), when incentive schemes are introduced. This is a relevant prediction, provided that it explains the impact of a (costly) organizational measure with potential benefits. Under this scenario, the model predicts that adopting an incentive scheme, when no incentives are initially implemented, causes an increase of about 121.5% in the probability of declaring both SAL = 5 and SAS = 5. Explicitly, the probability of reporting a high level of satisfaction would grow about 112.6%, whereas the increase in the probability of reporting a high level of learning would be equal to 79.5%. In other words, incentive schemes affect the level of satisfaction most. Another interesting result is obtained when analyzing a scenario in which people are not assigned any amount of time for studying. In fact, the model predicts that when specific time for study is assigned to people, the probability of reporting a high level of learning increases of about 80%.

5. CONCLUSION

E-learning does not automatically imply learning. In fact, it might actually reduce the level of learning, motivation and satisfaction of users, when it consists of just uploading course materials initially designed for use with classic methods. Organizational factors as e-learning course flexibility (Sun, P.; Tsai, R.J.; Chen, Y. & Yeh, D., 2008) or perceived self-efficacy (Liaw, S., 2008) were considered in previous research as key issues that influence e-learning satisfaction. As a consequence, a complete pedagogical system has been developing through recent years, which is aimed at establishing the educational process in the new technological environment. At the same time, companies are conscious that some organizational features may facilitate learning within the new model and thus, they could contribute to improve both its efficacy and efficiency.

Following this approach, this paper has considered the impact of a number of organizational features on satisfaction and learning of users. It is shown, firstly, that incentives schemes play a key role in augmenting the level of both satisfaction and learning. It should be noted, moreover, that by incentive we mean either monetary rewards, which are almost absent in our sample, or the expectation of advancements in the professional career within the firm. This circumstance may also help to a better understanding of the greater impact of incentives on satisfaction respect to learning, which we found out in our analysis. In other words, the organizational value of learning programs for employees, independently from the level of learning, not only causes their greater acceptance, but also it may also positively affect behaviors in such a way that favor learning. Efficacy of learning programs may thus improve by linking the human resources development policy with results obtained in e-learning courses.

Another interesting conclusion we obtained in the analysis is the importance of assigning a specific amount of time for attending e-learning courses. Explicitly, it is shown here that time available for learning has a positive and significant impact on learning, whereas it does not affect satisfaction. This implies a relevant question. In fact, one of the main advantages of e-learning is its temporal flexibility, which permit to work and study during the same period of time. Model predictions about the positive effects of a fixed amount of time available for learning are thus in apparent contrast with flexibility. On the other hand, it seems logical that focusing on study during a certain amount of time, without disturbance or interruption due to working activity, facilitates concentration and therefore, learning, and results about the importance of spatial flexibility support this conclusion. One possible solution we propose is to assign an amount of time to each student, who is then responsible to use it whenever he or she wants, thus obtaining some degree of flexibility together with reducing interruptions when studying.

On considering the limitations of our analysis, it should be noted that our sample includes users of e-learning programs within a single firm. The aim of this paper is the study of a particular case, and the generalization of results is beyond our scope. Results are in line with previous research mentioned in the introduction; we have considered how organizational factors, which are normally excluded from the analysis, have affected the implementation of e-learning at EVERIS. Moreover, we considered e-learning users independently from the program attended. In other words, an implicit hypothesis of this piece of research is that courses have homogeneous and comparable characteristics regarding their internal structure and therefore, users' opinion is assumed to be referred to the e-learning model rather than one particular program. This limitation suggests one step forward in the same line of research. It would be interesting, if data were available, to divide the sample and study the impact of organizational measures depending on both the type of course and the professional category of the student. In addition, comparing results with other firms that adopted the e-learning model would permit to analyze additional policies aimed at improving efficiency.

ACKNOWLEDGMENT

The authors sincerely wish to acknowledge the support of both managers and employees of EVERIS. Their collaboration underlined their involvement in the development of new organizational designs for increased effectiveness and satisfaction.

REFERENCES

Bose, K. (2003). An e-learning experience. *Campus-Wide Information Systems*, *20*(5), 193–199. doi:10.1108/10650740310507399

Box, K. (1999). *Human interaction during teacher training courses delivered via the Internet*. San Antonio, TX: Society for Information Technology & Teacher Education Conference.

Calhoun, C. A. (1989). *BIVOPROB: A computer program for maximum-likelihood estimation of bivariate ordered-probit models for censored data*. Working Paper, 38, Laxenburg, Austria: International Institute for Applied Systems Analysis.

Calhoun, C. A. (1991). Desired and excess fertility in Europe and the United States: Indirect estimates from world fertility survey data. *European Journal of Population*, *7*, 29–57. doi:10.1007/BF01796615

Calhoun, C. A. (1994). The impact of children on the labour supply of married women: Comparative estimates from European and U.S. Data. *European Journal of Population*, *10*, 293–318. doi:10.1007/BF01266566

Consulting Business Intelligence, S. R. I. (2003). *Quality and effectiveness of e-learning*. Learning on Demand.

Davis, R., & Wong, D. (2007). Conceptualizing and measuring the optimal experience of the e-learning environment. *Decision Sciences Journal of Innovative Education*, *5*(1), 97–125. doi:10.1111/j.1540-4609.2007.00129.x

Dawson, P., & Dobson, S. (2010). The influence of social pressure and nationality on individual decisions: Evidence from the behaviour of referees. *Journal of Economic Psychology*, *31*(2), 181–191. doi:10.1016/j.joep.2009.06.001

Ettinger, A., Holton, V., & Blass, E. (2002). E-learner experiences: Key questions to ask when considering implementing e-learning. *Industrial and Comercial Training*, *38*, 143–147. doi:10.1108/00197850610659409

Fernández Díez de Lastra, R. (2001). *La formación online y sus mitos*. Boletín Learnet, Marzo 2001. Instituto Universitario Euroforum Escorial. Retrieved from http://euroforum.cicei.ulpgc.es/learnet/bolMar_01/boletin.htm

Galagan, P. (2002). Mission e-possible: The CISCO e-learining story. In Rossett, A. (Ed.), *The ASTD e-learning handbook*. New York, NY: Mc Graw-Hill.

Greene, W. H., & Hensher, D. A. (2009. *Modeling ordered choices*. New York, NY: Mimeo, Stern School of Business. Kirby, E. (1999). *Building interaction in online and distance education course*. TX: Technology & Teacher Education Conference.

Liaw, S. (2008). Investigating students´ perceived satisfaction, behavioral intention, and effectiveness of e-learning: A case of study of the blackboard system. *Computers & Education*, *51*, 864–873. doi:10.1016/j.compedu.2007.09.005

McPherson, A., Homan, G., & Wilkinson, K. (2005). The implementation and use of e-learning in the corporate university. *Journal of Workplace Learning*, *17*, 33–48. doi:10.1108/13665620510574441

Netteland, G., Wasson, B., & Morch, A. I. (2007). E-learning in a large organization. *Journal of Workplace Learning*, *19*(6), 392–411. doi:10.1108/13665620710777129

Orbea, T. (2008). *UNE 66181. La calidad en la formación virtual*. Symposium 4th March, Ministry of Industry, Commerce and Tourism. Spanish Government.

Rahmandad, H., Repenning, N., & Stearman, J. (2009). Effects of feedback delay on learning. *System Dynamics Review*, *25*(4), 309–338. doi:10.1002/sdr.427

Redmon, J., & Salopek, J. J. (2000). A year in the life of an e-learning project. *Training & Development*, (September): 36–41.

Roy, K. (2006). The impact of learning styles on interactivity in asynchronous e-learning. *Performance Improvement Journal*, *45*(10), 21–26. doi:10.1002/pfi.4930451026

Sajaia, Z. (2008). Maximum likelihood estimation of a bivariate ordered probit model: Implementation and Monte Carlo simulations. *The Stata Journal*, *3*(2), 311–328.

Servage, L. (2005). Strategizing for workplace e-learning: Some critical considerations. *Journal of Workplace Learning*, *17*(5), 304–317. doi:10.1108/13665620510606733

Singh, H., & Reed, C. (2002). Demystifying e-learning standards. *Industrial and Commercial Training*, *34*(2), 62–65. doi:10.1108/00197850210417546

Snyder, K., Farrell, R., & Baker, N. (2000). Online mentoring: A case study involving cognitive apprenticeship and a technology-enabled learning environment. In J. Bourdeau & R. Heller (Eds.), *Proceedings of World Conference on Educational Multimedia, Hypermedia and Telecommunications 2000*, (pp. 1742-1743). Chesapeake, VA: AACE

Sun, P., Tsai, R. J., Chen, Y., & Yeh, D. (2008). What drives a successful e-learning? An empirical investigation of the critical factors influencing learner satisfaction. *Computers & Education*, *50*, 1183–1202. doi:10.1016/j.compedu.2006.11.007

Tynjälä, P., & Häkkinen, P. (2005). E-learning at work: Theoretical underpinnings and pedagogical challenges. *Journal of Workplace Learning*, *17*(5), 318–336. doi:10.1108/13665620510606742

Zhang, Z., & Jasimuddin, S. M. (2008). Toward a strategic framework of mobile knowledge management. *Knowledge and Process Management*, *15*(2), 87–96. doi:10.1002/kpm.300

Chapter 8
Organizational Commitment and Job Satisfaction:
An Empirical Analysis of Their Relationship in Private Teachers

Sandra Mª Sánchez Cañizares
University of Córdoba, Spain

Fernando J. Fuentes García
University of Córdoba, Spain

ABSTRACT

People are an essential source on the competitiveness of organizations. Numerous authors have recognized the importance of organizational commitment and job satisfaction as key attitudes related with work and influencing aspects like rotation, absenteeism or productivity. However, a clearly delimited posture on the relationship between both constructs does not exist. It is because of this that the present work theoretically analyses several models regarding this matter. Subsequently a proposed model is estimated and validated using the methodology of Structural Equations Modeling (SEM). This model confirms the influence of job satisfaction over organizational commitment and its moderating effect on the influence on the latter of various perceptual factors of the employee related to his/her work.

INTRODUCTION AND OBJECTIVES OF THE INVESTIGATION

People are an essential source on the competitiveness of organizations; employees meet the requirements of the Theory of Resources and Capabilities to build competitive advantage (Barney, 1991). Thus, a good HR management can make a difference because it will need to transform knowledge and creativity of people (intangible and difficult to imitate resources) into tangible results for the organization. But for this, the involvement and commitment of employees are imperative (De la Rosa and Carmona, 2010) so the analysis of these topics is mandatory under the framework of the new economy.

DOI: 10.4018/978-1-61350-207-5.ch008

Organizational commitment has been a construct of considerable interest in existing literature on the study of behaviour among employees. There is a great volume of work with respect to the topic, which is currently centred more and more on the relationship between the same and job satisfaction (Currivan, 1999; Gaertner, 1999; Linz, 2003). Authors such as Harrison *et al.* (2006), extending the work of Judge *et al.* (2001) consider that job satisfaction as much as organizational commitment acts as an indicator of an underlying global attitude towards work.

Numerous works connect a high level of commitment on behalf of the employee with lower rates of turnover and absenteeism, and improvements in productivity (Angel and Perry, 1981; Darden *et al.*, 1989; Dunham *et al.*, 1994). Likewise, there has also been an extensive study made from an economic viewpoint of the influence exercised by job satisfaction over these organizational indicators (Barraza and Ortega, 2009).

However, no unanimous agreement exists between the authors with respect to the link that unites these two intimately related constructs with the intangible value of human resources in the organization. The order of coincidence between satisfaction and commitment is a source of discrepancy in the investigations carried out. For this reason, this work hopes to clarify this point in certain measure through the application of Structural Equation Modelling (SEM) methodology to a proposed model that supports the relevant theoretical review.

In this way, initially, a theoretical exploration of organizational commitment and job satisfaction constructs is carried out. Furthermore, the review continues with the main models that analyse the mediating role played by job satisfaction between different variables about the perception of employees about their job and their loyalty to the organization. Subsequently, an analysis is made of the model proposed in this work and the methodology applied to the empirical investigation. Finally, the results reached are presented and their repercussions discussed for management of human resources.

THEORETICAL FRAMEWORK

Organizational Commitment: Definition, Dimensions, and Consequences

There are numerous definitions for commitment provided by specialised literature. Porter *et al.* (1974) understand organizational commitment as a global force of identification of the individual with the aims and values of the organization, a definition that has points in common with that of Blau and Boal (1987), who see commitment as a state in which the employee feels identified with the organization and its aims and who wishes to continue to be a part of it. For Price (1997), it deals with the loyalty experienced by the employee towards a social unity, considering that this construct as much as job satisfaction are variables which intervene in the process of the turnover of employees.

More recent definitions are those of Testa (2001), according to whom commitment reflects an emotional response of the employee in the face of a positive evaluation of their working environment, or that of Meyer and Herscovitch (2001, p.310), where the essence of commitment is defined as "an attitude which can adopt different forms and join the individual with a relevant course of action for a particular objective".

Among the abundance of definitions for organizational commitment, a common idea is detectable, according to which "committed" individuals believe and accept the objectives and values of the organization in such a way that they are prepared to remain at the heart of the same and afford it as a considerable effort (Mowday *et al.*, 1979). Organizational commitment then becomes a psychological bond which influences the employee to act in accordance with the interests of the organization (Porter *et al.*, 1974). In this way, it is understood that people can feel organizational commitment simultaneously and with different intensity, given that it deals with a multidimensional construct that can comprise of distinct dimensions and facets.

It must be said that the aforementioned multidimensionality is one of the questions that has generated more controversy. To this respect, Porter *et al.* (1974) consider various forms of commitment: the strong belief and acceptance of the aims and values of the organization; the will to make a considerable effort in the name of the organization; the definitive wish to continue as a member of the organization. However, despite the construct was defined making reference to three dimensions, in their well-known instrument OCQ, Porter et al. (1974, 1976, 2005) treated organizational commitment as a one-dimensional topic [1].

This line, which maintains the existence of more than one type of organizational commitment, had already been introduced by Becker (1960) in his Theory of Social Exchange, where he defines commitment to the organization as the bond which is established between the individual and the organization, the fruit of small investments made over a period of time. That is to say, it is thought that people continue in their firm because a change of situation would suppose sacrificing investments.

Subsequently, Meyer and Allen (1991) identify three forms of employee-organization unity:

- **Affective Commitment:** the strength with which individuals identify and implicate themselves in the organization. From the theory of affective bonding, commitment is defined as a component that makes individuals identify and implicate themselves with the values and aims of the organization in such a way that they declare their desire to continue within it.
- **Normative Commitment:** a feeling of moral obligation to continue as a member of an organization. Despite the fact that Porter *et al.* (1974) include this facet in the previous affective commitment. Meyer and Allen (1991) do distinguish between both forms. Insomuch that in normative commitment an obligation to loyalty exists; in affective commitment, only a desire is expressed.
- **Continued or Calculated Commitment:** the intention to continue in the organization because of the cost that leaving would suppose, or the rewards originated by remaining. This is linked to the definition of Becker (1960), and the possibility has been discussed of dividing it into two factors: "high personal sacrifice", associated with the cost of leaving the organization and "scarcity of perceived alternatives", related to the low odds of finding an alternative job to the current one (McGee and Ford, 1987). This breakdown has been empirically analyzed by Frutos *et al.* (1998) through a confirming factorial analysis which obtains a better fit in the model of four factors of commitment than in the model with three factors.

Despite the fact that this three-dimensional approach has been the most widely accepted in our time (Bergman, 2006; Herrbach, 2006), there does not exist, however, a general agreement over which are the dimensions that explain the phenomenon of organizational commitment. So, O'Really and Chatman (1986) and Harris *et al.* (1993) name them as the commitment of conformity, identification and internationalisation and Penley and Gould (1998), as moral, calculative and alienating. Work even exists where a greater number of dimensions or forms of commitment are defined, as in the model of Randall and Cote (1991), in which five are considered: protestant work ethic, job involvement, postponed commitment, affective commitment and career commitment.

With respect to the consequences derived from organizational commitment, the investigations have been centred mainly on the effects over turnover (Lin and Ma, 2004; Van Breukelen *et al.*, 2004) and the intention to leave (Gellatly, 1995; Powell and Meyer, 2004), as well as over productivity or the performance of employees committed to the firm (Leong *et al.*, 1996; Ward and Davis, 1995). The model Service-Profit-Chain (Figure 1) accepts that commitment of employees can lead directly or indirectly to an improvement in the performance and profitability of the organization. It suggests that some motivated employees remain loyal and will work close to their optimum level; indirectly, the commitment of employees, thanks to the quality of customer service, can lead to loyalty in the latter. At the same time, generating loyalty to the customer, greater profitability and development of the

In the empirical study conducted in this research, organizational commitment will be discussed from the point of view of the OCQ by Porter et al. (1974) since, despite being a fact the multidimensionality of the construct, there is a controversy about the possible overlap between affective and normative commitment (Cohen, 2003; Meyer et al. 2002) and a conceptual ambiguity recognized by Meyer and Allen themselves (1996) and other authors (Bergman, 2006; Swailes, 2002) in the continued or calculated commitment.

Job Satisfaction: Definition and Distinction of Organizational Commitment

Despite the fact that job satisfaction has been one of the most studied variables in the field of organizational commitment, no consensus exists as regards its definition, or to the existence of a comprehensive theory about what leads to satisfaction in the workplace (Gamboa, 2010)

However, despite the innumerable existing definitions, a common denominator can be perceived which makes it possible to categorize it from two different perspectives:

- On the one hand, we find authors who understand satisfaction as an emotional state, a feeling, an attitude or an effective response towards work (Bravo *et al*, 1996; Katzell, 1964; Larrainzar *et al*, 2001; Leal *et al*, 1999; Lee and Chang, 2008; Locke, 1969; Moynihan and Pandey, 2007; Price and Mueller, 1986; Robbins, 1994; etc).
- On the other hand, there are those who consider satisfaction as a result of a comparison or fit between expectations, necessities or current results of the work and the benefits that the job really offers to the employees (Holland, 1985; Mumford, 1976 and Michalos, 1986, among others).

Figure 1. Model service-profit-chain

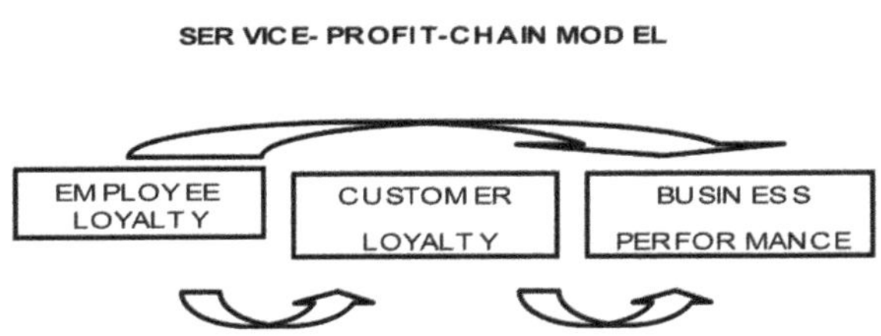

In this way, it could be considered that, uniting the two aforementioned positions, job satisfaction is:

A multidimensional and multidisciplinary concept which supposes the emotional state, attitude, sensation or degree of well-being that individuals experience towards their work as a consequence of the greater or lesser discrepancy which exists between their past and present expectations with regard to the rewards and the role which their job offers, and the measure to which these are fulfilled.

With respect to the existing difference between organizational commitment and job satisfaction, Mowday *et al.* (1982) argue that it deals with different constructs: while commitment is the reflection of an attitude of a more generic employee, satisfaction is the confirmation of a more fragile and changeable attitude. From an organizational point of view, commitment is a more global construct, which reflects a general affective response towards the organization understood as a whole, while job satisfaction is intrinsically bound to the job and certain aspects of work (Baker and Baker, 1999). Juaneda and Gonzalez (2007) hold that commitment is more stable in time, whereas satisfaction shows more immediate reactions to tangible aspects of the tasks in the working environment. Cohen (1993) proposes not to set the two concepts against each other; that is to say, not to understand them as opposites or as being in competition with each other, but to refer to them as being complementary.

Empirical Models of Causal Relationships between Job Satisfaction and Organizational Commitment

The present epigraph studies the relationship that other variables, with reference to the perception of the employee about certain aspects of their work, present over the level of job satisfaction and, as a consequence, over their organizational commitment.

Model of the International Survey Research (ISR)

In the case study carried out by ISR (2004), two levels of influence stand out, as much in job satisfaction as in organizational commitment of employees.

- **Perceptions about work experience:** It refers to opinions about the day-to-day and aspects such as supervision, the working team, the authority of the post, recognition or safety and hygiene at work. It tries to respond to the question "do I like my experiences in this job?"
- **Perceptions about the conditions of the organization:** It deals with the opinion at corporate level over the operation of the entity where employees carry out their work. It includes aspects such as leadership, communication channels, salary and received benefits, or the practices and policy of human resources. It would be the answer to the question "Do I approve of the way in which the organization works?"

Figure 2 shows a graphic representation of the model.

The analysis carried out over a telecommunications company shows results of an influence three times greater in the experience at work over job satisfaction than that over commitment to the job (the regression rate in the former is of 0.53, compared with 0.19 in the latter). With respect to the working of the organization, rates obtained show an influence six times greater at this level over organizational commitment (0.75) than over job satisfaction (0.14).

In conclusion, this investigation suggests that the measures of experience at work are critical when understanding the opinion of the employees and when measuring their job satisfaction. On the other hand, the questions related to the workings of the organization show special relevance when

Figure 2. ISR Model of job satisfaction

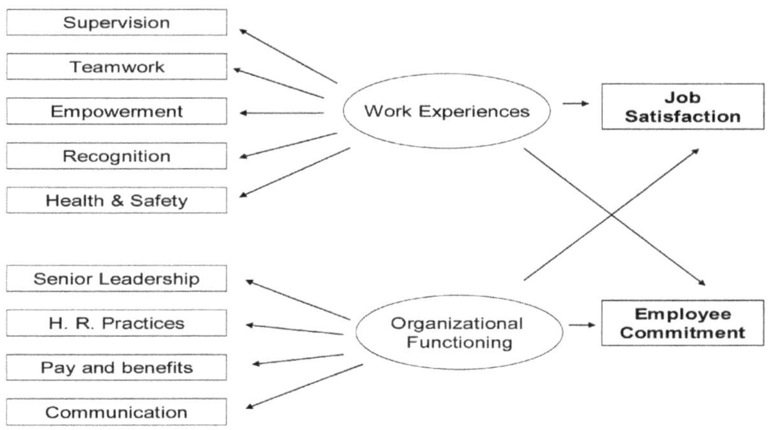

checking the commitment to the organization and to the job.

Nordic Employee Index Model

A different model, which rests equally over solid academic investigation in the field of satisfaction and commitment (Allen and Meyer, 1996; Baker, 1995; Clark, 2001; Eskildsen *et al.* 2002; Gaertner, 1999), is that used by the Nordic Employee Index. This model seeks an approximation to the measure of these constructs that could be applied to any organizational framework. The study, originally carried out in Nordic countries (Denmark, Finland, Norway and Sweden) has been extended to other member countries in such a way that currently it has become known as the European Employee Index. Its structure is based on the influence of five variables, as much over motivation as over satisfaction, which, in turn, act as intermediaries in the repercussion of the same over loyalty and commitment to the job. These five constructs are as follows:

- **Image of the job:** public perception and the pride of working for the organization.
- **Corporate leadership:** opinion of the employees over decision-making, communicative abilities and the ethics of the organization.
- **Immediate manager:** attitudes of the employees towards their immediate boss.
- **Cooperation:** mutual interaction and social climate at work.
- **Conditions of work:** including the perception of the employees with respect to three components: daily work (conditions of the work content, physical environment and job pressure), remuneration (salary, other benefits, job security), personal and professional development (training opportunities).

Loyalty to the position is measured by aspects such as the intention to leave, the search for another job, the recommending of the job to others, etc. The similarity of these items to those used to measure organizational commitment makes one think that both variables, with different denominations, rest over the same concept.

Figure 3 shows the graphic representation of the model.

In the empirical investigation carried out by the European Employee Index (2002), the construct "conditions of work" was that which showed a greater influence; with a regression rate of 0.60 over an estimate of Partial Least Squares (PLS), the rest having rates inferior to 0.25. On the other hand, motivation and satisfaction maintain a significant statistic relationship to loyalty, with a regression rate estimated at 0.65, which means that satisfaction linked to motivation leads to a higher organizational commitment.

Model of Ten Factors and Other Proposals

One model applied by Currivan (1999), more extensive than previous ones, considers the influence of ten determinant structures[2] (understood as characteristics of the job and guidelines of social interaction within the frame of work) over job satisfaction and organizational commitment, understood here as loyalty to the organization and, as such, similar to the previous model.

Three alternatives exist in the relationship of influence with diverse empirical bases of support.

The first possibility is that these determinant structures are directly related as much to job satisfaction as to organizational commitment (Kim *et al.* 1996).

The second version considers the influence of the structural determinants solely over job satisfaction, while its relationship to organizational commitment would be indirect via the same. This position is defended in the work of Price and Mueller (1986) and Williams and Hazer (1986).

The third position considers that different structural determinants have different effects over job satisfaction and organizational commitment. In this sense, Hom and Griffeth (1995) argue that certain determinants only show an impact over job satisfaction, while others exclusively influence organizational commitment.

The study carried out by Gaertner (1999), in which SEM methodology is applied, obtains low empirical support in the case of the first proposal, where all the structural determinants influence job satisfaction as much as organizational commitment in a direct way. The second possibility obtains an acceptable empirical fit, although in small samples the fit is not significantly greater than that of the first model. However, through an exploratory analysis, evidence is found of the direct influence of certain determinants over organizational commitment with a statistical relationship significantly greater to that shown over job satisfaction, which ultimately supports the theory of Hom and Griffeth (1995). However, the relationship of certain variables is detected, like

Figure 3. Model of the nordic employee index

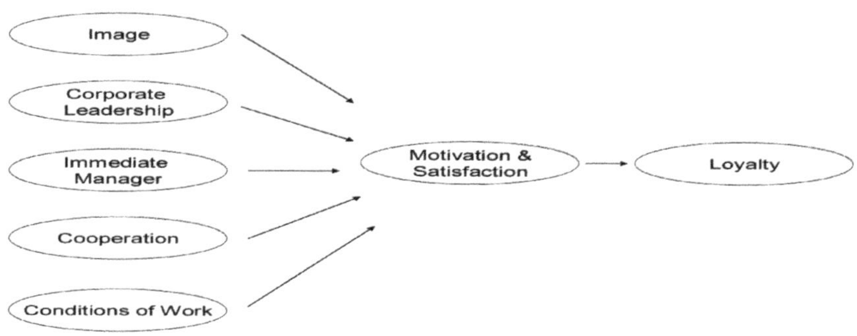

the support of the supervisor, the opportunities for promotion or distributive justice, as much with job satisfaction as with organizational commitment, while other determinants only show an influence over the first construct.

The work of Price and Mueller (1986) supports the theoretical base for the simplified model subsequently developed by Mathieu y Zajac (1990). The variable structures used are: autonomy, routine, social support (divided into peer support and supervisor support), stress at work (divided into role ambiguity, role conflict and workload) and pay. Only one characteristic of the individual is incorporated, involvement at work, which is defined as the measure in which the employee is prepared to develop his/her role at work.

The model specifically puts forward the positive influence of autonomy, the support of colleagues and the supervisor, as well as pay, over job satisfaction and organizational commitment. On the other hand, routine, ambiguity, role conflict and workload have a negative influence over both aspects for the employee.

The hypothesis of the positive impact of satisfaction over commitment is extended in such a way that it is also thought to have the same mediating effect between structural and individual determinants and organizational commitment (see Figure 4).

This model has been empirically applied by Currivan (1999) to an organization of North American State schools, carrying out a panel study in two stages to verify the effect of satisfaction over commitment. However, the conclusions obtained suggest that the relationship between the two constructs is equivocal, originated by common causes. The measure to which the determinants influence each one is different, given that only routine, social support in its two aspects and the workload show similar effects over satisfaction and commitment; the remaining variables show a different influence, depending on which aspect it is dealing with.

In short, the scientific literature unquestioningly supports the direct relationship between job satisfaction and organizational commitment (García del Junco and Bras, 2008; Jaik et al. 2010; Mendoza et al., 2010). But, indeed, both conceptual evidence and empirical research holds up the

Figure 4. Structural determinants and individual characteristics that influence job satisfaction and organizational commitment

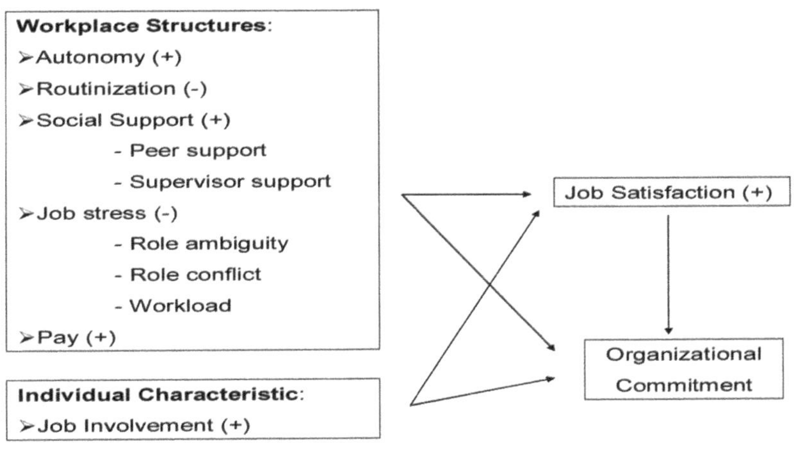

idea that satisfaction is an antecedent of commitment (Johnston et al., 1990, Álvarez, 2008).

METHODOLOGY OF THE INVESTIGATION

Site and Sample

The empirical analysis is directed to the personnel of private learning centres[3] in Andalusia. According to the data of the Ministry of Education and Culture for the academic year 2005-2006, it deals with 585 centres and 20,591 employees between teaching and administrative personnel. Using the population census offered by the Regional Ministry for Education in Andalusia[4], a list of these centres was obtained, which allowed the posting of the questionnaire to all of them.

The fieldwork lead to the receipt of 1,819 questionnaires from 55 centres, although, following a filtering process, the size of the definitive sample was fixed at 1,804 individuals.

Measure

The questionnaire used consists of the following parts:

- **General information:** learning centre to which one belongs and location.
- **Details of the job:** type of contract, dedication, antiquity, working hours, etc.
- **Teaching in Spain:** evaluation of teaching in recent years.
- **Job satisfaction:** advantages and disadvantages of the job; degree of satisfaction with different aspects linked to the profession according to the Likert scale of five points (1-very unsatisfied to 5- very satisfied); degree of global satisfaction on a Likert scale of five points; evaluation of different facets of the job compared with the same ones in public teaching.
- **Organizational commitment:** This part, as it was mentioned in the theoretical framework, is based on the Organizational Commitment Questionnaire (OCQ) of Porter et al. (1974) which consists of 15 items to answer the level of agreement or disagreement on a Likert scale of five points. Scores obtained from the answers to these items are added together and the average is calculated in such a way that the higher the result obtained, the higher the organizational commitment is. Six items exist (3, 7, 9, 11, 12 y 15) whose principles are drawn up negatively, in such a way that it is necessary to invert the scale when calculating the average scores. A great number of empirical studies guarantee the reliability and validity of OCQ[5]. According to the authors (2005, p. 177) this tool is easy to administer and can be used in different organizations and a wide variety of employees; the general support for this questionnaire comes from its good

Table 1. Technical details of the investigation

Geographical Limit	Regional (Andalusia)
Universe	Personnel of private learning centres
Size of universe	20,591 employees in 585 educational institutions
Method	Postal survey and structured and closed questionnaire
Size of sample	1.804
Period for collecting information	January-April 2006

psychometric properties (Commeiras and Fournier, 2001).

- **Sociological data**

Proposed Model

Following the analysis of the reviewed models in the previous epigraph, the model which is proposed in this investigation analyses the positive influence which six variables of the employee's perception about some aspects of his/her work exercises over job satisfaction as well as the relationship present between this and organizational commitment.

Applying SEM methodology, it is intended to determine the possible causal relationship between these variables and the higher or lower level of influence of each one of them over the rest. Its selection is founded on *ad hoc* literature, as well as its applicability to the empirical study carried out, and focused on private schools of the Autonomous Region of Andalusia. Variables are:

- **Perceptions of salary:** used as "payment and benefits" in the ISR model, included as "conditions of work" in the Nordic Employee Index model or as "salary" in the models extended to ten factors. It has been measured by items about the perception of the equity of the salary and the comparison with that of a professional of public schools.
- **Relationship with colleagues:** used as "teamwork" in the ISR model, as "cooperation" in the Nordic Employee Index model or as "peer support" in the model of ten factors. It has been measured by items covering the degree in which a good relationship with work-mates is maintained.
- **Relationship with the Headmasters of the centres:** is connected to the constructs "leadership" or "support of the supervisor" in the models reviewed. It has been measured by items covering the type of relationship maintained with those in charge of the management tasks of the learning centre to which they belong.
- **Perception of the prestige of the profession:** is connected to the variables "recognition" or "image of the post" of the ISR and Nordic Employee Index models respectively. It has been measured by items covering the way in which it is thought that society perceives the developed profession and the comparison with the prestige of the same profession when developed in a public school.
- **Perception of opportunities for promotion:** included in "conditions of work" in the Nordic Employee Index model or in "opportunities for promotion" in the model of ten factors. It has been measured by items covering the possibility of promotion or professional development.
- **Perception of the autonomy of the job:** is analysed in the model of ten factors and has been measured by questions about the degree to which the work can be freely developed, dependency when making decisions and the comparison with the autonomy of a professional in a public school.

From Figure 5 it can be gathered that the model supports *a priori* the chance order of satisfaction as a precedent to organizational commitment. The theoretical base on which this presumption is based is centred on the ideas of Lawler (1992). His theory about affective links provides a theoretical link to the most immediate positive emotions, like job satisfaction, and the lasting affective ties like organizational commitment. This theory assumes that positive emotional states come from affective links to collectives, acting as mediating factors between these and social structures.

Further on, Mueller and Lawler (1996) extend this supposition to the relationship between working conditions, job satisfaction and organizational commitment. They argue that working conditions produce emotions that can be negative

Figure 5. Proposed structural model

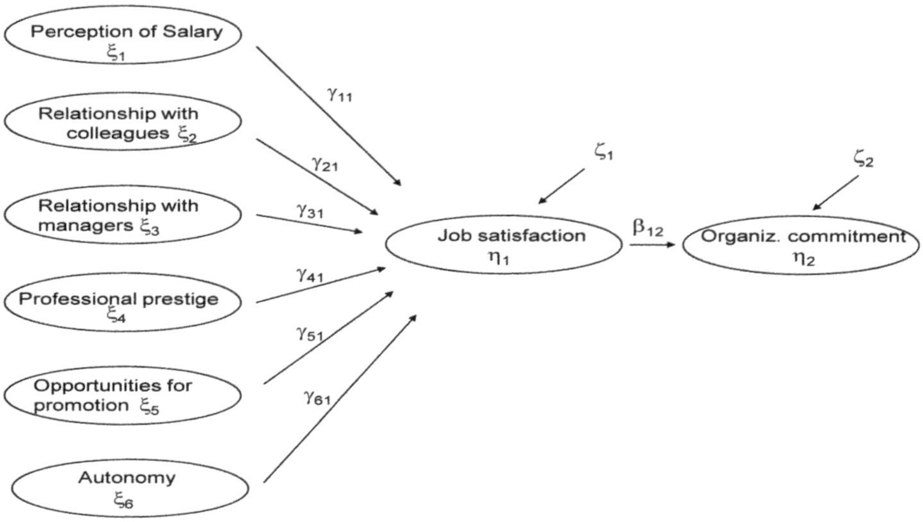

or positive (job satisfaction-dissatisfaction); the employees look for the source of those emotions, so that the organization starts to become the object of those feelings. If the firm becomes responsible for these positive emotions in the eyes of the employees, it is quite possible that it achieves this affective link on behalf of the same.

According to the basis of this theory, the proposed model shows job satisfaction in a dual position: on the one hand, it is important to determine organizational commitment and, on the other, it is a variable mediator between the individual – structural determinants and commitment.

However, although numerous authors defend the fact that job satisfaction constitutes the cause of organizational commitment (Kim *et al.*, 1996; Price and Mueller, 1986, Williams and Hazer, 1986), there is no lack of detractors to this theory (Farkas and Tetrick, 1989; Hom and Griffeth, 1995; Mathieu, 1991).

To go more deeply into this discrepancy over the chance order of the constructs, it can be observed that the dominant position in the literature considers that satisfaction is what causes commitment (Lincoln and Kalleberg, 1990; Mowday *et al.*, 1982; Mueller *et al.*, 1994; Price and Mueller, 1986). These investigators implicitly assume, on adopting this position that the orientation of the employee towards a specific job (satisfaction), necessarily comes from the orientation taken towards the organization as a whole (commitment). They consider that job satisfaction varies more directly and instantly with changes in the working conditions than with organizational commitment. Chen (2007) holds that the investigators usually understand satisfaction as preceding commitment and he supports this idea affirming that job satisfaction is related to a concrete working environment that can be prone to change, but supports the organizational commitment which maintains a stable and lasting growth. However, despite the existence of work where the statistical results defend that the increase in the satisfaction of an employee leads to greater commitment (Chen, 2007; Fletcher and Williams, 1996; Yavas and Bidur, 1999), the empirical investigation has yet to categorically confirm this chance order.

Different studies have detected that commitment is by chance previous to satisfaction (Bateman and Strasser, 1984; Pool and Pool,

2007; Vandenberg and Lance, 1992). These investigations argue that employees adjust their levels of satisfaction in such a way that they are consistent with their organizational commitment in the present moment. This position is supported by the psychological and social perspective that considers that individuals develop adequate attitudes towards the situations to which they feel committed. It is accepted that commitment increases the motivation of an employee, which has a direct impact on their level of satisfaction. Despite these arguments, the evidence for this second possible chance order is certainly limited in comparison with the first hypothesis.

Other studies conclude the existence of a reciprocal relationship between job satisfaction and organizational commitment (Farkas and Tetrick, 1989; Lance, 1991; Mathieu, 1991; Mottaz, 1988). However, the reciprocal relationship proves asymmetric as satisfaction shows a more potent effect over commitment than the opposite.

Finally, studies exist which suggest that the relationship between these two attitudes over work could be doubtful as both show multiple common causes (Currivan, 1999; Curry *et al.*, 1986; Dougherty *et al.*, 1985). This argument could be found attractive given the growing tendency in current investigations to consider that similar determinants affect the two positions in different measure, just as has been mentioned previously in this section. However, the empirical studies which support the inexistence of a relationship are very limited.

For all this, despite the position taken beforehand by the proposed model, SEM methodology will be applied to analyse the empirical value of the same, as well as of two alternative models, given that the priority is to identify the antecedents which gave rise to organizational commitment, with the objective to obtain the positive consequences derived from it.

Statistical Methods

The multivariable technique which will be used to statistically analyse the proposed model in the investigation is that known as Structural Equation Modelling (SEM), which has become an integral tool in academic investigation and it performs and comprises of a whole series of tools such as covariance structure analysis, analysis of latent variables, confirmatory factors analysis and LISREL analysis (a software program for the development of structural equations).

The proposed model will be specified with a series of equations that define the structural model that links the latent constructs.[6]

In the structural model, the causal relationship of the six latent exogenous variables over the different perceptions of work is analysed, with job satisfaction shown by the employees. Therefore, job satisfaction acts here as a latent endogenous variable. At the same time, a study will be made of how the construct "job satisfaction", analysed via its components, influences the variable "organizational commitment", which in this way represents the other latent endogenous variable of this model.

Figure 6 shows the graphic representation with a path diagram.

The method of estimation used is ADF (Asymptotically Distribution Free), given that its insensitivity to the lack of normality of the data makes it particularly applicable to this investigation.

RESULTS OBTAINED

Before the estimation of the model, an exploratory analysis of the data was carried out. Statistics obtained on typical errors and deviations with respect to the mean do not detect the existence of atypical values that could distort the sample.

With respect to the normality of the data, Kolmogorov-Smirnov contrast was carried out,

Figure 6. Estimation for the proposed structural model

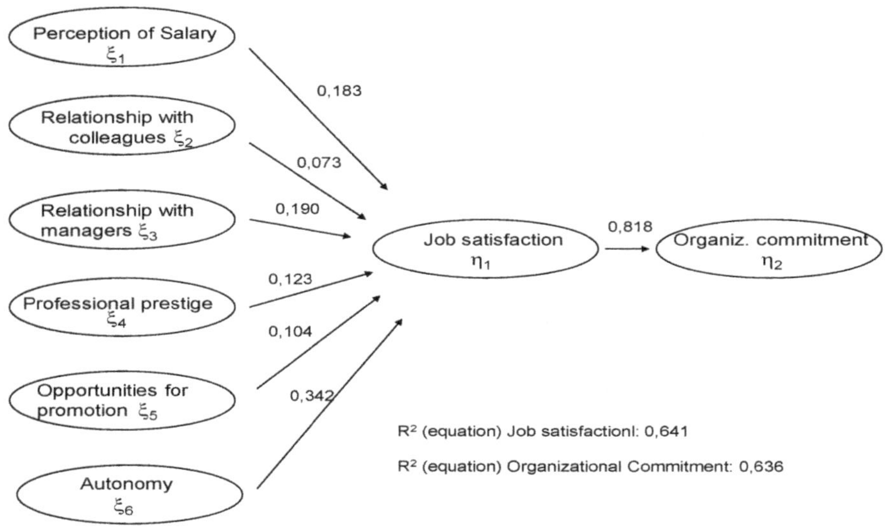

together with that of Shapiro-Wilk for the variables that form the measuring model. Obtained results lead to the rejection of the null hypothesis of normality in all cases. These circumstances, together with a big enough sample size, allow the application of the ADF estimation method, as has been previously indicated.

The model of structural equations proposed has been estimated with the LISREL program, version 8.7, which evaluates the identification of the model and shows the possible existence of errors. In this case, the estimation process has shown no restrictions through problems of identification.

No offending estimations were detected in the examination of the results of SEM (there are no negative variances of error, standard rate errors higher than 1, or high standard errors).

Figure 7 shows the result of the estimations obtained for the parameters of the structural model.

The interpretation of the estimation of the proposed model comes to the following conclusions:

- The weights of the exogenous constructs over job satisfaction are mostly significant and positively related.
- The determinant that shows higher weight (rate 0.342) is the sensation of autonomy at work. According to the measuring model, the indicator that shows most influence over this construct is the comparison with professionals from the public sector. In the opinion of those surveyed, the greater autonomy achieved in the carrying out of their work in a private centre over public one is valued very positively when expressing satisfaction with ones work.
- After autonomy, and carrying similar weight, the following determinants in influence are the relationship with management of the centre (equivalent to the relationship with a supervisor in the models used as a base) and the perception about the received salary. It is significant that salary is found in only third place of influence

Figure 7. Alternative model 1

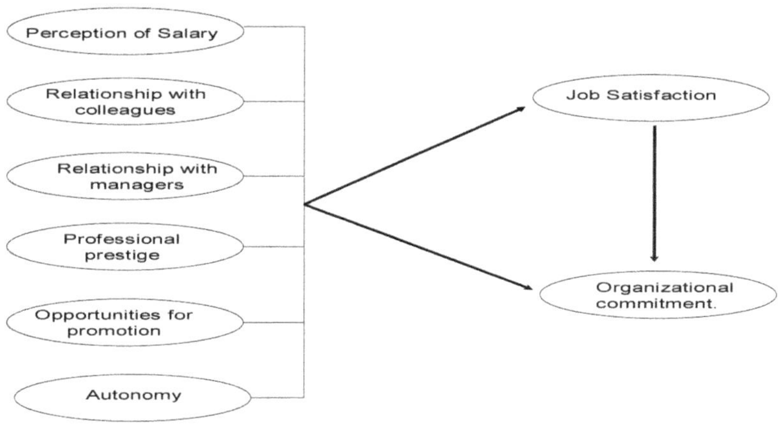

of the six exogenous determinants over job satisfaction. Empirical studies have shown that employees who maintain good quality relationships with their bosses have higher levels of performance, satisfaction and commitment (Liden et al., 2000; Lee, 2005; Ansari et al., 2007; De la Rosa and Carmona, 2010).

- With respect to the prestige of the profession and the opportunities for promotion, both obtain similar rates (0.123 and 0.104 respectively), which places them among the elements with less influence.
- Standing out in last place is the relationship with work colleagues, with a statistically insignificant rate of 0.073; this could be interpreted through the particularly individualistic nature of teaching and administrative work in a learning centre. The fact that one has a better or worse relationship with the rest of ones colleagues does not ultimately influence the satisfaction with the work carried out in this case. Possibly, this result depends in great measure on the type of work carried out, as well as the work philosophy of the organization. In businesses where group work becomes essential to the smooth working of the production process, this construct would possibly show a higher positive rate and have greater statistical significance.
- The structural equation that relates job satisfaction to organizational commitment shows a rate of 0.818, with an acceptable fit of 63.3%. Given that the parameter is significant and that it shows the appropriate direction (positive), it is possible to admit, with certain precaution, the influence that the level of the employee's job satisfaction has over his degree of commitment to the centre. The sense in this connection is preferred over other options raised by the rival models, with which satisfaction is established as a mediator in the connection of the exogenous constructs of the model over organizational commitment.

Table 2 shows some measures of fit for the validation of the model

The Satorra-Bentler Scaled chi square shows a level of significance below 1%, which means the existence of significant differences between the covariance matrixes. Nevertheless, the elevated number of indicators implies a greater sensitivity to this type of contrast, for which it will be necessary to analyse the other index.

Table 2. Measures of fit of the estimated model

Satorra-Bentler Scaled chi square (p-value)	46,893 (0,002)
Goodness of Fit Index (GFI)	0,8571
Root Mean Square Error of Approximation (RMSEA)	0,031
Adjusted Goodness of Fit Index (AGFI)	0,826
Normed Fit Index (NFI)	0,991
Non-Normed Fit Index or Tucker-Lewis (NNFI)	0,988

Source: Own Work

The Goodness of Fit Index shows a high value and, given that no established thresholds exist, its acceptability will be considered as marginal.

The Root Mean Square Error of Approximation shows a value of 0.031 (the confidence interval at 90% for this index shows an inferior limit of 0.018 and a superior limit of 0.043), for which, on finding both extremes below 0.05, it is possible to accept the hypothesis fit as more than acceptable.

With reference to the measures of incremental fit (AGFI, NFI y NNFI), apart from the AGFI index, the remaining measures go over the value of 0.9; this means an acceptable fit level.

Finally, a strategy of rival models will be followed for the comparison of the proposed model with a series of alternatives, with the objective of evaluating if, despite the fit within the acceptable limits of the proposed model, any other exists of a similar formation with a higher level of fit.

For the development of this strategy, two alternative models are shown:

- The first model sets out the direct causal relationship of the exogenous constructs, as much in job satisfaction as in organizational commitment, without the existence of mediation of the first in the influence caused by the said determinants over commitment. The causal relationship between the endogenous constructs, however, maintain the established order, satisfaction preceding commitment in the attitudes of the employee with respect to their work and the organization. This is the thesis defended by Kim *et al.* (1996) among others.
- The second model equally sets out the direct causal relationship of the exogenous constructs with the two endogenous ones. Nevertheless, in this case, the causal order is changed in such a way that the commitment becomes the antecedent and not the consequence of job satisfaction. Therefore, the position defended by authors such as Bateman and Strasser (1984), Pool and Pool (2007) or Vandenberg and Lance (1992) is analysed, which argued the fit of the employee's job satisfaction according to the commitment maintained with the organization at each moment.

As a comparison between the models, a series of goodness of fit measures are calculated (Table 3).

From the obtained measures we gather that the alternative model 2, which is unique in showing a GFI index higher than the proposed model, does not obtain a good fit in the other rates. This leads us to consider it as an unfeasible alternative and the previous supposed underlying theory of causality for organizational commitment is rejected.

With respect to the alternative model 1, the results are mixed, being favourable in the GFI and AGFI measures, but with a worse fit for the remaining than in the proposed model. If we limit the interest to the parsimonious index, with measures of comparison between models, the proposed one shows a more parsimonious fit and, ultimately, we accept it as the best, upholding the theoretical base on which it is founded and the empirical validation carried out.

Figure 8. Alternative model 2

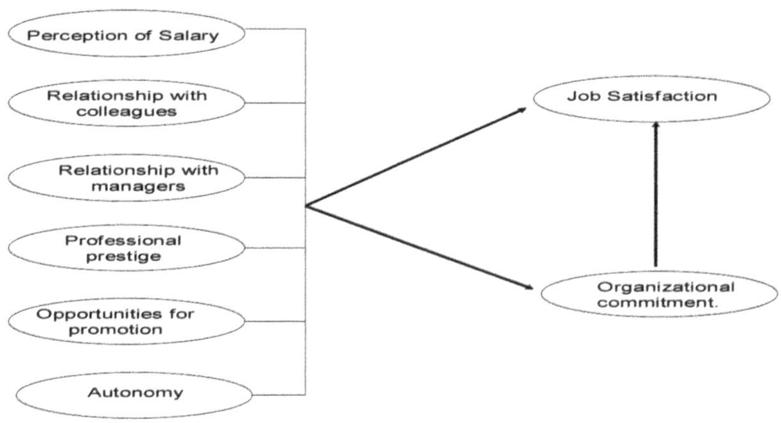

Table 3. Comparison of goodness of fit measures between the estimated model and rival models

Goodness of Fit Measures	Proposed model	Rival model 1	Rival model 2
Absolute fit measures			
Goodness of Fit Index (GFI)	0,857	0,993	0,893
Root Mean Square Error of Approx (RMSEA)	0,031	0,076	0,313
Incremental Fit Measures			
Adjusted Goodness of Fit Index (AGFI)	0,826	0,971	0,559
Normed Fit Index (NFI)	0,991	0,961	0,413
Tucker-Lewis Index (NNFI)	0,988	0,880	0,368
Parsimonious Fit Measures			
Parsimony Normed Fit Index (PNFI)	0,468	0,290	0,120
Akaike Information Criterion (AIC)	132,000	227,552	2013,06

Source: Own work

CONCLUSION AND FUTURE LINES OF INVESTIGATION

In an ever more competitive and global world, management of intangibles is becoming one of the key elements to business management. Moreover, given that human capital is one of the most important intangibles, especially when evaluating business growth, it becomes a source of competitive advantage to which special attention must be paid.

Management of human resources must evolve in the direction of new models that involve the workers in the philosophy and the objectives of the business. If the modern organizations, and more especially the service industries, need their values to be transmitted, they must count on workers who are professionally involved and who connect with the customer. In this situation of the market, the well-being of the employee becomes a strategy, and those businesses that become involved in the search for factors that lead them not only to the motivation, but also to the commitment of their employees, will obtain greater returns.

From this investigation, what is intended is a model which analyses the influence of six factors over the satisfaction of the employee and, secondly, over organizational commitment. Results reveal that elements such as autonomy when carrying out one's work and the relationship with management, together with a good view of the salary received (which is found in only third place) carry a lot of weight in job satisfaction. At the same time, the relationship between job satisfaction and organizational commitment has been considered in different alternative models, with better results being obtained in the model where the former is the cause of the latter and a mediator in the relationship with the six exogenous factors. Therefore, to gain a committed worker, the first fundamental step is to get a satisfied worker. But this, which is shown as a necessary condition, cannot be considered as a sufficient condition given that if satisfaction leads to commitment, a committed worker is not attained only by satisfaction, and it is the job of the organization to determine those additional elements that lead to the loyalty of their human resources.

In short, the analysis of the determinants of job satisfaction and organizational commitment is a priority given the strong impact that they have over determined behaviour such as absenteeism, turnover, productivity, etc. But locating the factors which influence commitment and to strengthen them seems to be still a challenge for Spanish businesses and work must be continued in this direction. It is necessary for utility services companies to become aware of the fact that their most important asset lies in their personnel; a redesign of management of human resources that considers an increase in the job satisfaction of the employees as a critical objective is needed. In a natural way this will lead to greater loyalty on their part towards the organization and, in short, be a guide towards a quality service which is ultimately reflected in customer satisfaction.

For future lines of investigation, the creation of models where more refined measures can be added is suggested, with a greater number of indicators or causal relationships re-specified for a greater convergence and fit of the model. Furthermore, organizational commitment measurement can be carried out by other instruments that take into account its multidimensionality, once it becomes clear the definition and distinction among the three facets (affective, normative and continued or free-will commitment) usually mentioned in specialized literature.

Equally, it is necessary to complete these types of investigations with longitudinal studies to demonstrate the preceding in time of the cause over the effect.

Furthermore, the use of mathematical methodology or econometrical alternatives of application in empirical studies is proposed, in such a way as to minimise the imprecision associated with the definition of the measures and indicators of the latent constructs such as those used in this investigation.

Finally, similar analysis should be carried out in other economic sectors (public teachers, tourism and hospitality industry employees, blue-collar workers, etc.) in order to be able to contrast the conclusions obtained.

REFERENCES

Akhtar, S., & Tan, D. (1994). Reassessing and reconceptualising the multidimensional nature of organizational commitment. *Psychological Reports*, *75*, 1379–1390. doi:10.2466/pr0.1994.75.3.1379

Allen, N. J., & Meyer, J. P. (1996). Affective, continuance and normative commitment to the organization: an examination of construct validity. *Journal of Vocational Behavior*, *49*, 252–276. doi:10.1006/jvbe.1996.0043

Alvarez, G. (2008). Determinantes del compromiso organizacional. ¿Están los trabajadores a tiempo parcial menos comprometidos que los trabajadores a tiempo completo? *Cuadernos de Estudios Empresariales*, *18*, 73–78.

Angel, H. L., & Perry, J. L. (1981). Empirical assessment of organizational commitment and organizational effectiveness. *Administrative Science Quarterly*, *26*, 1–14. doi:10.2307/2392596

Ansari, M. A., Kee, D. M. H., & Aafaqi, R. (2000). Fairness of human resource management practices, leader-member exchange, and intention to quit. *Journal of International Business and Entrepreneurship*, *8*, 1–19.

Ansari, M. A., Kee, D. M. H., & Aafaqi, R. (2007). Leader-member exchange and attitudinal outcomes: Role of procedural justice climate. *Leadership and Organization Development Journal*, *28*(8), 690–709. doi:10.1108/01437730710835443

Baker, J. G., & Baker, R. D. F. (1999). Perceived ideological differences, job satisfaction and organizational commitment among psychiatrists in a community mental health center. *Community Mental Health Journal*, *35*, 85–95. doi:10.1023/A:1018704327272

Baker, W. K. (1995). Allen & Meyer's 1990 longitudinal study: A reanalysis and reinterpretation using structural equation modelling. *Human Relations*, *48*, 169–186. doi:10.1177/001872679504800204

Barney, J. (1991). Firm resources and sustained competitive advantage. *Journal of Management*, *17*(1), 99–120. doi:10.1177/014920639101700108

Barraza, M. A., & Ortega, M. F. (2009). *Satisfacción laboral en instituciones formadoras de docentes. Un primer acercamiento*. Communication presented at the X Congreso de Investigación Educativa, Veracruz, México.

Bateman, T. S., & Strasser, S. (1984). A longitudinal analysis of the antecedents of organizational commitment. *Academy of Management Journal*, *27*, 95–112. doi:10.2307/255959

Becker, H. S. (1960). Notes on the concept of commitment. *American Journal of Sociology*, *66*, 33–42. doi:10.1086/222820

Bergman, M. E. (2006). The relationship between affective and normative commitment: Review and research agenda. *Journal of Organizational Behavior*, *27*, 645–663. doi:10.1002/job.372

Blau, G. J., & Boal, K. R. (1987). Conceptualizing how job involvement and organizational commitment affect turnover and absenteeism. *Academy of Management Review*, 290.

Boshoff, C., & Mels, G. (2000). The impact of multiple commitments on intentions to resign: An empirical assessment. *British Journal of Management*, *11*, 255–272. doi:10.1111/1467-8551.00165

Bravo, M. I., Peiró, J. M., & Rodríguez, I. (1996). Satisfacción laboral. In J. M. Peiró & F. Prieto (Eds.), *Tratado de psicología del trabajo, vol. I: La actividad laboral en su contexto.* Madrid, Spain: Síntesis

Chen, Y. J. (2007). Relationships among service-orientations, job satisfaction, and organizational commitment in the international tourist hotel industry. *Journal of American Academy of Business*, *11*, 71–82.

Clark, A. E. (2001). What really matters in a job? Hedonic measurement using quit data. *Labour Economics*, *8*, 223–242. doi:10.1016/S0927-5371(01)00031-8

Cohen, A. (1993). Work commitment in relation to withdrawal intentions and union effectiveness. *Journal of Business Research*, *26*, 75–90. doi:10.1016/0148-2963(93)90044-P

Cohen, A. (2003). *Multiple commitments at work: An integrative approach.* Hillsdale, New York: Lawrence Erlbaum.

Commeiras, N., & Fournier, C. (2001). Critical evaluation of Porter et al.'s organizational commitment questionnaire: Implications for researchers. *Journal of Personal Selling & Sales Management*, *21*, 239–245.

Currivan, D. (1999). The causal order of job satisfaction and organizational commitment in models of employee turnover. *Human Resource Management Review*, *9*, 495–524. doi:10.1016/S1053-4822(99)00031-5

Curry, J. P., Wakefield, D. S., Price, J. L., & Mueller, C. W. (1986). On the causal ordering of job satisfaction and organizational commitment. *Academy of Management Journal*, *29*, 847–858. doi:10.2307/255951

Darden, W. R., Hampton, R., & Howell, R. D. (1989). Career versus organizational commitment: Antecedents and consequences of retail salespeople's commitment. *Journal of Retailing*, *65*, 80–106.

De la Rosa, M. D., & Carmona, A. (2010). Cómo afecta la relación del empleado con su líder al compromiso con la organización. *Universia Business Review*, *2*, 112–132.

Dougherty, T. W., Bluedorn, A. C., & Keon, T. L. (1985). Precursors of employee turnover: A multiple-sample causal analysis. *Journal of Occupational Behaviour*, *6*, 259–271. doi:10.1002/job.4030060404

Dunham, R. B., Grube, J. A., & Castaneda, M. B. (1994). Organizational commitment: The utility of an integrative definition. *The Journal of Applied Psychology*, *79*, 370–380. doi:10.1037/0021-9010.79.3.370

Eskildsen, J. K., Kristensen, K., & Westlund, A. H. (2002). Measuring employee assets. *Business Process Management Journal*, *10*(5).

European Employee Index. (2002). *A benchmarking model for increasing the value of human capital.* Retrieved from http://www.ennovaconsulting.com/Default.aspx?ID=362

Farkas, A. J., & Tetrick, L. E. (1989). A three–wave longitudinal analysis of the causal ordering of satisfaction and commitment on turnover decisions. *The Journal of Applied Psychology*, *74*, 855–868. doi:10.1037/0021-9010.74.6.855

Fletcher, C., & Williams, R. (1996). Performance management, job satisfaction and organizational commitment. *British Journal of Management*, *7*, 169–179. doi:10.1111/j.1467-8551.1996.tb00112.x

Frutos, B., Ruiz, M. A., & San Martín, R. (1998). Análisis factorial confirmatorio de las dimensiones del compromiso de la organización. *Psicológica (Valencia)*, *19*, 345–366.

Gaertner, S. (1999). Structural determinants of job satisfaction and organizational commitment in turnover models. *Human Resource Management Review*, *9*, 479–493. doi:10.1016/S1053-4822(99)00030-3

Gallardo, E. (2008). Evolución en el estudio y medida del compromiso organizativo: Problemáticas y soluciones. In Pindado, J., & Payne, G. (Eds.), *Estableciendo puentes en una economía global* (p. 17).

Gamboa, E. J. (2010). Satisfacción laboral: Descripción teórica de sus determinantes. *Psicología Científica,* 1-7.

García del Junco, J., & Bras, J. M. (2008). Satisfacción profesional y compromiso organizativo: Un meta-análisis. *Revista Europea de Dirección y Economía de la Empresa*, *17*(3), 61–78.

Gellaty, I. (1995). Individual and group determinants of employee absenteeism: Test of a causal model. *Journal of Organizational Behavior*, *16*, 469–485. doi:10.1002/job.4030160507

Harris, S. G., Hirsfelch, R. R., Field, H. S., & Mossholder, K. W. (1993). Psychological attachment: Relationships with job characteristics, attitudes, and preferences for newcomer development. *Group & Organization Management, 18*, 459–481. doi:10.1177/1059601193184005

Harrison, D. A., Newman, D. A., & Roth, P. L. (2006). How important are job attitudes? Meta-analytic comparisons of integrative behavioural. *Academy of Management Journal, 49*, 305–325. doi:10.5465/AMJ.2006.20786077

Harrison, J. K., & Hubbard, R. (1998). Antecedents to organizational commitment among Mexican employees of a US firm in Mexico. *The Journal of Social Psychology, 138*(5), 609–623. doi:10.1080/00224549809600416

Herrbach, O. (2006). A matter of feeling? The affective tone of organizational commitment and identification. *Journal of Organizational Behavior, 27*, 629–643. doi:10.1002/job.362

Heskett, J., Sasser, E., & Schlesinger, L. (1997). *The service-profit-chain: How leading companies link profit to loyalty, satisfaction and value*. New York, NY: The Free Press.

Holland, J. L. (1985). *Making vocational choices: Theory of careers*. Englewood Cliffs, NJ: Prentice-Hall.

Hom, P. W., & Griffeth, R. W. (1995). *Employee turnover*. Cincinnati, OH: South Western.

ISR (International Survey Research). (2004). *Survey design and implications for driving job satisfaction and employee commitment*. Case study 901. Retrieved from http://www.isrinsight.com/pdf/insight/CaseStudy901.pdf

Jaik, A., Tena, J. A., & Villanueva, R. (2010). Satisfacción laboral y compromiso institucional de los docentes de posgrado. *Revista Electrónica Diálogos Educativos, 19*, 119–130.

Johnston, M. W., Parasuraman, A., Futrell, C. M., & Black, W. C. A. (1990). Longitudinal assessment of the impact of selected organizational influences on salespeople's organizational commitment during early employment. *JMR, Journal of Marketing Research, 27*, 333–344. doi:10.2307/3172590

Juaneda, E., & González, L. (2007). Definición, antecedentes y consecuencias del compromiso organizativo. In Ayala Calvo, J. C. (Eds.), *Universidad de La Rioja: Conocimiento, innovación y emprendedores: Camino al futuro* (pp. 3590–3609).

Judge, T. A., Bono, J. E., Thoresen, C. J., & Patton, G. K. (2001). The job satisfaction-job performance relationship: A qualitative and quantitative review. *Psychological Bulletin, 127*, 376–407. doi:10.1037/0033-2909.127.3.376

Katzell, R. A. (1964). Personal values, job satisfaction and job behaviour. In Borow, H. (Ed.), *Man in a world at work*. Boston, MA: Houghton Mifflin.

Kim, S. W., Price, J. L., Mueller, C. W., & Watson, T. W. (1996). The determinants of career intent among physicians at a U.S. Air Force hospital. *Human Relations, 49*, 947–975. doi:10.1177/001872679604900704

Lance, C. E. (1991). Evaluation of a structural model relating job satisfaction, organizational commitment, and precursors to voluntary turnover. *Multivariate Behavioral Research, 26*, 137–162. doi:10.1207/s15327906mbr2601_7

Larrainzar, M. J., Miñarro, J., Molinos, F. J., & Verdú, M. A. (2001). La satisfacción laboral en el Hospital de Sant Cugat de Asepeyo. *Capital Humano, 143*, 46–50.

Lawler, E. J. (1992). Affective attachments to nested groups: A choice-process theory. *American Sociological Review, 57*, 327–339. doi:10.2307/2096239

Leal, A. (Coord.), Alfaro de Prado, A., Rodríguez, L., & Román, M. (1999). *El factor humano en las relaciones laborales*. Madrid, Spain: Pirámide.

Lee, J. (2005). Effects of leadership and leader-member exchange on commitment. *Leadership and Organization Development Journal*, *26*(8), 655–672. doi:10.1108/01437730510633728

Lee, K. S., & Gao, T. (2005). Studying organizational commitment with the OCQ in the Korean retail context: Its dimensionality and relationships with satisfaction and work outcomes. *International Review of Retail, Distribution and Consumer Research*, *15*(4), 375–399. doi:10.1080/09593960500197495

Lee, Y., & Chang, H. (2008). Relations between team work and innovation in organizations and the job satisfaction of employees: A factor analytic study. *International Journal of Management*, *25*(3), 732–739.

Leong, C. S., Furnham, A., & Cooper, C. L. (1996). The moderating effect of organizational commitment on the occupational stress outcome relationship. *Human Relations*, *49*, 1345–1363. doi:10.1177/001872679604901004

Liden, R. C., Wayne, S. J., & Sparrowe, R. T. (2000). An examination of the mediating role of psychological empowerment on the relations between the job, interpersonal relationships and work outcomes. *The Journal of Applied Psychology*, *85*, 407–416. doi:10.1037/0021-9010.85.3.407

Lin, C. P., & Ma, H. C. (2004). Career commitment as a moderator of the relationship among procedural justice, perceived organizational support, organizational commitment, and turnover intentions. *Asia Pacific Management Review*, *9*, 519–538.

Lincoln, J., & Kalleberg, A. L. (1990). *Culture, control and commitment: A study of work organization and work orientations in the United States and Japan*. Cambridge, UK: Cambridge University Press.

Linz, S. J. (2003). Job satisfaction among Russian workers. *Journal of Manpower*, *24*, 626–645. doi:10.1108/01437720310496139

Locke, E. A. (1969). What is job satisfaction? *Organizational Behavior and Human Performance*, *4*, 309–336. doi:10.1016/0030-5073(69)90013-0

Mathieu, J. E. (1991). A cross-level nonrecursive model of the antecedents of organizational commitment and satisfaction. *The Journal of Applied Psychology*, *76*, 607–618. doi:10.1037/0021-9010.76.5.607

Mathieu, J. E., & Zajac, D. M. (1990). A review and meta-analysis of the antecedents, correlates, and consequences of organizational commitment. *Psychological Bulletin*, *108*, 171–194. doi:10.1037/0033-2909.108.2.171

McGee, G. W., & Ford, R. C. (1987). Two (or more) dimensions of organizational commitement: Re-examination of the affective and continuance commitment scales. *The Journal of Applied Psychology*, *72*, 638–642. doi:10.1037/0021-9010.72.4.638

Mendoza, M. I., Orgambídez, A., & Carrasco, A. M. (2010). Orientación a la calidad total, satisfacción laboral, comunicación y compromiso en establecimientos de turismo rural de la Sierra de Huelva. *Pasos. Revista de Turismo y Patrimonio Cultural*, *8*(2), 351–361.

Meyer, J. P., & Allen, N. J. (1991). A three component conceptualization or organizational commitment. *Human Resource Management Review*, *1*, 61–89. doi:10.1016/1053-4822(91)90011-Z

Meyer, J. P., & Herscovich, L. (2001). Commitment in the workplace: toward a general model. *Human Resource Management Review*, *11*, 299–326. doi:10.1016/S1053-4822(00)00053-X

Meyer, J. P., Stanley, D. J., Herscovitch, L., & Topolnytsky, L. (2002). Affective, continuance and normative commitment to the organization: A meta-analysis of antecedents, correlates and consequences. *Journal of Vocational Behavior*, *61*, 20–52. doi:10.1006/jvbe.2001.1842

Michalos, A. C. (1986). Job satisfaction, marital satisfaction and the quality of life: A review and preview. In F. M Andrews (Ed.), *Research on the quality of life*. Ann Arbor, MI: Survey Research Center Institute.

Mottaz, C. J. (1988). Determinants of organizational commitment. *Human Relations*, *41*, 467–482. doi:10.1177/001872678804100604

Mowday, R. T., Porter, L. W., & Steers, R. M. (1982). *Employee-organization linkages: The psychology of commitment, absenteeism and turnover*. New York, NY: Academic Press.

Mowday, R. T., Steers, R. M., & Porter, L. W. (1979). The measure of organizational commitment. *Journal of Vocational Behavior*, *14*, 224–247. doi:10.1016/0001-8791(79)90072-1

Moynihan, D., & Pandey, S. (2007). Finding workable levers over work motivation: Comparing job Satisfaction, job involvement and organizational commitment. *Administration Camp Society*, *39*, 803–832. doi:10.1177/0095399707305546

Mueller, C. W., Boyer, E. M., Price, J. L., & Iverson, R. D. (1994). Employee attachment and non-coercive conditions of work: The case of dental hygienists. *Work and Occupations*, *21*, 179–212. doi:10.1177/0730888494021002002

Mueller, C. W., & Lawler, E. J. (1996). *Commitment to different foci: The case of nested organizational units*. Iowa: Department of Sociology, the University of Iowa.

Mumford, E. (1976). *Work design and job satisfaction*. Manchester, UK: Business School.

O'Really, C. A., & Chatman, J. (1986). Organizational commitment and psychological attachment: The effect of compliance, identification, and internationalization on prosocial behaviour. *The Journal of Applied Psychology*, *71*, 492–499. doi:10.1037/0021-9010.71.3.492

Penley, L. E., & Gould, S. (1988). Etzioni's model of organizational involvement: A perspective for understanding commitment to organizations. *Journal of Organizational Behavior*, *9*, 43–59. doi:10.1002/job.4030090105

Pool, S., & Pool, B. (2007). A management development model. *Journal of Management Development*, *26*, 353–369. doi:10.1108/02621710710740101

Porter, L. W., Crampon, W. J., & Smith, F. J. (1976). Organizational commitment and managerial turnover: A longitudinal study. *Organizational Behavior and Human Performance*, *15*, 87–89. doi:10.1016/0030-5073(76)90030-1

Porter, L. W., Steers, R. M., & Mowday, R. T. (2005). Do employee attitudes towards organizations matter? In Smith, K., & Hitt, M. (Eds.), *Great minds in management* (pp. 171–189). Great Britain: Oxford University Press.

Porter, L. W., Steers, R. M., Mowday, R. T., & Boulian, P. V. (1974). Organizational commitment, job satisfaction and turnover among psychiatric technicians. *The Journal of Applied Psychology*, *59*, 603–609. doi:10.1037/h0037335

Powell, D., & Meyer, J. P. (2003). Side-bet theory and the three component model of organizational commitment. *Journal of Vocational Behavior*, *65*, 157–177. doi:10.1016/S0001-8791(03)00050-2

Price, J. L. (1997). Handbook of organizational measurement. *International Journal of Manpower*, *18*, 303–558. doi:10.1108/01437729710182260

Price, J. L., & Mueller, C. W. (1986). *Absenteeism and turnover among hospital employees*. Greenwich, CT: Jai Press.

Randall, M. D., & Cote, J. A. (1991). Interrelationships of work commitment constructs. *Work and Occupations*, *18*, 194–211. doi:10.1177/0730888491018002004

Robbins, S. P. (1994). *Comportamiento organizacional*. México: Prentice Hall.

Swailes, S. (2002). Organizational commitment: A critique of the construct and measures. *International Journal of Management Reviews*, *4*(2), 155–178. doi:10.1111/1468-2370.00082

Testa, M. R. (2001). Organizational commitment, job satisfaction and effort in the service environment. *The Journal of Psychology*, *135*, 226–236. doi:10.1080/00223980109603693

Van Breukelen, W., Van Der Vlist, R., & Steensma, H. (2004). Voluntary employee turnover combining variables from traditional turnover literature with the theory of planned behaviour. *Journal of Organizational Behavior*, *25*, 893–914. doi:10.1002/job.281

Vanderberg, R. J., & Lance, C. E. (1992). Examining the causal order of job satisfaction and organizational commitment. *Journal of Management*, *18*, 153–167. doi:10.1177/014920639201800110

Ward, E. A., & Davis, E. (1995). The effect of benefit satisfaction on organizational commitment. *Compensation & Benefits Management*, *11*, 35–40.

Williams, L. J., & Hazer, J. T. (1996). Antecedents and consequences of job satisfaction and commitment in turnover models: A reanalysis using latent variable structural equation methods. *The Journal of Applied Psychology*, *71*, 219–231. doi:10.1037/0021-9010.71.2.219

Yavas, U., & Bodur, M. (1999). Satisfaction among expatriate managers: Correlates and consequence. *Career Development International*, *4*, 261–269. doi:10.1108/13620439910279743

KEY TERMS AND DEFINITIONS

Job Satisfaction: Emotional state, attitude, sensation or degree of well-being that individuals experience towards their work.

Organizational Commitment: A part of the organizational behaviour that reflects an emotional response of the employee in the face of a positive evaluation of their working environment.

SEM: Structural Equation Modeling is a statistical technique for testing and estimating causal relations using a combination of statistical data and qualitative causal assumptions.

ENDNOTES

1. There is some empirical evidence that suggests the existence of two dimensions in the OCQ (Akhtar y Tan, 1994, Boshoff and Mels, 2000) but most researchers have applied this tool as a one-dimensional instrument (Cohen, 2003:4) and so it will be done in this research.
2. The determinants that are positively related to satisfaction and/or commitment are: salary, opportunities for promotion, distributive justice, support of colleagues, support of the supervisor and autonomy. Negatively related are: the workload, role conflicts, the ambiguity of the role and the routine.

3 Private schools have been chosen because their employees are subject to collective agreements and less arbitrary wages, which, ultimately, may lead to less distortion in the perception about their work, generating more reliable empirical research.

4 http://www.juntadeandalucia.es/educacion/scripts/rcentros/prcentros.asp

5 Many published studies have applied the OCQ, so it is associated with the development and practical applications of theories of organizational commitment (Harrison and Hubard, 1998; Gallardo, 2008). Currently, the OCQ is still used (Lee and Gao, 2005) and supported (Swailes, 2002) in empirical research.

6 A latent variable is an unobservable or immeasurable construct that can only be approximated by means of some kind of indicator.

Chapter 9
The Role of Internal Communication in Commitment Management

José Antonio Carazo
Capital Humano, Spain

ABSTRACT

The economic situation in the last few years has forced companies to adjust their workforces and reduce hierarchical levels. They have also had to undertake overhead cost and expense adjustment programmes, including salary freeze and the reduction of training budgets, amongst others; all of it within an environment where human capital has acquired special relevance. It has been shown that individuals represent enterprises' main asset, the element that can generate sustainable competitive advantages over time. Power has changed hands; it no longer lies in the traditional means of production, but in knowledge. And knowledge resides in people. Knowledge and people are the new engines of economy. The management of individuals, which is nothing but the management of their intangibles and the need to attract and retain talent, becomes especially important in this context. Commitment management is one of the most effective policies to achieve this aim. Managers must implement policies to commit employees to not only retain them, but also encourage them to do their best. The best tool to achieve this necessary commitment is internal communication. This chapter aims to offer a conceptual reflection on commitment management and analyses the state of internal communication in Spanish companies. The purpose is to establish a link between the emphasis placed by companies on achieving their employees' commitment and internal communication as a tool that can help to reach that goal.

INTRODUCTION

Managing people means managing expectations, aspirations, skills, attitudes, and perceptions. In other words, managing individuals is managing intangibles; hence the difficulty involved in this task. On the other hand, according to several authors, enterprises do not exist; it is individuals that exist. Companies are only groups of people oriented to reach a common goal. That is why it becomes so important to align the aforementioned intangibles (skills, attitudes) with the organisation's objectives making them compatible with personal goals (expectations, aspirations, perceptions, etc.).

DOI: 10.4018/978-1-61350-207-5.ch009

Knowledge is the engine of today's economy. It becomes necessary to identify and retain the best individuals, the most talented ones, those who can make a difference between competitors. 'Power' has changed hands. It no longer belongs to entrepreneurs (capital, technology) or managers (management, organisation). It resides in every person now. And I say every person because every single person plays, or must play, an essential role in each organisation's structure. Creativity and initiative reside in individuals. They permit to make qualitative leaps in the evolution of businesses. It is people who are the nearest to processes and customers, who know what works and what does not work: the main sources of continuous improvement. It is consequently obvious that individuals have become the most important item as far as management is concerned. How can objectives be achieved through people?

The way to do it is to ensure the commitment of every single member at each organisation. This task is complex for different reasons: the organisation's size and/or structure, the different ways in which functions and tasks are assigned, the fact that the strategy fixed by the management not always can be formulated openly, and finally the fact that not everything can be said to everyone in the same way, or in the same moment.

Commitment and Profitability

José Manuel Casado, CEO of 2C Consulting, points out that the magazine "Fortune" uses a measure called "The Great Place to Work Trust Index" to distinguish the top 100 U.S. companies in management commitment and, curiously enough, these are the ones which achieve better economic results (2011). And, in his view, "we must not forget that obtaining results in commitment management requires having been able to generate confidence in advance."

According to the dictionary of the *Real Academia Española* [Royal Spanish Academy (of Language)], *compromiso* [commitment] means an "incurred obligation" or a "word given;" whereas *lealtad* [loyalty] is understood as the "compliance with what is demanded by the laws of faithfulness and those dictated by honour and fine manhood." Finally, *fidelidad* [fidelity] is defined as "loyalty, observance of the faith which someone owes to someone else" (Espasa-Calpe, 2005).

Isidro Fainé, president of La Caixa, believes that "commitment is one step beyond motivation," and for Alfonso Jimenez, managing director of PeopleMatters, "commitment is related to motivation, values, attitudes, shared opinions and emotions. You want to do things in a certain way."

According to Carlos de Benito, vice president of Aedipe Centro[1], "loyalty is a quality of the human being and commitment is an expression of loyalty which appears in individuals who have a profound sense of duty. A loyal employee will always remain loyal, even if the enterprise does not deserve it. For that reason, the company is obliged to generate commitment" (de Benito, 2005).

In his view, "we need to draw a distinction between loyalty and commitment. Commitment is an expression of loyalty, or to put it in a different way, 'applied loyalty.' Loyalty is the cause; commitment is the effect. That is why we say that someone *es* [is] loyal [as a permanent quality] and *está* [is] committed [has decided to assume that stance at one particular point in time]." Commitment is an emotional kind of bond that becomes evident in a particular union or identification between one person and a personal or factual reality which is external to that person. Loyalty is the main source of commitment, but it is not the only, exclusive source. Commitment can be based on love, on shared or selfish interests, on pride or even on money alone. It must be stressed, though, that only positive motivation, whether it is altruistic or not, can generate commitment. Fear does not serve this purpose. The will moved by fear can prove to be very effective, but it will never be a committed will. In short, the quality of commitment, its strength and effects, are largely based on its motivation, on the reasons behind the will which triggers it. Commitment based on money and that based on loyalty are completely different matters" (2005).

A New Value System

A change of values has currently taken place within enterprises and more precisely in the management priorities of Human Resources Departments. The "engine" of economy has a new owner; instead of the entrepreneur who used to control the traditional means of production (organisation, capital and technology) the owner is now the employee, who is actually the person where a larger proportion of knowledge resides. In the words of Carlos de Benito, "the main driving force behind a positive attitude is personal commitment. The company appreciates commitment because it is aware that commitment can act as an engine to produce the results needed by the organisation" (2005).

Benito claims that commitment "is the motivation of the future. Now we are probably living a moment of transition, where the traditional motivation systems are changing. So far the company could obtain an acceptable outcome from motivating employees by means of two classical instruments: economic reward and fear, negative incentives; always powerful" (2005). He additionally explains that "the pressure on prices induced by pressure on costs has forced companies to adjust staff numbers and optimise organisational structures, reducing intermediate supervision levels and increasing the levels of responsibility and delegation. The traditional systematic and regulatory monitoring procedures turn out to be expensive and difficult to maintain within this framework, as opposed to relationships based on commitment, which require easier and less expensive monitoring systems", to which he adds: "Let us not deceive ourselves: the company mainly values results; it seeks commitment as the source to achieve them and appreciates loyalty as the source from where that commitment stems. We must make something clear, though: it is the commitment assumed by the competent individuals that matters; the loyalty of incompetent ones is not highly considered in the business world" (2005).

We have also said that commitment is a tacit contract between employer and employee, but not a lifetime contract. Pau Hortal [2], an expert in Human Resources Management, believes that "commitment is an increasingly rational agreement between the goals of employees and those sought by the employer, this agreement being further reinforced by the certainty that it is not possible to develop a whole professional life in a single organisation" (n.d.). This does not imply any disloyalty whatsoever. As Carlos de Benito sees it, "an employer who acts in good faith with an employee should not be surprised if that employee decides to leave after receiving an offer which gives him an opportunity for promotion and professional development outside the firm in question. He will not become a disloyal employee for that reason, unless he acts improperly" (2005).

The change in the power balance has given rise to new concepts: 'Power Employee', 'Value Proposal for the Employee,' 'Employee Branding'... Companies need the best employees and are willing to make an effort to hire them. And they also know that such policies are prioritised by the market. Hence their interest in having a good image as employers and fighting to improve working quality and professional development through the implementation of Human Resources Policies.

The Fight to Attract the Best Ones

Why do the best ones leave? According to Jose Manuel Casado, because they do not see an appropriate ratio between their performance and what they receive in return (Casado, 2003). They do not discern possibilities for growth, it seems to them that their superiors do not assign much value to their work, they think that their company is not making the most of their talent or have unclear expectations and do not tolerate a bad working atmosphere. Post-material values prevail at present, and people's priorities reflect a difficult balance between personal and professional aspects. The rising values are: professional development, talent appreciation, respect for different lifestyles and new family structures and everything that has to do with the conciliation of professional and personal life.

Nowadays people value their private sphere to a great extent and are unwilling to commit themselves to goals which can prove incompatible with their ethical standards. But employees can also be regarded as investors in time and knowledge. They will only stay in a company where they can obtain returns on their investment, where they can learn and develop professionally. A person who stops learning starts to become obsolete and, little by little, goes out of the labour market. If they do not obtain returns on their investment, they will invest somewhere else; they will find another job.

Furthermore, the engine of learning is twofold: on the initiative of the entrepreneur (or manager) and on the initiative of the employee. An increasingly high number of companies design training plans for all their employees, especially for those who occupy critical positions, and there are many which encourage their employees' initiative to design their own formative itineraries offering a wide range of courses which are usually oriented towards a twofold objective: improving employability and reaching corporate goals.

The new management and organisation models must adapt to the new value systems adopted by individuals. According to EADA Professor Jordi Assens[3], the model for what he calls 'enterprise integration' –from the perspective of people management– is characterised by the following ten principles:

1. Honesty in our approaches.
2. A value-transmitting leadership.
3. United human teams aligned with the mission, the vision and the values.
4. Work makes sense.
5. A strong sense of membership.
6. Clear guidance to harness and develop talent.
7. Ability to learn and adapt.
8. Its members' personal freedom is respected. Communication is open and transparent.
9. The person-company interdependence is experienced in a natural way.
10. People who leave the company maintain excellent relationships with it. (Assens, 2006).

Along the same lines, Pilar Jericó explains that, in her view, commitment is constructed by building trust, treating people fairly and acknowledging their efforts and performances. And all this must be communicated, i.e. made known. There are some companies where corporate culture encourages the creation and strengthening of commitment; these companies find their support on solid, clear and well-known values shared by everyone (Jerico, 2006).

Bancaja and DKV Seguros: Two Prototypical Cases

One of the companies that have pledged their commitment to its people is Bancaja. The *Caja de Ahorros de Valencia, Castellón y Alicante* [Savings Bank of Valencia, Castellón and Valencia] –Bancaja[4]– is one of the most important financial institutions in the *Comunidad Valenciana* [Valencian Autonomous Region]. Like nearly all Savings Banks, it experienced a strong growth outside its traditional natural boundaries a few years ago. This far-reaching expansion process outside the Valencian Region, the recruitment of professionals from other institutions, the need to become differentiated in a more open market through the creation of a solid corporate identity along with the need to unite the staff making an effort to avoid differences between branches and the head office (central services) led the Human Resources Department to further boost its policies in the areas of Conciliation, Compensation,

Development, Promotion and –of course– Communication.

Only a few months before, this financial institution had launched an advertising campaign with the slogan "The Bancaja Commitment: If it is not good for you, it is not good for us." The campaign materialised in a series of 'commitments' that Bancaja acquired with its customers in everyday commercial operations. Then they came up with another idea: this commitment with customers could be extended to the employees; this is how "The Bancaja Internal Commitment" was born (Carazo, 2007). The employees took part in the drawing-up of these 'commitments,' which were formulated as follows:

1. "Your queries have a single window."
2. "We guarantee a 2-year-stay in the same workplace."
3. "We inform about transfers 15 days in advance."
4. "We increase the number of women managers."
5. "We pay for the repair of your car in case of accident when it has been placed at the enterprise's disposal."
6. "15-days' leave when some relative is admitted to hospital."
7. "We do not summon you to formative activities during school-holiday periods."
8. "Assessment only has positive consequences."
9. "New employees receive 80 hours' training."
10. "Your opinion matters when it comes to undertake improvement actions."
11. "We create new communication channels between employees."
12. "Employees monitor internal commitment compliance."
13. "Two hours' leave to look after your children until they are 9 months old."
14. "You have advantageous conditions in other enterprises."
15. "We defer employee loans for 6 months free of charge in the event of death, divorce or separation of your partner."

The best way to prove an organisation's level of commitment to its people is to put it in writing, although we have already said at the beginning of this chapter that commitment is a moral 'contract.'

Codes of Conduct are another way to make explicit the extent to which an organisation is committed to its employees. The insurance company DKV Seguros[5] was one of the first pioneering enterprises to implement this type of policy.

A number of ideas underlie the Code of Conduct: the company should provide assistance for the human and professional development of its members; good professional work requires a certain level of ethical excellence when it is carried out; and this is a very suitable tool to reinforce the enterprise's culture. In the case of DKV, the Code affects the company's relationships with all stakeholders, no matter if they are contractual (customers, suppliers), inherent (employees, shareholders) or contextual (institutions), because "there is always a person behind them."

It has already been seen that commitment represents a tacit agreement made by two parties. In the words of Pilar Jericó, it is "a dance for two" (2008). You cannot make a pledge to yourself. It is consequently necessary to make this commitment explicit, not necessarily writing it down, but ensuring that both parties are aware of it and therefore know what each of them is obliged to do by virtue of the agreement. In order to achieve this, thorough information about the commitment must be provided within the organisation –in both parties' private environment. Internal communication, that flow of messages which circulates in all directions within an organisation, is the tool needed for this task. But, what is the state of internal communication in Spanish enterprises? Has it reached a sufficient degree of development, of maturity? Are the top executives committed to

Internal Communication? Do they support it? Do they think it adds value to their organisations?

The State of Internal Communication in Spain

The *Observatorio de la Comunicación Interna e Identidad Corporativa*[6] [Observatory of Internal Communication and Corporate Identity] constitutes a pioneering initiative in Spain which has as its aim the research, generation and dissemination of knowledge in enterprises' internal communication and corporate culture areas. It additionally detects the main trends and foresees their future development. It was created by the communication agency Inforpress, the review Capital Humano and the IE Business School 10 years ago with the desire to have a more in-depth knowledge about the characteristics of internal communication as a tool which generates value transmission, corporate identity, culture and internal brand management. Its activities are channelled through surveys and studies, experience exchange forums and best practice awards.

The *Observatorio* has conducted five studies on the state of internal communication within companies during the last two decades. The V *Estudio sobre la Comunicación Interna y la Gestión del Cambio en las Empresas Privadas Españolas y las Administraciones Públicas* [Study on Internal Communication and Change Management at Spanish Private Enterprises and Public Administrations], the last one carried out so far which –like the previous ones– had a national scope, collected data from 166 private enterprises and public administrations. Data-collection took place between July 2008 and August 2009. The results were contrasted with the views of 32 presidents, chairmen and women, chief executives, CEOs, Human Resources Managers and Communication Managers. A focus group was also held with representatives of public administrations. The main conclusions drawn are discussed below.

69.3% of the interviewed companies have a structured Internal Communication Plan, whereas 26.5% admitted that they did not have one. Furthermore, 43.2% of the enterprises have developed the plan on their own, as opposed to 45.2% of them which preferred to seek external support in aspects related to technology or advice.

The budget allocated by companies to Internal Communication activities is not high, although it must be remembered that many of the communication tools and actions cost nothing, have a very little cost or, sometimes, their cost is assumed by other activities. 28.9% of the companies allocate a budget of less than 50,000 euros and 26.4% have available more than 100,000 euros.

Regarding the allocation of human resources to these departments, there is only one person dedicated –full time or not– to this function in half of the companies (51.6%). 34.8% of the enterprises have two or three people and 13.5% have four or more people. The number of individuals is often determined by the resources available to the Human Resources or Communication Departments and the company's staff volume. In any case, we should not forget that internal communication is a task that affects all levels at organisations, and particularly top management and middle management levels. In fact, the responsibilities in this field are usually well defined in Internal Communication Plans.

Companies Have a Plan

This V study has analysed the extent to which internal communication influences various business management aspects. The results show that, above all, this exerts an influence when it comes to promote the awareness of the organisation's services and achievements, to support knowledge management, to improve the enterprise's public image and encourage savings and to achieve a better resource allocation, that is, when it comes to let all enterprise members know what their organisation places at their disposal, along with its achievements and objectives. This is a proof of transparency which is highly valued by people in order to give meaning to the tasks they have been entrusted with. Without a shadow of doubt, communication and commitment are closely related concepts. Communication, commitment and talent reappear as closely related concepts.

The Team Heads are Committed

Among the most highly developed Internal Communication aspects stand out, in this order: communication tools and channels and senior management's commitment, which comes to ratify the opinion expressed at the beginning of this chapter. The pending issue is the measurement of its impact on the enterprise's performance.

In 43.4% of the cases, Internal Communication is dependent on the Human Resources Department, and in 39.8% of them, it is managed jointly with External Communication. With residual percentages in almost 7.2% of the companies, internal communication is managed from several departments, and it is a staff department which reports directly to the Managing Director in 6% of them. Not all managers see it in the same way, but what really matters is the weight they attach to internal communication, regardless of its actual location.

The managers of Spanish companies are quite satisfied with the evolution of internal communication within their organisations. They give it an average mark of 6.1 out of 10, which slightly exceeds the mean in private enterprises (6.2) and also in those cases where internal communication is a staff department reporting to the Managing Director (6.7). 33.8% give it a B mark (above 7 out of 10) and only 15% give it a mark below the mean. Even though their perspectives may be different, the objectives sought by most of the companies which give importance to internal communication are similar.

The most widely used communication tools are, in this order: the Employee's Manual or Handbook (86.1% utilisation), Noticeboards (82.5%), the Employee's Portal (80.1%), random meetings (78.3%) and suggestions mailboxes (75.3%). It is worth highlighting the relevance acquired by the Employee's Manual in recent years. This handbook is a traditional Human Resources instrument which shows the close relationship between communication and people management that we have already referred to above. Noticeboards, which have lost importance in favour of technological tools lately, still play a significant role (e.g. in warehouses and labour intensive factories). Meetings have also increased their importance as means to transmit messages personally, which favours interpersonal relationships.

Internal television (29.5%) and corporate blogs (28.3%), which are making their way little by little, are among the tools which still have a residual use, either by the cost of the infrastructure required (television) or by the difficulty to control them and the uncertain interpretation by judges regarding the applicable law in case of conflicts (blogs). Traditional tools such as the internal review, videos or conventions reach a medium utilisation level.

As for the type of contents which are disseminated by means of the different internal communication tools, most of them (with 3.7 out of 4 in terms of frequency index) refer to enterprises' strategic issues (objectives, news, messages from the management), followed by operational contents (training, promotion, action guidelines), with 3.5, and emotional ones (awards, events), with 3.2. Purely financial contents (results, reports) occupy a much lower position in terms of priority. This is in keeping with the information gradually provided by the data and also with the aspects highlighted by the executives of the companies under study.

A Lever to Manage Change

Change Management concentrated much of the research attention in the *Estudio sobre la Comunicación Interna y la Gestión del Cambio en las*

Empresas Privadas Españolas y las Administraciones Públicas [Study on Internal Communication and Change Management at Spanish Private Enterprises and Public Administrations]. At a time when numerous companies have undergone changes, often profound ones, 88.5% of the respondents admitted using internal communication to support change management. The main areas which have suffered changes and required communication actions have been Strategies and Projects (77.1%), Organisational Structure (67.2%), Corporate Culture (66.4%), Corporate Image (54.2%) and Corporate Social Responsibility Integration (52.7%). Communication has been used to a lesser extent for matters such as mergers, transfers or employment regulation schemes, which are curiously enough the most delicate issues, and therefore the ones which should be the target of specific communication activities meant to maintain the information transparency policy and stop the circulation of rumours.

In any case, there is no doubt that most of the companies which took part in the study reveal that communication actions are needed to support the change processes which the organisation is going through. 48.8% of the enterprises do it with external actions and 40.2% with internal actions; and increasingly with a clear coordination of outward and inward messages. Concerning the moment or stage in which Internal Communication campaigns were undertaken to support change processes, previous or preparation campaigns have been carried out in 56.9% of the cases; campaigns have been undertaken simultaneously with the processes under way in 70.7% of the cases, and they have been maintained once these change processes have taken place in only 30.2% of the cases.

As for the channels and tools used to support change management, they vary depending on the type of messages, the audience they are addressed to and the company's technological sophistication level. The printed review (72.8%) is the most frequently used paper tool, followed by brochures and manuals. With regard to technological means, electronic mail –because it is so widespread and also for its immediacy– and the Employee's Portal are by far the most often used technological tools, well ahead of the online review, videos, microsites and blogs.

Finally, regarding in-house developed activities, meeting cycles (79%), training courses (49.2%) and conventions (46.8%) are those which have proved more effective, possibly because –as has already been seen– they are the ones which facilitate a direct contact between employees or (middle) managers and the senior management and among the employees themselves, regardless of their hierarchical level.

The Measurement of Results on the Debit Side

We have already pointed out before that research is one of the weaknesses shown by Internal Communication in Spanish companies. That becomes clear once again in the actions which are meant to support change management processes. Only 19% of the companies conduct research prior to undertaking those actions, and 9.1% admit that they do so afterwards. However, 45.5% declare that they carry out no research at all, neither before or after. Finally, over one fourth of the respondents (26.4%) reveal that they do research both a priori and a posteriori. It still represents a relatively small percentage, but it is gradually increasing.

Regarding the tools most often used to perform the corresponding research tasks, a majority of organisations choose satisfaction questionnaires (often included in Climate Surveys) with a 68.1% frequency rate, focus groups (47.8%) and in-depth interviews to executives at strategic positions (37.7%). The latter take longer and are less numerous (thus making it necessary to be very careful when selecting the prospective interviewees) but may provide more information and also of higher quality.

What cannot be denied is that not all messages are valid for all audiences. We cannot communicate the same things –and in the same way– to everyone. Companies segment their messages depending on the group they are addressing with the aim of modulating the different messages which need to be transmitted and achieving the highest possible effectiveness in that transmission process. 73.3% of Spanish companies segment their messages following hierarchical or functional criteria, while one in four enterprises admit that they do not carry out any segmentation whatsoever. 82.6% of them organise meeting cycles, 38.4% offer their employees specific training and 36% involve them in project creation processes.

CSR and People Management

A recent research study carried out within the *Estudio sobre la Comunicación Interna y la Gestión del Cambio en las Empresas Privadas Españolas y las Administraciones Públicas* [Study on Internal Communication and Change Management at Spanish Private Enterprises and Public Administrations] referred to the integration of Corporate Social Responsibility and its connections with internal communication, something very closely linked to actions in the areas of management commitment, membership pride and talent fidelisation.

CSR is run from Communication Departments in 48% of the cases, although this is a type of activity where other departments are usually involved. Only 10% have a specific CSR department. The main initiatives in the context of Social Responsibility ('business' or 'corporate' which, despite not being exact equivalents, are often used as synonyms) are, firstly, the actual communication of the actions performed (75.4%); secondly, the dissemination of environmental programs (72.3%) and then, actions associated with equality issues (69.2%), conciliation (64.6%) and the promotion of disabled individuals' integration (64.6%). These three aspects result from the change in values occurred both in the society as a whole and in individuals and its impact on the management policies adopted by companies.

As for the objectives sought with the implementation of the abovementioned actions, the most important one is clearly to improve relationships with the community or with the stakeholders (80.3%). Some objectives are somewhat less important, but still carry a lot of weight, namely: improving the working atmosphere (67.2%), brand creation –or branding– as an employer (62.3%) and talent fidelisation (60.7%). On a lower scale, we can find a number of secondary objectives such as innovation encouragement, new skills development or cost savings.

CONCLUSION

It is worth remembering that everything communicates. What we say and what we do not say; what we do and what we do not do, how we behave, the attitude we adopt before certain situations. There are eloquent silences as well as empty words. We cannot say that we trust people if we do not give them freedom, autonomy and responsibility. We will never be able to speak about commitment unless we pay attention to these individuals' development and concerns. We cannot ask for commitment if we do not commit ourselves.

At the present moment, when the economic situation has deprived us from the financial or material resources we were used to having, it becomes necessary to resort to intangibles. Moreover, the value scales have changed as well. People now assign more importance to their professional development, their growth as individuals or their free time than to other aspects. And human resources policies need to adapt to these changes.

People do not leave companies because they want to, but because they cannot stay any longer. The time has come to implement retention mechanisms and fidelisation policies. The time has come to 'sow' commitment. People need to commit themselves. We must lay the foundations so that, when better times arrive, people will want to stay even if they can leave. And if they do eventually leave, let us make sure that it is not due to neglect or lack of attention. We should at least try to ensure that they work with loyalty to the last minute. The key lies in making them work as if they had to earn their post every day, but with a long-term vision, as if they were going to retire in their current enterprise.

Internal Communication is one of the essential tools to achieve that commitment, which necessarily has to be mutual. Commitment is a tacit agreement, an unwritten pact where its elements must however be made explicit –with words and deeds. I insist: everything communicates. And we must be consistent in our actions and attitudes, messages and practices. It is time for us to design Value Proposals addressed to the Employee which are also useful to the company as a whole.

As for data, they show that internal communication at Spanish private enterprises and Public Administrations has reached a satisfactory development level. It has developed to a greater extent within those organisations which count on the wholehearted support of their top managers and see Internal Communication as an instrument which can bring consistency to corporate values and to the need to make messages fluent.

In most cases, internal Communication continues to be a competence of Human Resources Departments, although there seems to be an evolution towards the assumption of this responsibility by Communication Departments. This makes sense, since internal and external messages must be consistent.

Although the measurement of results is the aspect where some improvement still needs to be achieved, organisations are satisfied with their communication policies and give them a B^+ mark (at least 8 out of 10).

A large number of communication tools and channels currently coexist. Although technological tools have acquired great relevance in recent times, others which are typical of the Human Resources area have also developed, e.g. the Welcome Manual. Furthermore, the degree of sophistication in some communication policies makes it necessary to carry out a segmentation of messages depending on the audiences they are addressed to. Not everything can be said to everybody, and in the same way, if we want to reach our goals most effectively.

Finally, it is worth highlighting that internal communication also acts as a catalyst for change management and as a transmission belt for Social Responsibility policies.

REFERENCES

Aguilar López, J. (2011). *Gestión del Compromiso: De la continuidad por necesidad a la aportación libre de valor*. Capital Humano, No. 250. Wolters Kluwer, 2011.

Aguirre Barrera, V. (2011). *Engagement, más allá de un mero compromiso*. Capital Humano, No. 250. Walters Kluwer, 2011.

Assens, J. (2006). *Huevos con beicon. Cómo aumentar el compromiso de los empleados con la empresa y cómo gestionar su propio compromiso como empleados*. Barcelona: Granica.

Carazo, J. A. (2007). *Bancaja fideliza a sus empleados a través del compromiso interno*. Capital Humano, No. 205. Wolters Kluwer, 2007.

Casado, J. M. (2003). *El valor de la persona*. Madrid, Spain: Prentice Hall.

Casado, J. M. (2009). *El compromiso del miedo. Capital Humano, No. 237*. Madrid, Spain: Wolters Kluwer.

Casado, J. M. (2011). *El mantra del talento*. Barcelona: Ediciones Deusto.

De Andrés, E., & Primo, D. (2007). *Hacia una cultura de compromiso*. Capital Humano, No. 213. Wolters Kluwer, 2007.

De Benito, C. (2005). *Reflexiones sobre lealtad y compromiso en la empresa.* Wolters Kluwer. Capital Humano, No. 188. May, 2005.

Espasa-Calpe. (2005). *Diccionario de la lengua Española.* Madrid, Spain: Author.

Garcia Urbón, R. (2008). *El compromiso te hace libre: expectativa vs. compromiso.* Capital Humano, No. 232. Wolters Kluwer, 2008.

Hortal, P. (n.d.). *Personal blog.* Retrieved from http://pauhortal.net/blog/

Jericó, P. (2008). *La nueva gestión del talento. Construyendo compromiso.* Madrid, Spain: Prentice Hall.

Kofman, F. (2007). *Metamanagement.* Barcelona, Spain: Granica.

Observatorio De Comunicación Interna E Identidad Corporativa. (2009). *Estudio sobre la comunicación interna y la gestión del cambio en las empresas privadas y las Administraciones Públicas.* Retrieved from www.observatoriocomunicacioninterna.es

KEY TERMS AND DEFINITIONS

Attitudes and Skills: Attitudes are the pattern of the individual values formed from personal experiences and social experiences that are considered key factors in people management and teamwork. Skills relate to the ability or potential ability to perform some task or action but not yet learned that it is possible to get to learn.

Commitment and Psychological Contract: It is considered that informal, unwritten agreement between the employee and the company on what is reasonable can be expected of an employee to change what is in the employment contract.

Emotional Intelligence: It is the intelligent use of emotions, i.e. intended application involves the emotions with the goal of working for us, so help us guide our conduct and our thought processes to achieve better results.

Intellectual Capital: It is the possession of knowledge, applied experience, organizational technology, customer relationships and professional skills, all that could be developed but not to be apprehended, by the people who work for an organization.

Knowledge Management: It is the ability of organizations to acquire and exploit knowledge economically in a collective sense. That is, manage and exploit knowledge of the organization and its members as a source of value for customers.

ENDNOTES

1 *Asociación Española de Dirección y Desarrollo de Personas (Aedipe)* [Spanish People Management and Development Association]. It comprises 13 regional organisations, one of which is Centro.

2 http://pauhortal.net/blog/

3 Escuela de Alta Dirección y Administración (EADA), with campuses in Barcelona and Madrid.

4 http://bancaja.es/CAS/corporativa/clientes/compromisobancaja.aspx?urls=home

5 http://www.dkvseguros.com

6 www.observatoriocomunicacioninterna.es.

Chapter 10
Staff Restructuring in the New Economy

José Antonio Fernández Sánchez
University of Alicante, Spain

Encarnación Manresa Marhuenda
University of Alicante, Spain

ABSTRACT

In times of recession, survival for most businesses means to do more with fewer resources. It is necessary to encourage policies that will help a company to become more efficient, like staff restructuring. This policy encompasses many decisions, from laying people off to relocating employees to other positions as a consequence of the deployment of a more rational structure, which in turn answers to a change in strategy caused by the turn in the environment. This chapter explains the concept of staff restructuring in detail and offers several methods to carry it out. Afterwards, the potential problems that may arise from implementing this strategy are discussed, and, finally, some clues on how to avoid such problems, or, at least, mitigate their consequences, are presented. It is also made clear that a short-term approach in this regard will lead to a future lack of competitiveness, and therefore, endanger long-term survival.

CONTEXTUALISATION

Staff restructuring strategies are not a new fashion. Already in 1994, Cameron affirmed that it was an insufficiently studied topic (Cameron, 1994) and since then, research in this field has dramatically increased.

This issue is so relevant that the European Union held a conference especially dedicated to staff restructuring processes in Dublin in June 2003 [1] because, although these processes have been taking place, hundreds of experiences have been identified and many articles have been published for several decades, we still suffer from a knowledge deficit in this area and, what is worse, from a lack of action devices that have been tested, approved and shared by the different economic actors.

Firms that decide to implement a staff restructuring strategy do so aiming to considerably improve their performance in terms of net profit, total income, return on equity, productivity, and

DOI: 10.4018/978-1-61350-207-5.ch010

other items. In fact, there is supporting evidence of this phenomenon in Alicante province (Spain). Economic results improved according to subjective measures (managers' opinions) and objective measures (organisations' economic data) (Fernández Sánchez, 2011). However, most authors have come to the conclusion that staff restructuring processes fail to provide the desired results and that the situation tends to worsen instead of improving (Whetten et al., 1995).

All the above has encouraged us to delve into this topic with the aim of shedding some light on it, by means defining the term clearly, identifying those practices which provide the best outcome, describing available methods to try to alleviate the problems which can derive from these processes and, finally, looking for possible alternatives to restructuring.

Definition and Meaning of Staff Restructuring

Business restructuring processes have been studied by a large number of authors (E.g. Appelbaum, 1987, 1997, 2000, 2003; Gunnigle, 2012) mainly in the Anglo-Saxon context. Even though not many definitions of the term are available, and authors do not always reach a consensus on its meaning, this term is very often used to describe any modification of national policies, of capital structure and of management or running strategies, especially insofar as these affect social relationships.

At this point, staff restructuring can be defined as "the process by which each employee is assigned to the most suitable job so that they can jointly exploit the skills of individuals to the full; it may or may not imply changes in the organisational structure and a partial staff reduction." Along the same lines, Marr & García (1997) consider it a business rationalisation process, and other authors (Tomasko, 1990; Zehir & Keskin, 2003) speak about reorganising or redesigning the business firm.

Due to its external as well as internal consequences (Marr & García, 1007),staff restructuring processes are regarded as a very important issue –if not the most important one– within a company's staff policy, and it appears as one of the greatest challenges for decision-makers (Wilkowska, 2003).

As it is easy to infer, this process implies a total or partial reorganisation of the business structure and there are basically two reasons which can push an organisation towards the decision to restructure its staff (Band & Tustin, 1995): the competitive position that the company wants to adopt in the long run and maintaining or rapidly increasing its productivity level per employee. This decision may have been triggered by changes both in the national and international economic circumstances (globalisation, deregulation, innovation strategies undertaken by global competitors, etc.) (Wilkowska, 2003) and the internal conditions of the organisations themselves (for instance, strategies of innovation, differentiation, etc.); due to these causes, the staff costs cannot be compensated by their contribution to the company's added value.

According to the Resource-Based Theory, the growth of a firm will depend on the availability of idle resources that it tries to exploit in new markets and/or businesses. Staff restructuring or reduction can therefore be explained as a process of selective reorganisation or reduction of the resources available to an organisation (of human resources, in this case) with the aim of ensuring its survival and future (Suárez González, 1999).

Two basic kinds of problems may arise. On the one hand, when dealing with a partial staff reduction, this could mean the loss of certain resources and capacities needed by the firm to ensure its survival and success (Barroso & Casillas, 1998). Furthermore, there is also a risk of provoking the so-called "survivor syndrome" (Appelbaum et al., 1997; Appelbaum & Donia, 2000; Kim, 2003; Bashford, 2004), since the remaining workers might see their post in the firm threatened, experience feelings of uncertainty, anxiety, stress, and lack of confidence. This may lead to lower performance and productivity levels along with worse product or service quality. On the other hand, if we find ourselves in front of a transformation or reorganization, we bear the risk of underutilizing the available resources and capacities if things are not done properly.

This undoubtedly entails a series of strategic implications. As Drew (1994) pointed out, restructuring in general, and more specifically staff restructuring, should be accepted as a business strategy. A company should be able to anticipate events and design the strategy which suits best the situation which it is experiencing at that moment and have the capacity to adapt to change (Gómez-Mejía et al., 1998; De Miguel, 2001). Just the same as in any other type of strategy, foresight and planning are the best instruments to alleviate the problems derived from restructuring schemes.

Most authors associate this term with the general process which leads a company to reduce its size, with all the implications that this entails, essentially regarding the number of workers (Cameron, 1994; McKinley, Sánchez & Schick, 1995; Budros, 1999). As a matter of fact, it originally comes from the English term 'downsizing,' which literally means 'reducing size.' Nevertheless, there are some authors in whose opinion staff reduction is only one aspect within the wide range of processes covered by downsizing, understanding it as restructuring in a broad sense (Whetten et al., 1995; Pichault, Warnotte & Wilkin, 1998).

It is important to clarify that as a rule, a company does not only cut staff numbers (if it does) but it also reorganises individuals and jobs in accordance with its new dimension. Attending to Bowman & Singh (1993), Band & Tustin (1995) and Pichault et al. (1998), staff restructuring does not equal downsizing from a strict point of view, but all of them also recognise that staff reduction is one of the measures which normally accompany this type of processes.

Along the same lines, Freeman differentiated three aspects within this field already in 1994, pointing out that it does not always imply layoffs but can simply mean a relocation of individuals in other posts which has as its aim to rationalise the business processes and activities in the hope that this will ultimately improve the situation. She thus identified three types of downsizing depending on the strategy followed that could be defined as follows (Freeman, 1994):

- **Numerical restructuring:** it is a short-term strategy based on staff number reduction.
- **Functional restructuring:** in this case, a middle-term strategy is undertaken seeking to achieve process restructuring. Different methods such as task redefinition, level elimination, unit merging and removal and transformation of some functions can be used for this purpose.
- **Strategic restructuring:** a long-term measure which has as its aim to reshape the organisation and its parameters. Its characteristics would include: change of responsibilities, restructuring the workforce, global involvement, continuous improvement and questioning or rethinking the organisation's current system of values and regulations.

These last two types of restructuring activities refer to the same concept: reorganisation or redesign. Therefore, we will refer to two types of restructuring strategies in this paper:

- **Cut strategy:** restructuring based on reducing a part of the staff. This is a deliberate or voluntary reduction in the number of employees which goes beyond a mere decrease in the employment volume due to natural causes.

- **Reorganisation or redesign strategy:** restructuring based on the business functional and organisational reconfiguration or, to put in another way, on the relocation of individuals in different posts seeking an optimum exploitation of their skills. This type of strategy allows firms to achieve new forms of organisation thanks to which they will stand a better chance of meeting market demands successfully.

The main advantages, along with the drawbacks of these two types of strategy, are shown in Table 1.

Source: self-elaboration, based on the consulted bibliography

Finally, it is necessary to point out that the utilisation of the term 'strategy' in the case of staff restructuring must not be understood as something that implies a series of consistent, regular policies which are implemented being perfectly sure that the desired aims will be achieved and that they will correspond to carefully analysed critical success factors. Due to the actual nature of management and the uncertainty which characterises this environment, restructuring strategies are sometimes nothing but a set of proactive and reactive actions which include the responses to immediate crises. The specific restructuring strategies adopted vary depending on the country, the industry and the firm we are dealing with, and also according to the local imperatives, options and traditions (Winterton, 2003).

Once again, it is necessary to clarify that we are not talking exclusively about staff reductions or layoffs affecting part of the personnel, although that is the sense most frequently given to the expression and it is also the most common outcome in this kind of processes.

STAFF RESTRUCTURING PROCESS

A staff restructuring process like the one which is being defined requires an analysis similar to the one carried out in the selection process but much more complex, as more jobs will be examined (sometimes, all the posts available in the company). Such analysis would try to identify the skills needed to develop them conveniently and

Table 1. Advantages and drawbacks of staff restructuring strategies

	Advantages	Drawbacks
Cut Strategy	Short-term visible results	Possible loss of necessary resources and capacities
	Dramatic cost reduction	Traumatic both for the employees affected and for those who stay (demotivation)
		Potential cost increase due to layoff expenses
Reorganisation strategy	It can permit to achieve cost reduction avoiding massive layoffs	As a rule, results can only be appreciated in the middle or long run
	It favours efficiency	Potential workload intensification
		Demand for workers to have immediate availability when the enterprise needs them
		Possible downward revision of employees' levels and salaries
		Dualisation of the workforce (strategic employees and common employees)
		Lack of interest on the part of the enterprise in the professional development of employees regarded as 'ordinary' ones.

check how the current holders of those posts are carrying out their work, what skills are owned by those employees, and finally forecast which posts will still exist after the reorganisation and which individuals will hold them.

Band & Tustin (1995) propose a sequence of steps that lays out the path to a successful implementation of a staff restructuring process:

1. **Analysis of the current structure:** the adoption of a decision to restructure must be preceded by a previous analysis of the situation that will help to clarify the details corresponding to the remaining stages. First, the company's current competitive position should determined, as well as the staff structure required to, at least, maintain it. This means to identify the skills needed for that purpose and to check how many of them are already in the house. This analysis also entails an assessment of human resources practices in use and the identification of those critical areas bound to be affected by the process.
2. **Deciding whether to restructure:** once the previous analysis has been completed, the decision to restructure is made depending on whether the company plans to maintain or improve its competitive position. It is during this stage when a decision must be made regarding new workforce structure we want to equip the firm with. It is also necessary to draw possible alternatives that could be of use to restructure the abovementioned critical areas; and, finally, to consider the positive and negative aspects of each alternative.
3. **Formulation of the restructuring plan:** the way in which changes towards the new structure are going to be carried out; that is, which methods are going to be used and how long it is going to take.
4. **Implementation of the restructuring plan:** putting into practice the plan established in the preceding stage.
5. **Restructuring assessment:** the individuals in charge of the restructuring process must ask themselves if they have succeeded in achieving the structure which had been chosen during the second stage, always verifying that this 'ideal' structure has not varied due to both the internal and external changes which may have occurred in the time period elapsed since the process started.
6. **Analysis of the new structure:** once the 'ideal' structure has been achieved, the firm's competitive position has supposedly improved with respect to the initial position. From here, continuous efforts must be made in order to maintain that position, or improve it further if possible. However, it is necessary to keep in mind that certain problems are likely to show up as a result of this process. Besides, the company must beware of the risk of falling into a 'one-restructuring-after-another' dynamic that will ultimately lead to the disappearance of the company.

HOW TO CARRY OUT STAFF RESTRUCTURING STRATEGIES

Once the concept of staff restructuring has been discussed, and a plan has been laid out, it is time to carry it out. In this section it will be proposed a number of tactics to implement this strategy, along with their strengths and drawbacks.

Junior Employees

One of the most frequent restructuring techniques consists in removing from the staff those workers who have been hired more recently and have not been working for the company for long periods of time. Since layoff costs are usually higher for individuals in higher positions or have been working longer at the firm, this criterion entails lower layoff costs and, besides it is, so to speak, "socially accepted" (Devesa et al.1997). It is akin to the accounting term "first in, first out".

Nevertheless, staff reduction needs cannot be strictly done on the basis of legal regulations for the sole purpose of reducing costs and avoiding complaints by trade unions. Likely, the workforce behaviour will be affected by lower levels of satisfaction, reducing people's loyalty to the firm while increasing staff turnover –the lower the layoff costs, the higher the turnover- (Bobrow, 1996). Also, older people or people in a higher hierarchical position would stay if this approach was followed, which might have a negative effect on innovation capacity and work potential, these two elements being much more present among young and ambitious people (Marr & García, 1997). Therefore, it would be a great mistake to take layoff costs as the only or the main factor to decide who stays and who leaves the company.

Besides, as these authors also point out, layoffs are expensive and personnel costs, which were decreased after the downsizing process, could increase with layoff costs. This idea is shared by Requejo (1996) in a study where he reaches the conclusion that companies with a higher proportion of temporary workers have a greater likelihood of reducing their staff, which shows the relevance that layoff costs have for these decisions.

However, layoff costs should not be ignored, but rather valued in their own right, instead of simply becoming the basic element to adopt a final decision. It is necessary to analyse the ratio between layoff cost and worker performance, as it could eventually be more profitable to dismiss employees with higher posts or more years of service due to their poor potential performance. The criterion to let them stay should thus be the value provided to the business.

Early Retirements

Another common technique is to propose an early retirement to the oldest employees; that is, an agreement with the workers so that they accept to retire before reaching their legal retirement age. This can be deployed by means of a voluntary redundancy policy, or retirement with incentives. In this case, incentives are given for employees to stop working in the business from a certain age, adding a certain amount as an indemnity or economic compensation to what they should receive for their settlement in normal circumstances. This method is widely used both in Spain and abroad 8IESE, 2002).

Following the previous analogy, this would be similar to "last in, first out". Although this entails a higher cost in the short run, it will lead to sizable medium-term and long-term savings, if planned and implemented correctly, and if the company manages to remove from their jobs those individuals who are ultimately responsible for the critical situation (Appelbaum et al., 2003).

This method is based on a type of technique known as *rank and yank*. The idea is to establish a ranking or classification of the employees where three basic groups are distinguished: the best workers (which approximately includes 20% of the total number of employees), the worst ones (10%) and the rest of workers with a medium performance level (the remaining 70%). It is considered that the group formed by the worst workers has something so difficult to correct that the effort is not worthwhile. The best solution is, consequently, to remove them from the organisation.

However, although this method is widely used, it entails risks too, as in the case of the Ford Motor company, which ended up with lawsuits for discrimination (for reasons of age, amongst others) that finally forced the firm to abandon the strategy.

Other Techniques

According to evidence from Pin (2002), other practices in application in Spanish firms for staff restructuring processes are:

- **Functional mobility, recycling, and requalification:** it means a continuous update in terms of training which permits a greater staff turnover. In this way, workers can adapt to any other post.

- **Outplacement:** it basically means that that the company hires a specialised agency which can offer advice in the readaptation and relocation of those individuals who are going to lose their jobs. It is essentially used to reorient and relocate executives and managers, and the company usually dedicates part of the severance pay to cover the corresponding expenses.
- **Flexible contracts and temporary contracts:** modalities which replace ths usual full-time contract providing the organisation with higher flexibility.
- **Relay (or hand-over) contracts:** in this case, workers of advanced years who are about to leave the firm train and prepare the new employees who are going to occupy their post or a similar one.

In addition, the figure of *Expediente de Regulación de Empleo* [Labour Force Adjustment Plan] featured for the first time in 2005 in Spanish restructuring processes.

In 2005, the most commonly utilised practice in Spain, as per the evidence of a nationwide study, was the non-renewal of temporary contracts, followed by early retirements and voluntary redundancies with incentives supported by frozen recruitment policies. By 2011, even though these three tactics are still the most often used, their ranking has changed slightly: first, non-renewal of temporary contracts; then, policy of frozen recruitments; and, in third place, early retirements and voluntary redundancies with incentives (Fernández Sánchez, 2011b).

Regarding recent international data (Pulignano, 2011), research on 20 multinational firms in seven EU member states during the period comprised between 2006 and 2009 shows the predominant use of retirement with incentives, voluntary redundancy with incentives (or compensations), recruitment-layoff (that is, incorporating the suitable workers and retiring the unsuitable ones) and the mobility, recycling and requalification of employees.

The reason for the changes in preferences in time and over geographical locations is mainly the expected outcome between the positive and negative aspects of each technique. Table 2 shows the most important advantages and drawbacks associated with each one of these methods.

DEALING WITH PROBLEMS CAUSED BY RESTRUCTURING STRATEGIES

Implementing a staff restructuring strategy is going to give raise to a number of potential problems which should be avoided or, at least, their effects lessened to a certain extent. For starters, alignment of this strategy with the company's competitive strategy is crucial in the road to future success, this way the resources and capabilities needed in the organization would be the ones kept. Table 3 illustrates the relationship between competitive strategies and staff restructuring strategies.

Among those endangered competitive resources, the loss of human capital is the most direct consequence of a staff restructuring process. It can prove very costly when individuals leave the firm taking with them knowledge which cannot be detached from them or, which, being separable, is very valuable for a competitor (Simon, 1991; DiBella, Nevis & Gould, 1996). Large-scale layoffs and early retirement schemes sometimes cause precisely this loss of talent within an organisation. For this reason, an effort should be made to avoid this release of intellectual capital from the business insofar as possible, redistributing it instead for its better exploitation. For example, seeking to avoid the company's intellectual impoverishment after the loss of qualified professionals, the Spanish car firm SEAT reached an agreement that included partial retire-

Table 2. Advantages and drawbacks of staff restructuring methods

	Preretirement	Voluntary redundancy with incentives	Functional mobility	Outplacement (Relocation)	Flexible contracts	Temporary contracts	Relay contracts
Advantages	It compensates for the salary	Easily implemented	Employee polivalence	It softens the impact caused by layoff	Flexibility in time and shape	Quick resolution	Staff rejuvenation
	It is a compensatory measure	It minimises social impact	Constant update in training	It prevents demotivation	Savings for the enterprise	Flexibility in time and shape	Savings for the enterprise
	Freedom in its design		Improved employee qualification level	It improves the firm's labour climate and image			
	Staff rejuvenation		Freedom for the entrepreneur, flexibility	It reduces unemployment time			
	Non-traumatic and welcomed		Greater functional rotation among employees				
Drawbacks	It creates precedents and expectations which are difficult to maintain	It does not favour employee relocation	Considerable expense in training	Higher cost for the enterprise	Difficult implementation due to the Spanish culture	Precariouness for the employee	Need for collaboration on the part of the leaving employee
	High cost for the enterprise and for the Administration	High cost and it undercapitalises the enterprise		It may seem discriminatory		Situation of instability	
	Lack of guarantees	Traumatic					
	Legislative rigidity						

Source: IESE (2002)

ment schemes and relay contracts through which the most experienced professionals who were about to leave the firm trained the new employees before leaving the organization.[2]

Therefore, it is necessary to formulate a knowledge management policy in order to differentiate generic knowledge from specific knowledge, and preserve the latter within the company. Generic knowledge is easily obtained and copied and, therefore, it is the least interesting for the firm, because it does not distinguish that firm from its competitors. As for specific knowledge, it is not available in the market but develops within the company (Bueno Campos, 1998). This is the knowledge the organisation must try to cultivate and appropriate, to prevent its loss even if the employee leaves the firm voluntarily or is affected by the restructuring process.

Consequently, those employees who do not deal with specific management should be the first ones to leave the firm, for this will have a lesser impact on the value of the company's human capital. Besides, generic knowledge can be very easily gained again by means of temporary contracts and outsourcing.

Table 3. Relationship between business strategy and HR strategy

	Gómez-Mejía et al.		Porter			Miles & Snow		
HR STRATEGIC AREA	**GROWTH**	**STABILITY**	**COST LEADERSHIP**	**DIFFERENTIATION**	**HIGH SEGMENTATION**	**DEFENSIVE**	**EXPLORATORY**	**ANALYSING**
Staff Restructuring		Temporary contract non-renewal	Temporary contract non-renewal	Dismissal of the least qualified workers	Combination of both	Voluntary incentives	Layoffs, E.R.E. [Labour Force Adjustment Plan]	Combination of both
	Early retirement and voluntary redundancy with incentives	Recruitment freeze	Recruitment freeze	Dismissal of workers with a more generic training		Recruitment freeze	Early retirement and voluntary redundancy with incentives	
			Dismissal of those who have spent less time at the firm	Workforce renewal				
	Preferential re-recruitment does not exist	Preferential re-recruitment policy		Early retirement and voluntary redundancy with incentives		Preferential re-recruitment policy	Preferential re-recruitment does not exist	

Source: self-elaboration based on Gómez-Mejía, Balkin & Cardy, 1998; Valle Cabrera, 1995 & Armstrong, 1992.

As a complement to this knowledge management policy, the firm could deploy a sound information system to facilitate its acquisition and subsequent transmission to the other employees, thus ensuring that the business will not be significantly affected if a key employee leaves the company. It could also be of use to identify and promote critical practices and procedures, aiming to develop a learning system which permits to retain, keep and use the knowledge acquired by the organization's employees (Griggs & Hyland, 2003).

Business culture also comes into play here, as one of its main functions is to reduce the organization's dependence on a specific individual or group of individuals, by transforming corporative knowledge into "know-how", which may be re-utilised even if the worker leaves the organisation (Fiol & Lyles, 1985; Walsh & Ungson, 1991; Sáenz Blanco, 2001). This way, that person or team could be replaced at a relatively low cost (Fernández et al., 1998). The same authors mention other methods to appropriate knowledge and avoid the escape of human capital (legislation, salary compensations, pay based on length of service, high pensions which workers lose if they leave the firm, etc.) among which fixed or long-term employment stand out for their particular motivating importance (Auer & Cazes, 2000).

There is also a tool which has proved to be very useful to obtain information, whether if workers have decided on their own volition to leave their positions or if they were fired: the exit interview (Carvin, 2009). An exit interview is an interview conducted by an employer of a departing employee. This way, it is possible to obtain information which –if appropriate– may help adopting the correcting measures required to avoid future layoffs or undesired leaves (among which stand out relationships with partners, superiors and/or subordinates, their consideration of the business, the company to which they leave, the salary they have been offered, their opinions about what could be improved within the organisation, etc.). However, it presents a great drawback, namely that it needs the sincere and disinterested collaboration of the leaving worker, something that could turn out to be very difficult depending on the circumstances.

Furthermore, it is very important for the firm to pay attention to the different types of motivations, as this will influence employees' performance and, therefore, the organisation's results. In fact, different alternatives to cut strategies can be found as a combination of human resource management policies based on intrinsic (training, independency) and extrinsic (money, free time) motivation factors, as follows (Devesa et al., 1997; Menchero & Vale, 2002; Appelbaum et al., 2003):

- Suggesting a reduction of working hours with a proportional salary reduction.
- Establishing a permanent cost reduction policy.
- Implementing training and recycling programmes.
- Proposing a programme that includes early retirement and voluntary redundancy schemes with incentives, along with recruitment freeze.
- Encouraging workers to become self-employed so that the company can subcontract their services.
- Helping them to create a new business (spin-off).
- Reducing the fixed part of the salary while the variable part is increased.
- Favouring outplacement.

Finally, the introduction of information and communication technologies that are so pervasive in the new economy may help easing the unpalatable consequences of following a staff restructuring strategy by allowing, for instance, for teleworking positions or establishing virtual relationships with collaborators, suppliers, distributors and customers. Paradoxically, the solution to achieve savings in new technologies is more investment; since they make it possible to be more competitive while saving time, administrative costs, stock management costs, etc. Therefore, technology-based rationalization processes are due to be more efficient in the long term, and less damaging for the business.

CONCLUSION AND FINAL THOUGHTS

Summing up all that has been mentioned so far, it can be clearly observed that restructuring processes are very complicated, difficult and hard to rectify if some mistake is made.

In the absence of a clear strategic vision, as restructuring decisions have traditionally been based on legal and socially admitted aspects (age, length of service, type of contract, etc.) and on the staff reduction cost, the answers to these problems are generally short-term solutions where salary cost reductions are prioritised. For this reason, it is essential to plan things properly in order to achieve the expected outcome: that the firm's situation would improve, leaving behind the crisis, and becoming more efficient and rational in the deployment of its resources (human or not).

Finally, it would be interesting to insist on the convenience of researching alternatives to cutting jobs as a way to alleviate the problems provoked by an unavoidable restructuring strategy.

REFERENCES

Appelbaum, S. H., Delage, C., Labib, N., & Gouit, G. (1997). The survivor syndrome: aftermath of downsizing. *Career Development International*, *2*(6), 278–286. doi:10.1108/13620439710178639

Appelbaum, S. H., & Donia, M. (2000). The realistic downsizing preview: A management intervention in the prevention of survivor syndrome (part I). *Career Development International*, *5*(7), 333–350. doi:10.1108/EUM0000000005384

Appelbaum, S. H., Patton, E., & Shapiro, B. (2003). The early retirement incentive program: A downsizing strategy. *Journal of European Industrial Training*, *27*(1), 22–35. doi:10.1108/03090590310456500

Appelbaum, S. H., Simpson, R., & Shapiro, B. (1987). Downsizing: The ultimate human resource strategy. *Ivey Business Journal*, *52*(2), 52–60.

Auer, P., & Cazes, S. (2000). Fortaleza de la relación de trabajo duradera en los países industrializados. *Revista Internacional del Trabajo*, *119*(4), 419–450.

Band, D. C., & Tustin, M. (1995). Strategic downsizing. *Management Decision*, *33*(8), 36–45. doi:10.1108/00251749510093923

Barroso, C., 7 Casillas, J. C. (1998). Análisis del efecto de la reestructuración en las empresas españolas durante el quinquenio 90-94. *Revista Europea de Dirección y Economía de la Empresa*, *7*(3), 53–64.

Bashford, S. (2004). The survivor syndrome. *Human Resources (E-Journal of London)*, June, 43-44.

Bowman, E. H., & Singh, H. (1993). Corporate restructuring: Reconfiguring the firm. *Strategic Management Journal, 14*(Summer Special Edition), 5-14.

Budros, A. (1999). A conceptual framework for analyzing why organizations downsize. *Organization Science*, *10*(1), 69–82. doi:10.1287/orsc.10.1.69

Bueno Campos, E. (1998). El capital intangible como clave estratégica en la competencia actual. *Boletín de Estudios Económicos*, *53*(164), 207–229.

Cameron, K. S. (1994). Strategies for successful organizacional downsizing. *Human Resource Management*, *33*(2), 189–211. doi:10.1002/hrm.3930330204

Carvin, B. N. (2009). *Effective exit-interview strategies*. Human Resource Executive Online, September. Retrieved from http://www.hreonline.com/HRE/story.jsp?storyId=250192808

De Miguel Fernández, E. (2001). *Introducción a la gestión (management)*. Valencia: Universidad Politécnica de Valencia.

Devesa, J. E., Devesa, M., & Vidal, C. (1997). Viabilidad de los planes de prejubilación. *Dirección y Organización*, *18*, 5–13.

DiBella, A. J., Nevis, E. C., & Gould, J. M. (1996). Understanding organizacional learning capability. *Journal of Management Studies*, *33*(3), 361–379. doi:10.1111/j.1467-6486.1996.tb00806.x

Drew, S. (1994). Downsizing to improve strategic position. *Management Decision*, *32*(1), 4–11. doi:10.1108/00251749410050624

Fernández, E., Montes, J., & Vázquez, C. (1998). Los recursos intangibles como factores de competitividad de la empresa. *Dirección y Organización*, *20*, 83–98.

Fernández Sánchez, J. A. (2011a). *Uso de estrategias y sistemas de información en recursos humanos*. EAE.

Fernández Sánchez, J. A. (2011b). *Procesos de contratación y reestructuración de plantillas en España*. EAE.

Fiol, C. M., & Lyles, M. A. (1985). Organizational learning. *Academy of Management Review*, *10*(4), 803–813.

Freeman, S. J. (1994). Organizational downsizing as convergence or reorientation: Implications for human resource management. *Human Resource Management*, *33*(2), 213–238. doi:10.1002/hrm.3930330205

Gómez-Mejía, L. R., Balkin, D. B., & Cardy, R. L. (1998). *Gestión de recursos humanos*. Madrid, Spain: Prentice Hall.

Griggs, H. E., & Hyland, P. (2003). Strategic downsizing and learning organisations. *Journal of European Industrial Training*, *27*(2-4), 177–187. doi:10.1108/03090590310468994

Gunnigle, P. (2012). (in press). Weathering the storm? Multi-national companies (MNCs), human resources management (HRM) and industrial relations through the global financial crisis. *International Journal of Manpower*.

Kim, W. (2003). Economic crisis, downsizing and "layoff survivor's syndrome". *Journal of Contemporary Asia*, *33*(4), 449–464. doi:10.1080/00472330380000281

Marr, R., & García, S. (1997). *La dirección corporativa de los recursos humanos*. Madrid, Spain: Díaz de Santos.

McKinley, W., Sánchez, C. M., & Schick, A. G. (1995). Organizational downsizing: Constraining, cloning, learning. *The Academy of Management Executive*, *9*(3), 32–44. doi:10.5465/AME.1995.9509210276

Menchero, A., & Vale, V. (2002). Técnicas para cerrar el grifo. *Revista Emprendedores*, *55*, 60–66.

Pichault, F., Warnotte, G., & Wilkin, L. (Dirs.). (1998). *La fonction ressources humaines face aux restructurations. Trois cas de downsizing*. Paris, France: L'Harmattan.

Pin Arboledas, J. R. (2002). *Libro Blanco. Las mejores prácticas (best practices) en los procesos de reestructuración de plantillas*. Centre Telemàtic Editorial.

Pulignano, V. (2011). Bringing labour markets "back in": Restructuring international businesses in Europe. *Economic and Industrial Democracy*. Retrieved June 4, 2011, from http://eid.sagepub.com

Requejo, A. (1996). Reducción de plantilla y problemas de viabilidad financiera: El papel de la estructura de capital. *Investigacion Economica, 20*(1), 43–70.

Sáenz Blanco, M. T. (2001). Schneider Electric: El valor exponencial de las tecnologías aplicadas a la gestión de RR.HH. *Capital Humano, 148*, 28–36.

Simon, H. A. (1991). Bounded rationality and organizational learning. *Organization Science, 2*(1), 125–134. doi:10.1287/orsc.2.1.125

Suárez González, I. (1999). Decisiones de reducción de plantilla en la gran empresa española. *Investigaciones Europeas de Dirección y Economía de la Empresa, 5*(1), 79–100.

Tomasko, R. M. (1996). *Repensar la empresa. La arquitectura del cambio*. Barcelona, Spain: Paidós Ibérica.

Walsh, J. P., & Ungson, G. R. (1991). Organizational memory. *Academy of Management Review, 16*(1), 57–91.

Whetten, D. A., Keiser, J. D., & Urban, T. (1995). Implications of organizational downsizing for the human resource management function. In Ferris, G. R., Rosen, S. D., & Barnum, D. T. (Eds.), *Handbook of human resource management* (pp. 282–296). Cambridge, MA: Blackwell Publishers.

Wilkowska, I. (4-6 June 2003). *Corporate restructuring and human resource management: A case study of the banking sector in Britain*. Paper presented at the 7th Conference on International Human Resource Management, "Exploring the Mosaic: Developing the Discipline." University of Limerick, Ireland.

Winterton, J. (19, 20 June, 2003), *Un modèle heuristique de restructuration utilisé pour prévoir les pertes d'emploi et les futurs besoins de compétences*. Paper presented at the congress Responsible restructuring in Europe, Dublin.

Zehir, C., & Keskin, H. (2003). A field research on the effects of MIS on organizational restructuring. *Journal of African Business, 3*(1/2), 270–287.

KEY TERMS AND DEFINITIONS

Cut Strategy: Restructuring based on reducing a part of the staff. This is a deliberate or voluntary reduction in the number of employees which goes beyond a mere decrease in the employment volume due to natural causes.

Downsizing: To reduce in number or size; to dismiss or lay off from work; to make in a smaller size.

Redesign Strategy: Restructuring based on the business functional and organizational reconfiguration or, to put in another way, on the relocation of individuals in different posts seeking an optimum exploitation of their skills. This type of strategy allows firms to achieve new forms of organization thanks to which they will stand a better chance of meeting market demands successfully.

Staff Restructuring Strategy: A long-term measure which has as its aim to reshape the organization and its parameters. Its characteristics would include: change of responsibilities, restructuring the workforce, global involvement, continuous improvement and questioning or rethinking the organization's current system of values and regulations.

ENDNOTES

1. "Responsible restructuring processes in Europe" Dublin, June 19-20th, 2003.
2. "El nuevo acuerdo de SEAT: competitividad y flexibilidad". Expansión 12-12-2001.

Section 3
Competency Management for Sustained Competitive Advantage

Chapter 11
The Competency-Based Human Resource Management Model

Jorge Valdés Conca
University of Alicante, Spain

Susana de Juana-Espinosa
University of Alicante, Spain

ABSTRACT

It seems appropriate and coherent to start this study with a detailed description of the basic analysis unit in the human resource management model that constitutes the study object of this chapter: competency. With this aim in mind, the first section in this chapter is dedicated to three tasks the authors consider basic and introductory: (a) defining the concept of competency through the different approaches made by some of the most outstanding authors in this field; (b) dissecting that concept in its various elements or components so that it can be better interpreted; and finally (c) presenting a number of classifications thanks to which the treatment of certain competencies can be prioritised.

1. DEFINITION OF COMPETENCY, ELEMENTS, AND TYPOLOGY

Defining the concept of competency is a crucial aspect when it comes to implement a competency-based human resource management model; not only because it represents the analysis unit for the set of integrated policies within that functional area but also because a wrong interpretation or definition of the term might lead to fundamental errors with important consequences for the whole organisation. For example, an incorrect definition of the competencies that can describe a job hinders the promotion possibilities of the individual who holds that job (Kydd & Oppenheim, 1990). The definition of competency traditionally given by the literature presents it as an action tool but, in fact, the integration and development of the model into the firm can only be achieved if competency is understood as a decision tool, and not merely as an action tool (Rausch, Sherman & Washbush, 2001).

DOI: 10.4018/978-1-61350-207-5.ch011

Since this is a model that encourages strategic cohesion with the rest of the organisation, the negative consequences derived from a bad definition of the model would not only extend to the area of human resources –where all staff-related policies would be affected because the model is comprehensive– but also to the company as a whole.

Nevertheless, no unanimous agreement exists in this respect despite the importance given by all the literature to the definition of the competency concept.

Searching for the origins of this model and, therefore, for those on which the definition of the term 'competency' is based, allows us to state that, although McClelland pointed out already in 1961 that it was necessary to identify the attitudes of individuals that lead to business success, the first definition of competencies as achievement components associated with "groups of human behaviours" appeared in 1973, when the same author published his pioneering work "Testing competencies rather than intelligence" (McClelland, 1973). Of course, the aforementioned achievement components are not present in all individuals but only in those who are able to reach a successful performance in their professional activity. But, what are those components? It is important to highlight an idea in this respect: competency is not the sum of a series of factors but their synergic combination. In other words, these components are related to each person's reasons, features, values, skills and knowledge, but only their adequate combination can lead to the development of a competency at a certain level. A competency is relatively stable over time; that is, competencies do not usually evolve in a natural way –specific development programmes for those competencies are often needed instead. From this moment, these ideas represent the starting point for the preparation of 'ideal' competency-based professional profiles for the management of recruitment and training policies.

The methodology for the preparation of these 'ideal' professional profiles consisted in drawing a comparison between two types of individuals: those who reached a normal performance in their professional activity and those who had an excellent performance. As a result, competencies start to be defined as a set of knowledge, abilities, skills and other characteristics which distinguish medium performance from high performance.

The reflection made by Le Boterf (2000) on the conceptualisation of the term 'competency' deserves a special mention. In his opinion, those definitions which refer to competency as a sum of knowledge items –'knowing how to do' or 'knowing how to be'– and their application are weak. Such definitions cannot be the basis for the establishment of a good competency-based management system, as the fragmentation of competency in its definition inevitably destroys it. There is more into a competency than a mere addition of individual attributes. In short, competency is a multidimensional concept; it is not an isolated attitude, knowledge item or skill, but the integration of them all in the context of a specific profession and its sphere of action. Each observable behaviour taking place in the performance of a specific competency results from the combination of knowledge, skills and attitudes associated with it.

Along these same lines, Córdoba (2004) claims that the concept of 'competency' has suffered an evolution which makes it become an effective performance catalyst, as it permits to adapt the knowledge, professional experience and attitudes of individuals towards the company's goals, and not as a mere aggregation of elements, but as an integrated, intertwined and synergic management of those elements at the implementation of the different human resource practices.

Le Boterf additionally draws a clear distinction between the concepts of 'resource' and 'competency.' Whereas 'resources' would be the attributes or elements of competencies (knowledge, talents, skills, behaviours, etc.) and even the sum of them all, the 'competency' would be something more complex, since the different individuals build their competencies through the combination of the different attributes they own –which are not the same in all individuals and also evolve over time. In short–quoting this same author (Le Boterf, 2000)– "competency is a disposition and not an elementary gesture or an operation. A competency resides in the chaining or linking, combination and implementation of a sequence."

However, a conflict of interests arises at this stage regarding the definition of competency: unless a breakdown of the term is carried out, it will turn out to be too ambiguous and difficult to manage and, consequently, inoperative; and if, however, the fragmentation is excessive, we run the risk of applying a taylorist approach to competency. Therefore, finding that balance in the competency breakdown makes it absolutely necessary to deepen more into each one of the elements that it comprises. With that aim in mind, and taking as a reference the definitions previously formulated by different experts in this field, our conception of competency will from now on correspond to a set of patterns formed by characteristics which are underlying to individuals (knowledge, skills, disposition, conducts, etc.) and permit them to achieve an effective or higher performance in an activity or a job.

Once the term 'competency' has been defined, our next task will consist in dissecting this concept in order to identify the different elements it contains, so that they can allow us to clarify and specify the approach adopted for its definition.

In this respect, there are various typologies of elements and attributes which shape competency according to different authors. For instance, Pereda & Berrocal (2001) refer to knowing, knowing how to do, knowing how to be, wanting to do, and being able to do; for Arce, (2000) the elements or attributes in question are knowledge, attitudes and skills; in turn, Losey (1999) mentions intelligence, education, experience, ethics and interest; finally, Dirube (2004) speaks about a compendium of knowledge and skills, values and habits, and reasons. Nevertheless, most of them can be summarised in a single category on which nearly all the authors agree: knowing how to act, wanting to act and being able to act.

In principle, knowing how to act would include the knowledge (knowing) and the individual's experience (knowing how to do) and would consequently be a resource incorporated into the person. Knowing is a highly volatile element from the competency-based management perspective (Bacq, 1996), as the instruments used to validate this element (diplomas, degrees, certificates, etc.) reflect a variety of knowledge items and capabilities acquired at a particular moment which can become obsolete –and even turn out to be insufficient– with the passing of time if they are not recycled. On the other hand, having all the knowledge required to act competently becomes useless if the individual is unable to use it properly. For this reason, the organisation's main role regarding this first element would definitely consist in ensuring its development through different means, as is going to be explained later on in this chapter.

In other words, this knowledge and experience must be complemented by a second component of competency: wanting to act. Our focus at this point is going to be on a number of aspects which are inherent to the individual's personality and, therefore, much more subjective than the previous ones; in short, we are speaking about what is known by some scholars as 'knowing how to be.' However, this component appears as the most important one within the 'knowing how to act' context from the competency-based point of view, since it becomes the differentiating element between individuals who perform the same roles with the same knowledge and the same experience.

We can find the individual's 'attitudes' as an emotional element, along with 'behaviours' as a conduct-related element, within 'knowing how to be'. Attitudes are directly related to knowledge. The more we know about a competency, the more positive the attitude shown towards it. In turn, behaviours appear as a result of the combination between knowledge and attitudes; that is, we will adopt a behaviour that expresses greater acceptance or rejection depending on the degree of knowledge that we have about a certain issue and the attitude that this knowledge generates inside us (Arce, 2000). In other words, our way of thinking as well as our technical knowledge about the work to be performed will strongly influence our behaviour.

Therefore, as shown in Figure 1, there are specific circumstances where the configuration of individuals' behaviour will depend on their motivation and their conduct, the latter being in turn dependent on their attitude –as a personality feature– before certain situations as well as on their more or less thorough knowledge of those situations.

Other authors, such as Sagi-Vela (2004) add a fourth element –or rather, the combination of the preceding elements: the 'why' (Figure 2)– to the triangular competency system. This 'why' represents the purpose which gives sense to competent action on the part of the individual, and is closely linked to 'wanting,' since having a deep, detailed knowledge of the reason which leads us to perform a certain role and knowing what we expect to obtain with it will influence the individual attitude, which will eventually materialise in a specific behaviour. Therefore, even though the 'why' may indeed be linked to all three elements of competency, in our opinion, that link is closer with one of them: 'wanting.' On the other hand, this same author highlights that 'being able to act' does not constitute a competential element in itself, as it would be formed by the resources and instruments that the organization places at the individuals' disposal so that they can exercise their competencies and would thus fall outside the worker's voluntariness and personal action capacity. It is consequently an organisational element, and not a competential one.

After having defined the different competencies and analysed each one of their components, we are in a good position to classify them according to different criteria with the aim of relativising their importance and contextualising their application within the organisation. Our classification will start with one of the most frequently used typologies in the field of competency-based management, the one which draws a distinction between generic competencies and specific ones.

Generic Competencies vs. Specific Competencies

In this case, the classification criterion lies in the degree of relationship that the competency may have between different firms, sectors or even jobs. Generic competencies are those common to several organisational units. They coincide or are close to the concept of 'talents' used for selection and recruitment policies in the context

Figure 1. Configuration of behaviour

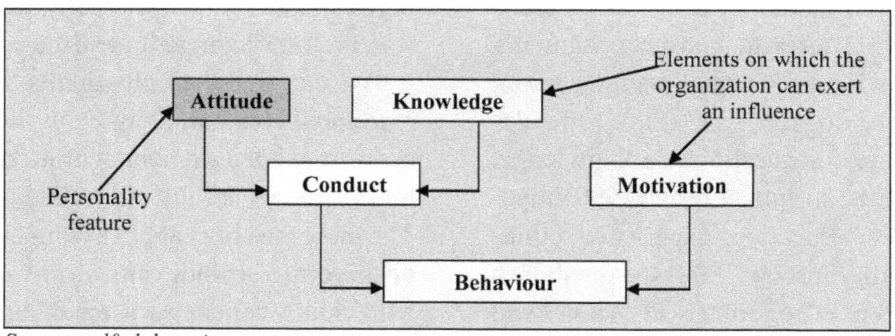

Source: self-elaboration

Figure 2. Elements of competency

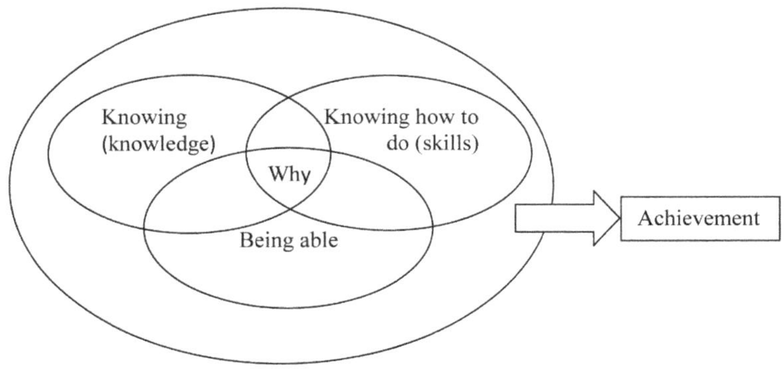

Source: Sagi-Vela (2004).

of traditional human resource management (Levy-Leboyer, 1997).

On the other hand, specific or 'technical' competencies comprise the professional knowledge and talents needed to carry out the technical contributions defined for a particular profession; they are in keeping with the correct performance of functions within an organisational unit.

Generic Competencies vs. Vocational Competencies

Other authors contrast the generic competencies described above with vocational competencies, the latter being derived from a specific training acquired by individuals during their educational stage. In this respect, Heijke, Meng & Ris (2003) establish a model where an organisation's human capital availability is determined according to the vocational and generic competencies of its individuals.

They infer from this model that the exercise of generic competencies is not confined to a particular context or work post; in fact it allows individuals –through their learning capacity– to develop the competencies required within a wide range of jobs or functions. On the other hand, vocational competencies favour the connection of individuals with the work posts that belong to the area of knowledge where they have been trained and on which they will be able to develop a competitive advantage. Generic competencies consequently permit the adjustment of vocational competencies to the specific job requirements. We are therefore referring to two complementary –and not substitutive– competencies (Heijke *et al.*, 2003).

Threshold Competencies vs. Differentiating Competencies

A third classification allows us to draw a distinction between threshold and differentiating competencies, according to the performance level stemming from the use of those competencies. Thus, threshold competencies permit a normal or adequate activity performance. They are typical of traditional staff selection processes and seek an adequate performance at the work post –not an outstanding performance.

Differentiating competencies are those personal characteristics which distinguish a normal performance from an outstanding or successful one. This type of competencies justify why one person with the same training and in the same circumstances as another is able to perform the tasks inherent to a job more satisfactorily.

Primary Competencies vs. Secondary Competencies

Following this typology, primary competencies would be formed by basic personality features of the individual, attitudes and talents which suffice to shape a competency for the efficient performance of the functions associated with a work post. Instead, secondary competencies would arise from the combination of several primary competencies; in other words, they would be constructs of individual basic competencies which enable a person to exercise a secondary competency –if they are developed at a certain level. For instance, whereas self-confidence and emotional stability are primary competencies, they could simultaneously form part of a secondary competency such as leadership capacity (Sánchez de Dios & Valldeperes, 1998).

Simple Competencies vs. Combinatory Competencies

We can finally distinguish between simple and combinatory competencies depending on the degree of development and complexity reached by the competency within the organisation. In this sense, simple competencies would be the individual ones, no matter if they are generic or specific, threshold or differentiating; as for combinatory competencies, they arise from the interaction of different simple competencies that result in new competencies which are much more complex and, consequently, much harder to copy or imitate.

2. INSTRUMENTS FOR COMPETENCY-BASED HUMAN RESOURCE MANAGEMENT

The second section in this chapter has as its main aim to study three of the main instruments –recognised by the authors who are well versed in this field– for human resource management and its different practices related to the concept of competency: the dictionary, the map and the balance of competencies.

The competency dictionary is a human resource management tool which contains both the definition that the company can conceive for all the competencies that it regards as being relevant and their progression at its different levels.

Competency levels represent a scale-based gradation in the individual's execution of a competency according to the achievement of its elements (knowledge, skills, attitudes, talents, behaviours, etc.). As the *Confederación Española de Organizaciones Empesariales* [Spanish Confederation of Business Organisations] (C.E.O.E., 2002) has explained, competency dictionaries will include the behaviours associated with those competencies which can be directly observed. Therefore, the different levels for each competency, usually between 4 and 7, fulfil a twofold purpose. On the one hand, they describe the levels of achievement reached by the individual in the competency under assessment and, on the other hand, they represent the demand profile for a specific work post in that competency.

In any case, the aim sought is the adjustment of levels between the candidate's evaluation and the demand profile for the post. The comparative assessment of these two elements (the candidate's profile and the demand profile for the post) will provide the basis for the subsequent development of different human resource systems, such as selection, promotion, training, career planning, pay, etc.; or to put it in another way, of the three competency-based management dimensions that will be presented in chapter 3, namely: acquisition, stimulation and development of competencies.

The competency dictionary undoubtedly appears as one of the most important instruments for model management. However, there are others which, despite not having that basic and elementary character, actually turn out to be very helpful when managing the model. In this respect, the importance assigned to the competency map derives from its geographical representativeness.

Competency maps are referential elements associated with specific work areas, posts, roles or occupations which become a highly useful reference framework and instrument both for the organisation and for the worker (C.E.O.E., 2002). It is advisable to show a small example that can help to clarify the process through which these maps are created.

The example shown in Figure 3 allows us to see the map of one competency, group leadership, which could belong to the human resource management role. In this map, the organisation includes the definition extracted from the dictionary, both for the competency and for the different levels used to grade it. Thus, the different jobs shaping the role –5 (A, B, C, D and E) in this case– are placed in a graph so that they can grasp the different competency levels required by the demand profile for the post. Finally, the relationships between the different jobs are established using arrows, thus clearly identifying the possible ways of promotion available to an individual within one particular role. This map–prepared for different posts within the same role under a competency– could also be adapted to the different roles that share a specific competency within the organisation.

The present section is going to finish with one of the instruments that have received the least attention in the literature on competency-based management: the competency balance. Despite the limited importance assigned to this instrument, its validity regarding previsional management and competency development matters is hugely relevant.

The purpose of this management instrument is to favour the identification and assessment of the competencies developed by individuals in the different jobs and roles that they have performed throughout their professional career.

The orientation towards individuals is one of the essential competencies for any human resource manager. Competency balances appear as one of the main orientative instruments for that purpose, since they permit not only to orient the professional career of individuals but also to carry out a periodical follow-up of that career (Joras, 1996). The balance consequently becomes a starting point (initial balance) for the design of a professional career, and a compass that permits to locate the individual's professional 'north' at all times, so that policies about training, promotion, pay, etc. can be oriented in the right direction (periodical balances) (Le Boterf, 2000).

3. APPROACHES TO COMPETENCY-BASED HUMAN RESOURCE MANAGEMENT

The specialised literature offers us two competency-based management approaches (Herranz & De la Vega, 1999): the Anglo-Saxon one and the Francophone one, the characteristics of both being presented in Figure 4. The most outstanding features in this dual approach are the focus of the model on generic or specific competencies depending on whether we adopt the Anglo-Saxon or the Francophone approach, respectively.

The first one of them is based on the assumption that, because generic competencies are the most widely used among companies, they will be more present in the strategic profiles of organisations than in the tactical or operational ones. This is so because the specificity of profiles increases as they go down the hierarchical ladder. Therefore, these are the competencies on which the attention of the model must be focused for the purpose of achieving the organisational flexibility that has

Figure 3. An example of a competency map for a specific role

Competency: Group leadership

Competency definition: Guiding and leading a group, as well as establishing and maintaining the group spirit needed to achieve the set aims.

Role: Human resource management.

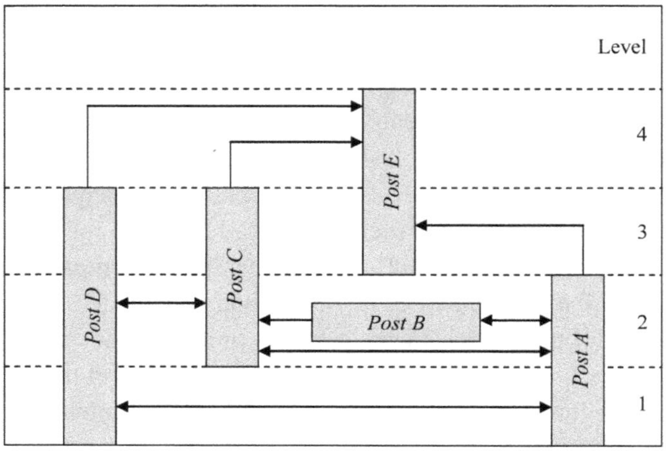

	Level definition
Level 1	Establishing the group's objectives in a clear and balanced way, especially when those objectives affect their organisation
Level 2	Encouraging everyone's participation in reflection and decision-making processes
Level 3	Paying attention to enterprise external factors with the aim of detecting opportunities
Level 4	Exerting an unquestionable influence on the whole organisation, being recognised as a qualified opinion for decision-making in any issues related to their area or even in issues from other areas. It is perceived as a 'guide' and inspiration for strategic decisions in the organisation as a whole.

Source: adaptation based on De Ansonera (1996).

been set as an objective. This Anglo-Saxon approach is also known as 'conductist approach,' as it starts from the idea that there will be no competency unless we can prove the existence of an effective and/or superior performance. The most representative supporters of this approach or school are McClelland (1973), Boyatzis (1982) and Spencer & Spencer (1993).

In turn, the Francophone approach–also known as 'contingent approach'– defends the idea that specific competencies are the ones which must have a strategic character and weight within the model, since these are the competencies which can eventually become a differentiating element and a possible source of advantage precisely because of this specificity which is likely to make them exclusive to the organisation. Therefore, according to this approach, firms have to design their competencies *ad hoc*, which would leave generic competencies –worthless because they have not been exclusively designed in accordance with the characteristics of the organisation and its environment– out of the picture. This approach –also referred to as 'constructivist approach' due to the belief that competencies can ultimately be built giving individuals an opportunity for development

Figure 4. Anglo-Saxon approach vs. Francophone approach

Anglo-Saxon approach	*Francophone approach*
The main objective is to increase individual or group performance.	Oriented towards the person. It acts as an audit element for individual capacity and the efforts made by the enterprise to keep its workforce in optimum employability condition.
It is based on qualitative measures.	Competencies are an inseparable mixture of work-related knowledge and experiences.
It is focused on generic and universal competencies, presupposing that any organisation in any country needs very similar competencies. Competencies represent the links which unite or connect individual conducts and organisational strategy.	Competencies must be developed *ad hoc* for each organisation; it is impossible to identify generic competencies.
The fundamental competency is flexibility.	Use of mega-competencies, such as learning capacity and the different learning strategies.
Focused on job contents, as well as on their relation to the global strategy.	It focuses more on learning processes than on people.

Source: Canós, Valdés & Zaragoza (2003).

even if they have not shown a successful performance yet– has its most representative supporter in Levy-Leboyer.

In this same sense, Mansfield (1996) draws a distinction between the two most common ways to develop and use competency-based management models: individualised for each job or common to all of them. However, in this author's opinion, neither of these approaches permits to differentiate the requirements for several posts or fit in the competencies for an individual profile with a wide range of jobs or tasks. In this way, a differentiation is made between the model based on specific competencies and that based on generic competencies.

- **The model based on specific competencies.** It was the first competency-based management model to be created and it is still quite widely used for its advantages, among which stand out the following:
    - It makes the key competencies in a post easier to identify and describe.
    - It informs workers about what they have to do to achieve better results.
    - The participation both of workers and of their superiors in the model favours their motivation.

Nevertheless, this model presents some inconveniences as well:

- Its cost –financially and also in terms of time– and the effort required to develop the model make it applicable to a small proportion of jobs in the organisation.
- Due to the constant changes suffered by work posts at present, the applicability of the model for a specific job does not usually exceed a two-year lifecycle.
- Sometimes it shows inconsistencies with the other organisational programmes implemented by the firm due to the emphasis placed by the model on the specific characteristics of each post.
- Also due to this individual character of the model, it is difficult to compare the compe-

tencies required for a post and those needed for another.

- ◦ **The model based on generic competencies.** It has as its main aim to identify and define a set of competencies for a large number of jobs within the organization. Its advantages are listed below:

- They involve a high number of employees.
- There is consistency in the description of the effective conducts that describe the competency.
- The model is in keeping with the values, the mission and other key concepts for the organisation.
- The competencies of individuals managed with this model can be compared.
- Its cost is modest considering its scope and impact.
- Because it is not based on individual posts, the model does not need to be updated every time a job is redefined.

But it has some drawbacks too:

- It does not clearly define the needs for a specific post.
- Workers have to see competencies as a presentation of values rather than as the skills needed to obtain results.
- It is not useful to guide the selection process for specific posts.
- It ignores the skills and knowledge which are essential for the fit between individuals and job tasks.

Therefore, seeing the advantages and drawbacks of a model based on specific competencies –according to the Francophone or contingent approach– and those of the Anglo-Saxon-inspired model focused on generic competencies, the authors of this chapter suggest a model which can cover a multiplicity of posts within the organisation. This 'model for a multiplicity of posts' permits the construction of different models from a common set of competencies. These sets, which are going to include both technical or specific competencies and generic ones, will make it easier to compare jobs for assessment purposes and also to achieve a better fit between individuals and competential profiles as well as more efficient training and development processes.

4. DEVELOPMENT OF THE COMPETENCY-BASED MANAGEMENT SYSTEM

4.1. Objectives and Uses of a Competency-Based Management System

Competency-based management permits to work from people, with people and towards the strategies and objectives determined by the organisation. One of the most significant advantages and contributions entailed by competency-based management is actually the achievement of a greater degree of proximity and convergence between the organisation's interests and those of its workers.

For Le Boterf (2000), this concurrence of interests constitutes a necessity rather than an organisational advantage. It is necessary to have more trust in workers –and less trust in procedures– because outcome achievement as well as the building of competitive advantages lies in the former, whereas the latter are mere instruments placed at the service of the organisation. Along these same lines, Bayón (1994) mentions three premises about the importance that human resource management has for the organisation: human capital is an firm's greatest asset, the decisive factor is the human team who applies the processes and not the processes themselves, the growing superposition of work as opposed to the capital. This same author implicitly refers to the main objective in the competency-based management model, the convergence and integration of interests between workers and the organisation, when he defines the human resource management function as "the one which has as its aim to adapt the company's interests and goals and those of the worker so that the unity and integration of attitudes, along with the work-related, professional, human and economic development can take place simultaneously (Bayón, 1994).

In short, if the model permits the convergence of interests between the staff and the company, the implementation of that model will supposedly imply a number of advantages, uses and requirements for both parts.

We can consequently distinguish three sets of benefits (Figure 5) brought to the firm by the adoption of the competency-based management: uses of a strategic nature, those related to human resource management improvement, and the ones associated with savings and cost reductions in processes.

As for strategic uses, we can firstly mention a greater strategic orientation towards the model, since it permits to identify and cover the key functions in an organisation successfully. Another of the strategic benefits brought by competency-based management is a greater capacity to anticipate and adapt to changes in the environment by means of a more proactive approach; this is so because the model favours this type of behaviours through continuous training, greater professionalism among the staff, competency development, etc. Thirdly, there is the necessary flexibility which has to accompany competency-based management in order to permit human resource mobility and the adaptation to the abovementioned changes in the environment. The strategic character adopted by the human resource function through the identification of behaviours which lead to excellent performances in functions is clearly one of the most important benefits that the organization can obtain from the model at a strategic level. Finally, a cultural change –along with a change in leadership style– becomes necessary for the implementation of the model. Indeed, the organisational culture must be more open and participative in order to permit the management and development of competencies as a strategic unit for the analysis of human resources. Similarly, the leadership within the organisation has to assume competencies in the areas of evaluation, motivation and development, but above all, in those of communication and dialogue with collaborators.

Regarding the uses provided by the model to the human resource management function, these are obvious because the model was born precisely to improve the management of this functional area and to give it a more strategic character. Although

Figure 5. Strategic, management-related, and process-related uses of the competency-based management model

Strategy	Management	Processes
Greater strategic orientation	Greater training, polyvalence and training	Time saving
A more proactive approach	A flatter organisational structure	Reduction of organisational costs
Higher flexibility	Achievement returns on labour costs	Reduction of 'non-quality' costs
Greater HR added value	Greater internal promotion	Savings in subcontracting
Cultural change and change in leadership style	Management and administration simplification	

Source: our adaptation from Sagi-Vela (2004).

these benefits are going to be repeatedly mentioned throughout this theoretical presentation, we will briefly comment on them here. Firstly, an improvement is achieved in training and motivation practices. Human resources can only become a strategic element in the organisation through an increase in their added value if they receive a kind of training that makes it possible to reach that performance level along with a subsequent development: a polyvalent training –thanks to the flexibility described above– which is also conducive to the stimulating professional acknowledgement of a successful performance. Internal promotion is another of the human resource policies bound to experience an improvement after the adoption of the model. Thanks to workers' polyvalence, together with their higher professional qualification, they will be ready to occupy posts of various kinds within the organisation with a broad business vision. On the other hand, exhaustive and rigid control mechanisms become less necessary thanks to the consideration of technical or specific competencies in an adequate combination with generic ones, thus favouring the increasingly common tendency to the reduction of hierarchical levels in the organisational structure of firms. Yet another improvement in human resource management would stem from making the most of labour costs, since the return on investment in this area increases, not because costs decrease but because there is an increase in productivity; a productivity which has to expressed in terms of added value on the part of the human factor –and not of productive volume– to the organisation's goals (Dalmau & Baixauli, 1989). The model ultimately permits a liberalisation of tasks in the human resource department, simplifying its management and administration, which in turn allows this area to focus its efforts on other functions with higher added value.

Finally, the implementation of the competency-based management model equally brings cost reductions and time savings in the corporation's activity processes. Thus, we can firstly observe a reduction in the process cycle due to the greater degree of professionalism and polyvalence among workers, which allows them to reduce wasted time periods, to simplify activities or to perform some of them simultaneously. Coordination costs will be reduced too, because the self-control that workers have in relation to a set of activities that they develop competently and with an overall vision of processes permits to simplify coordination mechanisms in the design of the organisation. For the same reason, the mistakes made due to the inability to reach a certain quality level in products and processes will be downplayed. Finally, it will be necessary to resort to the subcontracting of activities which are considered particularly relevant to the organisation. Workers will be sufficiently qualified to develop these processes efficiently, eliminating the lack of profitability that might have previously occurred if the organization had not decided to outsource those activities.

4.2. Design of Competency-Based Work Profiles and Model Implementation

In the first place, and as a step prior to the preparation of competential profiles for the different posts, it is important to carry out a good information and communication campaign addressed to employees (Pereda & Berrocal, 2001). This informative action pursues a double objective. On the one hand, achieving workers' involvement in the model design and implementation process, as they would represent the basic informative source both to identify and to define the organisation's generic competencies along with the specific competencies associated with their respective work posts; on the other hand, a good informative campaign undoubtedly helps to soften the environment before potential expressions of distrust among workers towards the new model, conceptualising it as a motivating instrument and not as a tool used to monitor their performance at the work post. A situation of rejection and lack of credibility may arise if the model is not accepted by all organisation members, both in relation to the model itself and regarding the work team that has implemented it, a situation that is very hard to correct in the short run –or even in the medium run.

After obtaining the acceptance and collaboration of workers, the first step in the design of the competency-based management model consists in identifying the competencies, describing them, cataloguing them and elaborating the reference profiles for the firm. It will be possible to establish these profiles for each professional role or even for each specific job (Gick & Pallares, 1997). These profiles comprise the set of competencies which the individuals occupying a work post have to own in order to perform their professional activity effectively, safely and satisfactorily (Pereda & Berrocal, 2001). Later on, workers will be evaluated with respect to the above-mentioned referential profiles, and the different human resource policies will be structured around the evaluation results, using the different instruments articulated by the company for these purposes, such as the competency dictionaries, balances and maps that we have referred to above.

As for the methodology used to define the competencies that will form part of each profile, two methods stand out as the most often used: top-to-bottom and bottom-to-top. In the first case, the decision-makers will choose the specific characteristics they want for each post; in the second, they will observe what the best people at each post are capable of doing and that information will serve as a reference to identify and define the competencies. In any case, working with a competency dictionary will make the task easier and will also permit to use more tools. Along these same lines, the III International Conference on Competencies, held in London at the end of 1998, highlighted the different methodologies for the identification and description of competencies and profiles depending on the management style applied by each firm. On certain occasions, competencies are actually defined by the top management; on others, they start taking as a reference what excellent employees do and associate their competencies with those corresponding to that post and that entity (control groups). In the first case, the imposition of competencies by the management over the rest of the organisation runs the risk of making them fall into clichés that will make their implementation more difficult. In the second case, the induced competencies may lack a future vision and fail to consider the foreseeable changes. Therefore, an intermediate methodology which takes both extremes into account and makes the most of the advantages brought by both is most likely to achieve the best person-post fit (Dirube, 1999).

A variety of techniques can be used to identify the competencies that the organisation really needs. Thus, we can mention the interview about critical incidents, panels of experts, labour climate studies, analyses of areas and functions, the Delphi technique, benchmarking and databases elaborated by external consultants specialised in these matters, (C.E.O.E., 2002), surveys, direct observation, self-evaluation, 360-degree evaluation, etc. (Parry, 1996). Of all those, Gick & Pallares (1997) highlight work groups, panels of experts, interviews about critical incidents, surveys, dictionaries and databases elaborated by external consultants as the most frequently utilised ones. Of course, the gradual development and acceptance of the model has led to the proliferation of competency identification and management service offers by consultancy agencies (Bartlett & Goshal, 1997).

In effect, the identification of competencies can be carried out through the definition of the workers' most outstanding characteristics, starting at the executive levels of the organisation and going down through the hierarchical structure, or resorting to the use of work or control groups. These work groups make it possible to analyse the conducts shown by excellent individuals at a specific work post when compared to those developed by other individuals who only achieve a normal performance. In any case, if the corporation aspires to achieve trust in the model on the part of individuals, to trigger the appearance of informal communication about the project and reach model efficiency, the best way to define competencies is to permit the participation of workers.

Now we would like to focus on studying two of the most widely utilised methodologies, depending on the case: the panel of experts and the interview about critical incidents.

In principle, the panel of experts may represent an alternative to control groups. If a work post is new or does not have a number of workers high enough for its analysis to be representative, the delegation of responsibility for the identification of competencies to a panel of experts seems to be the most advisable option. Furthermore, this methodology also permits the active participation and involvement of managerial levels in the competency identification and profile creation process.

In any case, a number of considerations must be taken into account when it comes to adopt this instrument. Firstly, it is necessary to ensure realism and functionality in the result obtained. In other words, the parameters according to which the panel of experts is going to identify competency profiles must be observable, measurable; otherwise, they will be impossible to assess and will therefore lose their functionality as instruments for the articulation of a comprehensive human resource management system. Similarly, if the profiles are excessively idealistic, they will create a demand for unreal performances with too high –or even unreachable– competential levels. On the other hand, this lack of realism in relation to competency profiles may derive from the lack of consideration towards the organization's values, beliefs, mission, objectives, etc. This means that, if the panel of experts does not take into consideration the firm's strategic plan and organisational culture, the outcome will be a set of competency profiles which remain distant both from reality and from the company's goals. For example, we cannot identify teamwork as a key competency for the organisation if the latter basically structures its pay policy around individual incentives.

Nevertheless, these inconveniences can be mitigated if the organization has previously created a dictionary from the observation and identification of the behaviour shown by excellent workers, or if it has acquired an external dictionary, always making sure that it is suited to the firm's organisational characteristics. In this way, the task for the panel of experts will be confined to choosing from the conducts and behaviours collected in the dictionary those which define the functions and responsibilities inherent to a job, but without needing to define and catalogue them.

As a final comment about the panel of experts' composition, it is worth highlighting that this panel can include a manager from the area corresponding to the competency profile that is going to be articulated, one or two direct superiors of the individuals who cover those profiles, three or four workers who develop their professional activity at the post around which the profile is going to be built, an expert or technician in human resource matters, a consultant, internal customers who receive the outcome produced by the job for which the competency profile is being designed and a representative of the works committee. The contributions made by each one of these panel members are the ones reflected in Figure 6.

Another of the most widely utilised instruments, not only for the identification of competencies but also in selection processes is the interview about critical incidents. The "Behavioural Event Interviewing" (B.E.I.) also known as "Competency-based interview" or "Critical Incidents Interview" is an instrument created by David McClelland in 1978 for the purpose of detecting the behaviours, feelings and thoughts which are typical of individuals and characterise their conduct (Donovan-Wright, 2002), avoiding opinions and hypothetical cases in order to prevent subjectivity-related biases in the information obtained. After all, the aim is to find out what leads people to act and not what they think or say that is en-

Figure 6. Composition for a panel of experts

Member	Contribution
Area manager	An overall vision and connection with the organisational strategy and culture
Direct superior	Limiting the competencies defined to real behaviours
Worker	Completing the realistic vision of the post with the direct experience of daily work
Human resource expert	Vision of the work post from the formal organisational point of view
Consultant	Methodology for the definition of competencies
Internal customer	Establishing the requirements to be fulfilled with competency performance
Union representative	Making the subsequent negotiation of the model easier

Source: our adaptation from Sagi-Vela (2004).

couraging them to act. These incidents are identified and registered by means of interviews, from which a pattern will arise that collects the most outstanding aspects in the performance of a competency that permit to distinguish a successful result from a failure result (Horton, 2000). With this aim in mind, the candidate goes through an in-depth interview structured around the individual's experience and training as well as the responsibilities assumed at the different work posts. The interviewees will have to specify three or four importance situations of successful results in their job, so that the competential elements related to 'knowing how' and 'wanting' to act competently can become visible, as well as others where the results obtained were unfortunately not as expected.

The conducts gathered through this type of interviews can be generalised by means of descriptions which reflect the main patterns, thus establishing a dictionary which facilitates a subsequent tabulation. However, if we already have at our disposal a good description of the competencies previously identified in the organisation, it will only be necessary to recognise the behaviours in the dictionary definitions, as it happened with the panel of experts. Moreover, as it allows the interviewee to make an open narration, the critical incidents interview makes it possible to reveal those competencies which, despite not being included in the dictionary, could turn out to be important for the organisation by complementing and improving this competency-based management instrument.

Although the competency-based interview and the traditional psychological interview are not mutually exclusive –as is pointed out by Alles (2000)– some differences do exist between them. The most important one lies in the fact that, whereas the traditional interview tries to elucidate the individual's behaviour through a battery of tests with the assistance of a psychologist, the competency-based interview extrapolates observable behaviours from past professional situations experienced by the individual to new work approaches, without the need for the intervention of a psychologist. In addition to this, traditional interviews do not permit to identify a person's competencies, since most individuals are not fully aware of the performance level they can reach at a work post, and not even of the behaviour they are going to develop in relation to those functions. On the other hand, when people face a traditional interview, they give answers about what they regard as socially desirable, about what they assume that the interviewer wants to hear, and not about what they really do.

Anyway, the identification of generic competencies for the company must take as a reference the definition of the organisation's strategic model, its mission, its values, etc., even though the behaviours described in each one of them will be different depending on the demands posed by the jobs. It is worth highlighting in this respect that not all generic competencies are equally relevant in all the corporation's functional units; not even in all the work posts belonging to the same functional area (Pereda & Berrocal, 1999).

The report prepared by C.E.O.E. (2002) indicates that, as the model evolves within the organisation, and as a result of successive revisions, there is a growing tendency to reduce the number of competencies and to simplify their content.

Returning once again to the methodology for the identification of competencies, it is worth highlighting the study carried out by Wustemann (2003) among various UK companies as an example. This research study revealed that work groups were the method most frequently used to identify the most representative behaviours in each competency. Seeking to achieve this aim, a decision was made to create transversal work groups formed by workers from different departments but with the same hierarchical position who were consequently going to share a considerable number of competencies in the performance of their functions.

Another method for the identification of competencies –though from a different perspective– can be mentioned: control groups. As we explained in the previous chapter, this methodology consists in drawing a comparison between two groups of individuals belonging to the same role, some with a medium performance level and others who achieve a high or excellent performance. This is the way in which the behaviours, attitudes, knowledge, etc. which serve to distinguish an outstanding function performance from a merely satisfactory one can be identified.

On the other hand, interview-based surveys among the organisation's collaborators are another widespread method for the identification of competencies, especially for specific competencies.

4.3. Competency-Based Management Dimensions

Seen from the perspective of the competency-based management model, the objective sought by the firm's human resource management is to have available the adequate competencies at the adequate level and at the right time and place. It is important to highlight that a specific competency level will exist for each post; high levels are not always advisable, what we really need is levels suited to the post circumstances. For instance, individuals who are excessively enthusiastic in the performance of their functions may prove to be particularly reckless in the decision-making process when they have to face highly complex and turbulent environments. That is why, although enthusiasm can be seen as a desirable primary competency for a managerial post, its development level will depend on the job to be occupied and its particular circumstances.

Competencies are a human resource management tool. If the organization is able to identify and define them in terms of conduct, they will make it possible to build a model that includes –under the same management system– the (internal as well as external) selection of people, the training and development of professionals, the preparation of succession plans, favouring a culture change, defining and assessing work posts, evaluating performance or even establishing pay criteria.

In order to achieve those aims, the company must pay attention not only to the staff acquisition or hiring for the preparation of a competency inventory but also to the set of policies that can contribute to their development, activation or deactivation depending on the organisational needs.

Competency-based management is therefore articulated in three basic dimensions which permit the integration of the different human resource management processes taking the concept of competency as the analysis unit (Ordóñez, 1995). Ramírez (1996) identifies a series of human resource management systems within each one of these dimensions which are collected in Figure 7.

We can place under the "*competency-based acquisition*" dimension those human resource management systems that we consider predominantly acquisitive; that is, which imply the selection of the competencies marked by the company's strategy (Cantera, García-Morán & Gómez, 1996). "*Competency stimulation*" –that is, how the preexisting competencies are mobilised and motivated seeking to reach the corporation's goals– has as its aim to focus on the management of performance and pay according to the competencies. Finally, "*competency development*" pursues as its ultimate aim to develop formative itineraries taking as a reference not the work post –as it has traditionally been done so far– but the prediction of future competential needs.

Regarding Figure 7, it is worth highlighting that the membership relationships between human resource management systems and the human resource management dimensions described in them are by no means unique and pure relationships. Each one of the systems can contain aspects from more than one of the dimensions. Thus, for instance, the staff selection system, despite being a predominantly acquisitive system, may consider the candidate's professional potential, which means that it would include nuances typical of the competency development dimension.

Gick & Pallares (1997) also want to make one thing clear about this table. In their opinion, not all competencies have the same capacity for development; whereas some can be used as selection criteria (acquisition), others are more appropriate for the detection and identification of training needs (development).

In turn, Ulrich *et al*. (1995) simplify all these human resource practices in 6 categories: selection and hiring, development, evaluation or assessment, pay, organisational design and communication. This same author explains in another of his works (Ulrich, 1998) that these 3 competency dimensions can be developed by means of 5 instruments: external recruitment, internal recruitment, 'outsourcing' through consultants, dissociation of those individuals who are unable to reach a certain performance level in the exercise of their competencies and retention of those workers who own critical competencies for the firm through motivation or stimulation policies.

Figure 7. Basic competency-based management dimensions

	Competency-based management		
	Acquisition	*Stimulation*	*Development*
Human resource management systems	Recruitment, selection and hiring	Pay and incentive policy	Training management
	Short-, medium- and long-term previsional management	Work post assessment	Information and communication
	Career plans	Performance assessment	Promotion system
	Succession planning	Motivation and integration	Assessment of potential

Source: Canós, Valdés & Zaragoza (2003).

Among all these policies, pay is without a doubt the most controversial one when it comes to implement a competency-based management model. Most authors think that linking workers' salaries to their competencies raises serious problems both from an ethical point of view and in terms of productive efficiency. For these reasons, this human resource policy has traditionally been relegated to the last place when the time has come to implement a competency-based management model, choosing instead others such as performance assessment, training or career planning. A large number of authors –among them Whiddett & Hollyforde (2003); Zingheim & Schuster (2003) or Rankin (2001)– see this discrimination of the pay policy within the overall competency-based human resource management model as something obvious, the reason lying in the fact that this policy is not usually considered essential for the design and implementation of the model. Nevertheless, as is pointed out by Blanco Prieto (2007), the necessary flexibilisation of organisations should be extended to pay policies nowadays. Going even further, Rábago López (2010) claims that, although most current organizations have not incorporated competencies into their pay policies yet for reasons related to caution, this position prevents them from taking advantage of all the potential offered by the comprehensive implementation of the competency-based management model.

On the other hand, as is suggested by Fernández López (2005), the pay pack should be broken down into three basic concepts: what the employee is; what the employee does; and what the employee achieves. The first one of these elements would possibly have to do with belonging to a specific professional group; the second one would be related to an occupational pay band; and the third one, to the variable part of pay.

In other words, performance assessment and pay are the most important human resource practices when it comes to trigger and reinforce desired behaviours among individuals and, consequently, to achieve the organisation's strategic goals. In fact, quite a few empirical studies have revealed the causal relationship existing between the human resource pay system and the strategic orientation of the firm (Collins & Clark, 2003). As is pointed out by Alles (2010), one of the main concerns among executives has to do with managing the company's organisational values so that they can be harnessed for practical purposes. In this sense, the adhesion of competencies to those values permits to turn them into real management indicators.

In any case, it is important to ensure that all these human resource practices, whether they are considered individually or grouped together around the three previous dimensions, are consistent with one another. On the one hand, this consistency will allow individuals to better understand what is expected from them and what they can obtain from the company and, on the other hand, it will give a response to the correlation of preferences, conducts and, of course, competencies shared by a group of individuals within the organisation, thus providing coherence to their treatment and favouring the formation of work teams (Baron & Kreps, 1999).

In short, employees' competencies are created and supported on these human resource practices (Wright, Dunford & Snell (2001), which contain important nuances in relation to the way of identifying and managing competencies.

REFERENCES

Alles, M. A. (2000). *Dirección estratégica de recursos humanos. Gestión por competencias.* Buenos Aires, Argentina: Ediciones Granica.

Alles, M. A. (2010). *Diccionario de preguntas. La trilogía.* Buenos Aires, Argentina: Ediciones Granica.

Arce, E. (2000). Desarrollo de competencias. *Training & Development Digest, 24*, 28–29.

Bacq, G. (1996). Décrypter les compétences. In J. M. Peretty (Ed.), *Tout DRH* (pp. 94-98). Paris, France: Editions D'Organisation.

Baron, J. N., & Kreps, D. M. (1999). Consistent human resource practices. *California Management Review*, *41*(3), 29–53.

Bartlett, C. A., & Goshal, S. (1997). The myth of the generic manager: New personal competencies for new management roles. *California Management Review*, *40*(1), 92–116.

Bayón, F. (1994). *Gestión de recursos humanos. Manual para técnicos en empresas turísticas* (2nd ed.). Madrid, Spain: Editorial Síntesis.

Blanco Prieto, A. (2007). *Trabajadores competentes. Introducción y reflexiones sobre la gestión de recursos humanos por competencias*. Madrid, Spain: ESIC.

Boyatzis, R. E. (1982). *The competent manager: A model for effective performance*. New York, NY: John Wiley & Sons.

C. E. O. E. (2002). La gestión por competencias en España. *Informe para el Observatorio Europeo de Gestión por Competencias.* Retrieved from www.ceoe.es/webceoe/ceoe/obseu/informe.pdf

Canós, L., Valdés, J., & Zaragoza, P. (2003). La gestión por competencias como pieza fundamental para la gestión del conocimiento. *Boletín de Estudios Económicos*, *48*(180), 445–463.

Cantera, F. J., García-Morán, R., & Gómez, G. (1996). Ingeniería por competencias. *Capital Humano*, *95*, 36–41.

Collins, C. J., & Clark, K. D. (2003). Strategic human resource practices, top management team social networks, and firm performance: The role of human resource practices in creating organizational competitive advantage. *Academy of Management Journal*, *46*(6), 740–751. doi:10.2307/30040665

Córdoba, A. (2004). *El reto de la gestión empresarial*. Barcelona, Spain: Deusto.

Dalmau, J. I., & Baixauli, J. J. (1989). *La productividad*. Valencia, Spain: Servicio de publicaciones de la Universidad Politécnica de Valencia.

de Ansonera, A. (1996). *15 pasos para la selección de personal con éxito: método e instrumentos*. Barcelona, Spain: Paidós.

Dirube, J. L. (1999). Competencias ¿impuestas o inducidas? *Training & Development Digest*, *17*, 40–42.

Dirube, J. L. (2004). *Un modelo de gestión por competencias. Lecciones aprendidas*. Barcelona, Spain. *Gestion*, ▪▪▪, 2000.

Donovan-Wright, M. A. (2002). How to capitalize on competencies. *Workforce online*. Retrieved from www.workforce.com/archive/feature/23/31/50/index.php

Fernández López, J. (2005). *Gestión por competencias. Un modelo estratégico para la gestión de recursos humanos*. Madrid, Spain: Prentice Hall.

Gick, A., & Pallares, T. (1997). Las competencias: ¿Qué está pasando en Europa? *Capital Humano*, *99*, 43–48.

Heijke, H., Meng, C., & Ris, C. (2003). Fitting to the job: The role of generic and vocational competencies in adjustment and performance. *Labour Economics*, *10*(2), 215–229. doi:10.1016/S0927-5371(03)00013-7

Herranz, A., & de la Vega, R. (1999). Las competencias: Pasado y presente. *Capital Humano*, *123*, 58–66.

Horton, S. (2000). Competency management in the British civil service. *International Journal of Public Sector Management*, *13*(4), 354–368. doi:10.1108/09513550010350508

Joras, M. (1996). Utiliser les bilans de compétences. In J. M. Peretty (Ed.), *Tout DRH* (pp. 165-172). Paris, France: Editions D'Organisation.

Kydd, C. T., & Oppenheim, L. (1990). Using human resource management to enhance competitiveness: Lessons from four excellent companies. *Human Resource Management, 29*(2), 145–166. doi:10.1002/hrm.3930290203

Le Boterf, G. (2000). *Ingeniería de las competencias*. Barcelona, Spain: Ediciones Gestión.

Levy-Leboyer, C. (1997). *Gestión de las competencias: Cómo analizarlas, cómo evaluarlas, cómo desarrollarlas*. Barcelona, Spain: Ediciones Gestión 2000.

Losey, M. R. (1999). Mastering the competencies of HR management. *Human Resource Management, 38*(2), 99–102. doi:10.1002/(SICI)1099-050X(199922)38:2<99::AID-HRM3>3.0.CO;2-T

Mansfield, R. S. (1996). Building competency models: Approaches for HR professionals. *Human Resource Management. Special Issue on Human Resource Competencies, 35*(1), 7–18.

McClelland, D. (1973). Testing for competencies rather than intelligence. *The American Psychologist, 28*(1), 1–14. doi:10.1037/h0034092

Ordóñez, M. (1995). *La nueva gestión de los recursos humanos*. Barcelona, Spain: Ediciones Gestión.

Parry, S. B. (1996). The quest for competencies. *Training (New York, N.Y.), 33*(7), 48–56.

Pereda, S., & Berrocal, F. (1999). *Gestión de recursos humanos por competencias*. Madrid, Spain: Editorial Centro de Estudios Ramón Areces.

Pereda, S., & Berrocal, F. (2001). *Técnicas de gestión de recursos humanos por competencias*. Madrid, Spain: Editorial Centro de Estudios Ramón Areces.

Rábago López, E. (2010). *Gestión por competencias. Un enfoque para mejorar el rendimiento personal y empresarial*. La Coruña, Spain: Netbiblo.

Ramírez, A. (1996). La gestión de las competencias. *Estudios Financieros, 160*, 87–100.

Rankin, N. (Ed.). (2001). *The IRS handbook on competencies. Law and practice*. London, UK: IRS LexisNexis Group.

Rausch, E., Sherman, H., & Washbush, J. B. (2001). Defining and assessing competencies for competency-based, outcome-focused management development. *Journal of Management Development, 21*(3), 184–200. doi:10.1108/02621710210420264

Sagi-Vela, L. (2004). *Gestión por competencias. El reto compartido del crecimiento personas y de la organización*. Madrid, Spain: ESIC.

Sánchez de Dios, L., & Valldeperes, P. (1998). Metodología de evaluación de competencias directivas. *Capital Humano, 110*, 20–30.

Spencer, L. M., & Spencer, S. M. (1993). *Competence at work. Models for superior performance*. New York, NY: John Wiley & Sons.

Ulrich, D. (1998). Intellectual capital = Competence x commitment. *Sloan Management Review, 39*(2), 15–26.

Ulrich, D., Brockbank, W., Yeung, A. K., & Lake, D. G. (1995). Human resource competencies: An empirical assessment. *Human Resource Management, 34*(4), 473–495. doi:10.1002/hrm.3930340402

Whiddett, S., & Hollyforde, S. (2003). *A practical guide to competencies: How to enhance individual and organisational performance*. London, UK: CIPD (Chartered Institute of Personnel and Development).

Wright, P. M., Dunford, B. B., & Snell, S. A. (2001). Human resources and the resource based view of the firm. *Journal of Management*, *27*(6), 701–721. doi:10.1177/014920630102700607

Wustemann, L. (2003). La comunicación a los empleados y su compromiso en el diseño del modelo de competencias. *Training & Development Digest*, *39*, 64–72.

Zingheim, P. K., & Schuster, J. R. (2003). Competencies and rewards: Substance or just style? *Compensation and Benefits Review*, *35*(5), 40–44. doi:10.1177/0886368703256932

KEY TERMS AND DEFINITIONS

Competency: The underlying characteristics of an individual (a motive, trait, skill, aspect of one's self image or social role, or a body of knowledge) which underlie performance or behavior at work.

Competency Model: Set of integrated Human Resource Practices (recruitment, assessment, compensation…) which takes the "competency" term as the basic analysis unit.

Human Resource Management (HRM): Is the function within an organization that focuses on recruitment of, management of, and providing direction for the people who work in the organization.

Chapter 12
Competencies 2.0:
A Theoretical Model for Defining and Managing a Presence Plan on the Web

Jaime Izquierdo-Pereira
NewMahwah - The Social Xperts Lab, Spain

Carmen Avilés-Palacios
Universidad Politécnica de Madrid, Spain

Joaquín García-Alfonso
Universidad Politécnica de Madrid, Spain

Manuel López-Quero
Universidad Politécnica de Madrid, Spain

ABSTRACT

Web 2.0 is changing the rules, reducing costs of access to technology, increasing the volume of cyber social contact through networking, both social and professional, making possible that any information is just a click away. This evolution causes the birth of new requirements, including the adaptation of people and, because organizations are made up of people, changing their map skills. This chapter presents a theoretical model for defining and managing a presence plan on the Web, emphasizing in those individual Competencies 2.0. It is defined from a Delphi method, involving experts in the use and development of Web 2.0. The strengths of this model are two: 1) It is a model that attempts to explain new tasks resulting from magazines not covered in sufficient depth from an academic and scientific perspective, though, they are been studied and discussed in forums 2.0; and 2) It explains the relationship between individual and organizational competencies 2.0, so that the first one can act on to modify the latter. On the other hand, this chapter has certain shortcomings, mainly two: 1) It has not been empirically validated, although research is underway to rectify this deficiency, and 2) There is no academic scientific knowledge enough, probably due to the novelty of this issue.

DOI: 10.4018/978-1-61350-207-5.ch012

INTRODUCTION

Web 2.0 is changing the rules. On one hand, a dramatic cost cutting has made a very complex technology accessible to practically all social classes (ADSL + Wi-Fi, mobile phones, IP telephony and video, digital cameras, GPS). Simultaneously, we have witnessed an exponential growth of network software tools (SaaS) with zero or very low cost, thanks to business models based on advertising or Freemium schemes (Free & Premium, Free & Fee). Finally, information (all the information) is just a click away.

As a result, over-informed people (not necessarily well informed) interact with each other through a universal network, continually finding new personal or professional uses which generate a collective value much greater than the sum of their individual contributions.

While technology has undergone a tremendous evolution, the uses that people make are clearly a revolution. The burden of change is not in the technology itself, but in people.

However, not all people have the ability or willingness to react to these changes in the same way. A new digital gap arises not only between areas of the globe with different technological development levels, but also between similarly developed areas and even inside certain areas with a high level of development.

Some experts such as Marc Prensky (2001), inventor of the terms digital immigrants and digital natives, considered this gap mainly generational, but from our point of view this approach assumes a gross simplification – very useful in certain scenarios but not always applicable.

METHODOLOGY

It has been made a thorough research in the academic literature relating to Web 2.0 in order to develop a model based on existing studies. As a result, began an exploratory study. Based on the method used to define competencies that could be labeled as traditional, it was performed a two-round Delphi in order to identify and define skills which were being developed through the use of Web 2.0 and also to describe the relationship between those individual and the organizational competencies. Two groups of experts were formed. All of them were users of Web 2.0; managers in organizations; and also they usually participate in forums and social and professional networks. The first group identified and defined 20 competences, both individual and organizational. The results were transferred to the second group who ordered them in order of the importance given. 8 competencies were ordered nearly similar by experts. In the following step of this research were defined the correspondences between this 8 individual competencies and the organizational ones. This paper aims to describe the 8 individual competencies 2.0 and pretend to model them.

THE WEB 2.0 IMPACT ON THE ECONOMY

First, the Internet's infrastructure suffered a new wave of evolution, in which distributed computers - according to the client-server model - are transferring their contents to the Web, or as it is often called in this context, the Cloud [1]. For businesses in the cloud, connectivity, data security and confidentiality become matters of prime importance, and management of service level agreements with their respective suppliers is taking apart CIOs from their traditional technological profile and converting them in pure service managers. IP technologies provoke huge impact on traditional industries. New giants like Google, Amazon and eBay offer their data centers at very competitive prices to companies around the whole world (Platform as a Service, PaaS). Telecom operators become unusually relevant.

Videoconference is putting business travel in serious jeopardy. Software is offered as a remote service (Software as a Service, SaaS).

Second, users find multiple and very easy ways to upload their own contents - and also replicate contents by others - to the cloud, producing a very substantial change on business models based on physical copies with well-known consequences in industries like press media, music, movies or videogames. People have switched the concept of copyright and are firmly committed to share contents in the cloud. Information becomes fully accessible and ubiquitous, ceases to be valuable itself, and is no longer classified by taxonomic but folksonomic[2] criteria - using labels, or tags, provided by the users.

Third, there is an unavoidable intrusion of the network in the lives of individuals and organizations, that now have a "digital twin" - not always easy to control. Digital identity, search engine rankings and customer's opinions become essential assets for individuals and organizations under the name "digital reputation". Companies take advantage of new network tools to "talk" with their customers, thus creating a new type of marketing which aims to capitalize the users' influence as a way to distribute products or services through the Web. Organization's hierarchical structures are also questioned, leading to flat, neuronal structures. Web recruitment processes become popular, putting in danger the traditional headhunting activity. Retention of talent is the new priority of Human Capital departments (Lookwood, 2007; Björkman & Smale, 2010; HR Management, 2006).

As a result, in a very short period of time we are struggling to deal with new forms of relationships, new business models, new job functions and new paradigms (such as the need of supranational laws to regulate global challenges); and it is far from obvious that "we" - as a whole - are prepared to face them.

NEW SKILLS FOR NEW CHALLENGES

At the first glance, Web 2.0 looks like a chaotic environment; however it evolves under continuous self control and self regulation. Therefore, we have in front of us the challenge to analyze and systematize the patterns of development of a new set of competencies – coexisting with the current or 'traditional' ones – that are essential to guarantee a pleasant and productive presence in Web 2.0:

- Pleasant, because Web 2.0 offers to both individuals and organizations countless possibilities to expand.
- Productive, because it is impossible to address all of them; therefore some control criteria are needed to find the optimal midpoint in each case.

Several factors complicate this task:

- Web 2.0 is essentially participatory and conversational.
- Brands or companies do not talk: people do.
- Matching personal and professional profiles is not always useful or suitable.
- Inconsistencies between the "natural" and digital identities (understood as the image projected on the network) are a high risk factor.

These behavioral patterns are translated into a new set of skills that we have called Competencies 2.0. Our investigation led us to a dictionary formed by eight personal competencies; eight corresponding corporate competencies; and two "frame" competencies. They will be described in detail along the following sections.

Each personal competency has an equivalent or corresponding corporate sister (Figure 1).

Our aim is to point that it doesn't make sense trying to achieve these competencies at a corporate level without encouraging the personal level ones. For instance, how could a brand start conversing with its consumers via Twitter, if the person that has to tweet does not have the minimum skills to communicate in 140 characters? As a temporary solution, a number of companies are outsourcing these services to specialized 2.0 communication agencies, but from our point of view this will prevent the enterprise capacity to "become 2.0" with a comprehensive vision, not only for marketing purposes. That's the difference between driving an airplane in the take-off track, and flying. This is a huge task for current and future HRM.

At the personal level, this model has four expansive competencies (Initiative, Connection, Global Vision and Contribution) and four control competencies (Time, Information, Transparency and Uncertainty management), looking for the right balance in order to have a pleasant and productive presence.

Trying to define a common behavior for all people is useless (unless 1984 become true at some moment in time). Each person has different vision and needs about how to socialize over the Internet. However, most people have not described - consciously - their goals when they register in a social network, or start blogging. Therefore we have two different problems here: first, there is a lack of definition of the objectives to be achieved; and second, when dealing with Competencies, a behavioral description is required, and it should be independent from the objectives. This is the difficult part of it.

THE NEED OF A PRESENCE PLAN

Very few people have consciously pre-defined their presence on the Web. For example, it is assumed by 2.0ers that a high number of followers is a good thing. However, a growing number of

Figure 1. Personal and corporate competencies

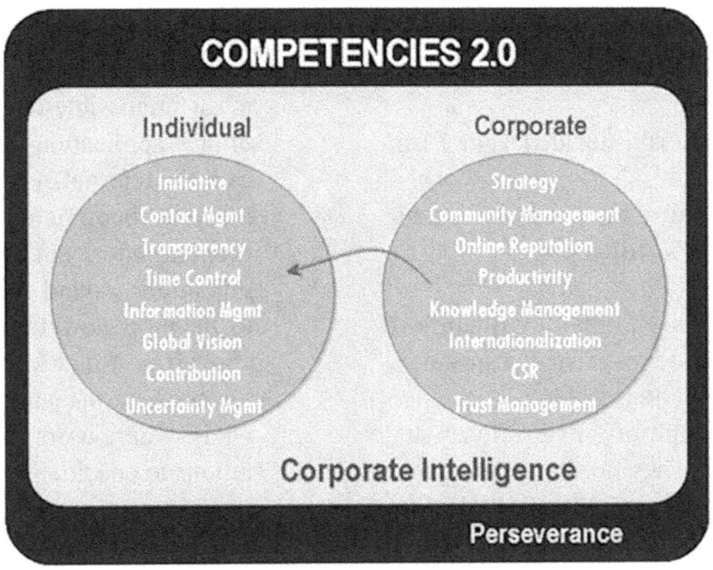

virtual contacts can become not only a very time-consuming task but also an unwanted obligation to talk with unknown people, something that not everybody is prone or ready to do. There are no thumb rules, each person is different and has different motivations.

The first thing to define is "why." Following the lessons learned from McClelland (1961), humans are driven by three basic motivations: achievement, affiliation, and power (a fourth motivation, curiosity, was added later to the list). Once we identify the main reason to start using Web 2.0, we can decide the targets we should contact with, the right tools to use, the presentation card we will show to others (most times in the form of a profile), and the valuable contribution that will make us attractive to them. The existence of such a plan allows measuring and managing the breach between expected and actual results. This evaluation can be performed by any external observer and results can be improved with behavioral patterns, thus making a competency model suitable. Accordingly, the behavioral patterns in our model are defined in such "ability to decide" basis.

Once we have decided how must be our presence in Internet, the Plan can evolve with our needs. The rules are:

- **Have a Plan.** The best reason to defend a Plan is having one.
- **Don't be afraid.** You decided your Plan, so follow it.
- **Be flexible.** If you're not happy with your Plan, change it. It's still yours.

Even if the Plan is not formally written, just thinking about it is a good exercise. For instance, parents concerned about their children's presence in the web have now a topic of conversation (actually eight) with them. They don't need to know the insights of MySpace or Facebook to discuss about the criteria to decide their virtual contacts and optimize their time and energy while connected.

THE MODEL: COMPETENCIES 2.0

1. Initiative

Readiness to Take Advantage of the Opportunities

This definition implies an ability to recognize the opportunities when they arise, which is not always easy for everybody; a proactive listening attitude is essential. Under this umbrella we also include anticipation and self-learning ability (especially for technology and tools). Let's explain all these items a little bit.

- **Listening attitude:** Ideas and opportunities pass in front of our eyes, waiting for someone to grasp them. This is an excellent time for entrepreneurship and collaboration. There are many people out there contributing to common knowledge, and information is one click away. Paying attention to what's going on can make the difference.
- **Anticipation:** Opportunities expire; in fact they only exist as long as they are not available to everybody. Only those who anticipate can take advantage of them.
- **Self-learning ability:** Technology evolves at an impossible-to-follow pace. Myriads of new applications and tools are launched over the Internet everyday, and even worse, many of them include an API (application program interface) to ensure that other programmers around the world will be able to develop new tools and enhancements above the initial idea. Manuals are being replaced by much more dynamic blogs and wikis, where users themselves provide solutions to specific problems.

The expression "take advantage" should be understood as getting some kind of benefit, not necessarily financial. For instance, if the main reason to sign up in a social network is finding new friends, the ROI of that action should be measured in terms of the number of new friends met.

Behavioral Checkout Examples: Initiative (Score 1=0% to 5=100%)

- Have you defined the reasons why you should use Web 2.0?
- Have you defined a Personal Presence Plan in Web 2.0? The plan should include:
    - A description of your desirable contacts' profiles,
    - A list of tools you understand essential to reach them,
    - The value proposition that will make you attractive to them, and
    - Your presentation to the rest of the world.
- How many of the tools named in point 2 above are you currently using?
- Do you keep track of all the tools and applications you ever used or registered in using a username and password, recording your experience for future reference?
- Do you consciously remove (or request for removal) your data from the tools you stop using?

From a corporate point of view, Initiative is not easy to implement, because it requires accepting some autonomy and error allowance. A level of autonomy looks unavoidable if we take in account that talent retention is one of the big challenges for HRM in the short term, and a certain error allowance is key to innovate. Organizations trend to flat structures for which soft control seems much more appropriate.

The correspondent corporate competency is Web 2.0 Strategy. Any organization should answer the same questions above (translated to the strategic level), i.e. should define a comprehensive Presence Plan and not only for marketing purposes, for which ROI measurement is easy in the short term.

After identifying the current web presence and positioning, a SWOT analysis is used to identify the strategic lines leading the Presence Plan. The stakeholders' list is matched up with suitable Web 2.0 applications in a target-tool matrix. Each cell in the matrix must be contrasted against the strategic lines and defined as an individual project with its own priority, goals, deadlines, and allocated resources. At the end, people involved in each project must be trained and empowered to accomplish their tasks, synchronizing corporate and personal plans.

2. Connection: Contact Management

Ability to Decide

- Number and quality of virtual contacts
- Depth, strength, and frequency of virtual relationships.

The human being is social. Generally speaking, we are not aware of the number of people we really know in the natural world. It is commonly assumed that the average path length between any two random human beings in the globe is six, after the experiments started by Stanley Milgram[3] in the later 60's. Social networks - as we understand them in Web 2.0 - are an excellent technology that allows staying in touch with other people despite the barriers of time and space. For instance, it allows families spread in different points of the globe share their pictures, comments, experiences, etc. as if they were living together. That is simply fantastic, and explains the unbelievable success of tools like Facebook (+500MM users) in such a very short time.

In this virtual world, however, whom should we connect to? Again, the answer to this question is not unique. Each individual decides, consciously or not, his or her virtual connections. The number of connections has become a kind of key success factor, leading to a strange run to beat a Guinness record. This could make sense from a branding/marketing point of view, but it doesn't seem a good idea for individuals trying to maintain true conversations. British anthropologist Robin Dunbar theorized that the maximum number of stable inter-personal relationships (about 150, known as the Dunbar's number) is a direct function of the neocortex size, with a small margin of error. He also stated that socializing with 150 people would require a 42% of the time. True or not, defining a prominent role for 150 stable contacts looks a good idea.

These are some example criteria used to define the profiles to connect with:

- Number of desired virtual contacts.
- Already connected offline (family, friends, colleagues), or completely new.
- Same background, or different. If innovation is an issue, connecting with different-minded people can be a good idea.
- Same or complementary interests.
- Decision level. Do you want to buy, or sell?
- Same country, or different.

Once identified the profiles of interest, the Plan must also take in account how to find them:

- Identifying the tools they use, knowing their specialization and user's average profile. As the number of users grows, social networks specialize in age (Tuenti), countries (Orkut, Xing, Viadeo), interests (LinkedIn, MySpace, Badoo)...
- Identifying which connections (hubs) can lead to them. New business models arise with this approach: for instance, bloggers with a high number of readers / subscriptors are given incentives to spread out the benefits of certain products and services, becoming a way for brands to reach a high number of targets with the minimum pay in.

This competency also implies the ability to say NO. Writing an assertive answer in advance is a good shield against unwanted or unrequested invitations to connect. The Plan itself becomes an excellent reason to decline such invitations.

Behavioral Checkout Examples: Connection (Score 1=0% to 5=100%)

- Have you defined your ideal number of virtual connections?
- How close is this figure to your connections database's actual size?
- Have you defined the aimed profile of the 150 first stable connections?
- How many of your contacts comply with the requirements?
- How often do you reject unwanted invitations to connect?

At corporate level, the corresponding competency is something equivalent to the well-known XRM (Extended Relationship Management, a more comprehensive view of CRM, which only takes care of Customers) or, using a common 2.0 marketing expression, Community Management. It has to do with the ability to actively manage the different groups of interests (or stakeholders) identified in the strategic plan and make them participate and provide valuable feedback.

The task is defining who in the organization will take care of each stakeholder, and how. In some cases, it could happen that Web 2.0 wouldn't be the most adequate vehicle to keep in touch. Regarding to customers, keeping the conversation alive and receiving feedback is becoming an essential issue, even if their opinions are negative.

3. Transparency Management

Ability to Decide

- Which self information should reside on the web
- Which contacts should have access to it.

Self information disclosure is probably the most common concern when dealing with social networks: "Why should I put my personal contact information available to other people?" "Apart of our family, who should care about our pictures?" "Doesn't it seem stupid showing party pictures when you're looking for a new job?" The bad news is that every possible answer is right and wrong at the same time: a common, universal, right answer for everybody simply does not exist. The difficulty relies in finding the right point where freedom and interests meet. And that point is essentially different for each single person.

Social networks are like a beach: commonly you dress as you like as long as you go there with people you know, but all other people around will also see you. And if you decide to give a walk along the shore for a while, the probability to meet other people who want to talk to you is never zero. However, each single individual in the beach is wearing the right clothes - of course, under his/her own point of view.

A certain level of transparency is key if you want other people know something about you; for instance, if you want to avoid unwanted advertising, or look for recognition in some area of expertise. This trend points towards a much more transparent world, as stated by the success of geolocation applications: many people are willing to tell the rest of the world where they have lunch just for fun (geotagging does not provide payback so far). On the other hand, other people with presence in professional networks like Linkedin are still discussing about the reasons to add or not their professional contact data in their profiles, which sounds equivalent to delivering a business card in blank.

An important fact to be taken in account is that Internet is technology, and technology fails sometimes. This is why we put in first place "which information should reside on the web". As the important paper documents, important information must be kept safe, and backed up. A risk management procedure should be set in place before uploading essential information to the net.

Regarding to the second question, there are no thumb rules. However, an acceptable level of transparency can be defined based on three variables:

The environment provides valuable information: the expected transparency. This is quite obvious: wearing a swimwear is not expected at the office, and its counterpart is also true (wearing a suit is not expected at the beach). In a professional network like LinkedIn, it seems obvious that individual's contact data are expected for professional reasons. In a social network like Facebook, personal information is commonly restricted to friends already known offline. These are good references; however there are exceptions to these general assumptions because of the other two variables.

Each individual's interests and likes define how far the individual would reach in his/her transparency in a no-constraints scenario. Some people are more conservative than others, in the street and in the Internet.

Each individual's degrees of freedom set the adequate constraints to prevent unnecessary suffering.

Aiming towards a reasonable convergence between both natural and digital identities seems a quite healthy behavior. Here it is where the expression unnecessary suffering makes sense: real life happens inside and outside the Internet, and a divergent image between both environments normally leads to a cost, of one the three following types:

- **Emotional:** shame, upsets
- **Personal:** an unpleasant personal contact, a harassment, or according consequences in the individual's environment: family, friends, etc.
- **Professional:** losing a job, or a deal

That cost can be dismissed if the individual is strong enough (considering each possible definition of strong) to pay it. If I won't feel any shame, and I don't care about loosing my job if my boss don't like them, why not upload my latest party pictures? However, if I need my job, I will not only think it twice before uploading any picture; I should also pay close attention to my friend's uploads and tags, just in case - regardless my own opinion about the pictures.

Therefore, Transparency Management starts with a good knowledge of all three variables in advance, to avoid unwanted bills. Teenagers (as many adults, by the way) are in jeopardy because they don't perceive their own limitations and dismiss the third variable. Strong people normally prefer changing the environment and staying in their tracks: whenever they don't feel comfortable they choose a different beach (i.e. a different social network) or a new job: since talent retention is key for the enterprises, talent gives one additional degree of freedom. However, do people have all the information they need to make the right choice? Nobody can answer this question.

Pictures are one easy way to explain transparency. However, there are many other places where individuals disclose (consciously or not) important self information, like their own presentation (or "bio"), tweets, blog posts or comments, geotags, etc. Friends, groups, and fan pages also provide enough information to a smart reader. Internet users must always keep in mind that their digital identity is influenced by all these factors.

Behavioral Checkout Examples: Transparency Management (Score 1=0% to 5=100%)

- Have you defined how you should appear on the Internet?
- Do you regularly inform your contacts about this?
- Are you aware of the information (about you) that actually is on the Internet?
- Have you identified different levels of information disclosure depending on your contacts' profiles?
- Do you know the different options and mechanisms provided by each of the channels and tools you are currently using to ensure that you share every piece of information with the right persons only?

The corporate competency equivalent to Transparency Management is Online Reputation. Keeping track - and trying to control - all the information published in Internet by different media, customers, partners, employees... about a certain name or brand can be a heavy task, however it is becoming more and more important for all organizations regardless their online activity. For instance, the arise of opinion platforms like TripAdvisor is changing tourism business model very rapidly, because other users' opinions can influence the digital identity of a certain hotel much more than any expensive advertising.

4. Time Control

Ability to Stay Connected Only as Much Time as Profitable

If transparency is the first concern about Internet, time consumption is surely second in the ranking. How much time should any person spend connected? Again, there are no universal answers: each person has different goals, skills, techniques, tools, requirements, and hobbies. As a consequence, the answer is different for each individual.

Time control has much to do with self-discipline. There are many different methods to gain productivity. One of the most popular is GTD (acronym of Getting Things Done), created by David Allen[4]. Allen's method recommends taking all inputs (of any nature) out of the mind by storing them in different lists, depending on their context and level of focus required, following a guiding flowchart. For instance, it recommends that any task taking less than two minutes should be done immediately, while other tasks require planning, other are desirable things to be done but without a milestone... Knowing the existence of the lists (you are taking care of everything) allows emptying your mind to concentrate in the execution of the tasks, one at a time. A weekly review is necessary to reorganize the lists. It sounds simple, but just think about the concept 'emptying your mind' and you will see it is not that easy. Any time control methodology without discipline will not work.

But what about being online? Staying connected is a complex task itself, and surfing one-thing-at-a-time is not usual at all. Is there any way to apply methods like GTD to the presence in the web? The good news is that any system or method like GTD is valid to manage those tasks - it is a crossing discipline. However it is important that individuals understand the concept of synchrony when communicating and participating in the web.

Let's take these two examples:

- Example 1:
  - E-mail is essentially asynchronous. Nobody expects an answer in seconds, maybe in minutes, or hours.
  - Instant messaging (IM) applications like Skype or MSN Messenger are synchronous. Whenever you type something and hit <Enter> you expect an immediate answer. Like in a telephone conversation, interruptions and delays become annoying.
- Example 2:
  - The essence of Twitter is its immediacy: a tweet sent two hours ago no longer exists.
  - Social networks like Facebook or Linkedin are meant to let you upload information which can be viewed later on.

An individual who understands the difference between both situations has an enormous advantage because he or she will be able to use methods like GTD to take apart those tasks which really don't need immediate answer, store them in different lists, and use the two-minute rule only when it is really necessary, accumulating similar tasks (like accepting contact invitations) in one single longer-than-two-minute task. The increase in productivity is then guaranteed.

A very simple advice to gain in productivity is reading e-mails only two or three times a day, of course planned in the agenda. Mobile devices and smartphones don't help at all, because they make everything seem immediate. However, smart users still know the difference and use them consciously. Money, not social networks, makes the World go round.

Behavioral Checkout Examples: Time Control (Score 1=0% to 5=100%)

- Do you understand the meaning of the term 'synchrony' in web applications?
- Do you apply synchrony to those Web 2.0 tools and applications you normally use?
- Do you use a time control method like GTD?
- Do you work with the same intensity and focus after an interruption?
- Are you able to manage more than one task at a time?

The corporate competency equivalent to Time Control is Productivity. As in Information Management, it is clearly accumulative: the most people in the organization have a better control of their individual time planning, the better the organization will perform, for obvious reasons.

However, organizations can help their employees to be more productive by providing them access to the Internet and the collaborative tools that Web 2.0 puts in its hands. Cloud applications have proven to be an excellent source of productivity and a catalyst for change. For instance, online shared calendars and office applications like Google Apps allow users to naturally share their documents and spreadsheets with their colleagues, and work in one single file at the same time. The key once again is knowing the limits and taking advantage of these tools whenever it makes sense. Empowerment and technical training must be provided together - a real challenge for a number of old-fashioned organizations.

5. Information Management

Ability to Prioritize and Retrieve Relevant Information Only

We live in an over-informed society. We cast our opinion about different items from the information found in thousands of sources: media, books, blogs, pages and advertising which bombard us everyday. The quantity of available information about any item grows continuously, and so does a sense of ignorance which sometimes leads to an anxiety status, a phenomenon described by Alfons Cornella (about ten years ago) with the term Infoxication: "we receive much more information than we can process"[5]. Some studies affirm that a standard person receives in one week more news than an 18th century person along a life.

Internet is the modern Encyclopedia: everything is there. A curious mind can spend a life looking for information, being an ideal victim of infoxication as well. But unfortunately being over-informed does not guarantee being well informed, especially when anybody can upload any information about any item. Democracy in content production has a clear risk of producing pieces of false or irrelevant information - however, it is commonly accepted that people's participation is based on goodwill and produces collective intelligence. This phenomenon is known with the term crowdsourcing[6], and is the aim of websites like Wikipedia, an encyclopedia created by internet users.

In such an ocean of information with thousands of sources, first thing to do is defining which information is really relevant to us. If we want to be productive, it is essential to declare a number of "things" that we need (or want) to be informed about. Keeping this number under control (five, or less) means that we can proactively look for information from diverse sources and formats (text, pictures, video...) about what we do really care. Feeding our mind - and curiosity - with information about any other items is perfectly acceptable, as long as we keep focused on the important "things".

Finding information and resources can be really challenging. However, the evolution of Internet search engine's technology is a Sci-Fi story since Google's birth - only eleven years ago when this book is being produced. In fact, a new feature called Google Instant (which provides results as long as you type your query) has been released a few days before I wrote these lines. Google broke all the expectations about querying the web with its relevance algorithm called PageRank, and has now a leading position in the search engine market with more than 90% visits share[7] in France, Germany, UK and Spain. Given the huge volume of results provided by Google, knowing the tricks-and-tips of advanced searching becomes essential.

On the other hand, Google's algorithm shows a list of links to a number of websites containing relevant information based on the words you typed in. For instance, if you type 'Capital City of Sweden' in Google you will get a list of web pages containing the words 'Capital', 'City', and 'Sweden' sorted by relevance. In our try, the first (thus most relevant) result is the Wikipedia entry about Stockholm, in which first line has exactly "Stockholm is the capital and the largest city of Sweden". However, there is a new and very interesting search engine called Wolfram|Alpha which goes further beyond: it tries to provide an answer to your specific question by analyzing the semantics of the typed-in text in order to understand the answer you really need, and completing it with additional useful information. Using exactly the same input, W|A result was:

- The name of the capital city of Sweden, of course: Stockholm
- Its location in a map, including the distances to other European cities
- A graph about its population's evolution
- Its current transit and weather conditions...

Of course, all this information can be found along the different websites proposed by Google. However, Wolfram|Alpha is interesting because of its approximation to the so-called semantic web or Web 3.0, a new and exciting technology jump by which the Internet will understand our natural speech. Apparently the universal development of this technology will take some time yet.

Other tools can help us obtain, store, retrieve and process required information with less efforts: RSS, alerts, e-mail filtering, and bookmarking. Acquiring good technical skills in these tools is not difficult, and they are an excellent way to help people stay focused.

Behavioral Checkout Examples: Information Management (Score 1=0% to 5=100%)

- Have you prioritized the five most important items you need to be informed about?
- Do you know how to use Google's advanced search features?
- Do you use at least two different search engines (with different relevance criteria) to query the web?
- Do you use RSS / bookmarking / alerts?
- Do you filter irrelevant e-mails?

The corporate competency equivalent to Information Management is Knowledge Management. This is understood in two ways: having the right information, and sharing it. In corporations, Information Management benefits are accumulative: the most people in the organization are able to obtain relevant information with less effort the better, for obvious reasons. However, sharing this information with other members in the organization is also extremely valuable for the organization - despite that sometimes it does not depend on technical aspects only. Web 2.0 provides several tools to help people share information, like blogs, wikis, and multimedia warehouses, three excellent ways to allow employees participate and share their knowledge.

Blogs are often understood as a good way to share knowledge with customers and partners. However, a growing number of organizations are using blogs for internal communications. By doing so, any member of the organization can subscribe (using RSS) only to those blogs dealing with relevant information for his/her job. Any post together with its comments conform an easy-to-follow conversation, replacing endless e-mail chains with CC to many individuals, also providing a desirable level of transparency in the communications. Further on, this solution is technologically easier and cheaper.

Wikis are websites created by several editors. They are excellent tools for workgroups and projects meant to store and share dynamic information within the workgroup or with other members of the organization.

Modern organizations also benefit from multimedia warehouses like Flickr, Picasa, YouTube, Vimeo, Slideshare, or Prezi, where employees can upload and share work-related pictures, videos or slide presentations with their colleagues.

6. Global Vision

Ability to Understand and Value Diversity as an Asset

The World is now connected as we never saw it before. Internet technology allows people from anywhere in the Globe share information, documents, pictures, videos, opinions... with people in any other place, in real time. An excellent example can be found in the expatriates and immigrants, who can now stay connected with their families and original cultures while integrating in a new country. Internet breaks time and space barriers, adding even more complexity to our already complex World.

Globalization is often seen as an easy way to bring new markets closer. When dealing about new or distant markets, Global Vision helps to understand those specific characteristics that can make the difference to succeed. This approach is valid, but limited: Web 2.0 is an excellent tool to incorporate diversity. The word incorporate does not mean being converted to any different culture, religion, laws, or customs, but taking in account all these different perspectives to help us understand the complex reality and take advantage, since not all people and organizations are able to realize its value. New opinions and points of view are often rejected as they will jeopardize the mission, vision and values which worked so far. Will this still be valid in a so-rapidly changing and global World?

It is commonly agreed that finding, retaining and developing talent is now one of the most important tasks for modern management. When trying to acquire talent as individuals, or managing collective talent as an organization, Web 2.0 can help us reaching global talent in both directions. This is the aim of Fédération Coséleam[8], a Swiss e-learning collaborative project for North African French-speaking universities, with more than 630.000 registered pupils. Talent and will, put together.

Web 2.0 is also rapidly developing in countries like Japan and China, because information is also global. Language will shortly stop being a problem: automatic translators are evolving at an extraordinary pace. For instance, technology has made it possible chatting in real time in different languages, which sounds astonishing. Translations are not perfect yet, but it is a matter of time and a right positioning can be, once again, essential.

A correct Global Vision also looks closer: too often we make huge efforts to integrate people from the other end of the world, while absolutely ignore our neighbor's opinions, especially if we know they have a different vote orientation. Local diversity is something we should also care about.

Behavioral Checkout Examples: Global Vision (Score 1=0% to 5=100%)

- Do you consider Global Vision is a competitive advantage?
- Do you promote diversity in your environment?
- Have you added global contacts to your timeline?
- Do you follow local contacts with different opinions than yours?
- Do you read blogs from remote countries when looking for information?

At corporate level, Internationalization would be the equivalent to Global Vision. However many organizations only take it under consideration for marketing and communication purposes, since they see the world as a market only. This means dismissing other competitive advantages as global information and global talent, which require a certain level of flexibility not that easy to reach.

The key is not only accepting, but promoting diversity. This means empowering, hiring and developing people with different cultures and points of view within the organization, while keeping the focus on its mission, vision, and values. This is probably the most discerning challenge for corporations when they face a "2.0" attitude and Google is probably the best example.

7. Contribution

Predisposition to Anticipate in Offering Value

All the previous competencies have to do with some abilities worth to develop from a self-centric point of view: how should I behave in my own benefit. However, the nicest part of Web 2.0 is that it gives a chance for people to participate, create, and give. The "2.0 attitude" has to do with Contribution: What makes me valuable to others? Why should they add me as a friend in Facebook, as a contact in Linkedin, or to their timeline in Twitter?

Dale Carnegie[9] explained in his book How to win friends and influence people, published in 1937: "You will win more friends in two months if you care about them, than in two years if you expect them to care about you". On the other hand, the following sentence is attributed to George Bernard Shaw: "If I have an apple, and you have an apple, and we exchange them, I will have an apple and you will have an apple too. But if I have an idea, and you have an idea, and we exchange them, I will have two ideas and you will have two ideas too".

Both thinkers died in the decade of 1950, much before the appearance of the Internet in our lives. However, both messages are even more valid now than they were eighty years ago. By the use of Web 2.0 technology, applications, and tools we can reach masses of people around the world in just a few clicks, but the answer to the question is still the same: if you want them to follow you, just be valuable to them. You must anticipate in offering your own value. You can participate - just do it.

Surprisingly, there are millions of people out there willing to collaborate and help by putting all their work and knowledge in the Internet for free. This model is completely new, and has very serious impact in the economy:

- Musicians publish their works in social networks and allow sharing as a form of quick mass distribution. What is the future of the old copyright model?
- If the cleverest algorithms are open sourced, which software should we pay for?

Wikipedia has 10 times more articles than the Encyclopedia Britannica. Yes, it could contain mistakes, but most of its contents are undisputable.

Wikipedia is the paradigm of Web 2.0: millions of people around the world creating contents about things they know, without any payback but the satisfaction of their contribution. As explained in Knowledge Management above, not many people are ready to understand this new paradigm after many years of living in an environment where 'information is gold'. The "2.0" attitude puts the value in other attributes, since we have in our hands countless free information resources provided by people and organizations. For instance, Instituto de Empresa, one of the most prestigious business schools in the world, publishes for free all the multimedia cases analyzed during its master classes[10]. Pupils no longer find value in the information itself: discussing the case, sharing their own opinions, and learning from other eager-to-learn colleagues save the real value to pay for.

Content supervision is probably #1 priority for the upcoming years, but millions of people providing knowledge are not a wave but a tsunami. Even if all this knowledge has mistakes, we should probably assume that it will become THE knowledge of our future generations.

Behavioral Checkout Examples: Contribution (Score 1=0% to 5=100%)

- How much do you consider contribution a competitive advantage?
- Have you decided your added value in Web 2.0?
- Do your presentation and bio show your contribution to others?
- Do you have a consistent presentation in all different Web 2.0 tools and applications you use?
- How often do you leave comments and opinions in other users' blogs?

According to the current model, organizations normally conceive, promote and even strain contribution within their own walls, but not along their relationships with the rest of the world. Some in-company social networking applications have appeared in the last two years to ease collaboration among employees as if they were acting in an open environment.

On the other hand, they suffer an increasing pressure to "give back" their earnings to the society somehow. This situation leads to effective contribution in two forms, the consolidation of the third sector (NPO, NGO, Foundations) and the appearance of Corporate Social Responsibility as a new and necessary activity close to marketing and PR in an exercise of Transparency Management. Terms like social accountability, sustainability, eco-responsibility, green management, etc. have become usual because private and public organizations have a real need to demonstrate any form of contribution. In both cases, social media tools have demonstrated to provide excellent channels to communicate with related stakeholders, and listen to their feedback - it will be cheap for them.

Jobs will evolve in the upcoming years, and learning to be useful to as many people as possible is a guarantee of success since Contribution, as a magnet, attracts interesting people at both private and corporate level.

8. Uncertainty Management

Ability to Stay Focused During Hard-Planning Times

We are living uncertain times. Many paradigms are changing (human relations, family), or falling (economy). A financial crisis hits strongly. Our perception about what is safe and what not has become irrelevant. New values and principles apply. New technology gives us the chance to participate, but many people are not able to take advantage of it. And, of course, we have no references from the past. Many people think we are living the rise of a new world. We are probably moving towards a new social and economic model, and scientists and philosophers have started drawing their crystal balls (Kelly, 2006). Planning is becoming harder than ever, because there are many new variables in the system and also the environmental conditions are changing. Does it make sense to stay focused under these conditions? And, is it possible?

Change has been a catalyst for progress in all times. However, as the Spanish sociologist Josep Burcet[11] says, previous changes happened along hundreds or even thousands of years, while current changes are happening in a period of time shorter than one full generation.

It is often said that "one man is so much intelligent as the number of uncertainties he can manage at the same time" (attributed to Immanuel Kant[12]). Yes, changes happen continuously, and being able to stay focused is a valuable characteristic. Remember, the knowledge generated using Web 2.0 tools is new, and it doesn't matter if it doesn't match what we now call "correct", because it will be (for sure) the only valid knowledge for future generations.

Let's summarize what can be done in order to stay focused and have competitive advantage using Web 2.0:

- **Contribute.** Be useful to others. The future new economy will make you rich if you make many people happy.
- **Promote diversity in your timeline.** Listen to everybody. Don't disregard unpleasant opinions.
- **Stay connected only as much time as required to be productive.** The use of Internet (not Internet itself) can be a time vampire.
- **Prioritize and look for the information you need only.** Old wise men had an advantage: they did not have Internet.
- **Do not tell or show about yourself more than you decide.** Learn to decide in advance.
- **Connect with as many people as you can manage.** Prioritize the first 150 contacts; you will have a thorough interaction with them. You deserve it.
- **Decide why you're using Web 2.0.** Have a plan. Carry it out. Change it whenever you're not happy. It's your plan.

Organizations should empower their employees to follow this list, or any other looking alike. It can be the seed of the so-called Social Media Guide of the organization. In fact, organizations need to develop at corporate level two more frame competencies:

Corporate Intelligence is the ability to empower the employees in the use of Web 2.0 and make them agents of the company/brand in social networks and, in general, Web 2.0 tools. This obviously requires a good Social Media Guide; a corporate presence plan; and a competency management system.

On the other hand, individuals can decide to connect or not, but once an organization incorporates Web 2.0 to its processes it must take in account Internet ubiquity to ensure an adequate response any time, any place.

HR professionals have the challenge of understanding, managing and incorporating, these Competencies 2.0 in order to place their organizations into the immediate future.

REFERENCES

Argyris, C., & Schön, D. (1996). *Organizational learning II: Theory, method and practice*. Addison-Wesley Publishing Company.

Björkman, I., & Smale, A. (2010). La gestión global del talento: Retos y soluciones. *Universia Business Review*, *27*, 30–43.

Kelly, K. (2006). *Next 5000 days of Internet*. Retrieved May 3, 2010, from http://www.ted.com/index.php/talks/kevin_kelly_on_the_next_5_000_days_of_the_web.html

Lookwood, N. R. (2007). *Leveraging employee engagement for competitive advantage: HR's strategic role*. Society for Human Resource Management Research Quarterly.

Management, H. R. (2006). Critical issues in HR drive 2006 priorities: N1 in talent management. *HRFocus*, *83*(1).

McClelland, D. (1961). *The achieving society*. D Priceton van Nostrand.

Presnky, M. (2001). Digital natives, digital immigrants. *On the Horizon, 9*(5). Retrieved June 5, 2010, from http://www.marcprensky.com/writing/Prensky%20-%20Digital%20Natives,%20Digital%20Immigrants%20-%20Part1.pdf

KEY TERMS AND DEFINITIONS

Connection-Contact Management: Ability to decide number and quality of virtual contacts, and depth strength and frequency of virtual relationships.

Contribution: Predisposition to anticipate in offering value.

Global Vision: Ability to understand and value diversity as an asset.

Information Management: Ability to prioritize and retrieve relevant information only.

Initiative: Readiness to take advantage of the opportunities.

Time Control: Ability to stay connected only as much time as profitable.

Transparency Management: Ability to decide which self information should reside on the web and which contacts should have access to it.

Uncertainty Management: Ability to stay focused during hard-planning times.

ENDNOTES

1 Definition at http://en.wikipedia.org/wiki/Cloud_computing

2 Definition at http://en.wikipedia.org/wiki/Folksonomy

3 Available at: http://en.wikipedia.org/wiki/Small_world_experiment

4 Available at: http://www.davidco.com

5 Available at: http://www.informaticaeducativa.com.ve/descargable/infoxicacion.pdf

6 Definition at http://en.wikipedia.org/wiki/crowdsourcing

7 Available at: http://www.atinternet-institute.com/en-us/search-engine-barometer/search-engine-barometer-may-2010/index-1-2-6-199.html

8 Available at: http://www.coselearn.org/fr/adherents

9 Available at: http://en.wikipedia.org/wiki/How_to_Win_Friends_and_Influence_People

10 Available at: http://openmultimedia.ie.edu

11 Available at: http://www.burcet.net/wh/agujero_blanco.htm

12 More information at http://en.wikipedia.org/wiki/Immanuel_Kant

Chapter 13
Competence-Based Profile to Characterize Successful Entrepreneurs

Lourdes Canós-Darós
Polytechnic University of Valencia, Spain

Cristina Santandreu-Mascarell
Polytechnic University of Valencia, Spain

ABSTRACT

In this chapter, the authors list different points of view about entrepreneurship and entrepreneurs. For this, they take into account the relationship between an entrepreneur and his or her environment. Thus, the chapter identifies the main competences that characterize an entrepreneur distinguishing between innate and/or learned competencies.

INTRODUCTION

Emergence of new businesses and entrepreneurial talent are fundamental in the current new economy, moreover if we are dealing with new technologies and dynamic environments. Entrepreneurship is a form of multidimensional and complex behavior which study has led to different theories. The question of what characterizes an entrepreneur has been widely debated (Hitt et al., 2002) based on two ideas (Levie and Muzyka, 1996; Chua et al., 2004): entrepreneur is born or made? Although there are many multidisciplinary empirical studies, there is no a general accepted theory (Amit and Schoemaker, 1993; Ripsas, 1998). These theories lead to three approaches: characteristics or traits approach, behavioral approach (Wooldridge and Jennings, 1995; Köllermeier, 1992) and a more modern perspective that considers creation of companies as a complex and dynamic process (Bruyat, 1993; Bruyat and Julien, 1999).

Entrepreneur is a concept based on some combination of competencies. An entrepreneur has the ability to create many different and new situations, in which confluence unusual factors. In this context, some options are more appropriated than other, setting up opportunity until the end. In other words, an entrepreneur tirelessly works

DOI: 10.4018/978-1-61350-207-5.ch013

in combinations of factors that create opportunities (notice that this is not the same as being an opportunist) and he or she finds the best way to be successful. Then, process starts again. This does not change the profile of the entrepreneur. Nowadays, new environment must be adapted to entrepreneurial needs from an academic point of view. Thus, future professionals have to acquire generic and specific skills, some of which promote and strengthen entrepreneurship (Casson and Wadeson, 2007; Tobin and Pettingell, 2008; Boone and Kurtz, 2010). Once the entrepreneur answers the question about what kind of opportunity he or she wants to pursue, it is figuring out how to assign material, technical or personal resources. Personal resources have to be used properly on time and in a good manner. Competencies can be used in different ways; indeed, lack or overabundance of some of them could also determine the characteristics of the opportunity to be set up and probabilities to implement it successfully (Zhao et al., 2010).

Entrepreneurs do not have to try to find combinations of competencies that ensure success, but rather to understand how to develop competencies in order to create opportunities. If launching a revolution is pursued, even implicitly, some exceptional talent and charisma competencies to achieve necessary support and assistance will be required, among other. For example, we can mention entrepreneurs who led the transformation undergone by IBM from a hardware company to a service solutions company; or entrepreneurs of some of the flagship companies of the early Internet era as e-Bay, who looked for the best talent, the more appropriate investments and the promotion of changes in consumer habits. On the other hand, if entrepreneurs identify a niche in a market already established, they also have to consider different solutions for the customer, the way to achieve profitable business models that retain higher value for the company, etc. (Shane and Venkataraman, 2007).

In this context, we can suppose that the common point in concepts of entrepreneur or entrepreneurship is innovation, regardless of theories or approaches (Lichtenstein et al., 2006). This is relevant in management studies and for the real situation of companies, because the need of continuous adaptation based on innovation, competencies, knowledge, etc. (Nonaka et al., 2000). Then, the profile of entrepreneur does not have to change, but to be focused to the environment. In this chapter we deal with this idea in order to know the main characteristics that identify an entrepreneur, showing common points studied in literature and highlight the more representative ones in the new economy.

MAIN PURPOSE

Proper use of resources requires balance to discern what is right in each moment and the adequate degree. Balance is applied to all competencies of entrepreneur: courage, but not reckless; towing capacity, but also the humility to listen and recognize his or her own mistakes; passion and perseverance, but identifying a ruinous initiative on time; creativity and improvisation, but also organization and awareness of time and budget limitations; planning, but without falling into paralysis; opened to new opportunities, but devoted to consolidate what has been achieved; promote teams and invest in training, but also know when to disregard any of them (Lambing and Kuelh, 1997). Is it possible to learn about balance? It depends on what we mean by learning. Experience tells us that there is not only theoretical knowledge to be learned in a classroom because an entrepreneur and his or her team also solve real life challenges (Birley and Stockley, 2000). Notice that there are no recipes even for current management so it is more difficult for entrepreneurs who are creating a way that may not exist. As noted by Bhidé (1994), there is no common measure to weigh inevitable pros and cons, and equally experienced and savvy entrepreneurs can easily disagree one to another. Thus, the key is not only to accumulate information (Bhidé, 1999). Obviously, theoretical concepts are important but, once acquired, the entrepreneur has to apply at the time and appropriate manner, and this can only be learned through practice and experience (Delmar and Davidsson, 2000).

In this paper we review some theories and approaches that characterize entrepreneurship, in concrete, the entrepreneur concept by following an analysis that includes changing environment factors and innovation. Given generalizable situation in international context in recent years, and on the basis of academic papers mainly from economics, study of entrepreneurship has emerged as a field of study focused on the business process that involves functions, activities and actions associated with perception and pursuit of business opportunities. Our purpose is to identify the main competencies that characterize an entrepreneur in current environment, distinguishing between different approaches that consider some innate and/or learned competencies.

ENTREPRENEURSHIP AND COMPETENCIES

First reference to entrepreneurship and the term entrepreneur was presented by Cantillon (Hayek, 1985), who defined the economic role of entrepreneurs. Nowadays, generally accepted definition by researchers is given in the 30s by Schumpeter (2000), who defines an entrepreneur as a person who composes new combinations of resources, which may be new products, processes, markets, organizational forms or supply sources. Therefore, entrepreneurship is the process of getting new combinations by breaking the balance of the market; its essence is another determinant of the innovation phenomenon (Weerawardena and Sullivan, 2006).

One obvious difference between competencies and personality traits is reported by Levy-Leboyer (2002). He considers that these two terms can characterize individuals and explain the change in their behavior in the execution of specific tasks. In this sense, competencies and personality traits are defined as differences between individuals (individual characteristics) (Snook and Khurana, 2004). While competencies refer to something broader and complex, personality traits and skills are acquired in order to fulfill a mission in the context of the company, closely linked to business activities and, more specifically, to the missions that are part of a job. Thus, there is a relationship between skills, abilities and personality traits, but with an important nuance, because competencies are more than individual characteristics and also allude to the knowledge acquired by individuals through experience. They are linked to a task or an activity that require real work practice to be developed. Fang et al. (2008) also establish a differentiation between competencies/skills and personality traits. For these authors, the first two concepts refer to aspects that differentiate people from each other. Levy-Leboyer (2002) also highlighted them as individual characteristics, showing a fundamental difference with respect to competencies: competencies underlie observable behaviors. However, competencies point to behaviors that are performed when put into practice knowledge, skills and personality traits (Sarasvathy, 2001).

At this point, we can make a first approach to the concept of competence affirming that they refer to implemented individual characteristics (skillls and personality traits) that facilitate the acquisition of knowledge, through real significant experiences, resulting in observable behaviors that can successfully lead to missions or tasks that a particular job requires. These ideas show that competencies are not only innate, but can be acquired through experience and practice in a real job (Iacobucci, 2002). Then, competencies can be either innate or acquired but almost all of them can be acquired even if they are not innate. Therefore, we can say that they are educable, that can be taught and, therefore, learned and developed.

There are different types of competencies. From the model suggested by Bunk (1994), four core competencies (technical, methodological, social and participatory) can be identified; they are integrated into what he calls competence of action. A similar model, although less developed, is the one from Vossio (2002) which, under different terminology, highlights common aspects to Bunk's by ranking competencies (specific, generic and essential). Those aspects or components of competencies allude to knowledge, attitudes and skills, among other. Finally, we want to refer to the work carried out by authors such as Coughlin and Lemke (1999) and Levy-Leboyer (2002), who classified competencies in generic and specific. First may be present in all the work of an organization and can be transferred easily from one job to another; they consist of knowledge, skills and personality traits and common behaviors. By contrast, specific skills are definite to a particular job or performance level within an organization; hence, it is not possible (or easy) to transfer.

ENTREPRENEURSHIP APPROACHES

There are three approaches to define entrepreneurs, who fit individual characteristics (skills and personality traits) and competencies to facilitate innovations. These approaches are: characteristics or traits approach, behavioral approach and a mix of both by considering creation of companies as a complex and dynamic process mainly because of the environment (Audretsch et al., 2002).

Characteristics or traits approach tries to determine basic characteristics or traits that distinguish entrepreneurs (Carland and Carland, 1991; Chell et al., 1991). Some authors consider it as one-dimensional and static approach (Gartner, 1990). The entrepreneur is the basic unit of analysis, trying to identify features that characterize individual entrepreneurs by comparing characteristics of entrepreneurs with those no considered entrepreneurs, and the matching of entrepreneurs who achieved success and those who failed (Timmons, 1997).

Neither from comparison nor matching a clear trait or characteristic has been concreted (Begley and Boyd, 1986; Brockhaus and Horwitz, 1986). However, we can create a list taking into account that is not quite complete but offers a number of innate traits and characteristics:

- Independence need (Veciana, 1999; Chell et al., 1991).
- Achievement or motivation need (Johnson, 1990; Begley and Boyd, 1986).
- Achievement motivation and entrepreneurial behavior (Shaver and Scott, 1991; McClelland, 1989).
- Internal control and self-determination (Jenssen and Kolvereid, 1992; Gibb, 1990).
- Propensity to take risks (Jennings, 1994).
- Tolerance for ambiguity (Gasse, 1982).
- Type A conduct (Chell et al, 1991; Stoner and Freeman, 1992).
- Other: integrity and credibility, training and management, innovation, creativity, market orientation, opportunities orientation, organizational capacity, ability to learn from experience (Harper, 1991; Timmons, 1997).

Behavior and conduct approach considers the creation of a business is a contextual event resulting of multiple influences in which the entrepreneur is a part. This approach is considered multidimensional and dynamic (Bygrave and Hofer, 1991). It consists of specify behaviors that make an entrepreneur effective in his or her goal of creating a new organization.

Behaviors are analyzed from the entrepreneur point of view about creating a company and also from the entrepreneur point of view in existing organizations. Collis and Moore (1979) indicate that entrepreneurial activities are independently undertaken within the organizational context.

About creating a new company we have to take into account three factors that affect this decision (Lyon et al., 2000; Lumpkin and Dess, 2001; Zahra, Neubaum and Huse, 2000). First, factors affecting the behavior of entrepreneur, as a need to express the unresolved problems in his or her childhood about autonomy, personnel difficulties, finding a sponsor, achieving basic negotiation, existence of an example or model of corporate performance, social marginality, role deteriorment, psychological or critical event displacement (Veciana, 1999). Second, influence of incubator organization, that is, an organization in which an entrepreneur has served. This influences the nature of the new company about location, size and preparation (Collins and Moore, 1979). Third, external factors belonging to environment:

- Factors favoring motivations for starting a business and its success: industrial and incentive system, training and corporate culture, flexible and clear legal framework, and a system of information and opportunities.
- Factors to stimulate creation of SMEs (Shapero, 1975; Shapero and Sokol, 1982): shares of attraction, generating new businesses or industrial ecology for survival of businesses.
- Other factors: examples of actions and entrepreneurial skills, societal attitudes, accessibility and availability of venture capital, business angels, availability of personnel and support services, customers and universities, opportunities to use consultants and economic conditions, among other.

In existing organizations, behavior of entrepreneurs plays an important role in organizational learning (Shane and Venkataraman, 2000) because they represent new opportunities (Lumpkin and Dess, 2001) and lead to new and different combinations of resources. These processes are influenced by internal and external determinants and strategic variables. Internal determinants are top management philosophy and values, resources and competencies, culture as an incubator for entrepreneurial organization and organizational characteristics (Veciana, 1999; Cooper et al., 1989). External determinants take into account skill and dynamism of environment, the degree of technological sophistication and the life cycle stage of the industry. Among strategic variables (Mintzberg, 1979), company's mission should include presence or absence of elements that may avoid innovation as a way of life of the firm (Grant, 2005). Managers have to develop business practices and competitive tactics that reflect the reality of innovation processes.

In the creation of companies as a complex and dynamic process, entrepreneur is considered as an additional factor (Shane and Venkataraman, 2000). In this context, company, environment, characteristics and dynamic process of change are studied. When refer to a continuous development, relationship and interaction are important. Several authors distinguish three types of phenomena that may or may not be interrelated (Teng, 2007; Loasby, 2007; Dess and Lumpkin, 2005):

1. Creating new businesses within an existing organization, defined as internal corporate venturing (Zajac et al., 1991) or internal entrepreneurship (intrapreneurship) (Pinchot, 1985).
2. Conversion of existing organizations through the renewal of key ideas, named strategic renewal (Guth and Ginsberg, 1990) or organizational renewal (Stopford and Baden-Fuller, 1994).
3. Innovation. We believe that both strategic renewal as corporate venturing suggest changes in strategy or structure within the organization, which may lead to innovation. The difference is that the former leads to

reconfiguration of an existing business and the second involves the creation of a new business (Sharma and Chrisman, 1999).

AN ENTREPRENEUR IS BORN OR MADE?

About required competencies of the entrepreneur and if they are innate or can be learned, we present a review that suggests that they have the same importance: entrepreneur has to combine them to lead innovations.

First studies on the origin of entrepreneur were based on innate traits and motivations that characterized him or her, that is, entrepreneur is born not made. From this perspective, McClelland, which together with Atkinson is considered the author of the so-called achievement level theory, provides important elements for researchers maintaining that entrepreneurs are made not born (Leite, 2000). McClelland does not ignore personality trait model (entrepreneur is born); however, he believes that social context influences the formative role of the entrepreneur. According to this idea, an entrepreneur has personality traits, which constitute his or her psychological profile, which drives him or her to initiate a transformative creative process by taking risks. Certain behaviors are associated to this profile, which can be trained in order to influence the effectiveness level or outcome of the venture.

Among the most recognized approaches to understand entrepreneurs' profile we find the one presented by Garcia et al. (2002). For these authors the key factors of entrepreneurial profile are: leadership, overcome motivation, opportunity obsession, creativity and self-confidence, risk tolerance, ambiguity and uncertainty, commitment and adaptability. On the other hand, entrepreneurship profile proposed by Leite (2000) involves high achievement need, high energy levels, take risks, self-confidence, solving problems, emotional detachment, status need and personal satisfaction need. In addition, Rodríguez (2007) considers as relevant the following variables to define entrepreneur profile: creativity and innovation, ability to take risks, responsibility, altruism, self-confidence, honesty, communication, leadership and coordination and negotiation.

We determine that for the above mentioned authors common characteristics to define entrepreneurial profile are creativity, risk taking, leadership, self-confidence and overcome motivation. Notice that competencies of the entrepreneurial profile are related to attitudes and behaviors associated with the entrepreneur's social development that has an influence because of its dynamism. Thus, competencies are essential to the course and outcome of any new initiatives. Can we identify key competencies? Are they innate? Is there an ideal entrepreneur? Can competencies be learned? Is the entrepreneur made? We can answer these questions through defining a model that determines the behavior of the entrepreneur (dynamic, changing, evolving) more than by using a static profile. Then, we can identify an individual who creates value with the help of a novel element in the environment that can be learned and developed. Entrepreneurship is not a conquest of intelligence but of will (Schumpeter, 2000); this is a special case in leadership social phenomenon.

There are three approaches to define entrepreneurs, who fit individual characteristics (skills and personality traits) and competencies to facilitate innovations. These approaches are: characteristics or traits approach, behavioral approach and a mix of both by considering creation of companies as a complex and dynamic process mainly because of the environment (Audretsch et al., 2002).

DISCUSSION OF RESULTS

We can often find entrepreneurial psychological profiles allegedly attributable to good entrepreneurs in textbooks and articles. Most of them emphasize features such as risk affinity, aggressiveness, vision, etc., considered as prototypes. These competencies reference to innate psychological traits or temperament, which lead us to conclude that there are only some valid psychological profiles to start a business or to introduce valuable innovations in it. In short, if an entrepreneur is not born with these traits, you can not do anything to acquire or develop them. Experience shows that this idea is far from certain. We have many counterexamples of entrepreneurs whose personality does not fit into that scheme. For instance, we can refer to sober and thoughtful decisions by analyzing Theo and Karl Albrecht brothers, founders of Aldi (German supermarket chain). Also, we can observe an extremely discreet entrepreneur as the founder of the fashion group Inditex, Amancio Ortega. These entrepreneurs demonstrate that they can succeed even without any of these supposedly essential competencies. Moreover, many studies confirm that there is so much variety of profiles as entrepreneurs.

If we focus on entrepreneur behaviour we are able to analyze what makes an entrepreneur. This approach allows us to extend this idea: exhibited entrepreneurial behavior by individuals starting a company can be also developed for all those that create opportunities wherever they operate: in a large organization, working in economic social or political initiatives. Thus, this behavior synthesizes a strong motivation to achieve something (a momentum that can have very different starting points): required competencies to perform it are entrepreneurial competencies.

According to this idea, three concentric dimensions (business, interpersonal and personal) of competencies were introduced by Cardona and Garcia (2005). The main advantage of this model is that it addresses all aspects of an entrepreneur understood as a decision maker, that is, as an individual embedded in a sphere of relationships with others, and, of course, as a businessman. From this basic classification, amalgamated with previous research, we can identify some competencies presented as a common denominator in entrepreneurial behavior. We can separate them into the three above mentioned levels. First one refers to all the knowledge and skills in the field of pure business. Second is about relations between entrepreneur and people involved in the project: entrepreneurial team, investors, customers, etc. Third focuses on the entrepreneur.

1. Business competencies.
    - **Business vision.** Special ability to recognize and exploit opportunities that affect competitiveness and effectiveness of business.
    - **Resource management.** Ability to use financial and material resources in the most appropriate, timely and effective way to achieve desired results.
    - **Networking.** Ability to develop, maintain and operate an extensive network of relationships with key individuals in company and industry.
    - **Negotiation.** Ability to generate satisfactory agreements to all involved parties, through discovering or creating elements that add value to relationships.
2. Interpersonal competencies.
    - **Communication.** Ability to listen and effectively communicate ideas by using appropriate channels and providing specific information to endorse comments and conclusions. Most entrepreneurs are good communicators in the sense that they are able to transmit business ideas to others, but communication competencies should also include the opposite direction.
    - **Implication (also known as charisma).** Ability to obtain commitment of colleagues and subordinates, inspiring confidence, giving meaning to their work and motivating them to

achieve objectives. Although many business ideas emerge from solitary individuals, its success lies in the ability of entrepreneurs to create and lead a project team.

- **Empowerment.** Ability to ensure that each team member has necessary information and resources to make decisions and fulfill their mission. Perfectionism lead many entrepreneurs to prevent their colleagues and subordinates to truly collaborate on their projects. Empowerment means delegate the execution of small parts, but not the ultimate responsibility.
- **Respect.** It is not a mere tolerance or acceptance of diversity, neither the mutual use of external resources. It is a recognition of qualities of others and mutual support based on the diversity of competencies, without exclusively comparing and measuring individual contributions. Respect is a requirement for empowerment and it also implies correctly relations with customers or investors, for instance.

3. Personal competencies.
   - **Creativity.** There is unanimity to consider creativity as one of the competencies that identify an entrepreneur. We understand it as a capacity to generate new ideas or concepts, or new associations between familiar ideas and concepts, which usually produce original solutions. Creativity does not have to coincide with an invention of a product or service.
   - **Determination.** It is about a determined and persistent attitude in the implementation and achievement of initiatives. Some authors unfold this competence in initiative and ambition. Initiative concerns the ability to start and drive necessary changes with necessary reliability. On the other hand, ambition means setting high goals.
   - **Integrity.** It means to do what is just and honest in any situation. Integrity is the foundation of trust of colleagues and subordinates. Eventually also secures relationships with suppliers, customers and other stakeholders.
   - **Tenacity.** Also known as discipline, self-control or willpower. This competence allows entrepreneurs to overcome obstacles and implement costly activities. Capacity of sacrifice to get something we pursue is one of the human unique skills.
   - **Emotional balance.** Ability to react to every situation with emotions and in appropriate mood. Emotional balance is a reflection of personal maturity. From an entrepreneur, maturity is joined by his or hers intelligence for business.
   - **Self-criticism.** Everyone agrees on the importance of learning to accept and get along with one's own limitations and mistakes. This assumption is not incompatible with the ability to learn, but quite the opposite: it requires an exercise of honesty with oneself and with others to identify errors and ways to overcome them. Against the arrogance that prevents learning, humility is the key to continuous improvement.

CONCLUSION

There is not a simple answer to the question: and entrepreneur is born or made? Although there are papers that show positive results for both options, almost all of them conclude that a mix of both facts is the ideal. Moreover, some authors propose as required some competencies to be a good entrepreneur, but there exist some examples in which this is not always true. In the context of the new economy, a more advanced profile for an entrepreneur has to be defined, based in business, interpersonal and personal competencies. Obviously, an entrepreneur has to have competencies related with innovation and flexibility, in order to adapt the company to the changes from the environment. Moreover, the analysis of some methods or techniques to develop or improve the key competencies for an entrepreneur in the new economy is open.

ACKNOWLEDGMENT

This work has been partially supported by the projects TIN2008-06872-C04-02/TIN and CORSARI MAGIC DPI2010-18243.

REFERENCES

Amit, R., & Schoemaker, P. (1993). Strategic assets and organizational rent. *Strategic Management Journal*, *14*, 33–46. doi:10.1002/smj.4250140105

Audretsch, D. B., Thurik, A. R., Verheul, I., & Wennekers, S. (Eds.). (2002). *Entrepreneurship: Determinants and policy in a European US comparison*. Boston, MA: Kluwer Academic Publishers. doi:10.1007/b109395

Begley, T., & Boyd, D. (1986). Psychological characteristics associated with performance in entrepreneurial firms and smaller business. *Journal of Business Venturing*, *2*, 79–93. doi:10.1016/0883-9026(87)90020-6

Bhidé, A. (1994). How entrepreneurs craft strategies that work. *Harvard Business Review*, *72*(2), 150–161.

Bhidé, A. (1999). Developing start-up strategies. In Sahlman, W. A., Stevenson, H. H., Roberts, M. J., & Bhidé, A. (Eds.), *The entrepreneurial venture* (pp. 121–137). Boston, MA: Harvard Business School Press.

Birley, S., & Stockley, S. (2000). Entrepreneurial teams and venture growth. In Sexton, D. L., & Landström, H. (Eds.), *The Blackwell handbook of entrepreneurship* (pp. 287–308). Oxford, UK: Blackwell.

Boone, L. E., & Kurtz, D. L. (2010). *Contemporary business*. Retrieved from http://books.google.es

Brockhaus, R. H., & Horwitz, P. S. (1986). The psychology of the entrepreneur. In Sexton, D. L., & Smilor, W. (Eds.), *The art and science of entrepreneurship* (pp. 25–49). Cambridge, MA: Ballinguer Publishing Co.

Bruyat, C. (1993). *Création d'entreprise: Contribution epistémologique et modélisation*. Doctoral dissertation, École Superieure de Affaires, Université Pierre Mendès France, Grenobles, France.

Bruyat, C., & Julien, P. A. (1999). Defining the field of research in entrepreneurship. *Journal of Business Venturing*, *16*(2), 165–180. doi:10.1016/S0883-9026(99)00043-9

Bunk, G. P. (1994). Teaching competence in initial and continuing vocational training in the Federal Republic of Germany. *Vocational Training European Journal*, *1*, 8–14.

Bygrave, W., & Hofer, C. (1991). Theorizing about entrepreneurship. *Entrepreneurship Theory and Practice*, *16*(2), 13–22.

Cardona, P., & García Lombardía, P. (2005). *Cómo desarrollar las competencias de liderazgo*. Pamplona, Spain: EUNSA.

Carland, J. A., & Carland, J. W. (1991). An empirical investigation into the distinctions between male and female entrepreneurs and managers. *International Small Business Journal, 9*(3), 62–72. doi:10.1177/026624269100900304

Casson, M., & Wadeson, N. (2007). The discovery of opportunities: Extending the economic theory of the entrepreneur. *Small Business Economics, 28*(4), 285–300. doi:10.1007/s11187-006-9037-7

Chell, E., Haworth, J. M., & Brearley, S. (1991). *The entrepreneurial personality: Concepts, cases and categories*. London, UK: Routledge.

Chua, J. H., Chrisman, J. J., & Chang, E. P. C. (2004). Are family firms born or made? An exploratory investigation. *Family Business Review, 17*(1), 37–54. doi:10.1111/j.1741-6248.2004.00002.x

Collins, O. F., & Moore, D. G. (1979). *The organizational makers: A behavioral study of independent entrepreneurs*. New York, NY: Meredith.

Cooper, A., Woo, C., & Dunkelberg, W. (1989). Entrepreneurship and the initial size of firms. *Journal of Business Venturing, 4*, 317–332. doi:10.1016/0883-9026(89)90004-9

Coughlin, E. C., & Lemke, C. (1999). *Professional competency continuum*. Retrieved form http://www.mff.org/pubs/ME159.pdf

Delmar, F., & Davidsson, P. (2000). Where do they come from? Prevalence and characteristics of nascent entrepreneurs. *Entrepreneurship & Regional Development, 12*, 1–23. doi:10.1080/089856200283063

Dess, G. G., & Lumpkin, G. T. (2005). The role of entrepreneurial orientation in stimulating effective corporate entrepreneurship. *The Academy of Management Executive, 19*(1), 147–156. doi:10.5465/AME.2005.15841975

Fang, C. H., Kang, Y. C., Chen, G. L., & Cheng, Y. S. (2008). Study on personality traits and key competence of R&D staff an empirical analysis in Taiwan. *Proceedings of the 2008 IEEE IEEM.*

García, E., Garza, R., Sáenz, L., & Sepúlveda, L. (2002). *Formación de emprendedores*. México: CECSA.

Gartner, W. (1990). What are we talking about when we talk about entrepreneurship? *Journal of Business Venturing, 5*(1), 15–29. doi:10.1016/0883-9026(90)90023-M

Gasse, Y. (1982). Elaborations on the psychology of the entrepreneur. In Kent, C., Sexton, D., & Vesper, K. (Eds.), *Encyclopedia of entrepreneurship* (pp. 57–71). Englewood Cliffs, NJ: Prentice-Hall.

Gibb, A. (1990). Enterpreneurship and intrapreneurship-exploring the differences. New findings and perspectives. In Donckels, R., & Miettinen, A. (Eds.), *Entrepreneurship*. Aldershot, UK: Gower Publishing Company Limited.

Grant, R. M. (2005). *Contemporary strategy analysis* (5th ed.). Blackwell Publishing.

Guth, W. D., & Ginsberg, A. (1990). Guest editors´ introduction: Corporate entrepreneurship. *Strategic Management Journal, 11*, 5–15.

Harper, S. C. (1991). *The McGraw-Hill guide to starting your own business*. New York, NY: McGraw-Hill.

Hayek, F. A. (1985). Richard Cantillon. *The Journal of Libertarian Studies, 7*(2), 217–247.

Hitt, M. A., Ireland, R. D., Camp, S. M., & Sexton, D. L. (2002). *Strategic entrepreneurship: Integrating entrepreneurial and strategic management perspectives*. Oxford, UK: Blackwell Publishers.

Iacobucci, D. (2002). Explaining business groups started by habitual entrepreneurs in the Italian manufacturing sector. *Entrepreunership and Regional Development*, *14*(1), 31–48. doi:10.1080/08985620110096636

Jennings, T. (1994). *Multiples perspectives on entrepreneurship*. Cincinnati, OH: South-Western Publishing Co.

Jenssen, S., & Kolvereid, L. (1992). The entrepreneurs reasons leading to start-up as determinants of survival and failure among Norwegian new ventures. In Birley, S., & MacMillan, I. C. (Eds.), *International Perspectives on entrepreneurship research*. Londres, UK: Elsevier Science Publishers.

Johnson, B. (1990). Toward a multidimensional model of entrepreneurship: The case of achievement motivation and the entrepreneur. *Entrepreneurship Theory and Practice*, *14*(3), 39–54.

Köllermeier, T. (1992). The East German transition. In Abell, D. F., & Köllermeier, T. (Eds.), *Dynamic entrepreneurship in Central and Eastern Europe* (pp. 65–86). Amsterdam, The Netherlands: European Foundation for Entrepreneurship Research.

Lambing, P., & Kuelh, Ch. (1997). *Entrepreneurship*. London, UK: Prentice-Hall International (UK) limited.

Leite, E. (2000). *O Fenómeno do Empreendedorismo* (2nd ed.). Brasil: Edicoes Bagaco.

Levie, J., & Muzyka, D. (1996). *New venture growth in France: Do local government policies make a difference? Frontiers of Entrepreneurship Research*. Wellesley, MA: Babson College, Babson Park.

Levy-Leboyer, C. (2002). *La gestion des competences*. Paris, France: Editions d'Organisation.

Lichtenstein, B., Dooley, K. J., & Lumpkin, G. T. (2006). Measuring emergence in the dynamics of new venture creation. *Journal of Business Venturing*, *21*(2), 153–175. doi:10.1016/j.jbusvent.2005.04.002

Loasby, B. J. (2007). A cognitive perspective on entrepreneurship and the firm. *Journal of Management Studies*, *44*(7), 1078–1106. doi:10.1111/j.1467-6486.2007.00729.x

Lumpkin, G. T., & Dess, G. G. (2001). Linking two dimensions of entrepreneurial orientation to firm performance: The moderating role of environment and industry life cycle. *Journal of Business Venturing*, *16*(5), 429–451. doi:10.1016/S0883-9026(00)00048-3

Lyon, D. W., Lumpkin, G. T., & Dess, G. G. (2000). Enhancing entrepreneurial orientation research: Operationalizing and measuring a key strategic decision making process. *Journal of Management*, *26*(5), 1055–1085. doi:10.1177/014920630002600503

McClelland, J. L. (1989). Parallel distributed processing: Implications for cognition and development. In Morris, R. (Ed.), *Parallel distributed processing: Implications for psychology and neurobiology* (pp. 8–45). New York, NY: Oxford University Press.

Mintzberg, H. (1979). *The structuring of organizations*. Englewood Cliffs, NJ: Prentice-Hall.

Nonaka, I., Toyama, R., & Konno, N. (2000). SECI, Ba and leadership: A unified model of dynamic knowledge creation. *Long Range Planning*, *33*(1), 5–34. doi:10.1016/S0024-6301(99)00115-6

Pinchot, G. (1985). *Intrapreneuring: Why you don't have to leave the organization to become an entrepreneur*. New York, NY: Harper & Row.

Ripsas, S. (1998). Toward an interdisciplinary theory of entrepreneurship. *Small Business Economics*, *10*(2), 103–115. doi:10.1023/A:1007975330428

Rodríguez, L. F. (2007). Emprendimiento y asociatividad como herramienta de desarrollo social para la educación media. *Cuadernos Latinoamericanos de Administración, 4*(2).

Sarasvathy, S. D. (2001). *What makes entrepreneurs entrepreneurial? Schumpeter, J. (2000). The theory of economic development*. New Brunswick, NJ: Transaction Publishers.

Shane, S., & Venkatamaran, S. (2000). The promise of entrepreneurship as a field of research. *Academy of Management Review, 25*(1), 217–226.

Shane, S., & Venkataraman, S. (2007). The promise of entrepreneurship as a field of research. *Entrepreneurship, 2*, 171–184.

Shapero, A. (1975). *Entrepreneurship and economic development. Entrepreneurship and enterprise development: A worldwide perspective*. Milwaukee, WI: Proceedings of Project ISEED.

Shapero, A., & Sokol, L. (1982). The social dimension of entrepreneurship. In Kent, C., Sexton, D., & Vesper, K. (Eds.), *Encyclopedia of entrepreneurship* (pp. 72–90). Englewood Cliffs, NJ: Prentice Hall.

Sharma, P., & Chrisman, J. J. (1999). Toward a reconciliation of the definitional sigues in the field of corporate entrepreneurship. *Entrepreneurship Theory and Practice*, Spring, 11-27.

Shaver, K. G., & Scott, L. R. (1991). Person, process, choice: The psychology of new venture creation. *Entrepreneurship Theory and Practice, 16*(2), 23–45.

Snook, S., & Khurana, R. (2004). Developing "leaders of character": Lessons from West Point. In Gandossy, R., & Sonnenfeld, J. (Eds.), *Leadership and governance from the inside out* (pp. 213–233). New York, NY: John Wiley & Sons.

Stoner, J. A. R., & Freeman, R. E. (1992). *Management* (5th ed.). New Delhi, India: Prentice-Hall of India.

Stopford, J. M., & Baden-Fuller, C. W. F. (1994). Creating corporate entrepreneurship. *Strategic Management Journal, 15*(7), 521–536. doi:10.1002/smj.4250150703

Teng, B. (2007). Corporate entrepreneurship activities through strategic alliances: A resource-based approach toward competitive strategy. *Journal of Management Studies, 44*(1), 119–142. doi:10.1111/j.1467-6486.2006.00645.x

Timmons, J. (1997). *New venture creation. Entrepreneurship for the 21st century* (revised 4th ed.). McGraw-Hill.

Tobin, D. R., & Pettingell, M. S. (2008). *The AMA guide to management development*. Retrieved from http://books.google.es/

Veciana, J. M. (1999). Entrepreneurship as a scientific research programme. *Revista Europea de Dirección y Economia de la Empresa, 8*, 2–10.

Vossio, R. (2002). Certificación y normalización de competencias. Orígenes, conceptos y prácticas. *Boletín Cinterfor, 152*, 51–73.

Weerawardena, J., & Sullivan, M. G. (2006). Investigating social entrepreneurship: A multidimensional model. *Journal of World Business, 41*(1), 21–35. doi:10.1016/j.jwb.2005.09.001

Wooldridge, M., & Jennings, N. R. (1995). Intelligent agents: Theory and practice. *The Knowledge Engineering Review, 10*(2), 115–152. doi:10.1017/S0269888900008122

Zahra, S. A., Neubaum, D. O., & Huse, M. (2000). Entrepreneurship in medium-size companies: Exploring the effects of ownership and governance systems. *Journal of Management, 26*(5), 947–976. doi:10.1177/014920630002600509

Zajac, E. J., Golden, B. R., & Shortell, S. M. (1991). New organizational forms for enhancing innovation: The case of internal corporate joint ventures. *Management Science, 37*(2), 170–185. doi:10.1287/mnsc.37.2.170

Zhao, H., Seibert, S. E., & Lumpkin, G. T. (2010). The relationship of personality to entrepreneurial intentions and performance: A meta-analytic review. *Journal of Management*, *36*(2), 381–404. doi:10.1177/0149206309335187

KEY TERMS AND DEFINITIONS

Capacity: Resources and skills that an individual has to develop a certain task or role.

Competence: Skills, abilities and qualification to develop a task in an excellent way.

Entrepreneur: A person who face difficult actions with resolution.

Professional Profile: Set of capacities and competencies that identify the qualification of a person to assume in optimal conditions the corresponding responsibilities, functions and tasks of a job.

Chapter 14
Cross-Cultural Competences in the New Economy

Irene Martin-Rubio
Universidad Politécnica de Madrid, Spain

Drew Rodgers
Oslo University College, Norway

Erik Døving
Oslo University College, Norway

ABSTRACT

Cross-cultural knowledge management in the e-environment results from interactions with others from different cultures that require new understanding and interpretations. The authors find that few studies have been conducted to address the objective of identifying the competences that are needed to promote learning and information processing in the 21st century. The authors' primary goal is to identify competences that will promote learning and information processing in global organizations. The chapter outlines a taxonomy of competences necessary for cross-cultural learning in the global e-environment based on Kolb's Experiential Learning Theory.

1. INTRODUCTION

The new economy is characterised by two pervasive trends: increasing globalization and a revolution in electronic communication. These render a number of key challenges for HRM. One particular challenge is the intercultural competencies of the company's workforce, as well as the specific skills related to electronic communication.

DOI: 10.4018/978-1-61350-207-5.ch014

General-purpose technologies, such as ICTs, have strong economic and lifestyle influences on all sectors and aspects of modern life. The new global e-environment has a profound impact on economy and the basic competences for the future (Şimşek et al., 2009). For instance, Autor et al. (2003) argue that computers substitute for workers in performing cognitive and manual tasks that can be accomplished by following explicit rules. Computers also complement workers in performing non-routine problem-solving and complex communications tasks.

Meanwhile globalization means that cross-cultural experiences occur in many forms: encounters with individuals of different cultures, overseas customers, suppliers, colleges, short visits to international divisions and emersion of a new host culture (Yamazaki & Kayes, 2004).

Moreover, the nature of the Internet, as a critical enabler of e-commerce and e-services, raises questions about the robustness of trust across cultures and usefulness and ease of use (Guo et al., 2010). Because all international business activity involves communication, competencies for intercultural communication allow employees to work effectively in international environments.

Whereas communication is a process, culture is the structure through which the communication is formulated and interpreted. Culture deals with the way people live. When cultures interact, adaptation must take place for people from different cultures to communicate effectively. With intercultural business communication, being aware of each culture's symbols, how they are the same, and how they are different is important.

At the heart of any successful intercultural team in the e-environment lies a group of individuals possessing competences to manage a complex, ambiguous and often stressful process. In this chapter we present the taxonomy of these competences from the perspective of ability to learn and improve each competence every day.

Our point of departure is that effectiveness, ability to deal with cultural uncertainty, and cultural empathy toward others can be enhanced through training (Matveev and Nelson 2004). For instance, Kayes et al. (2005a, 2005b) propose that organizations can improve the process of learning in a cross-cultural context through Experiential Learning Theory (Kolb, 1984). Experiential learning focuses on how individuals draw on direct experience with the world to solve problems, make sense of everyday events and create new knowledge. By integrating recent development on adult learning through Experience Learning Theory, and the latest research in cross-cultural learning and the definition of e-competencies, we outline a set of competences essential for communication between cultures.

In this chapter, we first consider different concepts in the literature of intercultural competence and digital competences. Cultural Competence, Cultural Intelligence and e-maturity are considered capabilities that can be measured. Instead, we are interested in the abilities to learn these last capabilities. To address this question, in the third section, we propose the four learning styles of the Experiential Learning Theory to understand the knowledge management process and the competences related in the e-global environment. In fourth section, we present the conceptualization and taxonomy of these competences to manage each phase of the learning process. We conclude the fourth section with the implications for HR Managers. Finally, some brief conclusions are offered.

2. MAIN PURPOSE

Communication is expressed through media, which take numerous forms from spoken and written messages to mediated ones (Schiller and Cui, 2010). Although face-to-face conversation still takes the primary role in our daily work communication, its dominant position is consistently being replaced by computer-mediated communication, such as e-mailing, teleconferencing, video conferencing and web conferencing.

The globalization of business has highlighted the need to understand diverse cultures, especially for multinational and trans-cultural organizations, and substantial progress has been made in describing and conceptualizing cultural differences (e.g., Hofstede, 2001). It has become clear that the essential competence for success in managing the global organization lies in acquiring appropriate skills and knowledge. To gain a better understanding of this process of learning organizations could improve their understanding of on-line collaboration and knowledge flux. We suggest that on-line cross-cultural collaboration requires valuing individual and cultural differences.

Schiller & Cui (2010) studied the positive influence of open communication in a collaborative group context. They look into communication medium and culture, two well-established factors that bring much insight to the understanding of communication openness. The choice of these two factors is also inspired by the call for action from other researchers who emphasized the importance of the interaction between culture and computer-mediated-communication (Stahl & Elbeltagi, 2004).

Ayoko (2007) define communication openness as the ease of talking to each other and the extent of understanding gained when talking to each other and has treated as synonymous to listening, honesty, frankness, trust, supportiveness and similar concepts in a variety of research studies (Rogers, 1987). Early researchers indicate that communication openness involves both message sending and message receiving behaviours (Redding, 1972). Rogers (1987) support this idea and add more dimensions to the concept by indicating that "communication openness incorporates the message sending and message receiving behaviours of superiors, subordinates and peers with regard to task, personal and innovative topics". Communication openness demands perceptiveness and adaptability.

Cross et al. (1989) consider that cultural competence is a set of congruent behaviours, attitudes, and policies that come together in a system, an agency or among professionals and enable that system, that agency or those professions to work effectively in cross-cultural situations. The word culture is used because it implies the integrated pattern of human behaviour that includes thoughts, communications, actions, customs, beliefs, values and institutions of a racial, ethnic, religious or social group. The National Center for Cultural Competence (NCCC, 1998), adds to Cross et al.'s (1989) definition, that cultural competence requires that organizations:

- Have a defined set of values and principles, and demonstrate behaviours, attitudes, policies, and structures that enable them to work effectively cross-culturally.
- Have the capacity to (1) value diversity, (2) conduct self-assessment, (3) manage the dynamics of difference, (4) acquire and institutionalize cultural knowledge, and (5) adapt to diversity and the cultural contexts of communities they serve.
- Incorporate the above in all aspects of policy-making, administration, practice, and service delivery; systematically involve consumers, families and communities.
- Cultural competence is a developmental process that evolves over an extended period. Both individuals and organizations are at various levels of awareness, knowledge and skills along the cultural competence continuum.

With regard to cross-cultural competencies in the e-environment, the key question is what makes a professional successful in the global e-environment. In other words, we are interested in finding out how people, who most often learned to be effective in their home culture, learned to be effective in other cultures through computer mediated communication.

Intercultural capability is captured by the notion of cultural intelligence (Ang & Inkpen, 2008), defined as a "person's capability for successful adaptation to new cultural settings, that is, for unfamiliar settings attributable to cultural context" (Earley & Ang, 2003:9). People having a high social or emotional intelligence are thought to be relatively more able to empathize, work with, direct and interact with other people. High social intelligence reflects a person's capacity to perform actions (such as problem solving) with and through others. High emotional intelligence reflects a person's capacity to understand and convey human emotion (Goleman, 2000).

From a cross-cultural perspective, however, the emotional and social intelligence approaches lack cultural context as they attempt to explain how and why people act as they do. Cultural intelligence captures this capability for adaptation across cultures and it reflects a person's capability to gather, interpret, and act upon these radically different cues to function effectively across cultural settings (Earley and Ang, 2003).

Cultural intelligence differs from social and emotional intelligence in other ways as well. Adaption across new cultural contexts requires that novel ways of dealing with others be discovered. Existing strategies must be adjusted, adapted or reinvented depending on the situation and culture. Thus, cultural intelligence places a heavy emphasis on metacognition, or "thinking about thinking". Likewise, the activities required in new cultures, may require people to develop and expand their behavioural repertoires. That is, cultural intelligence reflects a person's capability of developing entirely novel behaviour (e.g., speech sounds or gestures) if required.

At its core, cultural intelligence consists of three fundamental elements:

a. metacognition and cognition (thinking, learning and strategizing, monitoring awareness),
b. motivation (efficacy and confidence, persistence, value congruence and affect for the new culture, culture empathy)
c. behaviour: social mimicry and behavioural repertoire

An important capability in the e-environment is e-maturity (Adam, 2007). E-maturity is understood as an additional stage of development beyond e-confidence. The latter embodies high levels of ICT knowledge and skills, and a readiness to apply these to existing situations and new challenges. E-maturity is demonstrated when professionals apply ICT in strategic and discriminating ways, taking into consideration a balance of advantages and alternatives. E-maturity display values and attitudes, knowledge and understanding, and, finally, practice and skills that measure the e-maturity capability of the Professional Standards for Teachers (Adam, 2007).

Our focus is on the ability to learn and improve these capabilities. Competence means the proven ability to use knowledge, skills and personal, social and methodological abilities, in work or study situations and in professional and personal development in the context of the European Qualifications Framework. Boyatzis and Saatcioglu (2008) argue that to be an effective professional, a person needs the ability to use knowledge and make things happen, and these can be called competencies. We introduce a new conceptual framework for competences that enhance knowledge management in the organizations. At the heart of successful cross-cultural management effort lies an individual or group of individuals with the ability to learn in and manage a complex, ambiguous and often stressful process in the global e-context.

Our argument is that this approach provides a significant improvement on existing approaches for several reasons: a) it considers the way adults learns from experience, b) it provides an integrated approach with learning and knowledge management and c) it is built upon cultural adaptation through experience.

Few approaches to cross-cultural competences have considered learning as a primary driver of success (Yamzaki & Kayes, 2004; Kayes et al., 2005a, 2005b). The ability of learning, the translation of context specific experience into more generalized knowledge and back to context specific experience into more generalized knowledge and back to context specific knowledge, lies at the core of effective knowledge management (Kayes et al., 2005a, 2005b).

In sum, we find that, while studies in each of the areas mentioned above contribute valuable knowledge to our understanding of communication, few studies have been conducted to address the objective of identifying the competences that are needed o promote learning and information processing in the 21st century. Our primary goal is to identify competences that will promote learning and information processing in global organizations. We outline a taxonomy of competencies necessary for cross-cultural learning in the global e-environment based on Kolb's Experiential Learning Theory.

3. CONCEPTUAL FRAMEWORK: CROSS-CULTURAL COMPETENCIES IN THE E-ENVIRONMENT

In this chapter, we draw on a model of cross-cultural competences in the e-environment to identify the competencies for successful learning and performance across cultures. As the OECD (2005) remarks, the often unclear terminology combined with the existence of a multiplicity of definitions (e.g., new economy, e-economy, ICT sector) means that these concepts change depending the framework used. For the European Centre for the Development of Vocational Training (CEDEFOP, 2006), *skill* is "the knowledge and experience needed to perform a specific task or job". And *competence* is described as the "ability to apply knowledge, know-how and skills in habitual or changing situations". Thus we highlight that not only *knowledge* (tacit or explicit) is important, but also the capability to *apply* this knowledge in other situations.

The classification of skill proposed by CEDEFOP (2006) identifies *basic skills* as "skills and competences needed to function in contemporary society (e.g. listening, speaking, reading, writing and mathematics)" and *new basic skills* which "are information and communication technology skills, foreign languages, technological culture, entrepreneurship and social skills".

Boyatzis (1982) point out that competences are a catch-all term that designates the skills, abilities, knowledge, cognitive processes and communication techniques that allow an individual to be successful at specific task. This perspective has now been expanded to understanding and managing one's emotions and influencing others (Goleman et al., 2002). Thus cross-cultural work is primarily a process of interpersonal understanding, which in turn provides the basis for gathering new knowledge (Yamazaki & Kayes, 2004).

The literature on knowledge management has viewed individual learning from the perspective of information processing. Successful knowledge management is measured by the degree to which pre-existing knowledge structures become accurately represented as cognitive models. In this formulation, structure precedes language, where language acts as reflection of a pre-existing knowledge structure (Kayes et al., 2005)

We view learning from the perspective of experiential learning theory (Kolb, 1984), which describes learning as the process of transforming experience into new knowledge. Kolb's theory is inspired by earlier research that has given experience a central role in human learning and development. This theory defines learning as "the process whereby knowledge is created through the transformation of experience. Knowledge results from the combination of grasping and transforming experience" (Kolb, 1984:41).

In this model, there are four learning modes depending how is grasped the experience and how it is transformed into new knowledge. Experience is grasped in two dialectically related modes of learning:

- **Concrete Experience (CE):** associated to feeling, experiencing, and
- **Abstract Conceptualization (AC):** associated to thinking, conceptualizing.

In turn, experience is transformed in two dialectically related modes:

- **Reflective Observation (RO):** associated to reflecting, and
- **Active Experimentation (AE):** associated to acting.

Managing concrete experience requires interpersonal abilities, managing reflective observation requires perceptual and observational abilities, managing abstract conceptualization requires cognitive abilities and active experimentation requires behavioural abilities. Thus, individual learning involves moving through each of these phases in a cyclical fashion, calling on specialized abilities at each phase of the cycle.

Experiential learning, then, is a process of constructing knowledge that involves a creative tension among the four learning modes that are responsive to contextual demands. This model of learning comprises the whole human learning process, where experience is the foundation for four modes of learning: feeling, reflecting, thinking, and acting.

In sum, Experiential Learning Theory (ELT) as an integrative approach to learning that embodies a comprehensive set of abilities necessary for a variety of activities related to cross-cultural learning. The values underpinning ELT emphasize the ability to learn and develop in the face of cross-cultural experiences, and thereby place the learner in a host culture at the centre of the cross-cultural learning process. Because the ELT approach underlines the interaction between individual and environment, and offers a model to understand both the person and the relationship to the environment, this transactional approach helps describe the role of the individual learning as a process of adaptation to different cultures.

A Competence-Based Model of Cross-Cultural Knowledge Management in the E-Environment

This chapter provides a comprehensive model of the competences needed in the 21st century. By integrating recent developments in theory and the latest research in cross-cultural learning and e-competences, we outline an essential set of competencies necessary for interaction between cultures in the e-environment. We present a list of cross-cultural competences as well as e-competences that describe the learning process through the experiential learning theory presented above.

Yamazaki & Kayes (2004) and Kayes et al. (2005a,b) propose a model for cross-cultural management based on Knowledge Management that starts on experience. They suggest four sets of competencies in relation to Knowledge Management process:

1. Generating Knowledge includes creating new knowledge by remaining open to experiences
2. Organizing Knowledge involves putting existing knowledge into a framework or context by abstract conceptualization.
3. Gathering Knowledge involves identifying underutilized knowledge by reflective observation.
4. Applying Knowledge involves putting knowledge to solve problems by active experimentation.

According to Yamazaki & Kayes (2004), Kayes et al. (2005a,b), Kolb (1984) and Cobo (2009) we suggest that there are learning abilities related to the process of grasping experience and learning abilities related to the process of transforming experience. Therefore, cross-cultural knowledge management in the new economy requires the following set of competences:

A. Competences for grasping the experience:
   - (A.1) Generating knowledge by cultivating and remaining open to new experiences to create situations in which knowledge can be obtained. The generating process of knowledge management involves interpersonal abilities that have the following competences and skills:
     - two specific cross-cultural competences: valuing different cultures, and building relationships within the host culture, and

- one specific e-competence: strategic digital literacy.

(A.2) Organizing knowledge involves putting existing knowledge into a framework or context so that it can be actionable. Organizing involves abstract conceptualization and requires the ability to synthesize complex ideas into transferable knowledge. The organizing process of knowledge involves the following competences and skills:

- one specific cross-cultural competence: transforming complex information, and
- two specific e-competences: technological literacy and e-awareness.

B. Competences for transforming experience:

(B.1) Gathering knowledge involves taking account of existing opportunities and identifying existing sources of underutilized knowledge. Gathering may include reconfiguring existing sources of knowledge to fit new circumstances. The gathering stage requires the ability to manage large amounts of information while suspending judgment about its usefulness. The gathering process of knowledge involves

- two specific cross-cultural competences: listening and observing, and copying with ambiguity, and
- two specific e-competences: informational and instrumental digital literacy.

(B.2) Applying knowledge involves putting knowledge to work to solve problems, create opportunities or generate change. The Acting stage implies a degree of risk taking where unformulated ideas and their implications are tested out. The applying process of knowledge involves

- four specific competences: taking action and managing others, and
- One specific e-competence: media literacy.

C. Competences for grasping and transforming the experience: Flexibility and Managing Stress: Yamazaki and Kayes (2004) describe additional clusters that fit directly into the category of developmental learning competencies because they suggest a general ability to adapt to the multiple environmental demands of cross-cultural change. Unlike the previous skill clusters, which involve the specialization of one or two modes of learning, developmental competences entail the integration of multiple learning skills.

The combination of each dimension results in a knowledge management process along with corresponding competencies (Kayes et al.,2005a, 2005b). Taken together, the four stages of generating, gathering, organizing and acting constitute a full process of knowledge creation. Importantly, each stage of the learning cycle requires the use of abilities that are diametrically opposed to others, according to the type of learning involved. Generating knowledge requires a person to be open to new experiences and engage directly with the world. On the other hand, organizing knowledge requires a person to be abstract, becoming detached from concrete everyday experience to create generalized knowledge. Similarly, gathering knowledge requires a person to reflect, demonstrate patience and observe the world around them in order to gather all information and its possible meanings, acting on knowledge requires the ability to interact with others and test ideas uninhibited by the emergence of new knowledge. Being able to identify which stage of learning and which subsequent set of competences are needed is the crucial step to manage knowledge from individual learning point of view. The knowledge management process requires multiple repeated cycles

Figure 1. Cross cultural competences in the new economy

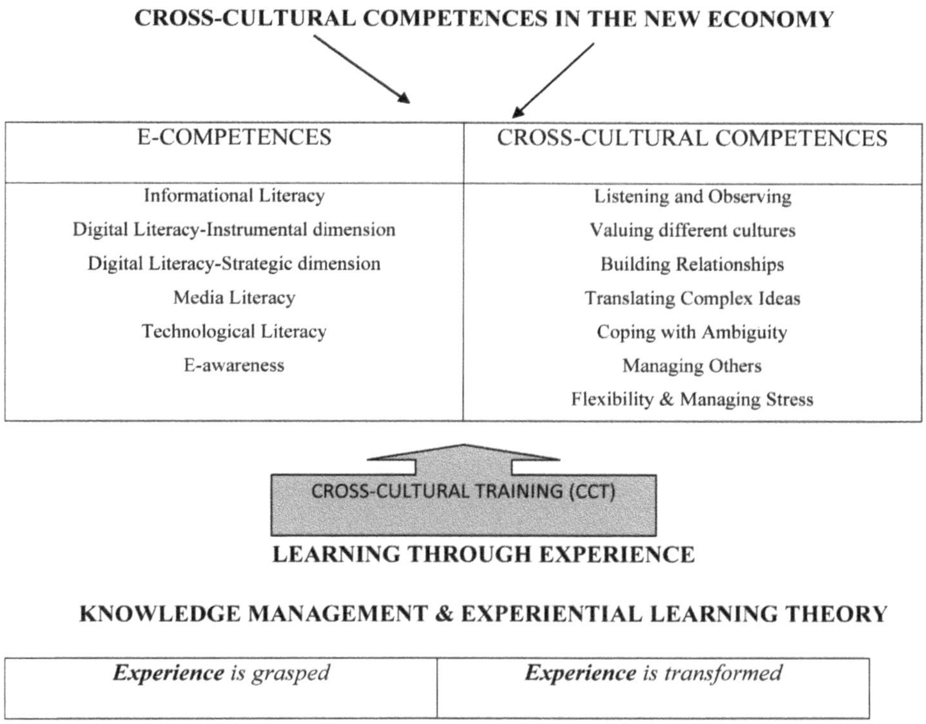

Source: Partially based on Kayes et al. (2005a,b), Yamazaki & Kayes (2004), Kolb (1984) and Cobo (2009)

that don't always follow the idealistic sequence we have suggested in Figure 1.

4. RESULTS: CONCEPTUALIZATION OF E-COMPETENCES AND CROSS-CULTURAL COMPETENCES

Conceptualization of E-Competences

Anderson et al. (1980) use the term computer literacy while Davis (1989) argues that users' reactions to computers are complex and multifaceted. Venkatesh et al. (2003) try to understand usage as dependent variable. They presented the theory of acceptance and use of technology adapted from the most relevant theories applied to computer and information system utilization: Theory of Reasoned Action, Technology of Acceptance Model, Motivational Model, Theory of Planned Behaviour, Combined Theory of Technology Acceptance and Theory of Planned Behaviour, Innovation Diffusion Theory and Social Cognitive Theory.

Morris (2010) described the extent and nature of the ICT skills and competences of the teaching workforce, and found that barriers to the use of ICT involve a lack of teacher awareness about what technologies are available and how ICT can be used to support the delivery of the curriculum. In this regard, the OECD (2005) distinguishes three categories of ICT competences:

1. ICT Specialists who has the ability to develop, operate and maintain ICT systems.

2. ICT Users who use advanced and often sector-specific software tools
3. Basic Users who are competent users of generic tools (e.g., Word, Excel, and Outlook) needed for the information society.

Cobo (2009) has reviewed the term of e-competences in order to propose a new conceptualization. Cobo defines e-competences as a set of capabilities, skills and abilities to exploit tacit and explicit knowledge, enhanced by the utilization of digital technologies and the strategies use of information. Below we outline underlying concepts that constitute the expression of e-competences.

Organizing

- **E-awareness:** This cognitive ability is characterized by a user's awareness of ICT's and appreciation of the relevance of ICT in the information based society. This understanding of the human, cultural and societal issues related to technology and their practice also includes legal and ethical behaviour (digital citizenship).
- **Technological literacy:** The confident and critical use of electronic media for the storage and management of information. The technologies involved in this definition evolve according to technological transformation. The ability to use these tools can be acquired in a formal environment like schools or informal ways such as at home or through friends.

Gathering

- **Informational literacy:** This is the ability to understand, assess and interpret information from all kinds of sources. It means the ability to read with meaning, to understand critically and to evaluate, connect and integrate different judgments about what is found on or offline, identifying the sources, authors and their diverse approaches. Being able to evaluate the reliability and quality of information is a key aspect in deciding what and when information is needed for specific audience. Two very important abilities related to information literary are:
    - **evaluation** (reflecting too make judgments about the quality, relevance, usefulness, efficiency, authority and timelines of the information),
    - **integration** (interpreting, summarizing, drawing conclusions, comparing and contrasting information from multiple digital sources).

Generating and Gathering

- **Digital literacy:** The proficiency to build new knowledge, based on the strategic employment of ICTs is termed digital literacy. The main aspects related to digital literacy are:
    - Instrumental dimension: how to get relevant information, and
    - Strategic dimension: how to manage and produce new knowledge.

Being digitally literate means using technology for information and knowledge in order to access, retrieve, store, organize, manage, synthesize, integrate, present, share, exchange and communicate in multiple formats, either textual or with multimedia. Digital literacy also means to understand that the management and sharing of new products of information could be enriched through networks of collaboration, just as open software communities do.

Taking Action: Applying Knowledge

- **Media literacy:** Media literacy has to do with understanding how the traditional mass media and the digital media are merging and evolving toward a new media landscape. This process of understanding and using mass media in an assertive way includes an informed and critical viewing or critical analysis of the media's nature. Media literacy competence facilitates the evaluation of the risk to put into new context the new information in an assertive way, managing events as they unfold and dealing effectively with the unintended consequences of action.

This set of literacies can be defined broadly or narrowly. Following Cobo (2009), we suggest understanding the term *e-competences* as a meta-competency that denotes the interaction of different skills and knowledge which are constituted by the five underlying concepts we reviewed. The proposed definition embraced cognitive abilities and also technical proficiencies. It encapsulates the idea that the development of e-competences is enriched by the continuous interaction (and connection) between knowledge and experience, and it evolves as new technologies and the labour market evolve.

E-competence involves confident and critical use of ICT. Reflective attitude towards available information and a responsible use of the interactive media is needed when engaging in e-work teams. The following cross-cultural competence complement the taxonomy of e-competence presented above. In the cross-cultural competence we find the abilities to improve collaboration in the global environment.

Conceptualization of Cross-Cultural Competences

Cultural toughness (Mendenhall & Oddou, 1985) or culture novelty refers to the degree to which the host culture differs from the employee's home culture in terms of numerous cultural indicators, such as values, assumptions about human life, or norms. Countries can be ranked in terms of their culture novelty from the occidental point of view and the implications for the acquisition of new cross-cultural skills. Cultural depth refers to the length and task complexity of the overseas assignment as well as the requirement to work closely with co-workers from different cultural backgrounds. The deeper the assignment, the more in-depth and detailed the cross-cultural training needs to be (Engle et al., 2001).

Yamazaki and Kayes (2004) propose a taxonomy of skills necessary for cross-cultural learning based on Kolb's Experiential Learning Theory (ELT). As Yamazaki and Kayes point out, the interaction between home and host cultures provides and alternative way to understand cross-cultural research by suggesting that specific learning strategies might be necessary in the face of specific cross-cultural learning situations. The cross-culture competences clusters are grouped through each learning skill dimension, following Experiential Learning Theory and Knowledge Management Process (Yamazaki and Kayes, 2005a,b).

Generating Knowledge: Interpersonal Learning Skills

- **Valuing different cultures:** This competence involves understanding the complexities of cultural norms and how they contribute to new knowledge. A number of cultural typologies exist to guide an initial understanding of cultural differences (e.g., Hofstede, 1980). House et al. (2002) suggest that culturally embedded values lie along nine dimensions: uncertainty avoidance, power distance, societal collectivism, in-group collectivism, egalitarianism, assertiveness, future orientation, performance orientation, and human orientation. Valuing different cultures goes beyond simply knowing the abstract differences between cultures to appreciating how these differences are expressed in day-to-day situations.

- **Building relationships within the host culture:** This cluster consists of competences related to fostering and developing human relationships. New knowledge is generated by coming into contact and, thus, creating new experiences with others. This involves putting a human face on culture.

Gathering Knowledge Observational Learning Skills

- **Listening and observing:** Absorbing knowledge across cultures requires individuals to carefully observe the host culture and its local practices and to understand the rationale behind the practices. Listening and observing requires patience in the face of pressing demands.
- **Understanding nonverbal cues:** is essential to successful cross-cultural knowledge management for employees. Watching how people interact, the closeness of their physical distance, the expressions on their faces and the manner in which they engage can provide clues. Listening and observation competences are among those most likely to be important in cross-cultural knowledge efforts according to Yamazaki and Kayes (2004). The Reflective Observation (RO) learning mode calls for understanding the meaning of ideas and situations by carefully listening and watching.
- **Coping with ambiguity:** Successful cross-cultural learning requires the ability to deal with ambiguities and lack of clarity that comes with working in a foreign culture. Successful professionals learn to cope with the lack of clarity around certain cultural interactions. This cluster of skills constitutes RO abilities, as dealing with ambiguity and making sense of unstructured and unpredicted situations requires patience, impartiality and thoughtful judgment, as well as the ability to reflective observe large amount of new and seemingly disparate data.

Organizing Knowledge: ANALYTIC Learning Skills

- **Translating complex ideas:** Translating complex ideas is central to the knowledge management process. Translation relies on a grasp and command of the local language and its meaning. The degree of proficiency required to effectively absorb and transfer knowledge remains in question; however, it is clear that without an in-depth knowledge of the host language, deep absorption and transfer of knowledge is less likely. Language is particularly important because it allows meaningful relationships to be developed. Yamazaki & Kayes (2004) conclude that translating complex information into another culture was primarily a language skill and was theoretically associated with the Abstract Conceptualization (AC) mode of learning.

Applying Knowledge: Action Learning Skills

- **Taking action:** The ability to deal with ambiguity brought about through cross-cultural interaction becomes essential for taking action. Even when the implications, details or cross-cultural applicability of knowledge transfer remain unclear, successful cross-cultural learning requires taking action. Successful employee learning involves appreciating differences between host and home culture, but also knowing when to move a project forward, despite these differences. The action and initiative skills cluster describes how people actively handle an intercultural situation. This orientation relates to the Active Experimentation (AE) learning mode, which involves the ability to actively affect people and influence situations.

- **Managing others:** This competence involves managing people from different cultures. The management skill serves to resolve conflicts between local employees and expatriates and to maintain close relationships between them. The management skills relate to action skills and the AE mode. The AE mode calls for actively affecting people and emphasizes both practical applications and pragmatic approaches.

Developmental Learning: Adaptability

Yamazaki and Kayes (2004) describe two additional clusters that fit directly into the category of developmental learning competences because they suggest a general ability to adapt to the multiple environmental demands of cross-cultural change. Unlike the previous seven skill clusters, which involve the specialization of one or two modes of learning, developmental competences entail the integration of multiple learning skills. Developmental learning constitutes higher order adaptive competences because these competences demonstrate an ability to respond to multiple circumstances, to exert flexibility in learning style, and to move out of one's preferred mode of learning.

- **Adaptability and Flexibility:** The ability to accept changes, setbacks and challenge and not be rigid. CE, AE and RO are critically for cross-cultural adaptation. Societal, intercultural and human skills may be more important than quantitative and analytical skills. The most adaptive employees tend to learn from his or her experience through proximate, specific situations.
- **Managing Stress:** Stress management skills constitute the ability to deal with the tensions inherent in complex learning situations. Working with people from different cultures can create high tension experiences, so this is an essential skill. This cluster of skills has been described in the literature as emotional maturity to tolerate stress, depression and loneliness.

Implications for Cross-Culture Training (CCT)

Black and Mendenhall (1990) examined the effectiveness of Cross-cultural Training (CCT) relative to three outcomes: cross-cultural skill development, cross-cultural adjustment and job performance. They also proposed a theoretical model that was largely absent in the empirical work reviewed. Cross cultural training is any procedure used to increase an individual's ability to cope with and work in a foreign environment (Tung, 1981). Further, individuals could avoid some role conflict through anticipatory socialization in which they obtained advance information about a new culture and adopted their anticipated roles.

Most of the scholarly work on CCT has focused on the discussion of methods of training and classifications of these methodologies, while less attention has been focused on frameworks able to determine which training method to utilize. Landis and Brislin (1983) have proposed a typology of methods based on a broad review of the cross-cultural literature. Their classification scheme is the following: They differentiate among information or fact-oriented training, behaviour training (attributions and cognitive behaviour modifications), cultural awareness training, experiential learning through active participation in other cultures. Finally, interaction training where trainees interact with natives or returned expatriates in order to become more comfortable with host-nationals and to learn from the first-hand experience of the returned expatriates.

Tung (1982) presented a contingency framework for choosing an appropriate CCT method. Tung argued for two key situational factors:

1. the degree of interaction required in the host culture and
2. the similarity between the individual's native culture and the new culture.

Mendenhall & Oddou (1986) in addition proposed three key elements related to training:

1. Training methods
2. Low, medium and high levels of training rigor
3. Duration of the training relative to degree of interaction and culture novelty.

Despite its strengths, the framework is limited with regard to training and learning process.

Social Learning Theory (SLT) is a complementary approach. SLT as described by its leading proponent (Bandura, 1977) implies that learning takes place by the effect reinforcement has on behaviour and by imitating the behaviour of others and symbolically or vicariously making associations between behaviour and consequence without direct, actual experience. SLT has four central elements for modelling and learning new behaviour: attention, retention, reproduction and incentives. Within SLT there are basically two modelling processes – symbolic and participative:

1. **Symbolic modelling** simply involves observing modelled behaviours. However, this observation can have two forms: reading and hearing. Specific CCT methods that generally exhibit this type of modelling include: lecture, books, verbal factual briefings, films, role modelling, demonstrations, and non-participative language training.
2. **Participative training** essentially means that in addition to observing the modelled behaviour, the trainee also participates in modelling the behaviour. The first form involves "verbal" participation when the trainee describes verbally what he or she would do. CCT methods that generally exhibit this type of participative modelling include case studies and culture assimilators. The second form of participative modelling involves more physical participation in modelling the behaviour being learned. CCT training methods that generally require this type of participative modelling include role plays, interactive language training, field trips, and interactive simulations. Trainees are more cognitively involved when they must physically as opposed to only verbally participate in modelling the behaviours being taught.

One technique that can help individuals value another culture is reading novels or stories written by and highly valued in that culture. These writings provide insight into the values of the host culture and provide an opportunity to show interest in the host culture when engaged in conversations with host culture members. Black and Mendehall (1990) provide a ranking in terms of rigor for a set of specific and common CCT methods:

1. **Experiential Training Methods:** These methods held high training rigor, are more rigours ad implies a complex modelling process. E.g.: Field trips, simulations, role plays, interactive language training. These methodologies develop active experimentation.
2. **Analytical Training Methods:** These methods imply medium training rigour. E.g.: Case studies, sensitivity training, culture assimilators, classroom language training, and films.
3. **Factual Training Methods:** These methods involves symbolic modelling process but not participative with low training rigour. These methods develop observational capabilities. E.g. Books, lectures, and briefings.

In general, the SLT literature and the CCT literature strongly suggest that the more rigorous the training the more effectively the trainee will be able to actually and appropriately execute the learned behaviour. The basic explanation for this relationship is that rigor increases the level of attention and retention thereby proficiency.

One could go beyond this view and argue that CCT should focus on training individuals to develop culturally consistent modes of managing impressions (Giacalone and Beard, 1994). Such a view would posit that anticipatory socialization of roles is not enough; employees need to be socialized at a more basic level and should understand the context and content of making a good impression from a fundamentally culture-based point of view. Thus, training employees in culturally consistent and specific impression management techniques provides the avenue for the employee to consciously switch from an automatic, home-culture mode to an impression management mode that is culturally appropriate and acceptable.

Over time, and with training, practice and patience, employees will tend to recognize and adopt different countries attitudes. It is irrelevant as to whether employees share the communication styles, traditions, values or desires of the people with whom they interact. What it is important, however, is that other employees, other people perceive that they share these cultural dimensions or at least respect these dimensions.

Understanding culturally consistent impression management would require a broad range of training approaches, where an experiential approach is preferable. For instance, field experiences in the other culture might provide insight into the deficiencies of the training as well as to sharpen the impression skills. Trainees should at least be given a working knowledge of how they can mitigate having said or done the wrong thing (protective strategies) as well as how they might enhance their reputation and perceived contribution (acquisitive strategies). Although the impression management strategies of excuses, justification, apologies, and disclaimers may all help an expatriate mitigate an offense he has created, the particulars of how these strategies are used contextually, the range of the specific actions within each strategy (for example, the varying types of excuses that one might use), as well as their overall acceptability, will no doubt vary from culture to culture.

Despite extensive research and critical reviews several barriers to understanding cross-cultural learning remain (Yamazaki and Kayes 2004). First, there is a lack of cross-disciplinary approach that integrates diverse findings. Second, reviews have focused on the skills and abilities necessary for training, success or job promotion, but have failed to account for how those skills are learned or developed. Therefore, Yamazaki and Kayes suggest the ELT framework to promote cross-cultural competence learning. Through the ELT framework cross-cultural training methods are evolving to a participative one where complex modelling behaviour is possible.

Case Study for Cross-Cultural Training

Case studies are shifting from teaching about culture to developing skills to manage across cultures. The former focuses on abstract knowledge, whereas the later focuses on developing interpersonal skills and increasing awareness in self and others. Through blended learning, employees come improve their e-competence and cross-cultural competences.

The case study method also develops communicative competence by putting the students at the centre of the action where they can use language actively and practice communication skills. As Rodgers, (1998b, p. xi) puts it "The role of the teacher becomes that of negotiations coach, facilitator, reference person and provider of feedback and the classroom becomes a workshop in communication and language skills and cross-cultural awareness development". In the active approach to learning fostered by this method, students will achieve communicative competence more efficiently. This method allows the students to expand their repertoire of communication skills by requiring them to develop presentations, team working, and networking skills, as well as critical/analytical problem-solving skills, all of which are required in a modern business context.

Gesteland (2005) try to reduce the confusion of cultural differences by classifying international business customs and practices into logical patterns. Chaney & Martin (2007) sensitize readers to the dynamics of international diversity by acknowledging both cultural and individual differences. They address a broad range of issues and skills that are crucial to effective intercultural encounters. Topics selected in Chaney & Martin's book include the nature of intercultural communication, contrasting cultural values, cultural shock, language, communication patterns, etiquette and customs, intercultural negotiation, and law.

Implications for HR Managers and Professionals

From the perspective of experiential learning, knowledge management is the process whereby new knowledge is created, transformed, absorbed and applied through new experiences. Cross-cultural knowledge management in the e-environment results from interactions with others from different cultures that require new understanding and interpretations. Language and communication serve as the basis for such learning, given that the communication instrument is Information and Communication Technology the challenges also become greater.

Once the particular stage of the learning cycle has been identified, the individual can work towards matching the appropriate skill necessary to manage the process. Enlisting others and identifying ways to obtain the skills necessary form managing different stages of the process may be the key of the process in knowledge management.

The four styles of learning (Concrete Experience CE, Reflective Observation RO, Abstract Conceptualization AC, and Active Experimentation AE) are related to four abilities of learning (Interpersonal Abilities, Observational Abilities, Cognitive Abilities and Behavioural (Action Learning and Adaptability) Abilities). In this way, human resource managers can take steps to improve competences. Effective knowledge management efforts require continual efforts (Kayes et al., 2005a,b). By viewing the process of knowledge management as four-stage cycle of learning (CE, RO, AC, AE), giving attention to generating, gathering, organizing and acting provide a comprehensive approach to knowledge management the cross-cultural e-environment.

There are three appropriate steps to develop the model proposed here. First, people should work towards identifying and developing general cross-cultural and e-competences. Cross-cultural Training provides a means for modelling and valuing different behaviours. The next step is to understand the nature of specific cultures and their specific demands through learning from experience. Finally, individuals should transform the experience into more complex and adaptive competences.

In the e-environment, HHRR professionals should consider the effects of the virtual distances managed through the e-competences and cross-cultural competences.

5. CONCLUSION AND FINAL THOUGHTS

This chapter has identified nine competences related to cross-cultural learning and five competencies related to e-competences. Experiential Learning Theory and Knowledge Management Model for organizing, gathering, applying and organizing knowledge provides a useful and comprehensive models to enlist these competences. In particular, we view knowledge management from an interpersonal perspective, where knowledge is gathered, processed and put into use through interpersonal relationships in the global e-environment.

In order to adopt a wider approach to ICT skills beyond the instrumental (and basic) use of technology, we propose that e-competences go beyond the use of any specific ICT, including the proficient use of information and the application of knowledge to work individually and cross-culturally collaboratively in changing contexts.

Organizations that support these competences, through training and development efforts, are more likely to realize the knowledge management process needed in the 21st century. Further, organizations that consider these competencies in the selection and evaluation of employees are more likely to foster knowledge management and increase flexibility.

In the context of cross-cultural training the primary implication is a shift from teaching about culture and language to developing skills to manage across cultures. The former focuses on abstract knowledge, whereas the later focuses on developing interpersonal skills and increasing awareness in self and others. Understanding a culture without understanding the language is difficult. Although languages often do have equivalencies in other languages, the social reality cannot be fully conveyed to a person who does not speak the language.

REFERENCES

Adam, N. (2007). *Workforce e-maturity-characteristics and development*. Nottingham, UK: NAACE. Retrieved January, 9, 2011, from http://www.naace.co.uk/448

Anderson, R. E., Krohn, K., & Sandman, R. (1980). *User's guide for the Minnesota computer literacy and awareness assessment*. St Paul, MN: Minnesota Computing Corporation.

Ang, S., & Inkpen, A. C. (2008). Cultural intelligence and offshore outsourcing success: A framework of firm-level intercultural capability. *Decision Sciences*, *39*(3), 337–358. doi:10.1111/j.1540-5915.2008.00195.x

Autor, D., Levy, F., & Murnane, R. (2003). The skull content of recent technological change: An empirical exploration. *The Quarterly Journal of Economics*, *118*(4), 1279–1333. doi:10.1162/003355303322552801

Ayoko, O. B. (2007). Communication openness, conflict events and reactions to conflict in culturally diverse workgroups. *Cross Cultural Management: An International Journal*, *14*(2), 105–124. doi:10.1108/13527600710745723

Bandura, A. (1977). *Social learning theory*. Englewood Cliffs, NJ: Prentice Hall.

Black, J. S., & Mendenhall, M. (1989). A practical but theory-based framework for selecting cross-cultural training methods. *Human Resource Management*, *28*(4), 511–539. doi:10.1002/hrm.3930280406

Black, J. S., & Mendenhall, M. (1990). Cross-cultural training effectiveness: A review and theoretical framework for future research. *Academy of Management Review*, *15*(1), 113–136.

Boyatzis, R. E. (1982). *The competent manager: A model for effective performance*. New York, NY: Wiley.

Boyatzis, R. E., & Saatcioglu, A. (2008). A 20-year view of trying to develop emotional, social and cognitive intelligence competencies in graduate management education. *Journal of Management Development*, *27*(1), 92–108. doi:10.1108/02621710810840785

CEDEFOP. (2006). *ICT skills certification in Europe. Dossier. European Center for the Development of Vocational Training*. Luxembourg: European Communities.

Chaney, L. H., & Martin, J. S. (2007). *Intercultural business communication*. New Jersey: Pearson, Prentice Hall.

Cobo Romani, J. C. (2009). *Strategies to promote the development of e-competencies in the next generation of professionals: European and international trends*. ERC Centre on Skills, Knowledge and Organisational Performance, Cardiff and Oxford Universities. SKOPE Monograph No. 13 ISSN 1466-1535. Retrieved December 27, 2010 from http://www.cardiff.ac.uk/socsi/research/researchcentres/skope/publications/monographs/monograph13.pdf

Cross, T., Bazron, B., Dennis, K., & Isaacs, M. (1989). *Towards a culturally competent system of care* (*Vol. 1*). Washington, DC: Georgetown University Child Development Center, CASSP Technical Asistance Center.

Davis, F. D. (1989). Perceived usefulness, perceived ease of use, and user acceptance of Information Technology. *Management Information Systems Quarterly*, *13*(3), 319–340. doi:10.2307/249008

Earley, P. C., & Ang, S. (2003). *Cultural intelligence: Individual interactions across cultures*. Palo Alto, CA: Stanford University Press.

Engle, A. D., Mendenhall, M. E., Powers, R. L., & Stedham, Y. (2001). Conceptualizing the global competency cube: A transnational model of human resource. *Journal of European Industrial Training*, *25*(6), 346–353.

Gesteland, R. R. (2005). *Cross-cultural business behavior*. Liber, Denmark: Copenhagen Business School Press.

Giacalone, R., & Beard, J. W. (1994). Impression management, diversity and international management. *The American Behavioral Scientist*, *37*(5), 621–636. doi:10.1177/0002764294037005004

Goleman, D. (2000). An EI-based theory of performance. In Goleman, D., & Cherniss, C. (Eds.), *The emotionally intelligent workplace: How to select for, measure, and improve emotional intelligence in individuals, groups, and organizations*. San Francisco, CA: Jossey-Bass.

Goleman, D., Boyatzis, R., & Mckee, A. (2002). *Primal leadership: Realizing the power of emotional intelligence*. Boston, MA: Harvard Business School Press.

Guo, C., Shim, J. P., & Otondo, R. (2010). Social network services in China: An integrated model of centrality, trust, and technology acceptance. *Journal of Global Information Technology Management*, *13*(2), 76–79.

Hofstede, G. (1980). *Culture's consequences: International differences in work related values*. Beverly Hills, CA: Sage.

Hofstede, G. (2001). *Culture's consequences: Comparing values, behaviours, institutions and organizations across nations* (2nd ed.). London, UK: Sage Publications.

House, R., Javidan, M., Hanges, P., & Dorfman, P. (2002). Understanding cultures and implicit leadership theories across the globe: An introduction to project GLOBE. *Journal of World Business*, *37*(1), 3–10. doi:10.1016/S1090-9516(01)00069-4

Kayes, D. C., Kayes, A. B., & Yamazaki, Y. (2005a). Essential competencies for cross-cultural knowledge absorption. *Journal of Managerial Psychology*, *20*(7), 578–589. doi:10.1108/02683940510623399

Kayes, D. C., Kayes, A. B., & Yamazaki, Y. (2005b). Transferring knowledge across cultures: A learning competencies approach. *Performance Improvement Quality, 18*(4), 87–100. doi:10.1111/j.1937-8327.2005.tb00352.x

Kolb, D. A. (1984). *Experiential learning: Experience as the source of learning and development*. Englewood Cliffs, NJ: Prentice Hall.

Landis, D., & Brislin, R. (1983). *Handbook of intercultural training* (*Vol. 1*). Elmsford, NY: Pergamon Press.

Matveev, A. V., & Nelson, P. E. (2004). Cross cultural communication competence and multicultural team performance. *International Journal of Cross Cultural Management, 4*(2), 253–270. doi:10.1177/1470595804044752

Mendenhall, M., & Oddou, G. (1985). The dimensions of expatriate acculturation: A review. *Academy of Management Review, 10*(1), 39–47.

Mendenhall, M., & Oddou, G. (1986). Acculturation profiles of expatriate managers: Implications for cross-cultural training programs. *The Columbia Journal of World Business, 21*(4), 73–79.

Morris, D. (2010). Are teachers technophobes? Investigating professional competency in the use of ICT to support teaching and learning. *Procedia - Social and Behavioral Science, 2*(2), 4010-4015.

NCCC National Center for Cultural Competence, Georgetowm University Center for Child and Human Development, Centers for Excellence in Developmental Disabilities. (1998). *Definitions of cultural competences*. Washington, DC: Georgetown University. Retrieved 10th January, 2011, from http://www.ncccurricula.info/culturalcompetence.html

OECD. (2005). *New perspectives on ICT skills and employment. Working paper on the information economy.* Directorate for Science, Technology and Industry, Committee for Information, Computer and Communications Policy. Paris, France: Organisation for Economic Co-operation and Development. DSTI/UCCP/IE. (2004). *10/Final*.

Redding, W. C. (1972). *Communication within the organization: An interpretive review of theory and research*. New York, NY: Industrial Communication Council.

Rodgers, D. (1998b). *Business communications. international case studies in English*. Cambridge, UK: Cambridge University Press.

Rogers, D. (1987). The development of a measure of perceived communication openness. *Journal of Business Communication, 24*(4), 53–61. doi:10.1177/002194368702400404

Schiller, S. Z., & Cui, J. (2010). Communication openness in the workplace: The effects of medium (F2F and IM) and culture (U.S. and China). *Journal of Global Information Technology Management, 13*(2), 37–75.

Şimşek, Ö., Altum, E., & Ateş, A. (2010). Developing ICT skills of visually impaired learners. *Procedia-Social and Behavioural Sciences, 2*(2), 4655–4661. doi:10.1016/j.sbspro.2010.03.745

Stahl, B. C., & Elbeltagi, I. (2004). Cultural universality versus particularity in CMC. *Journal of Global Information Technology Management, 7*(4), 47–65.

Tung, R. (1981). Selecting and training of personnel for overseas assignments. *The Columbia Journal of World Business, 16*(1), 68–78.

Tung, R. (1982). Selection and training procedures of U.S., European & Japanese multinationals. *California Management Review, 25*(1), 57–71.

Venkatesh, V., Morris, M. G., Davis, G. B., & Davis, F. D. (2003). User Acceptance of Information Technology: Toward a unified view. *Management Information Systems Quarterly*, *27*(3), 425–478.

Yamazaki, Y., & Kayes, D. C. (2004). An experiential approach to cross-cultural learning: A review and integration of competencies for successful expatriate adaptation. *Academy of Management Learning & Education*, *3*(4), 362–379. doi:10.5465/AMLE.2004.15112543

KEY TERMS AND DEFINITIONS

Competence as Abilities: A person's ability to adapt to specific environment or circumstance. Competences are developed into more concrete form that can be called skills, following Kolb's terminology.

Cross-Cultural Competence: A person's ability to adapt to new cultural settings.

Cross-Cultural Training: Classification of training methodologies to recognize and adopt different countries attitudes in work environments.

E-Competences: A person's ability to adapt to new digital environment.

Experiential Learning Theory: Kolb's Theory describes learning as the process of transforming experience into new knowledge. Knowledge results from the combination of grasping and transforming experience.

Knowledge Management: It represents the process for creating, gathering, applying and organizing knowledge.

Section 4
Leadership and Management Issues in the New Economy

Chapter 15
Leadership Talent Development in the New Economy

José Manuel de Haro
AGBAR Group, Spain

José Antonio Carrión
Opem Consultants, Spain

ABSTRACT

In this chapter, the authors analyze how it is possible to achieve the best results in the task of developing leaders. For this purpose, they first present as a starting point a framework based on strategic principles for managerial development which have to be assumed before taking action and, after which, they propose the operational means which allow to go beyond what has so far constituted the content of the organizational practices in this field. Intentions are as follows: reducing the level of complexity perceived about the concept itself and about its tools; including the most relevant managerial development contents; presenting an orderly, consistent process which can guide future interventions in this field. At present, and even though the authors think that the contribution made in this chapter will help introduce significant improvements in managerial development practices, there are still quite a few pending issues that will gradually have to be solved in order to make progress.

INTRODUCTION

Acting faster and faster, depending on shared knowledge and competing in complex, global and diverse scenarios are some of the *critical requirements of the new economy environment* where leaders will have to get by. In this complex and demanding scenario, those who occupy the posts involving a greater deal of responsibility inside the organizations will have to gradually assume new functions that permit those organizations' continuity and growth. Managerial development, integrated in the *human resources* function, is the tool that can contribute to prepare leaders so that they can acquire or develop the competencies required to meet the demands of these new roles.

The deep changes occurred in the economy on a global scale have followed a process of economic waves (Alvin Toffler, 1980, 1998). History can be divided into periods (waves), the

DOI: 10.4018/978-1-61350-207-5.ch015

first of which started in the Neolithic Age 8,000 years ago (*agricultural wave)*. The second one broke out halfway through the 17th Century: the *industrial wave*. The third one, the *technological wave*, which began in 1950 with the first computers, lasted until the 1990s, when the emergence of the Internet generated the fourth wave, the *Information and Communication Technologies Wave*, which is the one we are in at the moment, and the one where the new economy is located.

The "*New Economy*" is essentially supported on *knowledge,* due to the new breakthroughs in technology and to the economic globalization (Friedman, 2005). Three fundamental, highly-interrelated aspects characterize it: (1) Information and knowledge as the bases of organizations' production, productivity and competitiveness; (2) Globalization (the economy is interrelated and it is organized on a worldwide scale); and (3) the importance of the Net –through the Information and Communication Technologies (ICT)– in the organization, communication and commercial transaction system.

The elements in this new scenario are creating challenges and opportunities which will not only require new ways to establish relationships inside markets, new ways to communicate and new working patterns, but will also make many of the traditional jobs become obsolete; and they will equally demand *competencies* oriented to meet as quickly as possible the quality requirements of customers seeking the highest possible alignment with the organizational strategy. The new world of work is global, rapidly changing and customer-oriented; it expects quality and is intolerant with delays and excuses (McLagan, 2008). This situation is redefining the foundational concepts of time (speed: instantaneous), space (global connectivity), structure (networks), matter (dematerialization), and competition (increasing economic returns) on which the economy operates.

All these changes which shape this neo-economic environment are giving rise to trends that will have deep implications for the leaders of all organizations (Jaworski & Scharmer, 2000): vast globalization, the primacy of networking and connectivity, the steady de-materialization of products into knowledge, speed as the prime source of competitive advantage, and the widening dominance of winners, takes most modes of competition. The different executives and leaders in a company play a key role in its success. Research has proved that leadership affects organizational performance and accounts for more variance in performance than other organizational variables (Burke & Litwin, 1992). Its impact has become even stronger in present-day enterprises. The increased competitiveness has forced organizations to adapt, reducing costs and removing managerial layers. For those managers who stay, the objectives are to improve product quality, to reduce costs and to head increasingly numerous groups 2002). As a consequence, leaders and managers have to adapt to a whole host of new ground rules for business (Jaworski & Scharmer, 2000).

If we do not have agile and efficient mechanisms to develop the abilities of the leaders who demand this environment, the organizations will not be able to compete advantageously. The *human resources function,* which must lead the satisfaction of these demands, has to anticipate events and –in the light of the new strategic prospects– develop well-built practices and tools which permit the implementation of the best solutions inside the organization. The development area, and particularly the leadership development area, becomes especially relevant within this function, as it must be able to identify the criteria which define *high managerial performance* and to provide the best tools for the *learning of its critical competencies.*

Therefore, the development of managers so that they can perform new roles is more important than ever and, in our opinion, it has to be structured around the following five broad axes:

1. **Detecting the future; too much of the present.** Strategic thinking is what differentiates leadership in the best firms (HAY GROUP,

2010). Within this framework, we assign particular relevance to the conclusions drawn from the study by Jaworski & Scharmer (2000) on the principles and practices of the new economy leaders. According to these authors, if leaders want to be successful in this environment, they are going to need a new cognitive ability, the *ability to detect and update the emergent realities.* They see *anticipation* as the main competency. Looking into the future requires experience as well as creative and determined risk-takers. This ability, which will need to modify its orientation toward a more middle-term approach, can be developed.

2. **Focusing on results without forgetting people.** Executive leadership is concerned with creating organizations that have a high spirit of performance. We can group together the attributes of result-centered leadership that will permit to reach this spirit into four categories (Ulrich, Zenger & Smallwood, 1999): *setting the course* (vision, customers, future); *showing personal character* (habits, integrity, trust, analytical reasoning); *mobilizing individual dedication* (making others commit themselves, sharing power); and *generating organizational capacity* (creating teams, managing change). But the new requirements for the new economy managers in terms of competencies cannot come exclusively from outside, *the employees* also define how their leader should be: *know your followers, stand up for your beliefs, speak with passion, lead by example and conquer yourself (*Kouzes & Posner, 2003). Generating trust between leader and follower and making it possible to achieve extraordinary results requires the development of practices such as: *challenge the process, inspire a shared vision, enabling others to act, modeling the way and encouraging the heart* (Gordon, 2003). The leaders of this new age have to find their own voice and inspire others to find their voice as well.
3. **Innovating without losing sight of reality.** An organization's ability to *innovate* becomes the competitive edge in a global economy. The need to innovate is growing stronger as innovation comes closer to being the sole means to survive and prosper in highly competitive and globalized economies (David & Foray, 2003). Seeking ways to build flexible expertise that is the source of innovation becomes an increasingly important goal. To that end, it will be necessary to develop new competencies which are closer to the cognitive field, thinking originality, innovation, inductive and deductive reasoning, and complex problem-solving (Uhalde & Strohl, 2006). *But senseless innovation is useless.* Having 'practical intelligence' (Sternberg, 1996) or 'good judgment' (at the moment of making the right decision) is regarded as one of the essential qualities of leadership which develops over time and includes 3 areas: *individuals, strategy and crisis* (DeRose & Tichy, 2009).
4. **Being permanently in action without forgetting to stop to think.** The leaders of the future will have to embrace *complexity* and the skills needed to harness it. Complexity has to be addressed through collaborative and interdependent work. Managing unforeseen events, coping with diversity or acting globally are only a few examples of this complexity. In this scenario, we will definitely need more speed, the adoption of a multi-area approach and greater flexibility and resistance to handle unforeseen events. However, in order to ensure that actions are the most appropriate ones, this permanent action has to combine with thinking phases in which access is granted both to the rational and the emotional aspects. The option of networking thanks to the possibilities offered by the new information technolo-

gies will permit to learn faster and going from accumulating operational knowledge to sharing *tacit knowledge*. Possessing an *increasingly wide range of competencies* (it suffices to take a look at the repertoires of Day & Halpin (2001), Lombardo & Eichinger (2003) or Zenger & Folkman (2008)) will become a sine qua non condition to succeed as a manager.

5. **Creating and selling a leadership brand.** A leadership brand represents the identity and reputation of leaders throughout a company. Thinking about leadership as a brand offers insights into leadership effectiveness and into creating sustained and consistent leadership (Ulrich & Smallwood, 2004). Companies that consistently produce effective leaders who allow them to differentiate themselves from competitors generally have a clear leadership brand (*the identity of the leaders throughout an organization that bridges customer expectations and employee and organizational behavior* (Ulrich & Smallwood, 2007)). The integrating leader, seen as someone who breeds multiple perspectives, who connects these perspectives consciously and applies a variety of skills in order to establish new directions, options and solutions for the organization, can be a good model to start creating that leadership brand.

In this chapter, we are going to analyze how it is possible to achieve the best results in the task of developing leaders whose behavior is near these five axes. For this purpose, we will firstly present as a starting point a framework based on strategic principles for managerial development, after which we will propose the operational means for its implementation.

KEY PRINCIPLES (ASSUMPTIONS) FOR LEADERSHIP DEVELOPMENT

The concept of managerial development is not easy to define. The different ways to understand and conceptualize leadership make it difficult to specify its study object.

In some cases, these attempts have focused on the *development process* itself. We can classify them into four categories: *leadership training through personal growth; leadership development through conceptual understanding; leadership development through feedback; and leadership development through skill building that focuses on key leadership competencies that can be taught* (Conger, 1992).

In other cases, they have emphasized its evolutionary character, as something potential which is constantly moving. The orientations of McCauley, Moxley & Van Velsor (1998), who defined managerial leadership development as "the expansion of a person's capacity to be effective in leadership roles and processes", or that of Brungardt (1996), who understands it as "every form of growth or stage of development in the life-cycle that promotes, encourages and assists the expansion of knowledge and expertise required to optimize one's leadership potential and performance", are examples of this category.

There have also been authors who focused on the aims of leadership training, which must be: leader's knowledge, skills, and attitudes; training in success and effectiveness as a leader; and training and education on leadership styles (Bass, 1990).

Finally, some have tried to *differentiate it* from other concepts, e.g. management development (it is indeed often confused with management development (Kotter, 1990; Kouzes & Posner, 1995)), or have considered it a part of the education for leadership: *individual skill development, instilling organizational values that promote leadership, and strategic interventions* that encourage dialogue and effect change throughout an organization (Conger & Benjamin, 1999).

It was easier to find a consensus on the organizations' need to develop it and invest in it (Day & Halpin, 2001). The American Society of Training and development (ASTD, 1995) reported that 60% of the Fortune 500 companies surveyed in 1995 listed leadership development as a high priority up from only 36% in 1990.

Seeking to provide a properly structured framework that can help us guide future work in the area of managerial development, we are going to use this section to present the basic principles or assumptions which, in our view, should be included in the treatment of this topic as a previous step before the proposal of key action patterns that must guide our future interventions. Our intentions are: reducing the level of complexity perceived about the concept itself and about its tools; including the most relevant managerial development contents; presenting an orderly, consistent process which can guide future interventions in this field. With that aim in mind, we are going to make a tour through the elements that we regard as essential within the process shaping our conceptual and methodological approach.

Leadership Development Must be Based on Performance Improvement

Training is a form of intervention meant to develop new behaviors or improve the already existing ones through learning. However, its value as an agent of change should be revised (Rackman & Morgan, 1977) because its impact is very limited (Mena, 1989). When formal managers perceive a problem related to performance, training may be the solution, may be part of the solution or may have nothing to do with it (Harless 1989). In most cases, it is bound to apply only to problems caused by a behavior that someone "cannot" (i.e. is not able to) perform. If the problem is different, training is not the solution. Therefore, despite being a necessary requirement for change, training on its own does sometimes not suffice to make that change happen (Komaki, 1998). Environmental pressures may prevent the new skills learned in the process from being implemented.

Nevertheless, even if its contribution is limited, the purpose of training should be to improve performance. Training has to be oriented to solving performance-related problems and not only to improving the knowledge, skills or attitudes of employees. From this point of view, training or development are *independent variables* with effects that we want to check in the behaviors which are relevant to the post. On the other hand, the object that we wish to improve, i.e. performance, is the system's *dependent variable* which includes both what is done (behaviors) and what is achieved (results). Performance is defined as the outcome or result as well as the indicator of effort and achievement (Burke & Litwin, 1992). These outcomes included productivity; profits; service quality; and customer or employee satisfaction.

If we want to modify managers' *performances* –understood as a combination of their *behaviors and results*– it is necessary to work on their behaviors. If we manage to modify the behavior, we will succeed in changing performance, which in turn means producing a different outcome. Changes in performance are no longer monitored only through training but also through *the performance technology*, which includes all the possible interventions meant to reduce *or eliminate performance deficits.*

Leadership Development Considers Leadership a Kind of Talent

Managerial development is consequently a type of intervention which seeks to alter, in a positive sense, critical behaviors in relation to leaders' results, acting on them with the aim of producing a high organizational performance. The way to conceptualize leadership is important, as it influences development strategies. For us, leadership is a form of talent characterized by a certain combination of competencies.

The key element is *talent,* the outstanding mastery of systematically developed abilities known as *competencies* (knowledge and skills) in at least one field of human activity and to an extent which places individuals within the upper 10% of their age peers who are or were active in that field (Gagné, 2004). We can summarize this as follows (De Haro, 2009b):

1. Talent only exists *when it becomes evident.* Potential, even if it is high, is not talent yet. Giftedness is a potential requirement for high execution levels, but that giftedness is nothing but a feature indicating a subject's potential.
2. There are many *types of talent,* as many as fields where excellent performances can become visible: the academic world, language, science, strategy games, technology, mechanics, computing, arts, visual, drama, music, tutorials, sales or business (Gagné, 1991).
3. A specific *level of intelligence and creativity* is needed to guarantee the existence of high potential. Intelligence is a necessary but not sufficient condition for outstanding performance. A high IQ does not suffice to succeed in a sphere of life. After a certain limit, estimated by some at an IQ above 116 (Renzulli, 1986), other aspects such as creativity start becoming more important.
4. A number of *personal variables* associated with motivation and temperament intervene in the appearance and development of talent. Neither intellectual or academic excellence nor technical skills are enough to reach success. It is necessary to consider and quantify volitive and motivational components such as *initiative and creativity, interests, perseverance, adaptability, cognitive independence, self-esteem, values, style to cope with success and failure, motivation toward achievement, tolerance and stress, tolerance for ambiguity, emotional intelligence; and learning and working strategies* (Lorenzo, 2006).
5. High potential has a *biological substrate,* which in turn suffers changes with the appearance of talent. Links between the level of male hormones (testosterone) and talent (Kanazawa, 2003), or the intervention of myelin in the induction of talent, have been demonstrated.
6. The *social environment* where the individual moves is essential for the appearance of talent if the other psychobiological conditions exist. The macroenvironment and the cultural factors have an impact on talent, which develops within a specific society at a specific moment and is influenced by the family and the school, the labor context, the mentors or models, the climate, the social atmosphere and the community amongst other things.
7. Talent *may or may not* end up being produced from high giftedness. It depends on the psychological variables, the social determinants and the activity performed by the subject. Talent is not innate, but it is not necessarily made either. If a quality is not stimulated over time o is not integrated into the set of qualities talent is made of, it is pushed into the condition of a mere curiosity.
8. Talent *can be measured* using situational tests to evaluate the critical behaviors that according to what is determined through the corresponding validation process, form part of talent competencies; in other words, those associated with successful results (competencies).
9. Although there seems to exist a genetic endowment which sets a limit to the level it can reach, *talent can be developed.* From a situation of high potential, one can reach talent, technical mastery in a field by means of practice within a favorable environment.
10. A *working model* is necessary to be able to study talent inside organizations. An integrated one which includes four elements –engines, limiters, catalysts and results– has been proposed elsewhere (De Haro, 2009b).

The Subject of Leadership Development is an Adult (Leader)

Training in the classroom is not the best way to develop the potential to be a manager. According to the U.S. Labor Statistics and the Educational Development Center (1998), 70% of what employees knew about their job had not been learnt through formative actions, but through informal learning acquired in their workplace. Many organizations realize that the type of traditional, lecture-based, classroom training found in most formal leadership development programs is at best only partially effective at preparing leaders for 21st-century problems (Dotlich & Noel, 1998).

While training may even be a necessary element of leadership development, developmental experiences are likely to have the greatest impact when they can be linked to or embedded in a person's ongoing work and when they are an integrated set of experiences: "Put people in jobs before they are ready" (Chambers, Handfield-Jones & Michaels, 1998). Managerial talent is not taught in classrooms; it is learnt through professional experiences. Classroom training should not be the only part of a leadership development initiative, and it may actually the least critical part.

Therefore, the most effective learning is not acquired in courses but in the performance of certain types of jobs (McKinsey Quarterly Report, cited by Guadalupe Fernández, 2002) which imply assuming a new role with higher responsibility, taking over a new project with great visibility and importance or accepting a post which involves a great deal of responsibility in a foreign country.

Perhaps this has to do with our way to learn things inside and outside the classroom. In the classroom, we tend to use the *intellectual learning* (knowledge) that takes place in the brain cortex. Outside the classroom, when we want to learn new habits, we use more often *emotional learning* which, in addition to the cortex, involves the limbic system (the brain's most primitive area associated with our fast, instinctive reactions).

The principles of adult learning *(Andragogy)* differ from those regulating children's learning *(Pedagogy)*. The latter have been used in most of the formative interventions addressed to adults. Malcolm Knowles (1988) stated in his paper "*The art and science of adult learning*" that adults have their own rules of the game when it comes to learning: adult learners are pressed for time, are goal-oriented, bring with them previous knowledge and experience, have a finite interest in types of information, have different motivation levels, and have different learning styles too.

That is why the most effective programs must incorporate adult learning principles to enhance strategic thinking and other critical and complex learning capabilities (Collins, 2001).

Leadership Development Aims to Improve key Competencies (Those which Define Specific Leadership Talent)

As shown above, developing leadership is a type of intervention where a change is sought. Determining the objective of this change becomes essential in any type of intervention if we want to achieve positive results. In the case of interventions oriented to leadership development, this objective has *not* always been well or clearly defined. Increasing the decision-making process (Vroom & Yetton, 1973), developing self-awareness and self-integration (Shamir & Eilan, 2005), social intelligence (Connelly *et al.*, 2000) or tacit knowledge (Hedlund *et al.*, 2003) are just a few examples of this plurality in the purpose of the programs which have been carried out so far.

The diversity of methods used in the programs implemented, or the lack of an objective validation process, has hindered the task of creating a consistent framework which can guide the construction of an effective intervention. For this reason, the first thing we must do in order to make sure that we will obtain the desired results is to reflect on what can or cannot be the objective of managerial learning.

A first consideration derived from this analysis has to do with *the actual possibility of change.* Different characteristics have different margins of learning and change. Not everything can be changed and, of course, some things may be easier to change than others. For instance, the room for improvement in personality features is 50%, however, in the case of intelligence it is only 25% (Seligman, 1993). In this sense, many traditional training and development methods have failed because they focused on aspects such as personality, attitudes or motivation –which are hard to quantify and change– instead of behavior itself (Makin & Cox, 2004).

The second consideration is related to the *emotional or intellectual character* of the contents covered in managerial development programs. Although the contents included as objectives of change in the work with leaders are generally complex and have a multidimensional nature, we could group them together as belonging to one of these two broad categories: *emotional skills* and *cognitive skills.*

The first category above includes all the aspects associated with *emotional intelligence*, which is defined as a set of competencies that derive from a neural circuitry emanating from the limbic system and involve self-awareness and self-management (Boyatzis & Van Oosten, 2002). This set of variables have been included more often as the object of interventions in leadership development because it is closer to what experts and practitioners have understood as "managerial skills" and because it has been proved that they are connected with high performance at work. Effective leaders use emotional intelligence competencies every day more than others in leadership positions (Goleman, 1998). Higher levels of emotional intelligence are associated with better performance in the following areas: *anticipative management, putting people at ease, self-awareness, balance between personal life and work, straightforwardness and composure, building and mending relationships, doing whatever it takes, decisiveness, confronting problematic employees and change management.* There is now a considerable body of research suggesting that a personal ability to perceive, identify and manage emotion provides the basis for the kinds of social and emotional competencies that are important for success in almost any job.

The emotional intelligence programs derived from Goleman's socioemotional competencies model (Boyatzis, Goleman & Rhee, 2000) have shown their effectiveness working with competencies belonging to four large groups: *self-awareness, self-management, social awareness and social skills.* Thus, for example, Ballou, Bowers, Boyatzis & Kolb (1999) demonstrated that 45-to-55-year-old professionals and executives had improved in an statistically significant way in *self-confidence, leadership, help, definition of objectives and action skills,* which accounted for nearly 70% of the emotional intelligence competencies assessed in the study. Cherniss & Goleman (1998) establish up to 22 guidelines to orient the effective development of emotional intelligence in four stages: *preparation, training, transition to work, and maintenance and evaluation.*

The other great approach to the study of emotional intelligence (Mayer, Salovey & Caruso, 2000b) based on the concept of emotional intelligence as a general ability has also given rise to applications in leadership development through its model with the 4 branches of emotional intelligence assessed using the MSCEIT test (Mayer & Salovey, 2002). These authors consider that emotional skills training should emphasize raising awareness, broadening people's coping repertoire and helping people learn from experience, rather than drilling a narrow set of skills over and over again.

The second category, *cognitive skills*, has received less attention as a goal of leaders' training. Although the first approaches based on rational management (Kepner & Tregoe, 1965; Plunkett & Hale 1982) were initially quite solid, they progressively lost more or more weight because of the impression that they were perhaps not as effective as alternative approaches when it comes to explaining how managers behave. The influential studies carried out by Mintzberg (1973) on what managers do as opposed to what they are supposed to do, or what they say they do, showed that managers hardly ever used rational approaches when making decisions.

However, we can find other recent works like, for example, those on executive intelligence by Justin Menkes (2006), who focuses on 'critical thinking' when it comes to determining what needs to be developed in leaders. Its improvement affects three areas in a manager's effectiveness: *effectiveness with tasks, with people and with oneself.*

A promising paper along these lines is the one written by Mumford, Friedrich, Caughron & Antes (2009), who identified the conditions under which leaders' actions make a difference for their organization when it has to face complex issues. In their integrated model for the leader's behavior, leadership depends on: *the ability to solve problems and the social judgment and expertise capacity.* Their leader development model focuses on five processing activities which cause an impact on the leader's performance: *Scanning (the environment), analysis, forecasting, evaluation and idea generation and planning.*

In our opinion, this is not about choosing between one of these categories but about including the relevant contents in each one of them to implement managerial development programs, since they both impact on leaders' effectiveness. The term *competency,* which goes beyond academic results and the IQ (intelligence quotient), appears as the way to bring both fields together and to succeed in operationalizing talent so that it can form part of the practices developed by our human resources professionals. After all, there are socioemotional and intellectual aspects in every competency.

Leadership Development is Carried out within an Organization

One of the factors which determine the success of a managers' development program is taking into account the organizational environment where it is taking place, i.e. the *strategic requirements.* A review of the literature indicated that HRD practitioners must align managerial leadership development initiatives with the strategic needs of the organization (Koch & McGrath, 1996; Rummler & Brache, 1995; Swanson & Holton, 1999).

The most effective program for an organization is the one where the *strategic framework of the organization drives the content of the training program* (Collins, 2001). The alignment of programs with the organizational culture and strategy thus becomes a facilitator of learning. Being disconnected from the company's real life may lead to the failure of a training program for managers (Berry, 1990) by incorporating dimensions which do not suit the organization or failing to provide value to the corporate strategy (Swanson & Holton, 1999). In order to prevent this from happening, the training program must be preceded by a needs assessment to ensure that the program objectives meet the specific needs of the organization and are linked to the organizational strategy.

The type of atmosphere in which leadership training takes place equally affects these programs. Organizations with stronger *learning environments* demonstrate greater organizational effectiveness (Tannenbaum, 1997). Researchers believe that significant financial payoffs are found among companies that emphasize training and development (Huselid, 1995; Jacobs & Jones, 1995; Lam & White, 1998; Ulrich, 1997). It is necessary to *create an environment that supports the development* of talent and links business strategy with the kinds of experiences that people have if we want to lead a company and help it fulfill its mission (McCall, 2001).

Leadership Development is a Process that must be Assessed

Today's managers *recognize* the impact that evaluation has on the organization, but they rarely think of evaluation as an essential part of their strategy (Bassi, Benson, & Cheney, 1996). Evaluation of interventions is among *the most critical issues faced by the HRD field today* (Holton, 1996). According to Rummler & Brache (1995), evaluation is the primary tool for linking individual performance to that of the organization, for the purpose of taking improvement action. Managers and leadership development systems need to consider evaluation at all the phases of program design and implementation (Day & Haplin, 2001), as the ability to document the economic impact is closely linked to a leadership development program's credibility and influence. Without evaluation, performance cannot be controlled, managed or improved.

Despite this awareness of the importance of evaluation, the analysis of the most outstanding practices in management development indicates that *only high-performance* organizations always evaluate the impact of their managerial leadership development processes (Collins *et al.*, 2000; Fulmer & Wagner, 1999). Even when best practice organizations attempt to assess the impact of leadership development, many of these efforts are *informal and non-rigorous* (Day & Halpin, 2001). Few organizations utilize comprehensive evaluation strategies for management training and education (Saari *et al.*, 1988).

Although there is not too much *literature on managerial leadership* development interventions, some relevant works on this matter have actually been produced. The meta-analyses of Burke & Bay (1986) and the more recent study by Collins & Holton (2004) are two good examples. In the first one, it was found that *managerial training is, on average, moderately effective*. As for the second one, its authors checked that interventions with knowledge outcomes were found to be more effective than in Burke & Day's (1986) meta-analyses, with the most effective interventions using a single-group pre-test/post-test research design. Data suggest that practitioners can attain substantial improvements in both knowledge and skills if sufficient front-end analysis is conducted to assure that the right development is offered to the right leaders.

Most of the attempts to measure the effectiveness of training within organizations have been based on Donald Kirkpatrick's Four-Level Model. However, many researchers believe that the four-level evaluation model has failed the profession (Bassi, Benson, & Cheney, 1996; Swanson & Holton, 1999), primarily because of its emphasis on reactions versus the fundamental performance results of the host organization. Reaction measures cannot be used as surrogates of other measures (Alliger, Tannenbaum, Bennett, Traver & Shotland, 1997). For some authors, Kirkpatrick's evaluation model clearly did not contain the elements necessary to measure the impact of strategic leadership on organizational performance (Holton, 1996). For all these reasons, even though we recognize the value of the contribution made by the models based on levels, *we do not think they have helped clarify things much*. Work has been done to develop ways to measure levels, without analyzing in depth what these levels reflect, if they really refer to the same dimension, if they all have to be measured in a linear way or if they are totally independent (de Haro, 2009a).

OPERATIONAL PRINCIPLES FOR LEADERSHIP DEVELOPMENT IN THE NEW ECONOMY

Urgency in the workplace has had a dramatic impact on the velocity of business change. Companies in every industry, not just high-tech, need to accomplish more in less time. The response time available to fend off competitors is approaching zero. The best way for managers to add value is to reduce the time it takes employees to learn and apply the skills and information needed to compete in the new mercurial marketplace. *Leadership development must be accelerated.* The idea is to create and display 'expertise' more efficiently so that we can achieve the aims required by the increasingly demanding organizations. The traditional learning methods are no longer appropriate. The new economy leaders need to learn in different ways, deciding what to learn, when and how.

The *method used* to develop leadership within this display process becomes especially relevant. Training programs are commonly designed using multiple training techniques and multiple outcome measures (Conger & Benjamin, 1999). Among them, on-the-job training (93%) and external seminars (90%) have usually led the rankings of most frequently utilized methods (Saari, Johnson, Mclaughlin & Zimmerle, 1998). A wide variety of formal training programs are occurring in organizations: *formal training* continues to be the primary type of managerial leadership development intervention, while *job assignment* remains a close second (Collins, 2001). In recent times, development through *job experiences,* such as *on-the-job-training, job-performance evaluations and feedback programs, participation in special projects or task forces, coaching or mentoring, job rotation, succession planning, and career planning* have emerged as a powerful source of learning for managers (McCauley & Brutus, 1998). Other activities like action learning and 360-degree feedback are increasingly key elements of leadership development initiatives

Despite this variety of methods and their evident influence on leadership programs, it was not easy to quantify the effectiveness corresponding to each one of them. It even seems that the choice of a type of method is not the most important thing. Day & Halpin (2001) found that the best organizations combined these techniques in some way or other, so that the most effective system actually seemed to be a function of *the interdependence of several practices.*

In addition to the method itself, there are other variables which come into play when defining the effectiveness of a managerial development program. After analyzing the best programs created by organizations and business schools for leadership development, the Consortium for Research on Emotional Intelligence in Organizations –an independent body formed by experts from the academic world, the business world and the administration like Richard Boyatzis, Lyle Spencer or Daniel Goleman– concludes that managerial leadership is a process and as such it requires *time* and the need to consider *competencies* as the object of improvement. Furthermore, they construct a *decalogue* on what has to be done in order to achieve success in this field: (1) focusing on the competencies who have proved to be critical for success in the position; (2) evaluating a person's competencies through an external, objective vision like 360° assessment; (3) identifying the personal weaknesses that will act as the foundations of the development process, apart from the improvement opportunities; (4) taking into account that we adults only learn when we have a motivation to do so (achieving our personal aims); (5) ensuring that development is self-guided; (6) focusing on realistic development objectives; (7) defining a clear action plan; (8) assuming that improving competencies takes time (it is impossible to achieve sustainable changes before 6 months have elapsed); (9) providing external support which can reinforce or encourage the sustained practice of new behaviors; and (10) measuring changes in competencies and performance using valid methods.

As we said above, it is not so important to have a 'best practice' which leads to the successful development of leadership as to implement whichever it is, but in a consistent way. Moreover, the search for the key operational principles which enable us to perform managerial development strategies which suit the requirements of the new economy environment may start from the *demonstrated high-performance practices* which have been functioning until the present day, but it cannot stop there, *it must go beyond.* Therefore, it is necessary to introduce new formats within the learning methodologies which permit the accelerated acquisition of competencies that make it possible to go from strategy to action as quickly as possible. Acceleration of expertise can be achieved when training is designed on the basis of human psychological learning processes (Colvin, 2008). These learning accelerators revolve around the guidelines that we are going to describe below.

Focus on Results Adjusted to Organizational Requirements

A meta-analysis carried out by Conger & Benjamin (1999) revealed that managerial leadership development for the top levels of the organization was effective, and particularly those programs with *behavior or system outcomes.*

Ulrich, Zenger & Smallwood (1999) highlighted in their *outcomes-oriented leadership* model how important it is to connect learning practices with the business strategy so that a leadership development program can be successful. The authors developed *Results-Based Leadership* as a guidebook that bridges the gap between leadership theory and leadership skills, focusing on producing results that can be measured and *integrated into any business strategy or corporate culture.*

Leaders have to achieve their results but, at the same time, they must own the attributes which define the way to achieve them. Effective leadership depends *both on the attributes and on the results.* Ulrich and his collaborators believed that effective leaders know how to "connect their attributes with results" to improve performance. These desired results must be strategic and must align with the purposes of the organization (Ulrich *et al*., 1999). Kanter (1997) suggested that *a mission statement* is the motivator, as it enables people to recognize the importance of the work they perform.

The mechanism which shapes and integrates the organizational mission with the leader's personal features is *the action plan,* which represents one of the key aspects to guarantee that results are attained. In this plan are described the goals of change –that is, the desired performance– in a measurable, observable way. It is necessary to have some previous knowledge of the manager's current state or starting point (baseline). The plan provides a focus for their future efforts. It provokes them to think of how to use their day-to-day and week-to-week experiences as the basis for learning. In this sense, the plan helps each person turn their life and work activities into a learning laboratory.

Inside the plan, *goal-setting* serves to relate conscious intentions to performance at work. One goal is an antecedent because it acts as a stimulus which precedes the behavior and because it directs the attention and the action, mobilizing energy and effort meant to achieve the goals sought. Peter Druker (1999) highlighted the importance this has to 'hit the target' in plans, *"managing oneself in oneself".* It is necessary to choose the right objective that, for this author, includes knowing what we are already good at.

This is the phase in which behavioral changes are established, specifying what is going to be done in relation to behavior: *increasing, casting, diminishing, restricting or generalizing.* We speak about setting the objectives for the action plan in a specific way, because the mere wish or visualization when the time comes to achieve aims is not enough. The way in which the plan is formulated is not harmless; it has an impact on its level of success. There is consequently a proper way to design it (Wiseman, 2009), which includes amongst other things specifying the objectives, sharing the information with others or incorporating the positive aspects once they have been achieved.

We cannot forget that improvements have a purpose in this plan. When an improvement is achieved in an attribute, it is *'for something.'* Ulrich, Zenger & Smallwood (1999) make it very clear in their work, proposing up to 14 recommendations to develop leaders oriented toward results, which go from starting with an absolute concentration on results to serving as an example of the methods and fighting for the results that leaders want their respective groups to use and achieve.

The outcome from the correct formulation of these antecedents is to ensure, on the one hand, the *understanding of the objective sought through change* (indicators of the competency) and on the other hand, the *commitment needed* to initiate action (next phase), the influence of which on making the most of training has been proved.

Finally, we cannot forget either that *results are specific to each organization*. The configuration of the right behavioral indicators will make it possible to obtain the desired results, this has to be custom-made. The competencies for management are specific to the situation; they are directly related to the missions which have to be accomplished in a job. We have to start from the key success factors in the firm to derive competencies and indicators organized in a system of causal relationships which, starting from situations and attributes can allow us to reach the desired outcomes. The *criterion* in this respect is the starting point to define which excellent performance we are interested in (De Haro, 2005).

It is Necessary to Work Hard

Leadership development under the talent model we propose includes designing and performing the actions which allow us to pass from the initial state to a situation where we can exercise talent (mastery). *A process of active learning which produces increase in a specific expertise domain* is needed to reach a high level of achievement in the exercise of leadership. But leader thinking is a complex phenomenon, and leadership cannot be developed through a single, short-term intervention. It requires a progressive and sequential series of interventions.

Becoming competent in a field requires *a suitable intervention program* and a considerable *dose of work, effort and motivation for achievement and excellence* (Touron, 2001). In his change model, McClelland (1985) suggests going through at least six stages when an attempt is made to change a set of behaviors. From a behavioral perspective, specialists speak about going through at least five stops (Luthans & Kreitner, 1985).

The adoption of a *training approach* in which *effort and continuity* are combined can guarantee the level of dedication that will permit to reach mastery at a certain skill or a considerable improvement in highly demanding tasks (Tannenbaum, 1993). The knowledge about the field of *expert performance* provided by Professor A. Ericsson and his team, gives us the patterns to define how this type of training is going to be implemented. Studies in a variety of domains have provided us with lessons learnt from expert research (Colvin, 2008):

1. **Expertise requires extensive practice.** High levels of expertise demand years of practice. Productive talent in any area, and this includes leadership, becomes evident in *achievements that take years or months to be reached.* Someone has even quantified the amount of effort required. It is approximately *10,000 hours.* We need an intensive activity within a specific area where highly positive results and exceptional performances can be obtained, a full 8-to-10-hour working day for about 10 years (Ericsson, Krampe & Tesch-Romer, 1993).
2. **Expertise is domain-specific**. Fields of expertise are very narrow, because expertise relies on a large body of specific knowledge accumulated over time in memory. That knowledge stored in the memory is the basis for expertise.
3. **Expertise requires deliberate practice.** The key activity for the acquisition of expert performance is *deliberate practice*, which implies a *tiring, intensive routine* of training activities designed for the *sole purpose* of improving the current action

levels (Ericsson & Smith, 1991). The core assumption of deliberate practice is that expert performance is acquired *gradually* and that the effective improvement of performance requires the opportunity to find suitable training tasks that the performer can master sequentially-typically. The design of training tasks along with the monitoring of the attained performance is done by a teacher or a coach (Ericsson, 2006). The effects of mere experience (routine practice) differ to a great extent from those of deliberate practice. This is specifically *tailored* to improve their performance and requires a *concentration* that can be maintained only for a limited period of time.

4. **Experts see with different eyes,** and then choose the most appropriate strategies to solve problems or improve performance. Building expertise is training the brain to see problems through the eyes of an expert.
5. **Experts can get stuck,** can be inflexible, and they can have trouble adapting to new problems.
6. **Expertise grows from two intelligences.** Crystallized and fluid abilities must and can be developed, initially through direct instruction and thanks to cooperative learning by means of learning projects that emphasize abstraction (Neitfeld, Finney, Schraw & McCrudden, 2007).
7. **Challenging problems require diverse expertise.** Most of the problems faced by large organizations are complex enough to require diverse expertise. Innovation will increasingly depend on what psychologists call 'distribution cognition' (for example work teams). It is necessary to consider not only training but also other vehicles for the leveraging of diverse skills. The evolution of web 2.0 with social software such as wikis opens new channels for distributed expertise in organizations.

Leadership development must become a systematic process, not an event. Perhaps the most meaningful principle is that successful leadership development depends more on *consistent implementation* than on the use of innovative practices (Day & Halpin, 2001). Leadership development is an investment in the future and it is important to recognize that it may take years before dividends are realized. But only few managers are willing to maintain the effort. As is pointed out by Zenger & Folkman (2002), leaders initially go through a period of great learning, but once they have learnt the basic aspects, they no longer try to improve. They are satisfied with doing their job properly enough instead of wishing to become better leaders.

Cognitive and personal competencies modulate the speed of change within this tiring learning process. Among them, *self-regulation* processes largely contribute to producing permanence in behaviors leading to managers' goals. Self-evaluation and monitoring; goal setting and strategic planning; strategy implementation and monitoring, and strategic outcome monitoring constitute the 'steps' of self-regulated learning (Zimmerman, 1996). *Self-control and the ability to face the obstacles* appearing on the way equally form part of the behaviors which facilitate the improvement process. In this sense, specific practices about how to take *self-talks* into account (Makin & Cox, 2004); *overcoming procrastination* (Zeigarnik, 2007), *using training in self-instructions* or *double thinking* (Oettingen & Mayers, 2002) can also be helpful to reinforce self-regulated behavior and focalize training toward the outcomes sought.

What Happens Before (the Antecedents) is Important

The improvement process cannot overlook what happens before the act of learning. Improving performance by acting on behaviors can be done with work on two aspects: 1. on what occurs before it *(antecedents)*; or 2. on what occurs after it *(consequences)*. We are going to refer to antecedents in this section, leaving consequences for the next one.

Inside the organization, an antecedent is any person, place, thing or situation which arises prior to a behavior and drives us to execute that behavior. Antecedents may be of various types (Lopez Mena, 1989: indicators ('prompting'), modeling, goal-setting and participation; they serve to transmit information (about behavior and its consequences) and 'work' because they are associated with the consequences. In the case of managerial leadership, we would like to highlight two of these antecedents: one with an attributional character (*subject's dispositional states)* and another of a more social nature (*modeling)*.

The first one of them has a lot to do with *beliefs about change*. The subject's mental condition and other attributional variables are related to beliefs about change. A number of studies by Heslin, VandeWalle & Latham (2006) have shown that many managers do not believe in personal change. Some managers believe in natural talent and do not look for people with a potential to develop. However, good managers do believe in their collaborators' unlimited capacity to improve. Everything can be acquired with the right training (Buckingham & Coffman, 1999). For W. Bennis (1989), most great leaders believe that a leader is made, not born, and is made to a greater extent by themselves than through media that are external to them. Gilbert (1978) claims that the average manager's overcoming potential is 360%. Any person, of any age and in any circumstances, is able to carry out self-transformation (which is not the same as becoming a leader).

A lot of corporative training will have little value without a strong belief in the development of human beings. McCall (1998) believed that the real leaders of the future are those who have the ability to learn from their experiences and *remain open to continuous learning.* Carol Dweck (2006) has faithfully collected this idea in her research on 'mindset' when she draws a distinction between two groups of individuals: the *fixed mindset* and the *growth mindset* ones. The former think that intelligence is static, which leads to a tendency to avoid changes, to adopt a defensive attitude before obstacles, to see effort as something useless, to ignore the useful feedback from others, to feel threatened by other people's success and as a result of all this, to be unable to develop their whole potential. Instead, the latter believe that intelligence can be developed, which leads to a permanent wish to learn and a tendency to desire changes, to persist in overcoming hindrances, to see effort as the way to mastery, to learn from criticism and find inspiring lessons in other people's success. The most important lesson in this respect is that the 'growth mindset' can be taught to managers. Dr. Dweck has created workshops called "brainology" which have as their aim to develop the 'growth mindset.'

The second one of the antecedents we referred to above, *modeling,* has been used for the training of managers (Sims & Manz, 1982) and has proved its effectiveness as a form of learning. It is based on the fact that managers learn through the observation of other managers' behaviors (models) which constitute an important set of antecedents for behavior that is valuable for its frequency of appearance. Modeling is possible through four sets of learning activities: *modeling strictly speaking, role game, social reinforcement and training transfer or generalization* (Goldstein & Sorcher, 1974). The observation of the right behavior, its practice and the achievement of positive feedback has proved very useful in the development of management and problem-discussion skills.

Use Consequences to Reinforce what you Want to Get

Just like what happens before the behavior influences the way it starts, what happens afterwards is essential for its reinforcement and consolidation. A consequence is any event that follows behaviors and has an impact on how likely they are to occur in the future. If the antecedent serves as a cause of the ensuing behavior, the consequence serves to determine the behavior preceding it. If we want to influence the behavior before it happens, we will be using the antecedents. Instead, if we want to influence the behavior after it has occurred, we will be using the consequences. Consequences affect performance, improving or worsening it, and are the most effective tool for the modification of managers' performance.

In a study which reviewed more than 10 years of research, Daniels (1989) found that managers spend an average 80% of their time in activities that can be categorized as antecedents (defining tasks, creating procedures or giving instructions), as opposed to the rest of follow-up or consequence facilitation tasks. Nevertheless, the most effective leaders spend more time speaking to and asking their employees; in other words, providing consequences (Komaki, 1998).

The *feedback* process focused on managers' performance can help us utilize consequences for the purpose of introducing adjustments in their learning goals and persuading them to dedicate more time to the most productive managerial tasks. When trying to achieve this, feedback is one the most powerful intervention instruments which can be used within a behavioral approach to leadership, because people learn most effectively when they receive prompt, clear and objective feedback about their performance (Ericsson & Charness, 1994).

As a type of informative consequence, feedback becomes a key component within the learning process (Kuchinke, 2000) and improves performance because it can result in increased self-awareness and more dialogue between leaders and subordinates. *Self-Awareness* is the key to leadership development (McCall, 1998) and can be developed through the practice of seeking ongoing feedback.

The way in which feedback is provided affects its *effectiveness* in the reinforcement process. Effective feedback is characterized as being: *Positive, Immediate, Graphics and Specific* (the PIGS acronym). On the other hand, verbal feedback on a one-to-one basis and avoiding evaluations is more likely to be effective (Makin & Cox, 2004). What managers do with feedback also seems to matter. Recent research findings indicate that managers *who met with direct reports to discuss their* upward feedback demonstrated greater change in the form of performance improvement than managers who did not discuss their feedback (Walker & Smither, 1999). Another one of the strongest ways to provide feedback is the 360-degree feedback and multi-rater assessment, which is becoming one of the key components in any managerial leadership development (Church & Bracken, 1997). Its effectiveness is connected with other processes such as, for example, the degree to which it is linked to coaching actions like the ones we are going to comment on in the next section.

The Power of Personalized Support

Personalized support, understood as a 'professional relationship' of help with a third party, can adopt two main forms: coaching and mentoring.

(Executive) *coaching* as an action-learning process is a short-term intervention which seeks to develop specific leadership skills. *Mentoring* is typically defined as a committed, long-term relationship in which a senior person supports the personal and professional development of a junior person. It may be a formal program or a much more informal process.

Both of them try to achieve certain results in the manager's behavior, though differently. In this section, we will only focus on coaching, which has been used more often as a tool for the development of competencies for leaders and can be regarded as a more pure accelerator of managerial talent learning.

Coaching stimulates self-development and helps identify strengths and weaknesses, formulate improvement objectives and define and implement change plans. Taking as a reference the comparison between real and ideal performance and from the commitment of the person who wants to improve, the coach helps that person to carry out his/her action plans using communication strategies (mainly questions, listening and feedback). For executives or managers, carrying out such plans permits to fulfill the *most important objectives of coaching* which could be: developing their potential, for instance, to be successful in a new mission or role, adapting to new needs of the business or the environment, improving their contribution in terms of performance and objective compliance or identifying with the firm as far as its vision, mission, values, strategy or culture are concerned. In general, the objectives of coaching are focused on improving individual performance and personal satisfaction, and, consequently, on enhancing organizational effectiveness (Kilburg, 1996). Coaching may be used to improve individual performance, enhance a career, or work through organizational issues such as culture change (Katz & Miller, 1996).

Although the practice of coaching has become widespread recently, there is little published empirical research on the topic of executive coaching effectiveness (Kilburg, 1996). Most of the literature is anecdotal or provided by the practitioners or service providers themselves. Except for the use of tools like 360º-feedback, there has not been a great volume of empirical research on the aspects of executive coaching. Nearly all the studies have focused on studying the effectiveness of coaching, measured by the coachees' satisfaction with the treatment or their perception of progress, but it has not been checked whether or not a change of behavior has really occurred. The results obtained in this type of studies do not seem to be too optimistic, not even regarding the actual users' satisfaction. 73% of executives view informal feedback and coaching as essential or very important to development, but only 30% rate their company as excellent or very good at providing them. Fortunately, companies can nudge leaders to offer more feedback through '360º- feedback' programs (where contributions come from those above, below, and around an individual) and other formal mechanisms (Chambers, Handfield-Jones & Michaels, 1998).

Other significant results about its effectiveness have been summarized by Bayon, Cubeiro, Romo & Sainz (2006) in European organizations: a return on investment (ROI) equaling 6 times the cost of coaching in performance and quality improvements; performance increased by 88%, using coaching plus training as opposed to 22.4% using only training; reducing the rotation index from 35 to 16%; or improving the attraction of talent. The most admired enterprises according to the Fortune magazine differentiate themselves from the rest in the extent to which coaching is used (57% as opposed to 22%). In the 1,000 firms included in the sample, the use of coaching produced increases in productivity (53%), customer service (32%), profits (22%), improvement in keeping senior executives (32%) and cost reduction (23%). Executive coaching as a follow-up to a training program was shown to increase productivity by 88% in public sector managers (Olivero, Bane & Kopelman, 1997), which was a significantly greater gain compared with training alone. There are even studies on the ROI (Philips, 2005), which have demonstrated the effectiveness of coaching when it comes to creating value within an organization (Nations Hotel Corporation).

In the end, the review of empirical studies about coaching reveals great confusion regarding their results (Boyaztis, 2006), which has not prevented coaching from becoming increasingly prevalent as a formula for professional leadership, probably driven by the fact that it can be easily tailored according to the user's needs and thanks to its capacity to involve the user in the improvement process. Another of the explanations for understanding the how, why and what of effective coaching comes from the research on *implementation intentions* (Gollwitzer & Brandstatter, 1997). There is clear evidence that when people experience difficulty translating their goals into action, they can use situational cues to help make their responses stronger and more automatic using mental structures of the type "when a situation *x* arises, I will do *y*". The development of *trust relationships* which help and encourage the person at each step in the process is another of the reasons why coaches become so crucial to change. Coaches help people reflect and develop their dreams and aspirations; they look at the way they come across to others; they identify strengths and weaknesses and overcome blind spots in their leadership style. Coaches commit themselves in a collaboration alliance with the individual for the establishment and clarification of purposes and objectives and the later development of an action plan to accomplish those aims (Zeus & Skiffington, 2002).

Create Social Networks

Leadership learning does not take place individually; it constitutes a *social event*. Executives in best practice organizations commit the majority of their time to developing subordinates within the organization (Day & Halpin, 2001). The ability to show oneself as a cooperative, contributing, and constructive member of the group is critical for long-term career success.

A growing recognition that leadership development involves more than just developing individual leaders has now led to a greater *focus on the context in which leadership is developed*, thoughtful consideration about how to best use leadership competencies, and work/life balance issues. From this approach, everyone is considered to be a leader. Rather than asking oneself "How can I be an effective leader?" the pertinent question from the relational approach would be "How can I participate productively in the leadership process?" (Drath & Palus, 1994). The implications of this growing interest in relational aspects for leadership development are that the individual's ability will shift from taking charge to taking part, thus moving the focus toward interdependence. Direction will arise from the 'reciprocal-relatedness' of individuals, emphasizing the leadership ability of *work groups*.

Due to all this, in the context where the leader's learning develops, and from an organizational point of view, the *social environment* where managers can learn may be used to fight individuality and self-sufficiency, to develop their understanding of others and their ability to work with other people, as is required by managerial posts. The *creation of social environments* which permit the connection between the managers who find themselves in an improvement process is another accelerator of learning. The construction of this type of relationships within development programs through a *formally configured team* serves not only to develop competencies, but also for socialization, dynamization and mutual support, making it possible to create a safe environment for learning. An example of the use of these groups formalized for the improvement of managerial development is found in the dynamic groups proposed by Brenner (2010).

The *networking* initiatives within some organizations represent another example of the application of this principle to a manager's development. An important goal of *networking* initiatives is to develop leaders beyond merely knowing *what* and knowing *why,* to knowing *who* in terms of problem-solving resources. Networking is also about expanding one's definition of what and how through the exposure to other people's thinking, which can challenge basic assumptions about what we think we know. It is also a means of encouraging organization members to form commitments with others outside their immediate work group. In this way, networking is all about investing in and developing *social capital* with a primary developmental emphasis on building support. One type of networking involves the interaction of groups of managers and executives who have common training or job experiences.

One reason why networking is thought to be beneficial to professional and personal development is because it fosters peer relationships in work settings. *Peer relationships* offer unique value for development because of the degree of mutual obligation and the duration of the relationship. Research has shown that some peer relationships can span an entire 20-or-30-year career (Kram & Isabella, 1985), as compared with a typical mentoring relationship that lasts between three and six years (Kram, 1985) or an executive coaching relationship that generally lasts around six months (Levinson, 1996). Organizations should consider peer relationships as a potentially valuable component of an overall leadership development system. Networking opportunities build peer relationships across functional areas, leading to the creation of additional social capital.

Another of the implications derived from seeing social interactions as an accelerator of managerial learning is that it can be implemented in some of the custom-made training tools, e.g. action learning or job assignments. In *Action learning,* participants collectively construct social meanings and shared realities in a community of practice (Drath, 1998). In *Job assignments,* executives always prefer to have full control over everything they are responsible for, but jobs where they have to achieve results by influencing rather than controlling others contribute powerfully to their development.

Coaching, about which we talked above, also helps strengthen interpersonal relationships within the organization. The quality of an individual's relationship with an organization is positively associated with the effectiveness of coaching for the development of that individual.

Finally, networking is a prime means of enhancing social capital and favoring the transmission of *tacit knowledge* in an organization by means of "communities of practice." Tacit knowledge, which refers to the practical knowledge that is hardly ever described or taught formally, distinguishes experts from beginners and predicts managerial results (Wagner & Sternberg, 1985). Managers who build the kinds of networks that allow them to transcend the organization's formal structure –especially when they form non-redundant ties with people in other networks– are most likely to benefit in terms of information and entrepreneurial opportunities (Burt, 1992). When used in conjunction with other developmental practices, networking connects individual leadership development with collective leadership development.

Apply Learning Technologies

Just like the rest of areas in training, the area of leadership development has also benefited from the incorporation of new learning practices based on the information technologies. Apart from more traditional applications of the new technologies like e-learning, the advent of the Internet and mainly the World Wide Web as a medium of communication for the masses, along with the success of blogs, wikis or peer-to-peer technologies, have given the information technologies an increasingly social dimension which comes as a result of the evolution from Web 1.0 (only reading) to Web 2.0 (reading and writing). Because of the innovation they represent and also because they are very relevant in the field of managerial development, we are going to highlight two applications which, in our opinion, have greater significance as accelerators of leadership learning: *e-coaching* and *the use of social networks.*

E-coaching (Roujol, 2010) as a form of coaching based on using the new technologies through the Internet, through an intranet or using other types of electronic media, permits access to the 'coachee' seeking to facilitate the appearance of individual solutions, while at the same time they compose the process independently of time and place and the trainer's presence. Based on the application of a technological system which finds its support on the use of learning platforms (e-learning), it has also developed an ability to make the most of the incorporations coming from web

2.0 (a concept generated by O´Reilly (2005)), a real breakthrough in the development of e-learning platforms which has allowed the user to remove space and time barriers and therefore to be able to access a real knowledge management: *exchange of ideas, opinions, practices, experiences*. The use of this format, which has meant a cost reduction and has given the user greater space-time autonomy, has had its strongest impact on the development of formative actions for managers in the form of 'blended learning.' Other resources such as podcasts, videos or digital classes (teaching), synchronous and asynchronous tutorial systems, self-coaching or networked consultation, have been used as a way to reinforce the potentialities of online platforms.

Various studies have demonstrated its effectiveness as a tool (Goldsmith *et al*., 2003; Li-Bin Wang & Wentling, 2001). The research work carried out by Brookes (2009) where he used e-coaching with a group of executives indicates that it can be an effective method for leadership development programs, increasing the transfer of learning, permitting the access to a greater number of participants without affecting costs, and obtaining good results at the learning transfer. However, as is pointed out by Rossett & Marino (2005), e-coaching is an issue that goes beyond turning on the computer and establishing a link with a trainer.

As for *social networks*, they are possibly (along with some applications of e-learning) the most important contribution made by the information technologies to learning, using the relational aspects (Boyd & Ellison, 2007). Their use is growing progressively, due not only to the weight of the technological aspects but also to the weight that the current organizational culture assigns to learning and interpersonal relationships above the traditional system structured around control and authority (Keeys, 1994). The introduction of social networks into managerial development programs makes it possible to accelerate their completion. The use of the multiple services available on the Net (*Fotolog, LinkedIn, MySpace, Flickr, Facebook, YouTube, Bebo, Ning* or *Twitter*) will give users the possibility to articulate a list of other users with whom they will be able to share relationships (Boyd & Ellison, 2007). In turn, internal blogs and wikis will be used more often, as they make it easier to hold conversations, to share knowledge and to establish new internal links. The groups or debates opened in Linkedin or Xing in which managers take part will provide the framework to rehearse what has been learnt before applying it in the real world. However, the tools alone do not generate this '2.0' movement, rather they help us discover the concept of community and the influence that the way in which we share knowledge on the Internet has an influence on the Social Media (Boyd & Ellison, 2007), which will take us to another learning context where the individual will have more and more control as well as greater flexibility.

In short, networks make conversation easier, they permit to share knowledge and, using that knowledge, they generate learning in a permanent feedback process. In that respect, networks are also training in themselves (Celaya, 2008). But their power does not end here; they can equally contribute to achieving commitment through an increased participation in business policies and the democratization of their creation process, even if that means requiring leaders with values that are in keeping with this new working patterns (Backstrom *et al*., 2006) and leaders with other ways to learn, who will learn what they need to know, when they need it and from the best possible source.

Do not Forget the Learning Methods

Both the *organizational factors and the learning objectives* affect the effectiveness of the overall training process. Some methods are more suitable than others according to the type of learning content that is going to be developed. Therefore, the improvement will be more or less easy depending on whether what we want to change is knowledge, skills, competencies, attitudes, beliefs or values. Thus, for example, knowledge can be changed more easily with traditional techniques and resources. The most difficult part is leadership development per se; facilitating changes in attitudes and behavior brings those associated closer to effective leadership. If we want to develop skills, we will need to include practical activities such as challenge- and mission-based learning instead of conventional learning. The fact that they work with *real problems of the organization* means that managers can learn much more quickly (Revans, 1980). For instance, the 'action learning' methodology, which uses assessment in a formal way to select learning projects increases the quality of the experience producing a greater managerial development. Action learning goals that are aligned with individual goals result in meaningful developmental experiences.

Similarly, the use of *job assignments* for developmental purposes provides benefits that go beyond getting the job done and may even result in *competitive advantages for the organization* (Ohlott, 2004). *Job assignments* have been identified as particularly helpful to managers in learning about building teams, how to become better strategic thinkers, and how to gain valuable persuasion and influence skills (McCall, Lombardo & Morrison, 1988). When it comes to developing leadership, the most admired firms use the planned responsibility assignment, providing their high-potential employees with a wide variety of useful experiences for their personal and professional learning. Challenging job assignments is a powerful form of leadership development and provides many of the developmental opportunities in organizations today. These methods encourage development by providing managers with the opportunity to learn and acting as a motivator for learning (McCauley, Ruderman, Ohlott & Morrow, 1994). Leadership development is enhanced when assignments are matched with individuals' developmental needs (Day, 2001). Nevertheless, some types of jobs are easier to develop than others and different kinds of assignments for development produce different kinds of learning (McCauley & Brutus, 1998). Jobs that are more developmental include 'stretch' assignments that put a manager in a new situation with unfamiliar responsibilities, especially high-responsibility and high-latitude jobs.

Adopt a Quantitative Vision of Training and Development Assessment

It is necessary and possible to measure effectiveness in the acquisition of managerial talent. Considering the limitations of the level-based approach that we described in the first part of this paper, we are forced to think of a different way to conceptualize the assessment of training. This approach, which instead of using levels reflects on which way should be chosen when assessing training, has already been presented elsewhere (De Haro, 2009a). The two ways which can lead us to solve the problem of effectiveness in training are: *The short-range way and the long-range way.*

The first one of them, which we can call *internal effectiveness,* considers focusing only on the achievement of training objectives. According to this approach, the objective of training is *to produce learning.* Therefore, its effectiveness is only measured by the 'degree of achievement' of the objectives set by the formative action. It is necessary to measure the learning level before the action and then compare it with the learning level after the action in order to be able to evaluate that change.

The second way, linked to the technology of human performance and which we can call *external effectiveness* implies going beyond the actual formative action to check if it impacts –and if it does, to what extent– on the organization's results. An effective training is the one which has an influence on behaviors *that are later used at the workplace.* From this perspective, what matters is not only what has been learnt but what is done with what has been learnt and, of course, the economic value of this learning.

This economic value derives from the outcomes achieved, and it is a dependent variable the values of which have to be calculated like for any other variable from any design. Tools that forecast the economic return of training (Swanson & Gradous, 1988) and models related with evaluation exist. Other evaluation models used in the profession as described by Phillips (1997a) include Kaufman's five-level, CIRO, CIPP, Phillips five-level ROI, and the critical outcome technique, as well as Kaufman and Watkins' (1996) cost-sequence analysis instrument. And they have been applied with promising results too. For example, Philips & Philips (2005), in their paper *"best-practices case studies from the real world"* report positive results of three cases of ROI assessment in managerial development programs (Central Plains Rural electric Cooperative, BMW manufacturing Company and Global Car Rental).

CONCLUSION: CHALLENGES FOR THE DEVELOPMENT OF A NEW LEADERSHIP

Acting faster and faster, depending on shared knowledge and competing in complex, global and diverse scenarios are some of the *critical requirements of the new economy environment* where leaders will have to get by.

In this complex and demanding scenario, those who occupy the posts involving a greater deal of responsibility inside the organizations will have to gradually assume new functions that permit those organizations' continuity and growth: detecting the future but more on a medium-term basis; focusing on the results without forgetting people; innovating without losing sight of reality; being permanently in action without forgetting reflection; and generating a powerful framework image as a leader.

On the way toward the mastery of these *increasingly complex behaviors*, managerial development, integrated in the *human resources* function, is the tool that can contribute to prepare leaders so that they can acquire or develop the competencies required to meet the demands of these new roles.

In order to achieve this aim, we propose a *work setting* which operates on a double level. On the one hand, we should start from a set of general key principles about managerial development which have to be assumed before taking action and, on the other hand, we should perform some operational practices which allow us to go beyond what has so far constituted the content of the organizational practices in this field.

The *general principles* which, in our opinion, should be shared before taking action are: the consideration of leadership as a type of talent constituted by a series of competencies which have to be developed under a performance improvement approach. In this process where effectiveness has to be assessed from the very beginning, we act on adults with their own learning principles and inside a specific organization with its own determinants for behavior.

Regarding the *operational principles* which can allow us to implement specific development plans, we believe that with a clear orientation toward organizational results, measurement, and what has been scientifically proved, the best approach must be based on hard training, sustained over time, which takes into account the antecedents and consequences of behavior and finds its support on the accelerators of learning derived from the social, technological and human aspects.

At present, and even though we think that the contribution made in this paper will help introduce significant improvements in managerial development practices, there are still quite a few pending issues that we will gradually have to solve if we want to be able to make progress. Although the results reflected in the literature on the effectiveness of managerial development programs are somewhat contradictory and have left a number of unanswered questions, organizations must continue to explore new options to find ways to measure effectiveness in these programs. As final conclusions and integrating the future challenges with the still incomplete lines of work on which we think it is necessary to insist, we would like to highlight the following three ideas:

1. It is necessary to develop *new ways to think about leadership and about the managerial development methods and programs.* We have to suggest new ways to understand leadership, its purpose and its effectiveness. We must regard individual and social variables as moderating elements. Managerial development programs cannot be focused exclusively on individual development. They should also pay attention to the set of leaders' behaviors that *can transform* the whole organization (Conger & Benjamin, 1999). In the end, research has demonstrated that leadership effectiveness is impacted by personality characteristics, situational variables, and the interaction between them (Northouse, 2007). Developing the talent of their managers should not be only a responsibility of the organizations' human resources professionals either. It must be a *priority of the top management*, even in an aggressive way (Chambers, Helen Handfield-Jones Ed Michaels, 1998) if the organization wants to be competitive.
2. Overall there is a deficiency of scholarly knowledge about managerial leadership development, about what it is, and about its impact on the organization (Lynham, 2000). This is why *more* applied research needs to be done on the effectiveness of the different methods and the connection between the outcomes of interventions and organizational operation. Additional theoretical and empirical work is equally needed to better map the various dimensions and types of experience onto individual and organizational development. More research is needed to fully understand managerial leadership development (Collins, 2002). In doing so, research can help inform practitioners on how to match assignments and developmental needs on a more scientific basis. We must resort to what has been proved as a result of the research work, connect the academic world with the business world and link studies to the achievement of results. Although it seems that the best practice in evaluation processes is to test and analyze the relationships and linkages between performance measures at different levels of analysis (Collins, 2002; Swanson & Holton, 1999), *more research must be conducted* related to "how organizational capabilities and leadership competencies lead to and are connected to desired results" (Ulrich *et al.* 1999). Further research is necessary to determine when and what leadership activities influence organizational performance and the conditions under which these activities may have maximum effect (Svyantek & DeShon, 1992). More evaluative research needs to be conducted too (Burke & Day, 1986), particularly on linking managerial leadership development to performance improvement (Lynham, 2000).
3. It is necessary to create specific working models that can guide us in the search for the best results, which take into account the specificity of the situation, and which are contingent rather than generalist (of "best practices"). A significant difficulty in our ability to assess the impact of training may be the lack of a unifying model and theory of research (Dionne, 1996). The reflection on leadership and the analysis of research should allow us to create models which can integrate, in a balanced way, technology (2.0 ICT), cognitive and emotional competencies (e-leadership?), personal and social support,

the premises for the modification of behavior that guarantee us the transfer of what has been learnt to the job and the contradictory requirements of the globalized, complex environment in which we are operating. The 'competential command structure' (De Haro, 2005) is an example of this type of approach which allows us to use the concept of competency as a behavior/outcome association when it comes to defining the learning objectives and the assessment of effectiveness in the programs.

Knowing the ways to implement effective leadership not only is far from being an easy task but also constitutes the 'new bottom-line of organizations' (Meisel & Fearon, 1996). As we have highlighted in this paper, the *organizational discipline* to drive managerial development is possibly the most important principle to ensure its effective implementation. We hope to have helped encourage the development of this discipline with our research work. If that is the case, the continuous effort which stems from a real interest and from the commitment to leadership development will surely lead us to see new results very soon.

REFERENCES

Alliger, G. M., Tannenbaum, S. I., Bennett, W., Traver, H., & Shotland, A. (1997). A meta-analysis of the relations among training criteria. *Personnel Psychology*, *50*(2), 341–359. doi:10.1111/j.1744-6570.1997.tb00911.x

American Society for Training and Development. (1995). *National HRD executive survey*.

Backtrom, L., Huyyrnlovhrt, D., Kleinberg, J., & Lan, X. (2006). Group formation in large social networks: Membership, growth, and evolution. *Proceedings of the 12th International Conference on Knowledge Discovery in Data Mining* (pp. 44-54). New York, NY: ACM Press.

Backus, C., Keegar, K., Gluck, C., & Gulik, V. (2010). Accelerating leadership development via immersive learning and cognitive apprenticeship. *International Journal of Training and Development*, *14*(2), 144–148. doi:10.1111/j.1468-2419.2010.00347.x

Bass, B. (1990). *Bass and Stogdill's handbook of leadership: Theory, research, and applications* (3rd ed.). New York, NY: Free Press.

Bassi, L., Benson, G., & Cheney, S. (1996). The top ten trends. *Training & Development*, *50*, 28–42.

Bayón, F., Cubeiro, J., Romo, M., & Sanz, J. (2006). *Coaching realmente*. Madrid, Spain: Pearson Education.

Bennis, W. (1989). *On becoming a leader*. Reading, MA: Addison-Wesley Pub. Co.

Berry, J. (1990). Linking management development to business strategies. *Training & Development*, (August): 20–21.

Biech, E. (2008). *American Society for Training and Development handbook for workplace learning professionals*. Alexandria, VA: ASTD Press.

Bolleu, R., Bowers, D., Boyatzis, R., & Kolb, D. (1999). Fellowship in lifelong learning: An executive development program for advanced professionals. *Journal of Management Education*, *23*(4), 338–354. doi:10.1177/105256299902300403

Boyatzis, R. (2006). Intentional change from a complexity perspective. *Journal of Management Development*, *25*(7), 607–623. doi:10.1108/02621710610678445

Boyatzis, R., Goleman, D., & Rhee, K. (2000). Clustering competence in emotional intelligence: Insights from the emotional competence inventory (ECI). In Bar-On, R., & Parker, J. D. A. (Eds.), *Handbook of emotional intelligence*. San Francisco, CA: Jossey-Bass.

Boyatzis, R., & Van Oosten, E. (2002) *Developing emotionally intelligent organizations*. Retrieved April 10, 2006, from http://www.eiconsortium.org

Boyd, D., & Ellison, N. (2007). Social network sites: Definition, history, and scholarship. *Journal of Computer-Mediated Communication, 13*(1). Retrieved from http://jcmc.indiana.edu/vol13/issue1/boyd.ellison.htmldoi:10.1111/j.1083-6101.2007.00393.x

Brenner, M. (2010). Group dynamics can enhance leadership development. In *SHRM's 2010 HR Trend book. Special supplement to HR Magazine* (pp. 2–33). Alexandria, VA: SHRM.

Brookes, B. (2009). *Leadership development using online group executive coaching*. UMUC Working Paper Series. Number 2009-001. University of Maryland.

Brungardt, C. (1996). The making of leaders: A review of the research in leadership development and education. *The Journal of Leadership Studies, 3*(3), 81–95. doi:10.1177/107179199700300309

Buckingham, M., & Coffman, C. (1999). *First break all the rules: What the world's greatest managers do differently*. Sydney, Australia: Simon & Schuster.

Burke, M., & Day, R. (1986). A cumulative study of the effectiveness of managerial training. *The Journal of Applied Psychology, 71*(2), 232–245. doi:10.1037/0021-9010.71.2.232

Burke, W., & Litwin, G. (1992). A causal model of organizational performance and change. *Journal of Management, 18*(3), 532–545. doi:10.1177/014920639201800306

Burt, R. (1992). *Structural holes*. Cambridge, MA: Cambridge University Press.

Celaya, J. (2008). *La empresa en la web 2.0: El impacto de las redes sociales en la estrategia empresarial*. Barcelona, Spain: Ediciones Gestión 2000.

Chambers, E., Handfield-Jones, H., Hanking, S., & Michaels, E. III (1998). Win the war for top talent. *Workforce*, December, 50-56.

Cherniss, C., & Goleman, D. (2001). *The emotionally intelligence workplace*. New York, NY: Wiley & Sons.

Churh, A. H., & Bracken, D. W. (Eds.). (1997). Special issue: 360-degree feedback systems. *Group & Organization Management, 22*(2).

Collins, D. (2002). *The effectiveness of managerial leadership development programs: a meta-analysis of studies from 1982-2001*. Dissertation Submitted to the Graduate Faculty of the Louisiana State University and Agricultural and Mechanical College in partial fulfillment of the requirements for the degree of Doctor of Philosophy in The School of Human Resource Education and Workforce Development, West Virginia University.

Collins, D., & Holton, E. (2004). The effectiveness of managerial leadership development: A meta-analysis of studies from 1982 to 2001. *Human Resource Development Quarterly, 15*(2). doi:10.1002/hrdq.1099

Collins, D., Lowe, J., & Arnett, C. (2000). In Holton, E. F., & Naquin, S. S. (Eds.), *Developing high-performance leadership competency* (*Vol. 6*, pp. 18–46). Baton Rouge, LA: Academy of Human Resource Development.

Collins, J. (2001). *Good to great: Why some companies make the lead and other don't*. New York, NY: Harper Collins.

Colvin, G. (2008). *Talent is overrated*. USA: Penguin Group.

Colvin, R. (2008). *Building expertise. Cognitive methods for training and performance improvement*. San Francisco, CA: Pfeiffer.

Conger, J. (1992). *Learning to lead: The art of transforming managers into leaders*. San Francisco, CA: Jossey Bass.

Connely, M., Gilbert, J., Zaccaro, S., Threlfall, K., Marks, M., & Mumford, M. (2000). Exploring the relationship of leadership skills and knowledge to leader performance. *The Leadership Quarterly*, *11*, 65–86. doi:10.1016/S1048-9843(99)00043-0

Coyle, D. (2009). *The talent code*.

Daniels, A. (1989). *Bringing out the best in people*. New York, NY: McGraw-Hill.

David, P., & Foray, D. (2003). Economic fundamentals of the knowledge society. *Policies Futures in Education*, *1*(1), 20–49. doi:10.2304/pfie.2003.1.1.7

Day, D. (2001). Leadership development: a review in context. *The Leadership Quarterly*, *11*(4), 581–613. doi:10.1016/S1048-9843(00)00061-8

Day, D., & Halpin, S. (2001*). Leadership development: A review of industry best practices*. United States Technical report by Army Research Institute for the behavioural and Social Sciences, Alexandria.

De Haro, J. (2005). El cuadro de mando competencial, una herramienta para el desarrollo de los recursos humanos en las organizaciones. *Revista del Centro de Estudios Financieros*, *6*(262), 245–280.

De Haro, J. (2009a). La evaluación de la formación, ¿un problema sin resolver? *Training and Development Digest*, *75*, 28–36.

De Haro, J. (2009b). ¿Podemos predecir el talento? *Revista del Centro de Estudios Financieros*, *320*, 137–170.

De Rose, C., & Tichi, N. (2009). *Leadership judgement: Without it nothing else matters*. Leader to Leader Institute.

Dionne, P. (1996). The evaluation of training activities: A complex issue involving different stakes. *Human Resource Development Quarterly*, *7*(3), 279–286. doi:10.1002/hrdq.3920070308

Dotlich, D. L., & Noel, J. L. (1998). *Action learning: How the world's top companies are re-creating their leaders and themselves*. San Francisco, CA: Jossey-Bass Inc.

Drath, W. (1998). Approaching the future of leadership development. In McCauley, C., Moxley, R. S., & Van Velsor, E. (Eds.), *Handbook of leadership development* (pp. 403–432). San Francisco, CA: Jossey-Bass Publishers.

Drath, W., & Palus, C. (1994). *Making common sense: Leadership as meaning making in a community of practice*. Greensboro, NC: Center for Creative Leadership.

Drucker, P. (1999). *Sobre la profesión del management*. Barcelona, Spain: Apóstrofe.

Druker, P. (1993). *The rise of knowledge society*. Boston, MA: Butterworth & Heinemann.

Ducharme, M. (2004). The cognitive-behavioral approach to executive coaching. *Consulting Psychology Journal: Practice and Research*, *56*(4), 214–224. doi:10.1037/1065-9293.56.4.214

Dweck, C. (2006). *Mindset, the new psychology of success*. New York, NY: Ballantine Books.

E-Learning 2.0. (2010). *Wikipedia, la enciclopedia libre*. Retrieved August 16, 2010, from http://en.wikipedia.org/wiki/E-learning#E-Learning_2.0

Ericsson, A. (2006). The influence of experience and deliberate practice on the development of superior expert performance. In Ericsson, A., Charness, N., Feltovich, P., & Hoffman, R. (Eds.), *The Cambridge handbook of expertise and expert performance*. New York, NY: Cambridge University Press.

Ericsson, A., & Charness, N. (1994). Expert performance: Its structure and acquisition. *The American Psychologist*, *49*, 725–747. doi:10.1037/0003-066X.49.8.725

Ericsson, A., Charness, N., Feltovich, P., & Hoffman, R. (Eds.). (2006). *The Cambridge handbook of expertise and expert performance*. New York, NY: Cambridge University Press.

Ericsson, A., Krampe, R., & Tesch-Romer, C. (1993). The role of deliberate practice in the acquisition of expert performance. *Psychological Review, 100*(3), 363–406. doi:10.1037/0033-295X.100.3.363

Ericsson, A., & Smith, J. (Eds.). (1991). *Toward a general theory of expertise: Prospects and limits*. New York, NY: Cambridge University Press.

Friedman, T. (2005). *The world is flat: A brief history of the twenty-first century*. New York, CA: Farrar, Straus and Giroux.

Fulmer, R., & Warner, S. (1999). Leadership: Lessons from the best. *Training & Development, 53*(3), 28–33.

Gagné, F. (1991). Toward a differentiated model of giftedness and talent. In Colangelo, N., & Davis, G. A. (Eds.), *Handbook of gifted education*. Boston, MA: Allyn and Bacon.

Gagné, F. (2004). Transforming gifts into talents: The DMGT as a developmental theory. *High Ability Studies, 15*(2). doi:10.1080/1359813042000314682

Gilbert, T. (1978). *Human competence: Engineering worthy performance*. New York, NY: McGraw-Hill.

Goldsmith, M., Govindarajan, V., Kaye, B., & Albert, A. (2002). *The many facets of leadership*. New York, NY: Financial Times Prentice Hall.

Goldsmith, M., Lyons, L., & Freas, A. (Eds.). (2000). *Coaching for leadership*. San Francisco, CA: Jossey-Bass.

Goldsmith, M., & Morgan, H. (2003). *Increasing leadership effectiveness: A review of leadership practices development of 8 large corporations*. Retrieved from http://www.personal.marshallgoldsmith.com/html/articulos/increasing.html

Goldstein, A., & Sorcher, M. (1974). *Changing supervisory behaviour*. New York, NY: Pergamon.

Goleman, D. (1998). What makes a leader? *Harvard Business Review, 76*(6).

Gollwitzer, P., & Brandstäterr, V. (1997). Implementation intentions and effective goal pursuit. *Journal of Personality and Social Psychology, 73*, 186–199. doi:10.1037/0022-3514.73.1.186

Hall, S., & Scraton, P. (1981). Law, class and control. In Fitzgerald, M., McLennan, G., & Pawson, J. (Eds.), *Crime and society*. London, UK: RKP.

Harless, J. (1989). *Job and analysis, design and development*. Newnan, GA: Harless Performance Guild.

Hawkins, P., & Smith, N. (2006). *Coaching, mentoring and organizational consultancy*. Open University Press.

Hay Group. (2010). Study "best companies for leadership". In G. Fernandez (Ed.), *Entrevista con Mary Fontaine*, (pp. 3-5). En persona.

Hedlund, J., Forsythe, G., Horvath, J., Williams, W., Snook, S., & Sterberg, R. (2003). Identifying and assessing tacit knowledge: Understanding the practical intelligence of military leaders. *The Leadership Quarterly, 14*, 117–140. doi:10.1016/S1048-9843(03)00006-7

Heslin, P., Vandewalle, D., & Latham, G. (2006). Keen to help? Managers' implicit person theories and their subsequent employee coaching. *Personnel Psychology, 59*, 871–902. doi:10.1111/j.1744-6570.2006.00057.x

Hogan, R., & Kaiser, R. (2005). What we know about leadership. *Review of General Psychology, 9*(2), 169–180. doi:10.1037/1089-2680.9.2.169

Holton, E. III. (1996). The flawed four-level evaluation model. *Human Resource Development Quarterly, 7*(1), 5–21. doi:10.1002/hrdq.3920070103

Horton, W., & Horton, K. (2003). *E-learning, tools and technologies*. Indianapolis, IN: Wiley.

House, R., Hanges, P., Javidan, M., Dorfman, P., & Gupta, V. (Eds.). (2004). *Culture, leadership, and organizations: The Globe study of 62 societies*. Thousand Oaks, CA: Sage Publications, Inc.

Huselid, M. (1995). The impact of human resource management practices on turnover, productivity, and corporate financial performance. *Academy of Management Journal, 38*(3), 635–672. doi:10.2307/256741

Jacobs, R. L., & Jones, M. J. (1995). *Structures on-the-job training: Unleashing employee expertise in the workplace*. San Francisco, CA: Berrett-Koehler.

Judge, T., Bono, J., Llies, R., & Gerhardt, M. (2002). Personality and leadership: A qualitative and quantitative review. *The Journal of Applied Psychology, 87*, 765–780. doi:10.1037/0021-9010.87.4.765

Kampa-Kokesch, S., & Anderson, M. (2001). Executive coaching: A comprehensive review of the literature. *Consulting Psychology Journal: Practice and Research, 53*(4). doi:10.1037/1061-4087.53.4.205

Kanazawa, S. (2003). Why productivity fade with age: The crime-genius connection. *Journal of Research in Personality, 37*(4), 257–272. doi:10.1016/S0092-6566(02)00538-X

Kanter, R. (1997). *Frontiers of management*. Boston, MA: Harvard Business Review Press.

Kanter, R. (1997). Restoring people to the heart of the organization of the future. In Hesselbein, F., Golesmith, M., & Beckhard, R. (Eds.), *The organization of the future*. San Francisco, CA: Jossey-Bass.

Katz, J., & Miller, F. (1996). Coaching leaders through cultural change. *Consulting Psychology Journal: Practice and Research, 48*, 104–114. doi:10.1037/1061-4087.48.2.104

Keeys, G. (1994). Effective leaders need to be good coaches. *Personnel Management, 26*(11), 52–55.

Kepner, C., & Tregoe, B. (1965). *The rational manager: A systematic approach to problem solving and decision making*. New York, NY: McGraw Hill.

Kilburg, R. (1996). Toward a conceptual understanding and definition of executive coaching. *Consulting Psychology Journal: Practice and Research, 48*(2). doi:10.1037/1061-4087.48.2.134

Knowles, M. (1988). *Adult education. What works at work*. Minneapolis, MN: Lake- wood Books.

Koch, M., & McGrath, R. (1996). Improving labour productivity: Human resource management policies do matter. *Strategic Management Journal, 17*, 335–354. doi:10.1002/(SICI)1097-0266(199605)17:5<335::AID-SMJ814>3.0.CO;2-R

Komaki, J. (1998). *Leadership from an operant perspective*. New York, NY: Routledge.

Kotter, J. (1990). What leaders really do. *Harvard Business Review*, (May-June): 37–60.

Kouzes, J., & Posner, B. (1995). *The leadership challenge: How to keep getting extraordinary things done in organizations*. San Francisco, CA: Jossey-Bass.

Kouzes, J., & Posner, B. (2003). *Credibility: How leaders gain and lose it, why people demand it*. San Francisco, CA: Jossey-Bass.

Kram, K. (1986). Mentoring in the workplace. In Hall, D. T. (Ed.), *Career development in organizations*. San Francisco, CA: Jossey–Bass.

Kram, K., & Isabella, L. (1985). Alternatives to mentoring: The role of peer relationships in career development. *Academy of Management Journal, 28*, 110–132. doi:10.2307/256064

Kuchinke, K. (2000). The role of feedback in management training settings. *Human Resource Development Quarterly, 11*, 381–401. doi:10.1002/1532-1096(200024)11:4<381::AID-HRDQ5>3.0.CO;2-3

Kumar, R., Novak, J., & Tomkins, A. (2006). *Structure and evolution of online social networks*. In ACM SIGKDD, August 20-23, Philadelphia.

Lam, L., & White, L. (1998). Human resource orientation and corporate performance. *Human Resource Development Quarterly, 9*(4), 351–364. doi:10.1002/hrdq.3920090406

Levinson, H. (1996). Executive coaching. *Consulting Psychology Journal, 48*(2), 115–123. doi:10.1037/1061-4087.48.2.115

Libert, B., & Faulk, R. (2008). *Barack, Inc: Winning business lessons of the Obama campaign*. New York, NY: Prentice Hall.

Lombardo, M., & Eicheinger, R. (1989*). Eighty eight assignments for development in place: Enhancing the development challenger of existing jobs*. Technical report, 136, Greensboro, NC. Center for creative leadership.

Lombardo, M., & Eichinger, R. (1997). The HR function that consists in creating leaders with advantage on competency. In Ulrich, D., Losey, M. R., & Lake, G. (Eds.), *Tomorrow's HR management*. Wiley and sons.

Lombardo, M., & Eichinger, R. (2002). *The leadership machine*. Minneapolis, MN: Lominger.

Longenecker, C., & Neubert, M. (2005). The practices of effective managerial coaches. *Business Horizons, 48*, 493–500. doi:10.1016/j.bushor.2005.04.004

López Mena, L. (1989). *Intervención psicológica en la empresa*. Barcelona, Spain: Martinez Roca.

Lorenzo, R. (2006). ¿A qué se denomina talento? Estado del arte acerca de su conceptualización. *Intangible Capital, 11*(2), 72–163.

Luthans, F., & Kreitner, R. (1985). *Organizational behavior modification and beyond: An operant and social learning-approach*. Glenview, IL: Scott & Foresman.

Lyham, S. (2000). Theory building in the human resource development profession. *Human Resource Development Quarterly, 11*(2), 159–178. doi:10.1002/1532-1096(200022)11:2<159::AID-HRDQ5>3.0.CO;2-E

Makin, P., & Cox, C. (2004). *Changing behaviour at work: A practical guide*. Oxfordshire, UK: Routledge.

Martin, G., & Pear, J. (1996). *Behaviour modification: What it is and how to do it* (5th ed.). Englewood Cliffs, NJ: Prentice Hall.

Mayer, J., & Caruso, D. (2002). *The effective leader: Understanding and applying emotional intelligence*. The University of Ontario Ivey Business Journal, November/December.

Mayer, J., Salovey, P., & Caruso, D. (2000b). Emotional Intelligence. In *R. J. Sternberg (2000), Handbook of Intelligence* (pp. 396–421). New York: Cambridge University Press.

McCall, M. (1998). *High flyers: Developing the next generation of leaders*. Boston, MA: Harvard Business School.

McCall, M., Lombardo, M., & Morrison, A. (1988). *The lessons of experience: How successful executives develop on the job*. New York, NY: The Free Press.

McCauley, C., & Brutus, S. (1998). *Management development through job experiences*. Greensboro, NC: Center for Creative Leadership.

McCauley, C., & Hughes-James, M. (1994). *An evaluation of the outcomes of a leadership development program*. Greensboro, NC: Center for Creative leadership.

McCauley, C., Rudermar, M., Ohlott, P., & Morrow, J. (1994). Assessing the developmental components of managerial jobs. *The Journal of Applied Psychology*, *79*, 544–560. doi:10.1037/0021-9010.79.4.544

McClelland, D. (1985). How motives, skills and values determine what people do. *The American Psychologist*, *40*, 812–825. doi:10.1037/0003-066X.40.7.812

McLagan, P. (2008). Competencies and the changing world of work. In Biech, E. (Ed.), *ASTD handbook for workplace learning professionals*. Virginia: ASTD Press.

Meisel, S., & Fearon, D. (1996). Leading learning. In Calvieri, S., & Fearon, D. (Eds.), *Managing in organisations that learn* (pp. 180–209). Cambridge, MA: Blackwell.

Menkes, J. (2005). *Executive intelligence*. New York, NY: HarperCollins Publisher.

Michaels, E., Handfield, H., & Axelrod, B. (2001). *The war for talent*. Boston, MA: Harvard Business School Press.

Mullins, L. (2007). *Management and organizational behaviour*. New York, NY: Prentice Hall.

Mumford, M., Friedrich, T., Caughron, J., & Antes, A. (2009). Leadership development and assessment, describing and rethinking the state of the art. In Ericsson, A. (Ed.), *Development of professional expertise. Toward measurement of expert performance and design of optimal learning environments*. New York, NY: Cambridge University Press.

O'Reilly Y. T. (2005) *What is Web 2.0? Design patterns and business models for the next generation of software.* Retrieved July 3, 2010, from http://oreilly.com/web2/archive/what-is-web-20.html

Oettingen, G., & Mayer, D. (2002). The motivation function of thinking about the future: Expectations vs. fantasies. *Journal of Personality and Social Psychology*, *87*, 1198–1212. doi:10.1037/0022-3514.83.5.1198

Ohtott, P. J. (2004). Job assignments. In McCauley, C. D., & Van Velsor, E. (Eds.), *The Center for Creative Leadership handbook of leadership development* (pp. 151–182). San Francisco, CA: Jossey-Bass.

Olivero, G., Bane, K., & Kopelman, R. (1997). Executive coaching as a transfer of training tool: Effects on productivity in a public agency. *Public Personnel Management*, *26*, 461–469.

Olson, M. L. (2001). *E-Coaching, ASTD's source for e-learning*. Retrieved March 3, 2008, from http://www.learningcircuits.org/2001/sep2001/olson.html

Pastor, J. (2002). El liderazgo de personas en la organización. In Bonache, J., & Cabrera, A. (Eds.), *Dirección estratégica de personas*. Madrid, Spain: Prentice Hall.

Phillips, J. (1997a). *Return on investment in training and performance improvement programs*. Houston, TX: Gulf Publishing.

Phillips, J., & Phillips, P. (2005). *ROI at work: Best practice case studies from the real world.* Alexandria, VA: ASTD Press.

Plunket, L., & Hale, G. (1982). *The proactive manager*. New York, NY: Wiley.

Rackman, N., & Morgan, T. (1977). *Behavior analysis in training*. New York, NY: Mc Graw-Hill.

Renzulli, J. (1986). The three-ring conception of giftedness: A developmental model for creative productivity. In Sternberg, R. (Ed.), *Conceptions of giftedness*. Cambridge, UK: Cambridge University Press. doi:10.1017/CBO9780511610455.015

Revans, R. (1980). *Action learning: New techniques for managers*. London, UK: Blond & Briggs.

Rossett, A., & Marino, G. (2005). If coaching is good, then e-coaching is.... *Training & Development, 59*, 11.

Roujol, B. (n.d.). *Definición de e-coaching.* Retrieved 22 July, 2010, from http://www.coachingavenue.com/2/view_news.php?id_news=472

Rummler, G., & Brache, A. (1995). *Improving performance: How to manage the white space on the organization chart*. San Francisco, CA: Jossey- Bass.

Saari, L., Jonhson, T., McLaughlin, S., & Zimmerle, D. (1988). A survey of management training and education practices in U.S. companies. *Personnel Psychology, 41*, 731–743. doi:10.1111/j.1744-6570.1988.tb00650.x

Seligman, M. (1993). *What you can change and what you can't*. New York, NY: Knopf.

Shamir, B., & Eilan, G. (2005). What's your story? A life-stories approach to authentic leadership development. *The Leadership Quarterly, 16*, 395–417. doi:10.1016/j.leaqua.2005.03.005

Sims, H., & Manz, C. H. (1982). Social learning-theory: The role of modelling in the exercise of leadership. *Journal of Organizational Behavior Management, 3*, 55–63. doi:10.1300/J075v03n04_06

Sivera, S. (2008). *Marketing viral.* Barcelona, Spain: Editorial UOC.

Spencer, L., & Spencer, S. (1993). *Competence at work: models for superior performance*. Nueva York, NY: Wiley and Sons.

Sternberg, R. (1996). *Successful intelligence: How practical and creative intelligence determine success in life*. New York, NY: Plume.

Svyantek, D., & Deshon, R. (1993). Organizational attractors: A chaos theory explanation of why cultural change efforts often fail. *Public Administration Quarterly, 17*(3), 339–355.

Swanson, R., & Gradous, D. (1988). *Forecasting financial benefits of human resource development*. San Francisco, CA: Jossey-Bass Publishers.

Swanson, R., & Holton, E. (1999). *Results: How to assess performance, learning, and perceptions in organizations*. San Francisco, CA: Berrett-Koehler.

Tannenbaum, A. (1993). Giftedness: A psychological approach. In Sternberg, R. (Ed.), *Conceptions of giftedness*. Cambridge, UK: Cambridge University Press.

Tannenbaum, S. (1997). Enhancing continuous learning: Diagnostic findings from multiple companies. *Human Resource Management, 36*, 437–452. doi:10.1002/(SICI)1099-050X(199724)36:4<437::AID-HRM7>3.0.CO;2-W

Tassabehji, R., Wallace, J., & Cornelius, N. (2007). E-technology and the emergent environment: Implications for organizational form and function. *The Journal of High Technology Management Research, 18*, 15–30. doi:10.1016/j.hitech.2007.03.007

Toffler, A. (1980). *The third wave*. New York, NY: Bantam Books.

Toffler, A. (1998). *Interview on Viva Newsletter*. Retrieved 24 July, 2010, from http://www.cuentometro.com.ar/221.htm

Touron, J. (2001). *Igualdad, eficacia y excelencia. Retos del sistema educativo ante la sociedad del conocimiento.* Actas del IV Congreso de economía Navarra. Pamplona, Spain: Departamento de Economía y Hacienda.

Uhalde, R., & Strohl, J. (2006). *America in the global economy: A background paper for the new commission on the skills of the American workforce*. Washington, DC: National Center on Education and the Economy.

Ulrich, D. (1997). Measuring human resources: An overview of practice and a prescription for results. *Human Resource Management*, *36*, 303–320. doi:10.1002/(SICI)1099-050X(199723)36:3<303::AID-HRM3>3.0.CO;2-#

Ulrich, D., & Samllwood, N. (2004). Capitalizing on capabilities. *Harvard Business Review*, *82*(6), 119–127.

Ulrich, D., & Smallwood, N. (2007). *Leadership brand, developing customer-focused leaders to drive performance and build lasting value.* Cambridge, MA: Harvard Business School Press.

Urch, V., Sala, F., & Mount, G. (2006). *Linking emotional intelligence and performance at work.* London, UK: Lawrence Erlbaum Associates.

Vroom, V., & Yetton, P. (1973). *Leadership and decision making*. Pittsburgh, PA: University of Pittsburgh Press.

Walker, A., & Smither, J. (1999). A five-year study of upward feedback: What managers do with their results matters. *Personnel Psychology*, *52*, 393–423. doi:10.1111/j.1744-6570.1999.tb00166.x

Wang, L., & Wentling, T. (2001). *The relationship between distance coaching and the transfer of training.* Paper presented at the Academy of Human Resource Development, Tulsa, Oklahoma.

Warner, R., & Sternberg, R. (1985). Practical intelligence in real-world pursuits: The role of tacit knowledge. *Journal of Personality and Social Psychology*, *49*, 436–458. doi:10.1037/0022-3514.49.2.436

Watkins, R., & Kaufman, R. (1996). An update on relating needs assessment and needs analysis. *Performance Improvement*, *35*(10), 10–13. doi:10.1002/pfi.4170351005

Wikipedia. (2010). *Web 2.0.* Retrieved 20 August, 2010, from http://en.wikipedia.org/wiki/Web_2.0

Wiseman, R. (2009). *59 seconds, think a little, change a lot*. London, UK: McMillan.

Zeigarnick, A. (2007). Bluma Zeigarnick: A memory. *Gestalt Theory*, *29*(3), 256–268.

Zimmerman, B., Bonner, S., & Kovach, R. (1996). *Developing self-regulated learners*. Washington, DC: APA. doi:10.1037/10213-000

KEY TERMS AND DEFINITIONS

Development: The process of getting new skills or knowledge, specially so that you can do a job more effectively. Leadership development focuses on learning events, such as mentoring, training, self-study and job rotations to prepare employees with the skills required to lead an organization.

Leadership: The state or position of being a leader. The active process through which an individual seeks to identify courses of action and to guide and stimulate the actions of others toward goals desired by the leaders.

New Economy: Described the transition from heavy industry to a new technology based economy.

Talent: Exceptional performance, as a result of a total command of certain skills (competentes), which have to do with one or several activity fields. Talent is influenced by biological, psychological, and social variables.

Chapter 16
Management and Leadership of Innovative Work Teams

Santiago Gutiérrez
Castilla-La Mancha University, Spain

ABSTRACT

This chapter focuses on the analysis of two factors; the characteristics of new work teams and the management and leadership tools required for these innovative teams, in order to generate greater effectiveness. Firstly, the authors present the differences between a work team and a group and the basic characteristics of innovative work teams (multifunctional teams, self-managing teams, virtual teams and open-innovation teams). Then, the authors focus on the management systems of these teams with the aim of generating greater effectiveness. High-performance practices and management skills are presented as tools for increasing motivation and commitment to the company and its business culture. Finally, the authors analyzed the leadership and the effect of leadership has an impact on the team's results as well as on the company's innovativeness, making it an essential part of creating effective innovative teams.

INTRODUCTION

Companies in the modern era operate in highly volatile markets, with ever-increasing competition. In order to survive these increasingly complex environments and respond to these new situations, innovation is key (Dess and Pickens, 2000). The most advanced organizations are developing new models of organization, based on work groups, which aim to promote innovation and increase their response capacity (King and Anderson, 2002). By development these types of corporate knowledge assets modern companies are able to achieve a sustainable competitive advantage. Resources such as creativity, the capacity for innovation, the development of human talent, the management of teams or leadership, are increasingly used in business management because they generate an improved competitive position for those companies that best develop them and that are able to implement them successfully. (Drazin and Schoonhoven, 1996).Moreover some of them are giving to innovation a higher bonus. (Drazin and Schoonhoven, 1996).

DOI: 10.4018/978-1-61350-207-5.ch016

Table 1. Innovative teams and changing environments

Principal changes in the business environment	Innovative Teams
More complex environment	Multifunctional teams
Better educated and trained employees	Self-managing teams
Advances and developments in technology and systems of communication	Virtual teams
Increased competition and lack of creativity	Open-innovation teams

Source: The author

Not only Innovation and but also its conversion processes turning creative ideas into concrete actions have been analyzed by literature (Mumford and Gustafson, 1988) including some other factors such as estrategy (Hitt *et al.*, 1996; Parnell, Lester and Menefee, 2000; Montes, Moreno and Morales, 2005), structure (Pierce and Delbecq, 1977; Damanpour, 1998), organizational environment (Amabile and Gryskiewicz, 1989; Isaksen *et al.*, 2001), spreading practices (Abrahamson, 1991), group interaction (Mumford *et al.*, 2001) and individual capacities performance (Mumford et al., 1998). Most of these studies are centred in innovation at an organizational level as work groups, nets or individuals (King and Anderson, 2002) as it is a concept embracing different dimensions from exploration and generation of ideas to its further defense and application (de Jong and den Hartog, 2010).

Individuals in organizations have traditionally been grouped according to their specialty and always under a common basic function. In this traditional system, one person acts as leader and has a high level of authority over the internal working of the group and over external relations. Individuals usually remain at the same level within these groups for a significant period of time. However, the needs of the modern company, derived from increasingly complex environments, require the formation of new work teams that are more flexible and more adaptable to prevailing circumstances. These work teams have the potential to be better prepared when adapting to new circumstances than traditional ones do. Now more than ever, individuals must work together and coordinate their efforts with the other members of the team and the rest of the organization.

The creation of these new work teams is the result of certain changes in the business environment. These teams are better prepared to tackle the new challenges presented by today's markets (Tushman and O'Reilly, 1997; de la Florida, 2002).

Table 1 shows the most significant changes affecting organizations that have led to the creation of new work teams:

MAIN PURPOSE

This chapter will be focused on the analysis of two factors; the characteristics of these new work teams, and the management and leadership tools required for these innovative teams, in order to generate greater effectiveness. In order to achieve this, the chapter will be divided into three sections, explained below.

Firstly, the differences between a work team and a group will be established. Subsequently, the basic characteristics of innovative work teams will be presented, as well as the results of investigations carried out on these teams. There will be a particular focus on multifunctional teams, self-managing teams, virtual teams and open-innovation teams, each of which create specific leadership requirements.

The second part concentrates on the management systems of these teams with the aim of generating greater effectiveness. High-performance practices, also referred to as high-commitment practices, and management through skills, are presented as tools that reinforce the knowledge, abilities and skills of an organization's members, increasing their motivation and commitment to the company and its business culture. Studying each of these management techniques is an important part of being able to manage innovative teams with greater precision and, consequently, achieve better results.

The third section of this chapter focuses on leadership and the influence it exerts over the development of a cooperative and collaborative environment. The effect of leadership has an impact on the team's results as well as on the company's innovativeness, making it an essential part of creating effective innovative teams (Tierney, Farmer and Graen, 1999). Defining the concept of leadership, identifying false leaders and analyzing different leadership styles will help to establish a type of leadership that is best suited to the specific requirements of these work teams. Making sure that each member of the teams looks beyond individual needs to improve innovation is the main task of leadership. Knowing about skills, values and behaviour is a way of advancing the creation of highly innovative work teams.

This chapter of the book finishes with a series of questions for contemplation as well as recommended lectures for the topics covered, so that readers may study in-depth the main theme of the chapter.

INNOVATIVE WORK TEAMS: PRINCIPAL CHARACTERISITCS

As stated previously, both economic studies and real-life business experience today highlight the importance of teamwork in the development of innovation and the continued improvement of organizations. Understanding what a work team is and the basic characteristics it has is essential to guaranteeing its successful implementation. The problem for many companies is that neither high-ranking directors nor other employees truly understand what it means to work in a team and mistakenly believe that being physically present in the same room is enough to ensure a group functions correctly. Clearly that is not the case. The simple situation of having one individual with excessive authority, who imposes his/her ideas, prevents effective interaction between the other members of the group. Therefore, it is extremely important to differentiate between what may be called a work team and what is merely a group, in order to then be able to analyze the main characteristics of innovative teams.

Fundamentally, a work group is characterized by having members that interact to share information and take decisions, thereby facilitating one another's professional development within their area of responsibility. Members of the group are not required to, and do not have the opportunity to carry out collective work that requires the efforts of all, resulting in a lack of synergy in the work they each carry out as there is little or no interdependency between the different functions. A work team, however, is comprised of a reduced number of people with complementary technical and personal skills, that interact to achieve a common objective, the outcome of which is a collective result that is greater than the sum if the individual parts for which they are all responsible (Katzenbach and Smith, 1993; Kinlaw, 1991). All members are committed to the common processes, performance and objectives.

One important difference between a team and a group can be found in their management styles. In a work group the power to take decisions and to lead the work is clearly assigned to one individual. However, in work teams the capacity for taking decisions and the responsibility of coordination is distributed among its members. In most cases, innovative teams are able to collectively take decisions that were previously taken by individual managers.

Other psychological mechanisms present in work teams, and which are essential for correct functioning, are social identification and social representation. Social identification comes from the human tendency to see the world through a 'them and us' dichotomy, resulting in a feeling of belonging for some groups and exclusion from others. This mechanism helps an individual to understand what it is they are doing and how they can do it, whilst simultaneously provoking feelings of pride and belonging. Feeling good about belonging to a group builds self-esteem and motives us to work harder.

Social representation refers to the collection of accepted ideas of reality and carries out two functions. Firstly, it explains what is happening around us and gives it meaning; and secondly, it reflects power and relationships in a social context.

The main features of the best work teams found in the literature are reduced size (Curral, *et al.*, 2001), complementary skills and abilities, commitment of all the members (Hyatt and Ruddy, 1997; Kiffin-Peteron and Cordery, 2003) and shared responsibility for the results achieved (Álvarez de Mon, 2000). Therefore, among the most important factors for effective operation are commitment, confidence, the objective, the feeling of belonging, communication, involvement, and a working method or process adopted by all (tools, planning techniques, meetings, support, etc.) (Wellins *et al.*, 1991).

Some theoretical models of innovation in work teams (such as Anderson *et al.*, 2004; Janssen *et al.*, 2004; West, 2002 and West and Hirst, 2003), determine the presence of certain factors that intervene in the innovation process. These factors are fundamentally related, either to the situation in which they are developed (organizational) or to the work teams. The main problem to arise is that some of these factors also act as modulators of the relation between certain inputs and innovation (Fay *et al.*, 2006). Some of the organizational factors described by studies as influential in innovation are: climate, organizational structure, size, age, etc. (West and Hirst, 2003). Studies determining theoretical models as well as empirical investigations carried out to analyze factors related to work teams, establish the characteristics of the tasks, external demands, composition of the team, groups processes and leadership, as those that have the greatest impact on improving innovation (González-Romá, 2008) (see Figure 1).

It should not be forgotten that a work team is one which carries out a number of tasks of collective importance for the organization, through which the structure, composition and functioning of the group are defined (West, 2002). According to this author, these tasks include the autonomy of the team, the opportunities for learning, and the development and the level of participation between the members. Therefore, the end result is that the members of the team identify with the task and achieve greater intrinsic motivation that affects their level of creativity and innovation.

On the other hand, work teams also develop innovation by being influenced by external de-

Figure 1. Factors influencing the innovation process

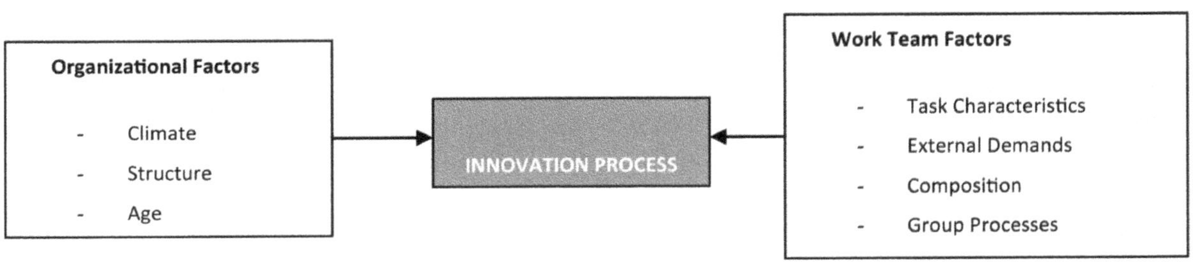

mands (West, 2002). Innovation is understood to be a strategy of competing when faced with threats in the business environment and other external pressures. The third factor to consider is the composition of the team, and particularly the issue of diversity (Jackson *et al.*, 2003; Van Knippenberg and Schippers, 2007). Although there are various types of diversity, the most analyzed type is functional diversity. Numerous studies have found a positive relation between functional diversity and the quality of innovation (Drach-Zahavy and Somech, 2001; Perretti and Negro, 2007), but the relation is not quite so simple. Some studies have even found contradictory and inconclusive evidence (Jackson *et al.*, 2003; Van Knippenberg and Schippers, 2007). Ultimately, it is argued that for diversity in the teams to contribute to innovation, this must be accompanied by group processes between the members (West, 2002). These group processes are measures related to the definition of clear objectives, shared by all members, incentivizing participation and cooperation, open and collective reflection on functioning, strategies and aims, support for innovation and conflict management.

As a consequence, team leaders play a crucial role in shaping the perceptions, tendencies and behaviour of the members (Schaubroeck *et al.*, 2007) and, therefore, in the performance of the teams.

Multifunctional Teams

Companies are constantly increasing the number of tasks that require the collaboration of different departments. To facilitate coordination, companies create multifunctional teams, incorporating members from several different areas, in order to develop and complete these tasks. The team is responsible for planning and carrying out a complex task that requires a high level of coordination and cooperation, and the collective resolution of any problems encountered. These teams are comprised of people trained in different areas, which gives them certain advantages over teams based only on people from one or very few different areas (Rochford and Rudelius, 1992). This allows the team to cover numerous wider sources of information to which the different specialists have access.

Multifunctional teams may be created independently from an organization to carry out different tasks and projects, or even to deal with certain clients[1]. The most common situation in which this type of team is used is for the development of a new product, in which the engineering, manufacturing and marketing departments are closely linked. Other examples are the implementation of new IT equipment, the planning of a publicity campaign, the development of a new training programme, etc.

Some multifunctional teams may be permanently incorporated into the formal structure of an organization, although the majority are temporary and are disbanded once the task they were created for has been completed. Membership of the group may be consistent throughout its existence or may change as the task progresses. Members may only work for the team on a part-time or full-time basis, although in the majority of cases the members of a multifunctional team are also part of a different functional unit within the organization. This diversity in the hierarchical position it occupies within the organization favours each member having different contact within and outside the company, thereby enabling the group to have a much wider range of available resources.

Multifunctional teams that also have cultural diversity should have a greater capacity to resolve problems derived from having a greater number of points of view. Some studies have shown a positive relation between a team's diversity and results, in terms of innovation and performance (Jackson *et al.*, 1995; Priem *et al.*, 1995).

It is important to be conscious of the fact that there is a limit to the diversity of a work team. It is often the case that larger groups do not have greater success, but they certainly do have higher administrative costs and increased communication problems. Furthermore, the greater the size, the greater the difficulty in creating a sense of group identity and the potential for what is referred to as *social laziness* is increased. Social laziness is produced when, due to an increase in the team's size, individuals feel that they will not receive due recognition for their contribution to the group effort, leading to less effort and commitment. Therefore, the average team size in the USA is 11 members (Devine *et al.*, 1999), although there is still considerable variance between them.

The main disadvantages are often related the problem of coordination and communication. Individuals tend to interact most with other individuals they consider to be similar. This tendency has been defined as *homophilia*. This preference is a result of the fact that it is easier and more comfortable to communicate with people that have the same dialect, mental processes and similar belief systems (Uzzi, 1997). This common dialect, confidence and familiarity make individuals more disposed to exchange information. Heterogeneous teams have greater difficulty in integrating different objectives and visions, which leads to conflicts and reduced cohesion within the group. This may be resolved by maintaining prolonged contact, so that the extensive interaction helps improve group processes (Watson *et al.*, 1993).

All these difficulties in creating committed group members and the existence of divided loyalties to protect the operational field to which they belong, increase the need to have one leader with considerable power and good personal skills. The majority of these teams have a formal leader, elected by a manager of superior rank. It must be kept in mind that a self-managing multifunctional team may lose much time resolving internal conflicts, not leaving sufficient time to complete the group mission. Studies reveal that self-managing multifunctional teams have a lower success rate (Cohen and Bailey, 1997). Multifunctional teams are not easily managed, that is the reason why during their early stages are time consuming till the time the group develops a kind of diversity related learning and team complexity.

Although the skills required by the leader of a multifunctional team are not clearly defined, many studies (Mumford *et al.*, 2002) suggest that leaders should have technical and cognitive abilities, interpersonal skills, and project management and political capacities. In this way the leader must promote interaction and trust (Robbins and Judje, 2009) in order to improve team performance.

Kichuk and Weisner (1997) explored five aspects of a leader's personality that affected a team's chances of success and found that those aspects that most favoured the success of a team developing new products were: being highly extrovert and highly sympathetic, and having a low level of neurosis.

Self-Managing Teams

Self-managing teams have been used more and more in recent years in the business environment (Kirkman and Shapiro, 1997; Kirkman *et al.*, 2001). Some studies (Lawler *et al.*, 1995) explain that 68% of the 1000 companies on the Fortune 1000 list make use of this type of team.

Self-managing teams are non-hierarchical groups of individuals with different and complementary experience and knowledge. Their objective is to develop a type of collective knowledge that requires the pooling of individual knowledge (Grant, 1997; 2001). No individual working alone could carry out the same tasks or produce the best innovations. For this reason companies turn to a combination of individuals with different and complementary skills and perspectives, deriving from this cooperation the achievement of these improvements and innovations.

Any type of team may be self-managed, but this type of management is usually reserved for teams that carry out the same type of functional task on a regular basis, with members that remain constant over a long period. Tasks are usually related or interdependent and they are used fairly regularly in manufacturing processed, although recently they can also be found in the provision of services. The team members offer similar functional experience and usually take turns in carrying out the different tasks for which the group is responsible.

The advantages of self-managing teams are team members more committed to the work, an efficient management of labour issues, greater efficiency and job satisfaction, less employee turnover and less absenteeism. However, their greatest advantage is without doubt the improved knowledge of the group (Lazear, 1998). The company's competitive advantage is better protected than a situation in which knowledge rests with just one individual (Wright *et al.*, 1994). When the members learn to carry out multiple tasks the team's flexibility increases when resolving problems, the work becomes more interesting and there is an opportunity to learn new skills.

However, there are also criticisms and negative aspects of this type of work, as it does not always produce the desired improvements in performance (Chaston, 1998). Certain resistance has been noted among individuals when it comes to sharing knowledge, which reduces performance.

Taking as a reference the theory of social dilemmas (Cabrera, 2002), one of the main problems created by this type of work group is that when it involves the voluntary contributions of different members of a collective, everybody benefits regardless of whether or not they have contributed to the success (Olson, 1965). Non-contribution may become the dominant strategy, meaning that no collective knowledge is gained (Cabrera, 2002), which would explain the resistance of other members to sharing their knowledge. This type of opportunist behaviour is based on lies, deception and calculated effort, known as the *free-rider problem*, whereby team members benefit from the group effort (Lazear, 1998).

In self-managing teams the members must complete their own tasks, control the result obtained and take responsibility for the innovation achieved. Therefore not only are suggestions made but also responsibility for implementing the chosen solution and its results is a collective responsibility. The organization in charge defines that mission, the scope of operations and the available budget. All other responsibilities delegated to the team vary from one organization to another. Sometimes, self-managing teams are also responsible for personnel decisions within the team, such as working hours, the selection and contracting of members, dismissal, and even determining salaries.

Despite the fact that trade magazines are full of articles describing the successful applications of self-managing teams (Cordery, Mueller and Smith, 1991), some precautions must be taken. The results of these teams on the investigations carried out are not always consistent and do not reflect great improvements in performance (Banker *et al.*, 1996; Cohen and Ledford, 1994). The difficulty in implementing self-managing teams makes having a specific situation and competent leadership all the more important.

Most studies (Cohen and Bailey, 1997; Kirkman and Rosen 1999; Pearce and Ravlin, 1987) outline the following conditions as favouring better results:

- Establishing clear objectives
- Complex and important tasks
- Consistent members
- Access to relevant information
- Sufficient recognition and remuneration
- Strong support from superior management
- Excellent interpersonal skills between members
- High level of discretion over work processes
- Competent and instructive leadership

When going into greater detail about the role played by the leader, it is important to distinguish between the roles of the external and internal leader. It is common to have internal leaders, elected by the members, a position which may even be rotated periodically. However, the most important responsibilities are shared by the members so that their principal responsibility is coordination and facilitating the taking and implementing of team decisions. In order to these teams work successfully, the leader must develop not only his self-reinforcement to be able to identify and evaluate a well-done job, self-criticism to detect inadequate behaviors and correct them, but also the establishment of his own goals to find key process areas, self-observation and self-evaluation to control his own behavior's consequences (Gil and García, 1996).

The external leader takes on responsibilities that have not been delegated to the team. When a team is created, the external role of the leader is normally to advise, moderate and assist the team to get started. This includes assisting members to develop skills and above all to promote self-confidence among the team members. Depending on how the team evolves, the role of leader is increasingly focused on advising new members. Another important role the leader carries out is obtaining information, resources and support from the rest of the organization. The leader acts as a link between the team and the company. Creating and maintain cooperative relations and an efficient exchange of information is essential. Finally, the external leader must act as the team's defender and lawyer.

Virtual Teams

Previous team types do their job "*on site*" but due to new technologies, other kinds of teams allowing people collaboration on line emerge (Robbins and Judge, 2009). During the last years, the number of these type of teams, called "*virtual teams*" have increased. Moreover, some authors foresee a sharp increase in the number of these teams in organizations and even expect them to revolutionize the workplace in the future (Townsend *et al.,* 1998).To what end that nowadays, nearly all teams make at least a part of their job remotely. There has been a sharp increase in the number of virtual teams in organizations and some authors even expect this type of team to revolutionize the workplace in the future (Townsend *et al.*, 1998). Virtual teams are teams in which members may be physically a great distance apart from one another, but are able to carry out an intense collaboration through the use of advanced information technology, such as email, video conferencing, IT programmes, groupware, mobile telephones, etc.

This allows individuals with unique skills and who are in different places to work on the same project. They are able, therefore, to collaborate without incurring travel costs and without interrupting their lives (Joinson, 2002). These types of teams are used frequently in global companies, but they also raise some interesting questions.

Proximity and frequent interaction favours communication and collaboration. In virtual teams, channels of communication are much less rich and present more problems when establishing norms and dialects than in direct or face-to-face forms of communication. It is for this reason that any new personnel selected must be comfortable with using new technologies and have solid interpersonal skills (Kirkman *et al.*, 2002). Their members should be able to work independently and have a strong work ethic, and individuals that tend to seek out interaction with others rather than avoid it must be selected. These types of teams also face challenges in developing confidence, resolving conflicts and exchanging knowledge.

As with self-managing teams, any type of team may be virtual, but the most common ones are similar to multifunctional teams. A virtual team may exist temporarily to achieve a specific task, or something more long-lasting to resolve technical problem, coordinate diverse activities, maintain external coordination with suppliers, clients, etc.

Virtual teams are most likely to comprise members with different cultures, work hours, and organizations, which poses new challenges to the leaders of these teams. The lack of contact creates difficulties for tasks such as supervision, influencing members, developing confidence, etc. But above all the biggest challenge is creating an identity for all the team members. The main objective of the leader is ensuring the commitment of all the team members.

Studies have started to analyze differences between dispersed and co-localized teams (such as Bell and Kozlowski, 2002) but it is likely that the same leadership roles are relevant for both the dispersed and virtual teams. These studies have shown that physical separation negatively impacts performance. However, in some cases the effect of dispersion may be beneficial, depending on the quality of the processes related to the tasks (Siebdrat *et al.*, 2009).

In order to make these teams effective, the leader should not underestimate the importance of short distances. For example, team located on different floors of the same building are not considered as dispersed teams and underestimate the barriers to collaboration, achieving worse results. On the contrary, groups that are in different areas of a country or even a continent, are aware of the situation and make a greater effort to improve vital processes such as coordination and communication linked to the tasks.

The leaders are also required to emphasize the capacities of the work team as the aim is to generate greater knowledge from dispersed members. A certain amount of self-leadership is required on the part of the team members in order to resolve collaboration difficulties and find effective means of overcoming different obstacles. Having efficiency as its main goal, the leader must get the trust between its members so that the team can make a progress keeping their aim in mind. Besides, it is important both to emphasize their efforts and spread their products making the team visible for the organization (Malhotra, Majchrzak and Rosen, 2007).

Open-Innovation Teams

As opposed to the types of teams analyzed previously, open-innovation teams are characterized by adding external sources of knowledge to accelerate and improve the innovation process (De Jong *et al.*, 2008). New network communication technologies have allowed us to experience phenomena that would have been impossible just a few years ago.

They are based on the fact that organizations do not have all the necessary talent to improve their innovations, and therefore require the collaboration of people from outside the company (Chesbrough, 2006). However, it is important to realize that open innovation has its limits with regard to its application, particularly in the type of task and the industrial sector to which it applies (Chesbrough *et al.*, 2006). It is not always possible to put into practice, nor would it always be desirable to do so.

Studies on these types of teams are at an early stage, meaning that there are not yet sufficient findings on which to make valued judgments. For this type of team there are great enthusiasts and sceptics in equal measure. There is greatest consensus in notion that open innovation is supported by external professionals as a source of ideas and as a way of marketing them.

In order to manage this open innovation, Boudreau and Lakhani (2009) propose two related models; though collaborative communities or through competitive markets. Collaborative communities are characterized as having social norms and 'weak' rules in order to promote open access to information, transparency and joint development and intellectual property. The best examples are perhaps the Linux operating system, from the Linux Foundation, the web browser Mozilla Firefox, the web server Apache, contributions to Wikis, as well as other open code software initiatives.

Competitive markets act in a different way, as instead of collaborating, external innovators compete amongst themselves. For example, Innocentive.com[2] is a website created in 2001 that connects 'seekers' (companies), who present scientific or technical problems so that 'solvers' (some 200,000 scientists, engineers, inventors, businesspeople, chemists and other professionals from a great variety of disciplines and countries) can resolve them. A problem is presented, a time-frame is established for resolution and a monetary reward is stipulated. Those interested in solving the problem do so isolated from others and from the searcher. At the end of 2008, 80 companies had presented 700 challenges of which approximately one third were resolved.

Leadership of these types of external innovators must pay special attention to their members' motivation. Although various studies are shown that the motivations of external professionals devoted to open innovation may be very similar, a large number of them may be grouped in two categories: extrinsic and intrinsic. While competitive markets usually prefer the first, collaborate communities are more oriented towards the latter. Members of collaborative communities are often happy to work without financial reward. The simple pleasure of innovative task itself, the challenge it represents, personal and professional identity, autonomy, as well as learning and the development of skills, are often much better motivation than things such as money, the needs of the user, or their own professional career.

MANAGEMENT OF INNOVATIVE TEAMS

In each one of the analyzed kind of teams, we have observed how unity among its members, confidence and compromise with their goals are essential for a good performance. In order to make leader tasks easier, some kind of practices, such as the so called high performance ones as well as skills management are proposed to make the results improve.

High Involvement Practices

High involvement practices (Huselid, 1995) or high performance practices (Lawler, 1986; Walton, 1985; Wood and Albanese, 1995), are also known by some authors as 'soft' managements models of human resources (Guest, 1987; Truss *et al*., 1997). These terms essentially refer to the planning of human resources management oriented towards the commitment of the employees, involving the active participation of these in decision making and providing the necessary organization support and resources (Roca *et al*., 2002). It stems from the idea that the effect produced between capabilities and motivation has a multiplying effect on the value created in the organization (Ulrich, 1998).

There are two principal advantages for organizations in implementing systems of human resources management oriented towards performance: firstly, it enables the organization to be more effective (Lawler, 1992) and secondly, it increases the social acceptance of this model in comparison with the traditional taylorist style based on strict control and subordination of the employees (Sashkin, 1984).

Researchers have attempted to determine which best human resources practices have generated the best results by grouping non-traditional practices that have been increasingly used in companies. Ichniowski *et al*., (1996) cite work flexibility, teamwork, contingent remuneration, empowerment, job security, etc., as practices that generate the greatest level of commitment among employees.

Marchington and Grugulis (2000) presented a model based on the seven practices identified by Pfeffer (1994, 1998) as generating success: job security, selective contracting of personnel, teamwork, contingent compensation, extensive training, reduction of status differences and transfer of information.

The main problem presented by recent studies is that there is little empirical evidence concerning which practices comprise a coherent system of human resources management oriented towards performance (Ferris *et al*., 1999; Roche, 1999). However, it can be noted that there is high amount of concurrence at a theoretical level between different authors when it comes to determining certain high performance practices, outlines in Table 2.

Table 2. Principal high performance human resources practices

HIGH PERFORMANCE PRACTICES	DESCRIPTION	PRINCIPAL AUTHORS
Creating positions with wide scope	Jobs with a wide scope, low horizontal and vertical specializations.	Schuler and Jackson (1987a); Capelli and Crocker-Hefter (1996)
Selective recruitment	Defined recruitment and selection criteria appropriate to the demands of the position. Analysis of past, present and future conduct.	Guest (1987), Arthur (1994), Pfeffer (1994), Huselid (1995); Youndt et al., (1996); Roche (1999); Bayo and Merino (2001); Ordiz and Fenández (2003); Céspedes et al., (2005); Beltrán et al., (2008); Wood and Menezes (2008).
Fixed contracting	Job stability to avoid temporary work.	Pfeffer (1994); Bayo and Merino (2001); Ordiz and Fernández (2003); Wood and Menezes (2008)
Extensive training	Great importance and investment in the training and development of the employee. Continuous and evaluated training of all employees. Incentivise skill acquisition.	Schuler and Jackson (1987a)
Performance evaluation	Evaluation schemes based on results.	Guest (1987); Pfeffer (1994); Huselid (1995); Huselid et al., (1997); Capelli and Neumark (2001); Bayo and Merino (2001); Richard and Johnson (2001); Gutrie et al., (2002); Roca et al., (2002); Mohr and Zoghi (2008); Beltrán et al., (2008); Wood and Menezes (2008)
Contingent compensation	Salary policies based on individual and group results. Reward above the market average. Tangible, intangible and flexible incentives.	Guest (1987); Arthur (1994); Pfeffer (1994); Huselid (1995); Roche (1999); Bayo and Merino (2001);Capelli and Neumark (2001); Gutrie et al., (2002); Ordiz and Fernández (2003); Céspedes et al., (2005) Zatzick e Iverson (2006); Beltrán et al., (2008); Wood and Menezes (2008)
Greater discretion	Award the position and all its responsibilities to the employee in order to evaluate conduct and norms in the job. Decentralization and less vertical specialization.	Schuler and Jackson (1987a and b); Arthur (1994); Pfeffer (1994); Huselid (1995); Roche (1999); Bayo and Merino (2001); Richard and Johnson (2001); Gutrie et al., (2002); Ordiz and Fernández (2003); Roca et al., (2002); Mohr and Zoghi (2008).
Internal promotion	Provide opportunities for professional development in the company through a formal system of professional careers.	Arthur (1994); Pfeffer, (1994); Huselid (1995); Roche (1999); Bayo and Merino (2001); Richard and Johnson (2001); Guthrie et al., (2002); Roca et al., (2002); Ordiz and Fernández (2003); Mohr and Yoghi (2008)
Teamwork	Incentivize and promote teamwork, employee commitment and Little job rotation.	Arthur (1994); Pfeffer (1994); Huselid et al., (1997); Bayo and Merino (2001); Capelli and Neumark (2001); Richard and Johnson (2001); Guthrie et al., (2002); Zatzick e Iverson (2006); Mohr and Zoghi (2008)
Shared information	Open and shared information systems for all members of the organization.	Guest (1987); Arthur (1994); Pfeffer (1994); Huselid et al., (1997); Roche (1999); Capelli and Neumark (2001); Bayo and Merino (2001)Richard and Johnson (2001); Guthrie et al., (2002); Ordiz and Fernández (2003); Roca et al., (2002); Zatzick e Iverson (2006); Mohr and Zoghi (2008) Wood and Menezes (2008)
Job security	High levels of job security. Specialized staff to improve job security.	Schuler and Jackson (1987 a and b)

Source: The author, based on Martí (2008)

Implementing all of these human resources measures has a profound impact on the employee as it increases his/her potential, motivation and commitment to the company (Sanchez, 2004), all of which are fundamental elements for improving innovation.

The commitment generated within the company makes the employee act in a positive manner, which translates into an effective way of achieving organizational objectives and leads to improved business results (Ordiz and Fernández, 2003; Mohr and Zoghi, 2008); Beltrán *et al.*, 2008).

Skills Management

The skills management model is able to resolve the problems posed by traditional approaches focused on evaluating work positions, utilizing such variables as regulated and outdated training. When it comes to analyzing the qualities required to occupy certain simple positions of work, it is relatively easy for experts to stipulate which aptitudes and personality traits are needed. But this is not the case when it comes to describing more complex positions, with greater responsibilities and functions, as is the case nowadays.

This model is based on what people can bring to organizations, in other words their competencies, and to do this it studies a worker's development looking for behaviour deemed appropriate due its relation to an organization's objectives and mission. Directors should provide their employees with opportunities to develop and put their new skills and knowledge into practice. The skills management model treats human abilities as valuable assets to invest in, and it consists of a full system of evaluation and improvement of organizations through the development of the people it comprises (Oliveros, 2006).

This model has arisen from the field of psychology at work and in organizations, which has been growing since the late 1960s and early 1970s (Spencer and Spencer, 1993). It started to be used following studies done by McClelland (1973) during the 1970s, when it attempted to identify variables to explain performance in the workplace.

Although there is not general agreement on the current use of the concept of professional competency, the Delors report of the European Union (1996) highlighted the fact that it are increasingly used by business leaders in their companies. The term competency refers to the unique make-up of each individual and covers the overall capacities and qualifications they have obtained throughout their technical and professional training, social conduct, aptitude for teamwork and even individual initiative. One skill can apply to various aptitudes and personality traits (Levy-Leboyer, 1997). The overall effect of this is the characterize individuals and explain differences in conduct when carrying out specific tasks. It is a more widespread notion than the concept of 'qualification', but it is nonetheless more complete.

When defining the concept of competency, two theories exist. Firstly, Levy-Leboyer (1997) defines it as a collection of observable behaviour that enables a specific job to be performed effectively and efficiently in a specific organization. This theory is based on the already classic Iceberg Model (Spencer and Spencer, 1993). In this, a profile is configured by a series of observable behaviour, linked to given organizational, strategic and cultural structures and a specific job. Within the same theory, Zarifain (1999) notes that it is necessary to introduce the concept of 'situation' to understand the concept of 'competency'. In this sense, one is competent in relation to a specific situation. Competencies affect the full implementation of aptitudes, traits and knowledge, which are acquired within a company with a specific culture and strategy.

The other option is to consider competency to be an underlying characteristic of an individual and one which determines conduct. In order for a person to be able to able to conduct themselves appropriately, the following components are required: knowing (overall knowledge), knowing how to do (being able to apply knowledge to solve problems), knowing how to be (behaving according to the norms, rules and culture of an organization), wanting to do (displaying interest and motivation), and being able to do (having the necessary means and resources to carry out the required tasks) (Pereda *et al.*, 2002).

Competency can be defined as complex 'knowing how to do', and is the result of mobility, integration and adaptation to knowledge, skills and attitudes used efficiently in different situations. It should also be noted that one fundamental point is that an outside expert is needed to make a competent person explain their conduct. A competent individual can display their competency but it is a big ask to request that they verbalize or that they teach others though a written report, rather than through continuously observing their conduct.

The skills management model proposed by Levy Leboyer (1997) is split into the phases; identification, evaluation and development of skills. It is a theoretically simple model but one that encounters certain difficulties when it comes to putting it into practice.

During the identification phase one must be aware that there are no clear analysis methods like those used for analyzing positions of work (Pearn and Kandole, 1988). Consequently, the task of identification is increasingly complex due, in the main, to the reasons (Pearn and Kandole, 1988): first, no position has a fixed content in time; second, the same position may be filled in a different manner by different people; and third, attention should not be paid to the title of a position as different skills may be required for positions with the same name. There are lists of skills that may make this task easier but it should be remembered that these are general skills and they may not always be adapted to the needs of each company.

In the literature there are different classifications of skills. One common theme is an appreciation for the suitability of differentiating between what are referred to as generic or transferrable skills and specific skills. Generic skills refer to a collection of common abilities or aptitudes for all types of know-how. For example: the selection and use of information, organization capacity, professional ethics, etc. Specific skills reflect theoretical knowledge and specific procedures for each profession. They are unique to a particular professional profile and are expressed through the basic abilities of each professional practice.

Boyatzis (1982) carried out one of the first studies of directive competencies and asked a group of directors which specific types of conduct had helped them to be efficient in their jobs. The result was a total of 21 types of skills, grouped in five clusters: management of actions and objectives skills, leadership, human resources management, management of subordinates and of relations with others (Abad and Castillo, 2004).

Other authors, such as Aubert and Gilbert (2003) describe four categories of skills: intellectual, interpersonal, business, and development.

The European Space Agency (1998-2003) has established a model of generic skills for its employees, divided into the following categories: cognition (thought), relations (interaction), corporate identity (way of being) and actions (way of doing).

Bennet (2002) carried out in the United Kingdom an analysis of the transferable skills listed in 1,000 adverts for graduate jobs. Ordered from most to least frequent were the following: communication, information technology, organization, teamwork, interpersonal relations, analytical thinking, self-confidence, numerical reasoning, initiative, presentation, foreign languages, leadership, and adaptability.

Feist and Gorman (1998) found among the most creative scientists a great range of common personality features including among others, goal oriented motivation, autonomy, open-mindless, flexibility, cognitive complexity, self-confidence, self-control and emotional stability.

It should not be forgotten that despite the existence of these lists, each organization must create a list of key skills for their own business, taking into account their strategy and thereby improving their competitiveness. Therefore there are other methods of identifying essential skills that can are best adapted to the needs of each company. Observation, questionnaires, semi-structured interviews, interviews for critical aspects, and group discussions (Krueger, 1988), are techniques that are more adapted to a specific organization. The problem that arises is that none of these techniques are completely valid and viable, meaning the best approach is to work with a combination of them, which allows the strengths and weaknesses of each of them to be balanced out (Blanco, 2007).

The phase of evaluating skills also throws up certain problems that must be resolved. It is not only a question of measuring to what extent a skill is acquired, but also evaluating the aptitudes and personality traits. This is done through signs, samples and references. In order to carry out an evaluation of a worker's skills it is useful to establish the different levels of completion or performance in order to compare the results with the required level (Oliveros, 2006).

Finally, the skills development phase follows the specific needs of an individual for their professional development and the needs of the company to make the most of talent and abilities and be able to develop new skills. This involves developing those skills considered most efficient when carrying out a job. Among the most commonly used methods in training programmes are role playing, company games, case studies, group dynamics and, more recently, mentoring and coaching. Boyatzis (1982) places particular emphasis on the need to develop emotional competencies in order to achieve better business results.

Although it does have limitations, this model is used quite frequently in the business world. If we study the work carried out by Sala (2006), in which the financial institutions that had implemented the skills management system were analyzed, we discover that they stick rigidly to this management model. It is worth mentioning the importance of all actors participating in the definition and implementation of skills measures, following the recommendations of the national system of qualification and the work of the national institute of qualifications. Furthermore, it also warns that the signs of operational skills should be transferable in order to not fall into 'a body of detailed experiences" (Sala, 2006). All companies define skills profiles and compare behaviour and levels of competency. It is also observed, in terms of competency, how and evaluation leads to the promotion of employees no longer being linear and increases exponentially, leaving all responsibility in the hands of the employee, making them feel more a part of the organization.

THE EFFECT OF LEADERSHIP

The literature on human resources establishes a series of factors required to achieve innovative results and continuous improvement. It is not enough to implement sophisticated management models such as those analyzed above, but rather it is necessary to create an appropriate environment among the team members if those models are to be effective. The work environment, called *high involvement* by Lawler (1992), or *high care* by Krogh (1998), generates a multiplying effect on team results and on innovation.

The study carried out by Zarraga and Bonache (2005) highlights that the figure of a committed leader involved with the teams exerts great influence over the environment of collaboration. Therefore, leadership has a substantial impact along the whole innovative process (Byrne *et al.*, 2009).

Social events and informal communication networks also favour the sharing of knowledge. These situations, created by the company, are more effective than other initiatives as they could provides incentives or even training (Zarraga and Bonache, 2005). It demonstrates that 'soft' initiatives have more influence than 'hard' initiatives, favouring mutual confidence and access to support in the work team (Szulanski, 1996). The lack of availability is one of the main problems facing work teams in organizations (Von Krogh, 1998), something that leadership must reduce or eliminate.

A qualitative study made among managers of different organizations came to the conclusion that leadership is a very important factor encouraging employees' creativity and innovation (De Jong and Den Hartog, 2010).

However, a problem arises when it comes to studying leadership. Despite it being a well-investigated topic, there is no consensus over its definition and even less regarding how it should be put into practice. The claim made by Burns (1978: 1), who said that "leadership is one of the most observed and least understood phenomena on earth", is still entirely accurate today.

Leadership has been defined as the process by which a person has the capacity to influence and motivate a group of people (followers) so that it contributes to the improvement and achievement of established objectives and to the success of the organizational project (House *et al.*, 2002). However, different studies of leadership, rather than definitions, have attempted to theorize it. With regard to an analytical perspective it is possible to group the vast majority of these studies into three categories, called *traditional*: the focus on characteristics, the focus on behaviour and the one focused on context and contingency. Most of these theories are cantered on the person as leader, leaving aside other aspects of the leadership process. Only the focus on context and contingency considers aspects beyond the leadership figure, such as the type of task, the level of pressure and the support of the group, the experience and maturity of collaborators, etc. Through these aspects it is possible to improve the effectiveness of leadership and generate better understanding.

The focus on characteristics and behaviour supposes that certain features and personal skills allow us to predict is a person will be able to take on leadership responsibilities and if he/she will perform effectively. These features include different personal characteristics, including aspects related to personality, character, needs, values and motivations. Most studies compared leaders with non-leaders, and even included physical characteristics, personality traits, attitudes, etc. Stodgill (1948) revised 124 studies carried out between 1904 and 1948 and concluded that the pattern of the results showed the concept of a leader who acquired their status by showing a capacity to help a group achieve its goals. The most important features included intelligence, awareness of the needs of others, understanding of the task at hand, initiative, persistency to overcome problems, self-confidence, extroversion, high emotional maturity, high levels of energy, a willingness to accept responsibility and occupy a position of power and control. However, it did support the basic premise that person should have a certain number of attributes in order to become a successful leader. Depending on the circumstances, some characteristics were more important than others. Not all characteristics were identified as to the extent they were necessary.

In 1974, Stodgill revised 163 studies of characteristics carried out between 1949 and 1970, in which he identified some characteristics and skills that were different to those studied previously. The results were more detailed but still didn't provide empirical evidence of universal leadership characteristics. Some characteristics increased the probability of a leader being efficient, but did not guarantee effectiveness.

Subsequently, situational and contingency models ruled out contextual variables influencing the efficiency of leadership (Hersey and Blanchard, 1982). Theories such as situational leadership, path-goal theory, cognitive resources theory of the theory of normative decisions, analyzed such variables as the maturity of subordinates, available resources, organizational capacity, effort, quality and level of acceptance of decisions, etc., but including many situation-related factors made it difficult for the theory to be contrasted. As a result, there was a very ambiguous definition of the theory making it difficult to elaborate specific propositions, which were subsequently contrasted. Therefore, the majority of studies only offered a partial contrasting of the theories. In any case it is important to be aware of the great difficulty a leader has in establishing the optimum conduct for each type of situation.

More recently, leadership has gone from being perceived as a single person phenomenon (traditional and classical focus) to being understood as the result of interpersonal relations (Guillén, 2006). During the 1980s, researchers began to study the relationships between leaders and followers, paying special attention to emotional and symbolic characteristics, and paving the way for two new focuses on leadership; transitional and transformational.

Leadership has generally been understood to be a transactional process or a process of exchange, in which leadership is deemed to be a transaction of interests between leader and collaborator, whereby a leader either motivates his/her followers through promises, praise, and rewards, changes their conduct through negative feedback, disapproval, threats or action. Some authors do not consider this a type of leadership, but rather a type of management.

Although the results of good transactional leaders are considered acceptable, the result may still be improved by establishing a different type of relationship that does not reduce it to a formal system of expected conduct (Cardona, 2000). This is the case with transformational leadership, whose development is considered to be an expansion of transactional leadership. The principal difference is that the transformational leader inspires followers to go beyond original expectations (Bass, 1985) by improving the leader-follower relationship. Instead of basing the relationship on economic aspects, it is instead focused on aspects of social character (Cardona, 2000).

Cascio (1995) demonstrates the relevance that this type of leadership has in modern day organizations, as it promotes greater interdependency, interconnection and cultural diversity, and achieves greater creativity, imagination and more effort on the part of the followers. There is strong empirical evidence to suggest that transformational leadership is more effective, productive, and innovative, and provides greater satisfaction for followers than transactional leadership (Lowe *et al.*, 1996).

Transformational leadership calls to the moral values of the followers in order to increase their awareness of ethical questions, thereby motivating them to make improvements to their institutions. According to Bass (1985), the transformational leader motives his/her followers, making them more conscious of the importance of task results, and encouraging them to look beyond their own interests and focus on the good of the organization and the team, fulfilling their hierarchical needs.

There has been no clear explanation of the process of influence used by this type of leadership, but a personal identification is seemingly implied as influence ideally affords the leader charisma in the eyes of the followers. Other conduct is at work in this process, such inspirational motivation (optimistic vision) and individual consideration (advising). Intellectual stimulation may increase the creativity of individual followers and of teams.

We do not have to forget the fact that the own nature of innovating people has important implications for the leadership. An immediate consequence is that leaders must structure the work in a way that is participative and with a high intrinsic motivation (Byrne, *et al.*, 2009), since, as previously seen, members of the innovating equipment have a high motivation towards work and this is related to creativity and innovation (Mumford and Hunter, 2005).

According to Bass (1998), transformational leadership can be efficient in any situation or culture. Empirical evidence supports the conclusion that in most situations, if not in all, some aspects of transformational leadership are relevant. However, this does not imply that is as efficient in all cases and situations. There are certain circumstantial variables that may increase the probability of this type of leadership being effective. These variables are the following: dynamic and changeable environments, encouraging change and giving leaders the power to be flexible and innovative. All of these variables are present in innovative teams.

In order to conclude and as a summary, we can say that the leader qualities an innovative equipment needs are, in the first place, specialized knowledge providing to him the power of being able to influence the people (Yunkl, 2006). In addition it requires of a high motivation, autonomy and a strong identity with the creative people, since all this allows him to represent the group, to communicate openly, to evaluate the necessities of the followers and to use and to foster the experience of them (Mumford *et al.*, 2002). Finally the leader needs creativity and ability to solve problems, capacity to provide knowledge, to select suitable projects, to evaluate in effective way the ideas propose and to provide the group with information (Basadur, Runco and Fertile valley, 2000; Mumford *et al*., 2000). We do not have to forget that these leaders will have to control actively the internal environment as much as the external one, since they will have to include commentaries of clients, suppliers, market studies, competitors, technological monitoring, cooperation with other companies, etc. (Mumford, Eubanks and Murphy, 2007).

DISCUSSION OF RESULTS

The complexity and dynamism of the environments in which organizations now operate increase the possibility of creating innovative work teams that have very different characteristics to traditional work teams.

These new teams share the same intent to obtain better results by making more and better use of the knowledge, skills and creativity of each of their members, regardless of the position they occupy, where they are located, the technology they use and the decision-making capacity they have.

In order for these teams to be effective, organizations need to incorporate new management tools that guarantee the integration of the members of teams and of the teams in their organizations. High performance practices or high commitment practices and skills management are seen as management techniques that have the potential to develop the individual skills of the different members of the organization and achieve increased integration and commitment.

Furthermore, these teams require a type of leadership focused more on interpersonal relations than on working methods and norms. This requires leaders to have better communication, conflict resolution, coordination, listening and self-control skills, etc. Essentially, more emotional competencies, which the leader must put into practice.

In light of the findings of this chapter there are certain relevant questions that must be asked:

- Do today's organizations have the capacity to incorporate and manage innovative work team in their staff?
- Do the new demands of the current situation favour companies that are well consolidated in their respective markets or, on the contrary, do they create new opportunities for new companies or even countries to compete?
- Are organizations aware of the new requirements of effective leadership?
- What type of training is most appropriate, for both team members and for leaders, in order that greater innovation is achieved?

REFERENCES

Abad, I. M., & Castillo, A. M. (2004). Desarrollo de competencias directivas. Ajuste de la formación universitaria a la realidad empresarial. *Boletín Económico del ICE, 2795*, 29–41.

Abrahamson, E. (1991). Managerial fads and fashions: The diffusion and rejection of innovations. *Academy of Management Review, 16*, 586–612.

Agencia Espacial Europea. (1998-2003). *ESA-Careers at ESA-Core skills.*

Álvarez de Mon, S. (2000). *El mito del líder. Profesionales, ciudadanos, personas: La sociedad alternativa.* Madrid, Spain: PriceWaterhuse Coopers.

Amabile, T. M., & Gryskiewicz, N. D. (1989). The creative environment inventory. *Creativity Research Journal, 2*, 231–253. doi:10.1080/10400418909534321

Anderson, N., De Creu, C. K. W., & Nijstad, B. A. (2004). The routinization of innovation research: A constructively critical review of the state-of-the-science. *Journal of Organizational Behavior*, *25*(2), 147–173. doi:10.1002/job.236

Arthur, J. B. (1994). Effects of human resource Systems on manufacturing performance and turnover. *Academy of Management Journal*, *37*, 670–687. doi:10.2307/256705

Aubert, J., & Gilbert, P. (2003). *L'evaluation des competences*. Sprimont, Belgium: Mardaga.

Banker, R. D., Field, J. M., Schroeder, R. G., & Shina, K. K. (1996). Impact of work teams on manufacturing performance: A longitudinal field study. *Academy of Management Journal*, *39*, 867–890. doi:10.2307/256715

Basadur, M., Runco, M. A., & Vega, L. A. (2000). Understanding how creative thinking skills, attitudes, and behaviors work together: A casual process model. *The Journal of Creative Behavior*, *34*, 77–100.

Bass, B. M. (1985). *Leadership and performance beyond expectations*. Nueva York, NY: Free Press.

Bass, B. M. (1998). *Transformational leadership: Industry, military and educational impact*. Hillsdale, NJ: L Erlbaum.

Bayo, A., & Merino, J. (2001). Quality management and high performance work practices: Do they coexist. *International Journal of Production Economics*, *73*(3), 251–260. doi:10.1016/S0925-5273(01)00089-5

Bell, B. S., & Kozlowski, S. W. (2002). A typology of virtual team. *Group & Organization Management*, *27*(1), 14–49. doi:10.1177/1059601102027001003

Beltrán, I., Roca, V., Escrig, A., & Bou, J. C. (2008). Human resource flexibility as a mediating variable between high performance work systems and performance. *Journal of Management*, *34*(5), 1009–1044. doi:10.1177/0149206308318616

Bennet, R. (2002). Employer's demands for personal transferable skills in graduates: A content analysis of 1000 job advertisements and an associated empirical study. *Journal of Vocational Education and Training*, *54*(2), 457–475. doi:10.1080/13636820200200209

Blanco, A. (2007). *Trabajadores competentes*. Madrid, Spain: Esic Editorial.

Boudreau, K. J., & Lakhani, K. R. (2009). How to manage outside innovation. *MIT Sloan Management Review*, *50*(4), 69–77.

Boyatzis, R. (1982). *The competent manager*. Nueva York, NY: John Willey.

Burns, J. M. (1978). *Leadership*. Nueva York, NY: Harper & Row.

Byrne, C. L., Mumford, M. D., Barrett, J. D., & Vessey, W. B. (2009). Examining the leaders of creative efforts: What do they do, and what do they think about? *Creativity and Innovation Management*, *18*(4), 256–268. doi:10.1111/j.1467-8691.2009.00532.x

Cabrera, A. (2002). La gestión del conocimiento. In Bonache, J., & Cabrera, A. (Eds.), *Dirección estratégica de personas: Evidencias y perspectivas para el siglo XXI*. Madrid, Spain: Prentice Hall.

Capelli, P., & Neumark, D. (2001). Do high-performance work practices improve establishment level outcomes? *Industrial & Labor Relations Review*, *54*(4), 737–775. doi:10.2307/2696111

Cappelli, P., & Crocker-Hefter, A. (1996). Distinctive human resources are firm's core competencies. *Organizational Dynamics*, *24*(3), 7–22. doi:10.1016/S0090-2616(96)90002-9

Cardona, P. (2000). El liderazgo relacional. In *Raíces éticas del liderazgo*. Pamplona, Spain: Ediciones Universidad de Navarra.

Cascio, W. (1995). Whither industrial and organizational psychology in a changing world of work? *The American Psychologist, 50*, 928–939. doi:10.1037/0003-066X.50.11.928

Céspedes, J. J., Jerez, P., & Valle, R. (2005). Las prácticas de recursos humanos de alto rendimiento y la capacidad de aprendizaje organizativo: incidencia e implicaciones. *Cuadernos de Economía y Dirección de la Empresa, 24*, 29–56.

Chaston, I. (1998). Self-managed teams: Assessing the benefits for small service-sector firms. *British Journal of Management, 9*, 1–12. doi:10.1111/1467-8551.00071

Chesbrough, H. (2006). *Open business models: How to drive in the new innovation landscape*. Boston, MA: Harvard Business School Press.

Chesbrough, H., Vanhaverbeke, W., & West, J. (Eds.). (2006). *Open innovation: Researching a new paradigm*. Oxford University Press.

Cohen, S. G., & Bailey, D. E. (1997). What makes teams work: Group effectiveness research from the shop floor to the executive suite. *Journal of Management, 23*, 239–290. doi:10.1177/014920639702300303

Cohen, S. G., & Ledford, G. E. Jr. (1994). The effectiveness of self-managing teams: A quasi-experimental. *Human Relations, 47*, 13–43. doi:10.1177/001872679404700102

Cordery, J. L., Mueller, W. S., & Smith, L. M. (1991). Attitudinal and behavioral effects of autonomous group working: A longitudinal field study. *Academy of Management Journal, 34*(2), 464–476. doi:10.2307/256452

Curral, L. A., Forrester, R. H., Dawson, J. F., & West, M. A. (2001). It's what you do and the way you do it: Team task, team size, and innovation-related group processes. *European Journal of Work and Organizational Psychology, 10*, 187–204. doi:10.1080/13594320143000627

Damanpour, F. (1998). Innovation type, radicalness, and the adoption process. *Communication Research, 15*, 545–567. doi:10.1177/009365088015005003

de Jong, J. P. J., & den Hartog, D. N. (2010). Measuring innovative work behaviour. *Creativity and Innovation Management, 19*(1), 23–36. doi:10.1111/j.1467-8691.2010.00547.x

De Jong, J. P. J., Vanhaverbeke, T., Chalet, T., & Chesbrough, H. (2008). *Policies for open innovation: theory, framework and cases*. Helsinki, Finland: Research project funded by Vision Era-Net.

de la Florida, R. (2002). *The rise of the creative class*. New York, NY: Basic Books.

Dess, G. G., & Pickens, J. C. (2000). Changing roles: Leadership in the 21st century. *Organizational Dynamics, 28*, 18–34. doi:10.1016/S0090-2616(00)88447-8

Devine, D. J., Clayton, L. D., Philips, J. L., Dunford, B. B., & Melner, B. S. (1999). Teams in organizations: Prevalence, characteristics, and effectiveness. *Small Group Research, 30*, 678–711. doi:10.1177/104649649903000602

Drach-Zahavy, A., & Somech, A. (2001). Understanding team innovation: The role of team processes and structures. *Group Dynamics, 5*(2), 111–123. doi:10.1037/1089-2699.5.2.111

Drazin, R., & Schoonhoven, C. B. (1996). Community population, and organizational effects on innovation: A multilevel perspective. *Academy of Management Journal, 39*, 1065–1083. doi:10.2307/256992

Fay, D., Borrill, C., Amir, Z., Haward, R., & West, M. A. (2006). Getting the most out of multidisciplinary teams: A multi-sample study of team innovation in health care. *Journal of Occupational and Organizational Psychology, 79*, 553–567. doi:10.1348/096317905X72128

Feist, G. J., & Gorman, M. E. (1998). Psychology of science: Review and integration of a nascent discipline. *Review of General Psychology, 2*, 3–47. doi:10.1037/1089-2680.2.1.3

Ferris, G. R., Hochwarter, W. A., Buckley, M. R., Harrell-Cook, G., & Frink, D. D. (1999). Human resources management: Some new directions. *Journal of Management, 25*, 385–415. doi:10.1177/014920639902500306

Gil, F., & García, M. (1996). *Grupos en las organizaciones*. Madrid, Spain: Ediciones Pirámide.

González-Roma, V. (2008). La innovación en los equipos de trabajo. *Papeles del Psicólogo, 29*(1), 32–40.

Grant, R. (1997). The knowledge-based view of the firm: implications for management practise. *Long Range Planning, 30*(3), 450–454. doi:10.1016/S0024-6301(97)00025-3

Grant, R. (2001). Knowledge and organization. In Nonaka, I., & Teece, D. (Eds.), *Managing industrial knowledge: Creation, transfer and utilization*. London, UK: SAGE Publications Ltd.

Guest, D. E. (1987). Human resource management and industrial relations. *Journal of Management Studies, 24*(5), 503–521. doi:10.1111/j.1467-6486.1987.tb00460.x

Guthrie, J. P., Spell, C. S., & Nyamori, R. O. (2002). Correlates and consequences of high involvement work practices: the role of competitive strategy. *International Journal of Human Resource Management, 13*(1), 183–197. doi:10.1080/09585190110085071

Hersey, P., & Blanchard, K. H. (1982). *Management of organizational behaviour: Utilizing human resources* (4th ed.). Englewood Cliffs, NY: Prentice-Hall.

Hitt, M. A., Hoskisson, R. E., Johnson, R. A., & Moesel, D. D. (1996). The market for corporate control and firm innovation. *Academy of Management Journal, 39*, 1084–1196. doi:10.2307/256993

House, R., Javidan, M., Hanges, P., & Dorfman, P. (2002). Understanding cultures and implicit leadership theories across the globe: An introduction to project GLOBE. *Journal of World Business, 37*, 3–10. doi:10.1016/S1090-9516(01)00069-4

Huselid, M. A. (1995). The impact of human resource management practices on turnover, productivity, and corporate financial performance. *Academy of Management Journal, 38*, 673–703. doi:10.2307/256741

Huselid, M. A., Jackson, S. E., & Schuler, R. S. (1997). Technical and strategic human resource management effectiveness as determinants of firm performance. *Academy of Management Journal, 40*(1), 171–188. doi:10.2307/257025

Hyatt, D. E., & Ruddy, T. M. (1997). An examination of the relationship between work group and performance: Once more into the breech. *Personnel Psychology, 50*(3), 555. doi:10.1111/j.1744-6570.1997.tb00703.x

Ichniowski, C., Kochan, T. A., Levine, D., Olson, C., & Strauss, G. (1996). What works at work: overview and assessment. *Industrial Relations, 35*, 352–332. doi:10.1111/j.1468-232X.1996.tb00409.x

Isaksen, S. G., Lauer, K. J., Ekwall, G., & Britz, A. (2001). Perceptions of the best and worst climates for creativity: Preliminary validation evidence for the situational outlook questionnaire. *Creativity Research Journal, 13*, 171–184. doi:10.1207/S15326934CRJ1302_5

Jackson, S. E., Joshi, A., & Erhardt, N. L. (2003). Recent research on team and organizational diversity: SWOT analysis and implications. *Journal of Management, 29*(6), 801–830.

Jackson, S. E., Mayo, K. E., & Whitney, K. (1995). Understanding the diversity of dynamics in decision making teams. In Guzzo, R. A., & Salas, E. (Eds.), *Team effectiveness and decision making in organizations* (pp. 204–261). San Francisco, CA: Jossey-Bass.

Janssen, O., Van de Vliert, E., & West, M. (2004). The bright and dark sides of individual and group innovation: A special issue introduction. *Journal of Organizational Behavior, 25*(2), 129–145. doi:10.1002/job.242

Joinson, C. (2002). Managing virtual teams: Keeping members on the same page without being in the same place poses challenges for managers workplace trends. *Human Resource Magazine, 47*(6), 68–72.

Katzenbach, J. R., & Smith, D. K. (1993). *The wisdom of teams* (pp. 21, 45–48). Cambridge, MA: Harvard University Press.

Kichuk, S. L., & Weisner, W. H. (1997). The big five personality factors and team performance: Implications for selecting successful product design teams. *Journal of Engineering and Technology Management, 14*(3-4), 195–222. doi:10.1016/S0923-4748(97)00010-6

Kiffin-Peteron, S. A., & Cordery, J. L. (2003). Trust, individualism, and job characteristics of employee reference for teamwork. *International Journal of Human Resource Management, 14*(1), 93–116. doi:10.1080/09585190210158538

King, N., & Anderson, N. (2002). *Managing innovation and change: A critical guide for organizations*. London, UK: Thomson.

Kinlaw, D. C. (1991). *Developing superior work teams* (pp. 3–21). Lexington, MA: Lexington Books.

Kirkman, B. L., Gibson, C. B., & Shapiro, D. L. (2001). Exporting teams: Enhancing the implementation and effectiveness of work teams in global affiliates. *Organizational Dynamics, 30*(1), 12–29. doi:10.1016/S0090-2616(01)00038-9

Kirkman, B. L., & Rosen, B. (1999). Beyond self-management: Antecedents and consequences of team empowerment. *Academy of Management Journal, 42*(1), 58–74. doi:10.2307/256874

Kirkman, B. L., Rosen, B., Gibson, C. B., Tesluk, P. E., & McPherson, S. O. (2002). Five challenges to virtual team success: Lessons from Sabre, Inc. *The Academy of Management Executive, 6*(3), 67. doi:10.5465/AME.2002.8540322

Kirkman, B. L., & Shapiro, D. L. (1997). The impact of cultural values on employee resistance to teams: toward a model of globalize self-managing work team effectiveness. *Academy of Management Review, 22*(3), 730–757.

Krueger, A. R. (1988). *El grupo de discusión. Guía práctica para la investigación aplicada*. Madrid, Spain: Ediciones Pirámide.

Lawler, E. E. (1992). *The ultimate advantage: Creating the high-involvement organization*. San Francisco, CA: Jossey- Bass.

Lawler, E. E., Mohrman, S. A., & Ledford, G. (1995). *Creating high performance organizations: Practises and results of employee involvement and total quality management: Practice and results in Fortune 1000 companies*. San Francisco, CA: Jossey-Bass.

Lazear, E. P. (1998). *Personnel economics for managers*. Jonh Wiley & Sons, Inc.

Leonard, D., & Sensiper, S. (1998). The role of tacit knowledge in group innovation. *California Management Review, 40*(3), 112–131.

Levy–Leboyer, C. (1997). *Gestión de las competencias. Cómo analizarlas. Cómo evaluarlas. Cómo desarrollarlas,* 1st ed. Barcelona, Spain: Ediciones Gestión 2000.

Lowe, K., Kroeck, K. G., & Sivasubrabramanian, N. (1996). Effective correlates of transformational and transactional leadership: A meta-analytic review. *The Leadership Quarterly*, *7*, 385–425. doi:10.1016/S1048-9843(96)90027-2

Malhotra, A., Majchrzak, A., & Rosen, B. (2007). Leading virtual teams. *The Academy of Management Perspectives*, (February): 60–70. doi:10.5465/AMP.2007.24286164

Marchington, M., & Grugulis, I. (2000). Best practice human resource management: Perfect opportunity or dangerous illusion? *International Journal of Human Resource Management*, *11*(6), 1104–1121. doi:10.1080/09585190050177184

Martí, N. (2008). *High commitment human resource practices in Spanish call centres. Nature, Sources and effects on voluntary turnover*. Thesis Doctoral Universitat Rovira i Virgili.

McClelland, D. (1973). Testing from competence rather than intelligence. *The American Psychologist*, *28*, 1–14. doi:10.1037/h0034092

Mohr, R. D., & Zoghi, C. (2008). The high-involvement work design and job satisfaction. *Industrial & Labor Relations Review*, *61*(3), 275–296.

Montes, J. L., Moreno, A. R., & Morales, V. G. (2005). Influence of support leadership and teamwork cohesion on organizational learning, innovations and performance: An empirical examination. *Technovation*, *25*, 1159–1172. doi:10.1016/j.technovation.2004.05.002

Mumford, M. D., Eubanks, D. L., & Murphy, S. T. (2007). Creating the conditions for success: Best practices in leading for innovation. In Conger, J. A., & Riggio, R. E. (Eds.), *The practice of leadership* (pp. 107–197). San Francisco, CA: Jossey-Bass.

Mumford, M. D., Feldman, J. M., Hein, M. B., & Nago, D. J. (2001). Tradeoffs between ideas and structure: Individual versus group performance in creative problem-solving. *The Journal of Creative Behavior*, *35*, 1–23.

Mumford, M. D., & Gustafson, S. B. (1988). Creativity syndrome: Integration, application, and innovation. *Psychological Bulletin*, *103*, 27–43. doi:10.1037/0033-2909.103.1.27

Mumford, M. D., & Hunter, S. T. (2005). Innovation in organizations: A multi-level perspective on creativity. In Yamarino, F. J., & Dansereau, F. (Eds.), *Research in multi-levels issues* (*Vol. 4*). Oxford, UK: Elsevier. doi:10.1016/S1475-9144(05)04001-4

Mumford, M. D., Marks, M. A., Connelly, M. S., Zaccaro, S. J., & Johnson, T. F. (1998). Domain based scoring of divergent thinking tests: Validation evidence in an occupational sample. *Creativity Research Journal*, *11*, 151–164. doi:10.1207/s15326934crj1102_5

Mumford, M. D., Marks, M. A., Connelly, M. S., Zaccaro, S. J., & Reiter-Palmon, R. (2000). Development of leadership skills: Experience, timing, and growth. *The Leadership Quarterly*, *11*, 87–114. doi:10.1016/S1048-9843(99)00044-2

Mumford, M. D., Scott, G. M., Gaddis, B., & Strange, J. M. (2002). Leading creative people: Orchestrating expertise and relationships. *The Leadership Quarterly*, *13*, 705–750. doi:10.1016/S1048-9843(02)00158-3

Mumford, M. D., Scott, G. M., Gaddis, B., & Strange, J. M. (2002). Leading creative people: Orchestrating expertise and relationships. *The Leadership Quarterly, 13*(6), 705–750. doi:10.1016/S1048-9843(02)00158-3

Oliveros, L. (2006). Identificación de competencias: Una estrategia para la formación en el Espacio Europeo de Educación Superior. *Revista Complutense de Educación, 17*(1), 101–118.

Olson, M. (1965). *The logic of collective action*. Cambridge, MA: Harvard University Press.

Ordiz, M. (2002). Prácticas de alto rendimiento en recursos humanos: Concepto y factores que motivan su adopción. *Cuadernos de Economía y Dirección de la Empresa, 12*, 247–265.

Ordiz, M., & Fernández, E. (2003). High-involvement practices in human resource management: Concept and factors that motive their adoption. *International Journal of Human Resource Management, 14*(4), 511–529. doi:10.1080/0958519032000057565

Parnell, J. A., Lester, D. L., & Menefee, M. L. (2000). Strategy as a response to organizational uncertainty: An alternative perspective on the strength-performance relationship. *Management Decision, 38*, 301–392. doi:10.1108/00251740010352811

Pearce, J. H. II, & Ravlin, E. C. (1987). The design and activation of self-regulating work groups. *Human Relations, 40*, 751–782. doi:10.1177/001872678704001104

Pearn, M. A., & Kandole, R. S. (1988). *Job analysis: A practical guide for managers*. Londres, UK: Institute for Personnel Management.

Pereda, S., Berrocal, F., & López, M. (2002). Gestión por competencias y gestión del conocimiento. *Revista de Dirección. Organización y Administración de Empresas, 28*, 43–54.

Perretti, F., & Negro, G. (2007). Mixing genres and matching people: A study in innovation and team composition in Hollywood. *Journal of Organizational Behavior, 28*(5), 563–586. doi:10.1002/job.464

Pfeffer, J. (1994). Competitive advantage through people. *California Management Review, 36*(2), 9–28.

Pierce, J., & Delbecq, L. A. (1977). Organizational structure, individual attitudes, and innovation. *Academy of Management Review, 2*, 27–37.

Priem, R. L., Harrison, D. A., & Muir, N. K. (1995). Structured conflict and consensus outcomes in group decision making. *Journal of Management, 11*, 691–710. doi:10.1177/014920639502100406

Richard, O. C., & Johnson, N. B. (2001). Strategic human resource management effectiveness and firm performance. *International Journal of Human Resource Management, 12*, 299–310.

Robbins, S. P., & Judje, T. A. (2009). *Comportamiento organizacional* (13th ed.). México: Pearson Educación.

Roca, V., Escrig, A. B., & Bou, J. C. (2002). Compromiso con los empleados y estrategia competitiva: un análisis intersectorial de su repercusión en los resultados. *Cuadernos de Economía y Dirección de la Empresa, 12*, 267–289.

Roche, W. K. (1999). In search of commitment-oriented human resource management practices and the conditions that sustain them. *Journal of Management Studies, 36*, 653–678. doi:10.1111/1467-6486.00153

Rochford, L., & Rudelius, W. (1992). How involving more functional areas within a firm affects the new product process. *Journal of Product Innovation Management, 9*, 287–299. doi:10.1111/1540-5885.940287

Sala, G. (2006). La gestión por competencias en el sector bancario español. *Trayectorias*, *8*(22), 21–33.

Sánchez, I. (2004). The creation of employee commitment in Spanish hotels through cultural practices. *Management Research*, *2*(2), 175–192.

Sashkin, M. (1984). Participative management is an ethical imperative. *Organizational Dynamics*, *12*(4), 5–22. doi:10.1016/0090-2616(84)90008-1

Schaubroeck, J., Lam, S. S. K., & Cha, S. E. (2007). Embracing transformational leadership: Team values and the impact of leader behaviour on team performance. *The Journal of Applied Psychology*, *92*(4), 1020–1030. doi:10.1037/0021-9010.92.4.1020

Schuler, R. S., & Jackson, S. E. (1987a). Linking competitive strategy with human resource management practices. *The Academy of Management Executive*, *1*(3), 207–219. doi:10.5465/AME.1987.4275740

Schuler, R. S., & Jackson, S. E. (1987b). Organizational strategy and organization level as determinants of human resource management practices. *Human Resource Planning*, *10*(3), 125–141.

Siebdrat, F., Hoegl, M., & Ernst, H. (2009). How to manage virtual teams. *MIT Sloan Management Review*, *50*(4), 62–69.

Spencer, L. M., & Spencer, S. M. (1993). *Competence at work. Models for superior performance*. New York: John Wiley Sons, Inc.

Stogdill, R. M. (1948). Personal factor associated with leadership: A survey of the literature. *The Journal of Psychology*, *25*, 35–71. doi:10.1080/00223980.1948.9917362

Stogdill, R. M. (1974). *Handbook of leadership: A survey of the literature*. New York, NY: Free Press.

Szulanski, G. (1996). Exploring stickiness: Impediments to the transfer of best practice within the firm. *Strategic Management Journal*, *17*(Winter Special Issue), 27-43.

Tierney, P., Farmer, S. M., & Graen, G. B. (1999). An examination of leadership and employee creativity: The relevance of traits and relationships. *Personnel Psychology*, *52*, 591–620. doi:10.1111/j.1744-6570.1999.tb00173.x

Townsend, A. M., DeMarie, S. M., & Hendrickson, A. R. (1998). Virtual teams: Technology and the workplace of the future. *The Academy of Management Executive*, *12*(3), 17–29. doi:10.5465/AME.1998.1109047

Truss, C., Gratton, L., Hope-Hailey, V., McGovern, P., & Stiles, P. (1997). Soft and hard models of human management: A reappraisal. *Journal of Management Studies*, *34*, 53–73. doi:10.1111/1467-6486.00042

Tushman, M. L., & O'Reilly, C. A. (1997). *Winning through innovation*. Cambridge, MA: Harvard Business School Press.

Ulrich, D. (1998). Intellectual capital = Competence x Commitment. *Sloan Management Journal*, *4*, 15–26.

Uzzi, B. (1997). Social structure and competition in interfirm networks: The paradox of embeddedness. *Administrative Science Quarterly*, *42*, 35–67. doi:10.2307/2393808

van Knippenberg, D., & Schippers, M. C. (2007). Work group diversity. *Annual Review of Psychology*, *58*, 515–541. doi:10.1146/annurev.psych.58.110405.085546

Von Krogh, G. (1998). Care in knowledge creation. *California Management Review*, *40*(3), 133–153.

Walton, R. E. (1985). From control to commitment in the workplace. *Harvard Business Review*, *63*, 77–84.

Watson, W., Kumar, K., & Michaelsen, L. K. (1993). Cultural diversity's impact on interaction process and performance: comparing homogeneous and diverse task groups. *Academy of Management Journal, 36*, 590–602. doi:10.2307/256593

Wellins, R., Byham, W., & Wilson, J. M. (1991). *Empowerment teams: Creating self-directed work groups that improve quality, productivity and participation*. San Francisco, CA: Jossey-Bass.

West, M. A. (2002). Sparkling fountains or stagnant ponds: An integrative model of creativity and innovation implementation in work groups. *Applied Psychology-An International Review-Psychologies Appliquee-Revue Internationale, 51*(3), 355–387.

West, M. A., & Hirst, G. (2003). Cooperation and teamwork for innovation. In West, M. A., Tjosvold, D., & Smith, K. G. (Eds.), *International handbook of organizational teamwork and cooperative working* (pp. 297–319). Chichester, UK: Wiley. doi:10.1002/9780470696712

Wood, S., & Albanase, M. T. (1995). Can we speak of a high commitment management on the shop floor? *Journal of Management Studies, 32*, 215–247. doi:10.1111/j.1467-6486.1995.tb00341.x

Wood, S., & de Menezes, L. M. (2008). Comparing perspectives on high involvement management and organizational performance across the British economy. *International Journal of Human Resource Management, 19*(4), 639–682. doi:10.1080/09585190801953673

Wright, P. M., McMahan, C., & McWilliams, A. (1994). Human resources and sustained competitive advantage: A resource-based perspective. *International Journal of Human Resource Management, 5*(2), 301–326. doi:10.1080/09585199400000020

Youndt, M. A., Snell, S. A., Dean, J. W., & Lepak, D. P. (1996). Human resource management, manufacturing strategy, and firm performance. *Academy of Management Journal, 39*, 836–866. doi:10.2307/256714

Yunkl, G. (2006). *Leadership in organizations*. Englewood Cliffs, NJ: Prentice Hall.

Zarifain (1999). *Objectif Competente, pour une nouvelle logique*. Paris, France: Éditions Liasions.

Zárraga, C., & Bonache, J. (2005). Equipos de trabajo para la gestión del conocimiento: La importancia de un clima adecuado. *Cuadernos de Economía y Dirección de la Empresa, 22*, 27–48.

Zatzick, C. D., & Iverson, R. D. (2006). High-involvement management and workforce reduction: Competitive advantage or disadvantage? *Academy of Management Journal, 49*(5), 999–1015. doi:10.5465/AMJ.2006.22798180

ADDITIONAL READING

Álvarez de Mon, S. (2000). *El Mito del Líder. Profesionales, ciudadanos, personas: La sociedad alternativa, Cap. IV*. Madrid: PriceWaterhouse Coopers.

Blanco, A. (2007). *Trabajadores Competentes*. Madrid: Esic Editorial.

Charan, R. (2010). Apprentice yourself for leadership, *Leader to Leader*, 48, 39-44. Edición en castellano: Aprenda a desarrollar su propia capacidad de liderazgo. *Harvard Deusto Business Review, 187*, 14–18.

Ordiz, M. (2002). Prácticas de alto rendimiento en recursos humanos: concepto y factores que motivan su adopción. *Cuadernos de Economía y Dirección de la Empresa, 12*, 247–265.

Pereda, S., Berrocal, F y López, M. (2002). Gestión por competencias y gestión del conocimiento. Revista de Dirección, Organización y Administración de empresas, 28, 43-54.

KEY TERMS AND DEFINITIONS

High Performance Practices: Are innovative practices of human resource management which improve the commitment and the performance of the company.

Innovation: Use of a new product, service or method in business practice immediately subsequent to its discovery and it´s good for the competitiveness.

Innovative Teams: It´s a work group focus on innovation and development technique for improving its performance and attitudes.

Leadership: Person who provides vision and direction for the team and teaches his values.

Skills Management: This model is based on people´s competencies and looking for provide their employees with opportunities to develop and put their new skills and knowledge into practice. It consists of a full system of evaluation and improvement of organizations through the development of the people it comprises.

Work Teams: Some persons with complementary technical and personal skills, that interact to achieve a common objective, the outcome of which is a collective result that is greater than the sum if the individual parts for which they are all responsible.

ENDNOTES

1 To further about training and development aspects of teams you can see: Gil, F. y García, M. (1996), chapter 3, pp. 39-52.

2 http://www.innocentive.com/about-us-open-innovation.php

Chapter 17
Listening and Leadership

Javier Pagán Castaño
Portsmouth University, UK

Dolores Garzón Benítez
Polytechnic University of Valencia, Spain

ABSTRACT

This era is both characterized by continually evolving and changing market conditions and the relevance of knowledge to adapt to the new environment.

In this chapter, the authors will focus on three ideas: the connection between the company results and the leader's listening skills; people as the firm's most valuable resource; and therefore, internal communication as the key for success and full potential of the company. However, the survey concludes that internal communication is scarce and not always effective since one way communication seems to be the most common form of interaction, even though leaders know the importance and value of their human resources.

INTRODUCTION

In the era that we live in, where the market conditions are continually evolving and changing, and organizations are struggling to adapt to the new conditions, leaders must understand that what they really manage in an organization is people. House (2004) defines leadership as "the ability of an individual to influence, motivate and enable others to contribute toward the effectiveness and success of the organizations of which they are members", The DePree´s (1992) idea of leadership, a position of servant hood, compliments with Drunker (1992) observation that the Japanese leader recognizes that rank does not confer privileges but rank entails responsibility. For Nelda Spinks and Barron Wells (1995) to make people feel as though they are accomplishing something and not just "putting time", a leader must work with people and not against them. Leaders must understand that people are the most important components in an organization and that having an effective communication with them is key to enhance performance. But reality seems to be different since leaders do not put in practice, during their daily interactions, the elements that

DOI: 10.4018/978-1-61350-207-5.ch017

Figure 1. Ideal and simple communication process (Source: Own elaboration)

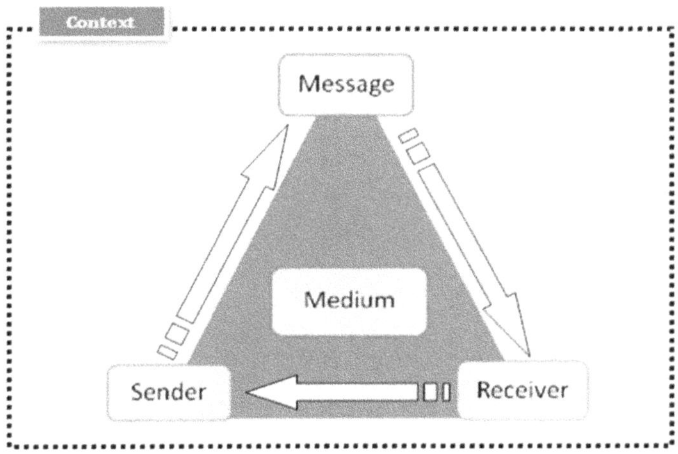

make good communication possible. Trusting and open environments are the exception in an overall organization's atmosphere characterized by judgment, interruptions, and misunderstanding of people's needs.

LITERATURE REVIEW

The Key Leadership Skill

Communication is the most important skill a leader must possess; since it occupies 70 to 90 per cent of his/her time (Mintzberg 1973, Ecles & Nohria, 1991). If that same study were done today including e-mail, cell phones, and text messaging, the percentages yielded would be even higher. The total amount of time managers dedicate to communicate emphasizes the importance of having strong communication skills to advance to leadership positions. Mastering leadership communication should be a priority for managers wanting their organizations to consider them as leaders (Deborah J. Barrett, 2006). Without effective communication, a manager accomplishes little. Without effective communication a manager is not an effective leader, and through effective communication, a leader leads.

Communication is the transmission of meaning from one person to another or to a group of people, whether verbally or non verbally. Leaders must pay attention to four elements to become effective communicators: the sender, the medium or channel, the receiver and the context (Figure 1).

This triangle shows what would be an ideal and simple communication process without miscommunications or misunderstandings. The person, who sends the message, understands the audience (receiver) and the context, selects the most appropriate medium and sends a clear message. On the other hand, the person who receives it understands the message as the sender intended. But in reality, communication resembles some variations, as in Figure 2 (Barret, 2006).

The complication in communication comes from the interruptions or interferences in that transmission, whether the sender causes them or the receiver. Leadership communication necessitates anticipating all interruptions and interferences through audience analysis and then developing a communication strategy that controls the rhetorical situation and facilitates the effective transmission of the message.

The purpose of the sender is to communicate a thought or idea formulated within their mind. Even though he or she clearly sees the idea, fac-

Figure 2. Real and complex communication process (Source: Barret, 2006)

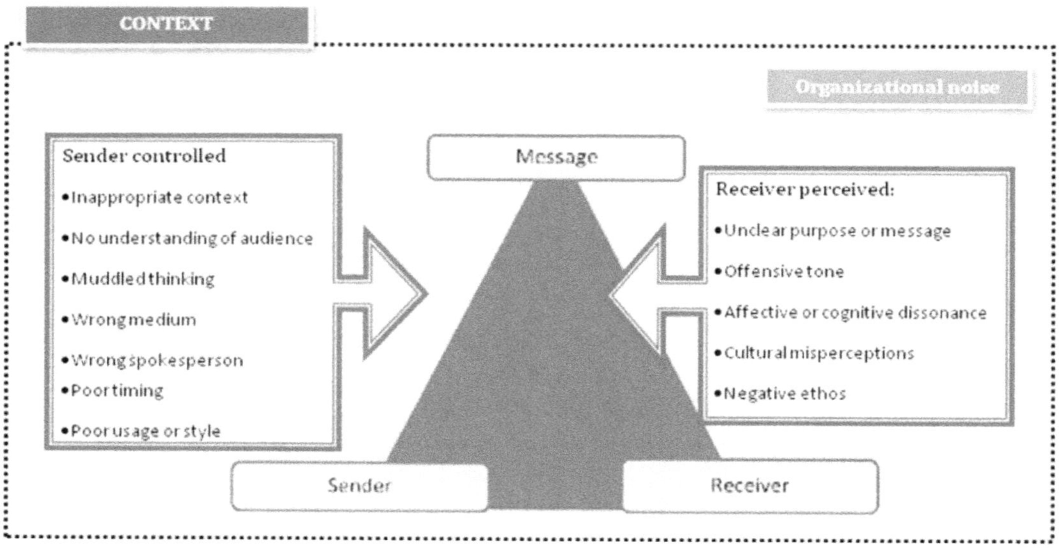

tors like the sender's current frame of mind, level of communication skills or familiarity with the receiver influence the communication outcome.

A channel of communication is selected, whichever written or oral. Written communication is exposed to different interpretations of symbols and words, and to poor production. Even though the clarity of the written message, it can be neglected or ignored because of improper spelling/grammar skills, the incorrect format or inappropriate length. In the same way, oral communication is vulnerable to poor listening or distracting voice. To gain effectiveness, the sender must consider the velocity, the tone and the body language used during the communication process. It is mandatory for the messenger to read the receiver's body language to evaluate the degree of understanding. Once the message arrives to the receiver, it is submitted to physical and emotional conditions, predetermined notions and different levels of receptivity.

New Communication Environment

In the past, interactions between individuals used to be face-to-face. Today, an ever-increasing number of encounters are mediated by technology. This new way of interacting involves different types and levels of risk, and they are only possible if users trust each other and the systems they use to communicate. As Checkland (1999) pointed out, any technical system that is brought to an organization can only work efficiently as part of a larger socio-technical system, i.e., the organization and its human actors. Some authors claim that the benefits associated with the implementation of new technology in organizations fail to yield the expected increases in productivity because of its inability to build social capital (Launder, 1996; Resnick 2002). Trust can be formed by informal exchanges, but if new technology contributes to make these exchanges archaic though mechanization, trust could be unavailable when it is required. Interactions that used to be face-to-face and based on long-established personal relationships are now carried out over distance by automated systems, a process called dis-embedding (Giddens, 1990). From this point of view, conducting more interactions using new technology rather than face-to-face can deprive leaders of opportunities to build trust. To evaluate the potential of new technologies for enabling new forms of interaction without considering these undesirable consequences, trust and the conditions that influence it, can undermine the opportunities for development.

A clear and extreme example of a new form of professional interaction is working at home. The benefits from teleworking programs, understood as any work agreement in which employees can work at any time or place, and can perform their tasks in an effective and efficient way, are not as easy as they might seem. Successful teleworking programs require, according to Chaudron (1995), the "right employee", "the right job" and "the right reasons":

- Employees whose personal traits will be suitable for teleworking.
- Jobs involving individual versus team contributions
- Management teams who do not just view teleworking as a benefit for their employees but also as an opportunity to enhance productivity.

Baruch and Nicholson (1997) research study reveals that ninety four percent of sampled teleworkers declared they wish to continue working at home, but less than ten per cent affirmed a desire to continue on that situation for the rest of their working lives. These results show the inner conflict experienced by the at-home worker. Anxiety, communication gaps, isolation, guilt, negative spillover between family life and work and reduced productivity are some of the feelings they can experience.

Participants at an ACM forum (1995) declared that teleworking does not facilitate the social and professional interaction of the employee, killing his/her most human quality. At-home workers, can feel left out of interactions from office rumors to changes in company procedures, activities or policies. On the other hand, managers and at-home workers who are part of a program to interact informally, develop interpersonal organizational networks and create synergistic relationships have shown reduced feelings of isolation (Kurland and Cooper, 2002).

Huws (1984) identified in his survey that 60 per cent of respondents felt isolation as the major disadvantage of working at home. This physical separation reduces not only the amount of feedback that teleworkers receive from clients, supervisors or coworkers, but also reduces its quality because the non-visual communication reduces the nonverbal cues (Norman et al., 1995). Hamilton (1987) stated that at-home workers miss the motivation of exchanging experiences and ideas with colleges

As it happens in the in office environment, the out-of-office one also needs good communication between managers and workers to maintain business and personal relationships and to accomplish tasks. But in the out-of-office environment communication channels are mostly electronic, document handling and e-mail being the most commonly used communication methods.

In Weiner and Hill (1995) qualitative study findings, the negative influence that at-home workers had on peer interaction and communication with managers and coworkers contrasts with the view of organizations as social systems whose basic structure is made of relationships among individuals (Fritz, Narasimhan and Rhee, 1998). Employees sharing the same physical location develop alliances, learn through communication and foster creativity through communication with coworkers through casual encounters around the coffee machine and elsewhere (Allen, 1977) Kraut et. Al. (1998) study showed that using e-mail to socialize was positively related to social isolation, depression and loneliness, indicating that e-mail is not as good communication channel as it is face-to-face. Weick (1987) suggests in his research that face-to-face or organic systems involve more managerial information, two way communication and advice rather than instructions, decisions and orders than electronic or mechanical systems. A leadership style closer to consultation than command (Courtright, Fairhurst and Rogers, 1989).

Communication and Leadership

Geneen (1984) identified in his study that good leadership, which inspires people to excel, contributes as much as 80 – 90 per cent of organization's success. In the opinion of Wells B. (1992) this leadership can only be exercised through quality communication. As DePress (1992) explained, quality communication is the avenue by which leaders clarify their visions and foster participative management within the organization. In this context, Bennis (1989) determined that leaders are people who are able to express themselves fully. They know what they want, why they want it and more important, how to communicate what they want to others to gain co-operation and support. In the words of Tracy Goss (1996) "language is the only leverage for changing the context of the world around you. This is because people apprehend and construct reality through the way they speak and listen". Robert Bolton (1998) expressed, it is how we manage our conversations and have a discipline about conversations that we can achieve change and transformation in organizations and consequently the achievement of breakthrough results.

The fast and efficient exchange of information is undeniably a competitive advantage in business. And for communication to be effective, everyone must be up to date. A balance flow of information reduces the frequency of misunderstandings and limits expensive mistakes. Employees feel a greater sense of connection with their supervisor and co-workers. They are able to adapt to changes more rapidly and they can offer targeted insights, ideas and perspectives. As a result, supervisors and staff work smarter, not just harder. The effect of quality information flows on quality performance (Forza 1995). Management practices are information and people dependent despite all the technological advances of our time. Organizations are still managed and operated by people and it is their knowledge and skills what determines success (Teresa Brannick et al., 2002). But unfortunately, in the rush to adapt to market needs, many organizations sacrifice effective communication. As priorities shift, executives may impart only the immediate goals and fail to share the rational and strategies behind them (Max Messmer, 2004). Managers must be so focused on results that they neglect to praise and reward their staff. And most of all, they may forget that communication is a two way exchange, one part dialogue, one part listening.

Poor communication is today's number one problem at work, at home and in the work place. People communicate through a set of filters shaped by history, sense of identity, beliefs about what is true, and values about what is right, as well as perceptions and interpretations of what is going on. When someone else communicates with us, we squeeze the message through our own personal filtering system to understand (Shelle, R. Ch., 1995). Every person has a certain number of filters by which they let in certain parts of the real world.

The Role of Listening

Leadership communication uses a full range of communication skills and resources to overcome interferences and to create and deliver messages that guide, direct, motivate or inspire others to action. It is estimated that leaders dedicate as much as 45 per cent of his/her communication time in the form of listening. For Burley Allen (1995) communication breaks down as in Figure 3.

Even though the importance of listening, few individuals are trained in the development of listening skills. Leadership and decision making are as much the art of listening as they are the art of doing (James T. Scarnati (1998). We tend to assume that listening is basically the same as hearing, leading us to believe that effective listening is instinctive. People must seek first to understand, and then to be understood (Covey, 1989).

The word listening comes from the Middle English word *listnen*, which derives from Old English *hlysnan* (Hlud means loud), which simply

Figure 3. Communication skills and resources (Source: Burley Allen, 1995)

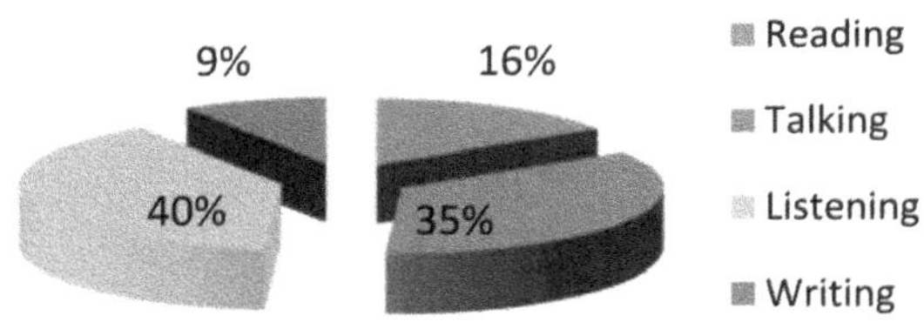

refers to paying attention to sound. In present literature, listening entails a more refined definition: listening is an in-depth process beyond hearing (Shoho, Woods and Smith, 2006). Webster's dictionary (Agnes 2003) defined listening as to make a conscious decision to hear. For the International Listening Association (ILA) listening is "the process of receiving, constructing meaning from, and responding to spoken and/or nonverbal massages" (Emmert, 1996,p.2) Burley – Allen (1995) and Bakhtin (1993) consider listening as a process that entails a sophisticated mental model that requires discipline and energy. Bakhtin (1993) considered that every word in a conversation replies to another word. Spears (1995) described listening as the active recognition and acceptance of employees′ ideas, opinions and suggestions. Bass and Avolio (1994) described it as the leader's motivation to entertain even the most eccentric ideas from followers, noting that this behaviour enhances follower's commitment. Bechler and Johnson (1995) came across with the relationship between leadership effectiveness and listening skills and with the strong positive connection between leadership emergency and listening skills (Johnson and Bechler, 1998). Kramer (1997) tested interpersonal skills and discovered a link between transformational leadership effectiveness and listening.

But it was not until the work of Ralph Nichols that listening was considered a study focus as an element of verbal communication. Lewis and Nichols published in 1954 a guide to effective verbal communication emphasizing the relevance of listening in becoming an effective speaker. In 1957, Nichols co-authored the book *Are You Listening?* The first book dedicated to listening. Shortly, business people and scholars became interested in Nichols′ work as the significance of communication between subordinates and supervisors became more apparent (Orick, 2002).

After Nichols′ work various factors have been linked to the concept of listening. Brandt, Emmert and Emmert (1992) discussed in their work six specific elements of listening, as shown in Figure 4.

And adopted these variables to develop the Listening Practices Feedback Report (LPFR). In their work, 860 people representing 22 firms in a variety of industries and business participated in the study. The survey asked them to make a list with the names of three colleges they considered to be good listeners and another list with the names of three people they considered to be poor listeners. The participants were also asked to write five habits or qualities they will use to identify an effective listener and five more to describe a poor one. Even though there was not a major difference between the group of good listeners and the group of poor ones, the attained results were used to group specific characteristics into general concepts (open mind, empathy, memory, attention, respect, and response) to develop the LPFR.

Open mind refers, according to Brandt Management Group (1999), to be free from personal biases. It is considering logic and content and it is not being critical of appearance, vocabulary, delivery, etc.; it is about not becoming defensive or emotional when confronting a difficult situation, keeping an proper balance between listening and talking; and staying away from emotion-packed clichés, phrases or words (Brandt Management Group, 1999).

Figure 4. Six specific elements of listening (Emmert and Emmert, 1992)

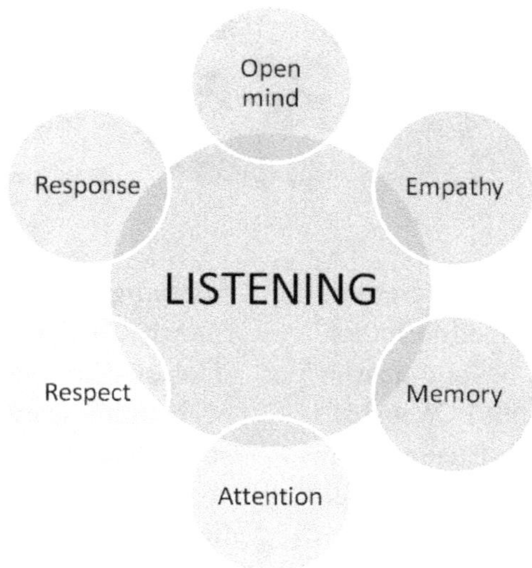

Attention describes the use of the listener's body, eyes and face to show the speaker that he or she is the most important thing occurring at that specific moment (McEwan, 2003). It entails the listener's non verbal communication as well as the degree the listener creates a private listening environment (McKewan, 2003).

Empathy refers, according to Covey (1989), to a person desiring to understand what the other person feels and thinks. It is the person's willingness to understand the paradigms of the other person although it is different from his or hers. It involves seeing the word from the other person's ayes and willingly being in the other person's shoes. Covey (1989) believed that empathy was key to find a common ground and to create a productive relationship. For the Brandt Management Group (1999) a person demonstrates empathy by summarizing, paraphrasing, or repeating the comments of the other person in order to ensure understanding by the listener, by giving confidence to others to give their views on subjects under discussion as well as reflecting on the subject under discussion before responding, and by putting oneself in the speaker's situation and understanding their feelings and concerns.

Memory is related to the result of listening. Good listening produces an outcome aligned and consistent with the guidelines or instructions agreed ahead, the precisely remembered comments and the notes taken when appropriate (Brandt Management Group, 1999).

Respect can be measured in numerous ways. It can be seen as the ability to be confident and listen honestly without just going through the actions. Respect can be articulated by having patience during a conversation and recognizing others′ ideas and words in spite of economic, business or social status (Brandt Management Group, 1999)

Response is related to the listener's reaction. A good response happens when the listeners ask relevant questions for clarification of points that are misunderstood or technical. Response also refers to the appropriateness of verbal and non verbal replies, such as facial expressions or nodding (Brandt Management Group, 1999).

The question that can arise at this moment is "Why should leaders listen to their people?" Osterman (1993) indicated that Covey (1989) could have answered this question when he settled that "the greatest need of a human being is psychological survival-to be understood, to be affirmed, to be validated and to be appreciated" (p.241) Listening carefully to another person gives "psychological air" to him/her. And it is only when this need is satisfied when the leader can focus his/her attention on problem solving or influencing.

Bennis (1997) talked about the four characteristics that people look for in their leaders:

- Sense of trust,
- Sense of purpose,
- Optimism and
- The capacity to achieve results

"One of the best ways to build trust is by listening. It is the most powerful dynamic of human interaction when people feel that they are being heard. Listening does not mean agreeing, but it does mean having the empathetic reach to understand one another" (p. 75).

In Peter Druker's effectiveness model (2002) executives must carefully choose (as illustrated in Figure 5).

The model relies on the listening abilities of the leaders, especially steps one, two and five. Listening is required during step one to understand the employees´ plans and improvements to avoid control of everything, therefore annulling step one. Druker's conviction that organization effectiveness depends on subordinate's contribution is emphasized in step two. It is only when subordinates and managers express their contributions that everyone feels ownership and a role in the organization effectiveness. The importance of searching for views and questioning about the basis for these opinions to make decisions and put plans into action is emphasized in step five as opposed to the alternative of centring the evolution on pre-arranged ideas. It requires listening with genuine curiosity to persuade others to voice

Figure 5. Listening effectiveness model (Source: Peter Druker 2002)

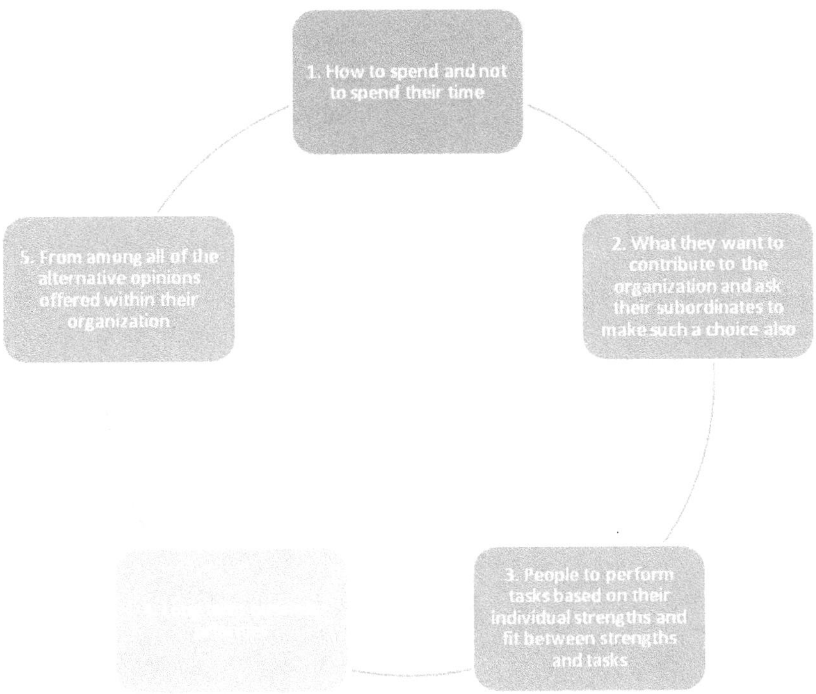

and sustain their opinions, instead of just setting forth proposals hoping form a general agreement.

For Kouzes and Posner (1987) pioneer leaders feel comfortable jumping into the unknown. They identified five common practices to successful leaders, shown in Figure 6.

They emphasize, in these five leadership practices, the power of encouragement, and noted that encouragement is achieved through listening and recognition.

On the other hand, Burley Allen (1995) mentioned the consequences of not developing listening skills:

- Individuals create unnecessary problems for themselves.
- Misunderstandings.
- Hurt feelings.
- Confused instructions.
- Loss of important information.
- Embarrassment
- Frustration

Listening is a potent force for reducing stress and tension, since true listening builds (Burley Allen, 1995):

- Team work
- Trust
- A sense of belonging to a group

When people know they are talking to a listener instead of someone who sits in judgement, they openly suggest ideas and share thoughts. The listener has an opportunity to respond to the person's concerns and needs that otherwise might have gone unnoticed. Factors such as the information not sounding good, not making sense or not ringing true act as filters because they are the result of values and past experiences. These filters result in a closed mind. Listeners with closed minds focus on (Burley Allen, 1995):

- Disagreeing.
- What is not right
- What is wrong

This negative process can be discouraging for the talker. A listening barrier exists when someone hears what he wants to hear, not what is really communicated. The James T. Scarnati (1998) formula: open your mind, your ears, your eyes, and your heart while closing your mouth is a winning combination for listening.

Many people are poor listeners, because listening is difficult and because it is usually more satisfying to be on the offensive. Listening in fact is often more tiring than talking. It demands intellect effort. The average person speaks at rate of 150 words per minute, whereas humans have the capacity to listen at the rate of over 1.000

Figure 6. Five leadership practices (Kouzes and Posner, 1987)

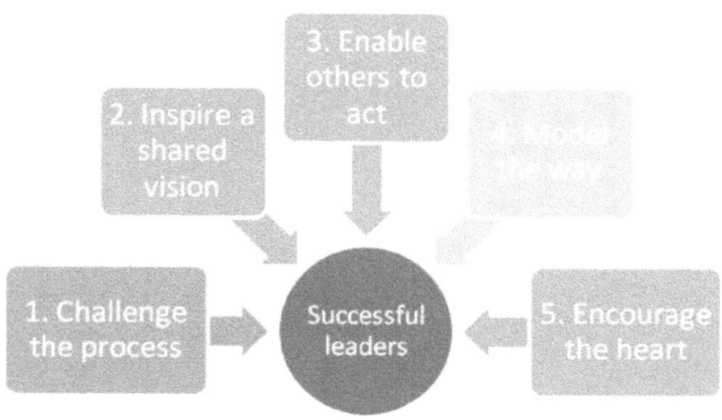

words per minute (Robins 1989). This leaves a mental deficit equivalent to a "distraction zone" of about 850 words per minute. Be a productive listener takes a great deal of mental effort. Successful leader will have not the loudest voice but the reediest ear (Bennis 1993).

Active Listening

Active listening provides a climate for decision making that can best impact strategic change. Nothing will kill genuine dialogue and the motivation to pursue new initiatives faster than a senior manager who is unwilling to listen and who demonstrates little patience or respect for what people have learned and the knowledge they want to transfer. Active listening by top management is a catalyst for change because It overcomes some of the distance created by the authority they hold (Kenneth J. Hatten and Stephen R. Rosenthal 2000). Leaders who practice active listening sign a genuinely open door, and demonstrate care and respect for those doing the talking.

In mastering the role of leadership, leaders must discern between behaviour modification and attitude modification. Behaviour modification implies that people change their outward action, while their attitudes, beliefs, feelings and opinions remain the same. On the other hand, attitude modification implies that followers change their inward thinking hence, their attitudes, feelings, and opinions are brought into line with those of the leader. In opinion of Wells B. (1992) attitude modification is more important.

Another element to take into consideration is the managers´ general assumption that all they are supposed to do is to issue directives and to make decisions. As Sayles (1979) points out, managers often do not know how to react when they encounter a situation that requires endurance, patience, spontaneous compromises continuous interactions and negotiation. Just because executives know their aspirations does not mean that they know how to attain them. Managers have, within themselves, the ideas, capacity and vision to become more effective and to accomplish their goals. However, because most directors need the support and assistance of other people to accomplish their objectives, they need to learn what to do in order to motivate others to perform relentless tasks. In this context, the charismatic and transformational approaches are ideal for innovating, initiating, or energizing others´ behaviour, but fail for directing and maintaining their behaviours.

Leaders responsible for maintaining performance on a day-to-day basis have to pay attention to how they interact with their followers. Letting people occasionally know what they liked or disliked about their jobs and frequently communicating, in a casual way, their impressions about their performance through the day make followers aware of how leaders perceive and value their performance and understand different ways to improve it. The managers´ agreements and disagreements resulting from their observation of the followers' performance can be called performance consequences. These consequences provoke a manager's response that is succeeded by a follower reaction:

- *Positive contingent consequences:* The manager compliments followers for an outstanding job.
- *Positive non-contingent consequences:* The manager praises followers when they did poorly and well.
- *Negative contingent consequences:* The manager reprimands followers when their performance was bellow the standard.
- *Negative non-contingent consequences:* The manager makes followers accountable for things over which they had no control.

Collaborators were more satisfied with their managers when they felt that contingent consequences were delivered by their supervisors. The majority of the problems are not based on peoples´ insincerity, but because they seem to be insincere.

So, in some cases, it is not a question of content, but a question of style; and in others, it is about a manager who does not have or know the information on which to support a sound judgement, even a positive one.

Another important aspect is the frequency of managers' acquisition and communication of information.

To avoid non-contingent consequences, managers must pay attention to the quantity and quality of their information gathering efforts. Frequently monitoring performance provides natural opportunities to provide consequences on a daily basis. Landy, Barness and Murphy (1978) found, after surveying employees of a large manufacturing organization, three aspects connected to the employees´ attitudes about the accuracy and equity of their evaluations:

- The supervisor's frequency of appraisals
- The supervisor's communication of the goals that the employee should try to attain in order to eliminate weaknesses
- The superior's knowledge of employee's level of performance and job duties.

Monitoring and consequences are two important elements to improve and sustain performance. Effective managers collect information about how their workers were doing and, who, based on this monitoring, present consequence that shows that bosses know what is going on. This back and forth information exchange must be an interactive process performed together by managers and workers in a sort of communication symphony. Even though their melody does not always sound pretty and peaceful, it is "in tune" in the sense that participants listen to each other.

FINDINGS

Communication is the most important skill in life since the ability to communicate well is critical for effectiveness. Having a clear vision and good communication skills are critical to becoming a successful leader since they are the only tools available to interact with the team members in order to understand their personal needs, values, objectives and intentions, to align them with the vision of the firm and to create a powerful team spirit.

Poor communication is today's number one problem. 59 per cent of the people participating in the study claim that communication in their work place is tense, disorganised, non-directional, and insufficient. When asked what they understand by communication, 42 per cent of them consider it as a one way process used to transfer information from one person to another. Since two monologues do not make a dialogue, perceiving communication as alternating monologues is a big problem. Communication suffers when the parties involved focus their attention on what to say, how to say it and how to communicate it better, because the speaker focuses his/her attention to achieve his/her conversation goals and forgets the other end of the message. For communication to exist, the receiver has to interpret and to understand the sender's message in the same way the sender intended it.

Leaders must understand that when communicating, people see things not as they are but as they perceive them to be. The communication process is complex and it is vital to have a clear understanding of it if a person wants to succeed.

For communication to be completed the receiver must understand the message and the sender must be sure that it happens. It seems it is an exception since more than 40 per cent of respondents described communication in their work place as a one way process. The Neuro-Linguistic Programming (NLP) presupposition of communication is that the meaning of communication is the response one gets. The response a person gets gives him/her valuable points about what to do next. People know if their message has been properly received by two way communication or feedback. It is surprising that more than 60 per cent of survey respondents understand feedback as something to be given not as something that is part of the communication process. The purpose of feedback is to understand the impact that the message causes on the other person in order to change or adapt it so the new message is better understood and communication can succeed. Communication is an exchange, not just an emission. All parties must participate to complete the information exchange.

People in business organisations must communicate clearly if they want to reach their basic purpose, to make a profit. Successful communication is based on the ability of practicing the act of listening. This "simple" act of listening to each other opens the door to connection, understanding and transformation and improves the individual's conversation capabilities. When two people are truly listening to each other they are not only present but also they are present to something beyond their individual selves. It is what James E. Sullivan defines as the "*dying the self*" and describes as the giving up, at least temporarily, of the individual point of view. This exercise requires putting individual perceptions aside for a moment in order to be present in the world of the other. It is to see what the other person sees, and to feel what the other person feels. It is this type of conversation that can transform a firm.

The reality shown in the survey indicates that people are constantly interrupting each other in order to make a point or convince someone to do something, since more than 60 percent of respondents did not feel respected or understood during the communication process. When speaking, people hear what they already know but when they listen they may learn something new. For a conversation to be effective participants can disagree among themselves but they must be able to see, respect and understand the other person's point of view. When listening to somebody it is essential to fully understand the other person's perspective. The listening process has to start with an attitude that presumes that the other part has sound reasons for why they are saying what they are saying. From the understanding of the topic it could be possible to reflect the importance of active listening in the Figure 7.

For communication to exist, someone, the speaker, must be willing to communicate something to someone, the receiver. In order for this process to be effective, the listener must be receptive to what is going to be communicated by showing some respect towards the speaker. The survey reveals that more than 80 per cent of respondents have regular bad communication experiences when communicating with their superiors, a circumstance that sure influence their approach in future conversations and that sure contribute to destroy the ground for a good communication process. On the other hand, a positive attitude, allows the listener to better understand not only the message but also the feelings involved in the message. The speaker's perception of this positive attitude of the receiver will make him/her feel comfortable and safe and will make him/her willing to go deeper into his/her thoughts and feelings. This circumstance will provide the receiver more and better information about the message formerly presented. The result of this circular process is not only a better understanding of the idea and feelings expressed by the speaker and a better recommendation given by the receiver but also the start of a long term trusting relationship. It is remarkable that 44 of the surveyed people felt interrupted, not listened, inferior and not comfortable when speaking with their superiors. The result of that is a low quality relationship and trust between them. The employees´ level of commitment and motivation decreases, dissatisfaction appears and staff turnover rises. The end result will be a company with a personnel working as individuals not as a team.

From the findings it is possible to graphically describe the result of a leader interaction depending on his/her listening skills as in Figure 8.

Figure 7. Importance of active listening (Source: Own elaboration)

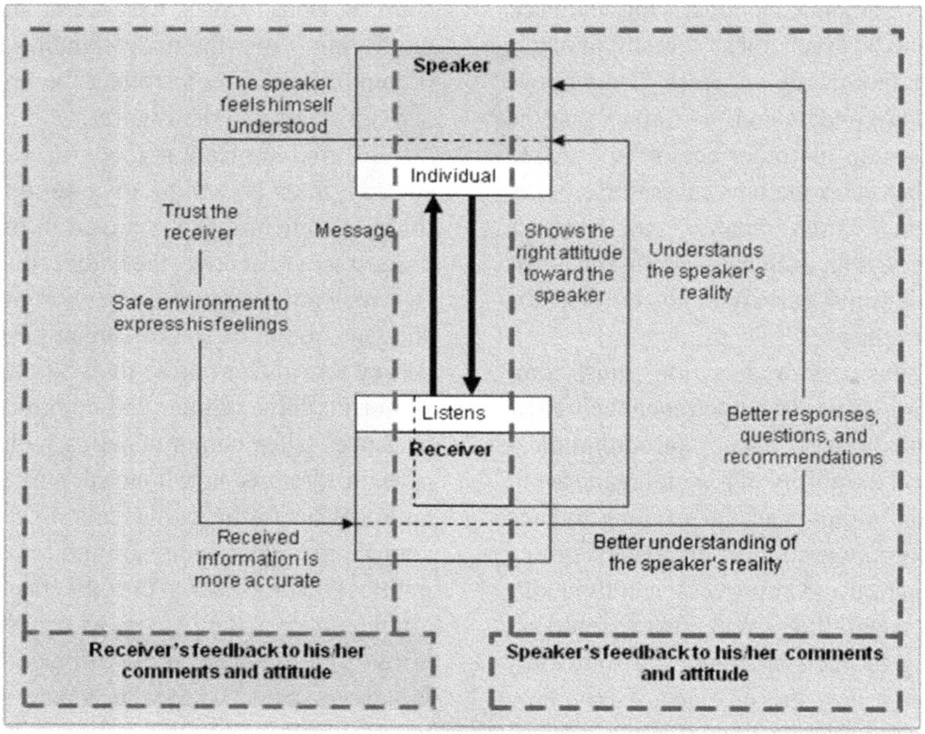

Having good listening skills is connected with having good interpersonal communication which facilitates professional growth. The better understanding derived from good listening skills avoids misunderstandings, hurt feelings, confused instructions, loss of important information, embarrassment and frustration, and facilitates career satisfaction. When people realise that they are in front of someone who listens instead of someone who judges, they openly share thoughts and suggest ideas. Listening leaders have the opportunity to respond to the follower's needs and concerns that otherwise might have gone unnoticed. The principle behind this is "the better the understanding, the better the response". But this principle is far from being present in the surveyed firms since more than 40 per cent of respondents claim that the communication with their superiors is a one way process.

The survey gives the idea that managers are inclined to show power by demonstrating that they know everything. They believe listening is not needed and they share the thought that if they are not talking they are not leading. The consequence of this behaviour is managers blocking openness to new ideas and concepts. Since new perspectives and ideas are not needed, why listening? Richard de Vos, chairman of the Orlando Magic NBA franchise, gives a good answer to this question:

"Listening provides the foundation of productive human reaction without which society cannot function. If we do not listen to each other, we cannot learn from each other. Listening is a vital skill. It is specially important in view of the enormous amount of information and misinformation that is exchanged in today's world"

Figure 8. Leader interaction depending on listening skills (Source: Own elaboration)

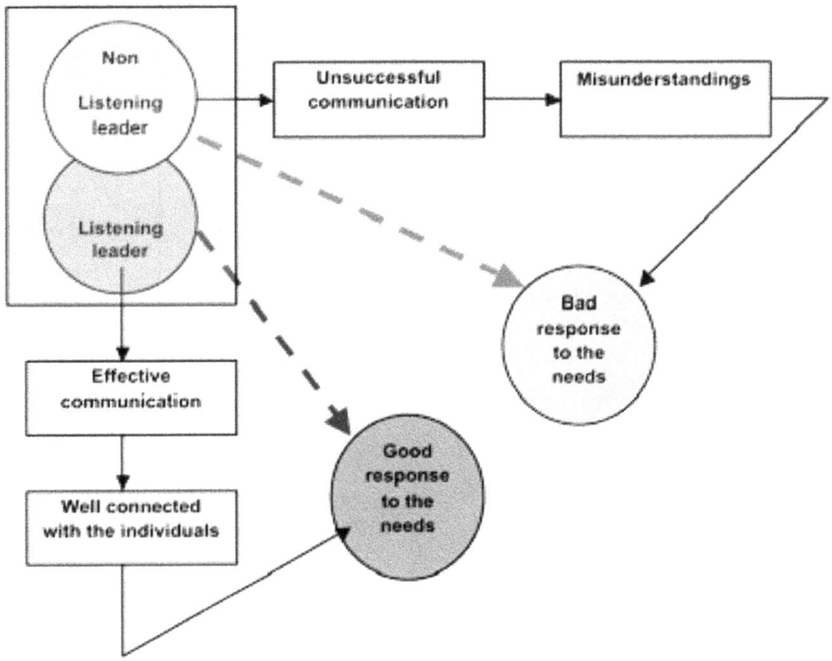

Information is power. Effective listeners are able to concentrate and find the most valid information in whatever they hear. Effective listeners are powerful people.

A listening barrier exists when someone hears what he wants to hear not what is really communicated. From the survey results it can be possible to draw the map to represent the listening barriers shown in Figure 9.

The purpose of listening, building a relationship, is specially challenging in the case of managers because they have a dual function. On one hand, they must show trust in their people allowing them to explore, make decisions and become increasingly independent, but on the other hand, they must also provide the boundaries, the safety lessons, the directions, the values and the beliefs that the employee needs to become self sufficient.

The required balance between these two responsibilities can be achieved creating an excellent communication environment, which will be possible thanks to the adoption of an active listening attitude.

The process of creating this ideal communication environment is not easy since there are inherent blocks to making real emphatic listening connections work. The human relations involved in the process are challenging because of the different goals of the people involved. Bosses and employees often have difficulties hearing each other when their interests are in apparent conflict.

During the employees´ development stage managers can judge rather than listen. The employee's inexperience at this stage will keep the manager/leader from listening because of his/her perception of the situation as a period of threat and will react projecting his/her experience onto the employee to guarantee success.

An analogy can be drawn between employees and children, the former learn from their leaders as children learn from their parents. They all absorb the values their mentors represent and there are no

Figure 9. Listening barriers (Source: Own elaboration)

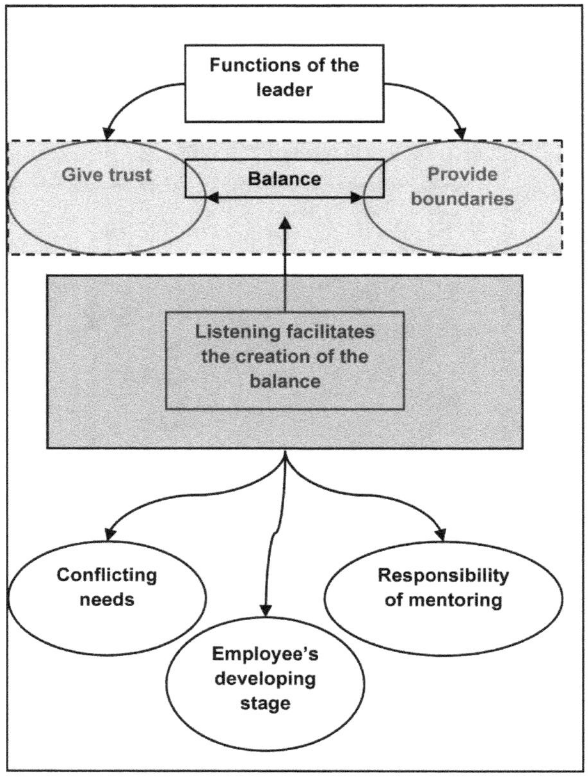

shortcuts in values education. "*Do what I say not what I do*" is a doomed message. Despite signs to the contrary, employees are always watching their leaders.

CONCLUSION

There is a connection between the company results and the leader's listening skills. Even though leaders understand the inefficiency of directive models to manage people they keep using a variant of it since listening seems not to be a priority on their day to day activities. Internal communication is scarce and not always effective since the one way communication seems to be the most common way of interaction among the interviewed people.

Internal communication is the key for success. Without it the survival of the firm is at risk or its developmental capabilities are not being used at its full potential. It could be reflected the importance of a leader's listening skills in relationship to achieving organizational performance as shown in Figure 10.

The objective of most firms is to make a profit in order to survive and the objective of the leader is to manage the company's resources in order to make the firm's objective possible.

In the knowledge era in which we live, 'people' is the firm's most valuable resource. So how the leader manages it makes the difference between success and failure.

To succeed, the leader's main responsibility is to create an open and trusting environment where everyone feels safe to express feelings without the threat or fear of being judged and to use this information to build a team whose members are included and involved. It becomes clear the role

Figure 10. Importance of leader's listening skills in relationship to achieving organizational performance (Source: Own elaboration)

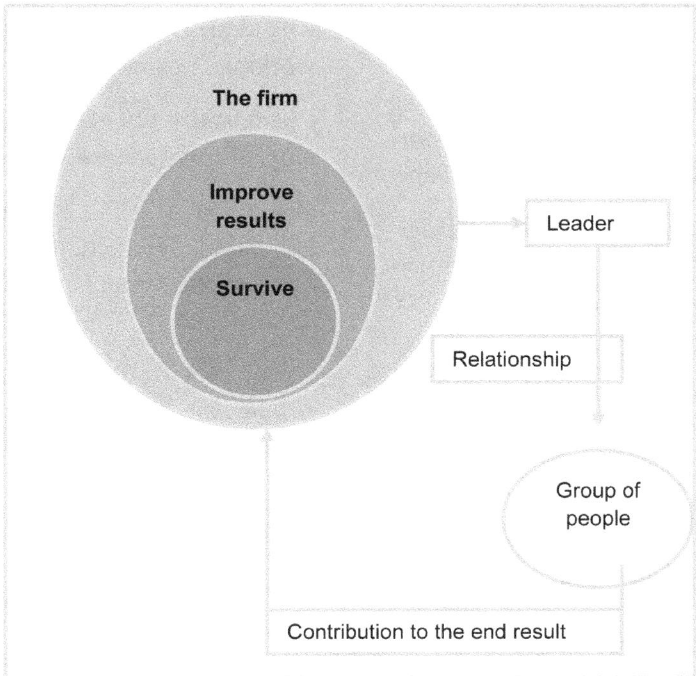

of communication in the leaders understanding of people's needs.

If the group of people decides to work a team, individuals working for the team instead of for themselves, the possibilities for the firm and the leader to succeed increase exponentially.

From the information collected in the survey it is possible to end up with the impression that leaders know the importance and value of their human resources but apparently they do not act accordingly. They continually blame failure on their employees' behaviour and seem not to know what to do differently to overcome the situation. But paradoxically when employees are asked about how to improve their present situation at work they demand more time and empathy from their managers. If they listen they understand the situation and the alternatives.

REFERENCES

ACM. (1995). Considerations for the virtual office. *Communications of the ACM, 38*(12), 13–15.

Agnes, M. (2003). *Webster's new world dictionary*. Indianapolis, IA: Wiley Publishing.

Allen, T. (1977). *Managing the flow of information*. Cambridge, MA: MIT Press.

Bakhtin, M. M. (1993). *Toward a philosophy of the act*. Austin, TX: University of Texas Press.

Barret, D. J. (2006). A leadership communication: A communication approach for senior level managers. In *Handbook of business strategy* (pp. 385–390). Emerald Group Publishing.

Barrett, D. J. (2000). Strong communication skills a must for today's leaders. In *Handbook of business strategy*, (pp. 385-390).

Baruch, Y., & Nicholson, N. (1997). Home 82Sweet work: Requirements for effective homeworking. *Journal of General Management*, *23*(2), 15–30.

Bass, B. M., & Avolio, B. J. (1994). *Improving organizational effectiveness through transformational leadership*. Thousand Oaks, CA: Sage.

Bechler, C., & Johnson, S. D. (1995). Leadership and listening: A study of member perceptions. *Small Group Research*, *26*, 77–85. doi:10.1177/1046496495261004

Bennis, W. (1989). *On becoming a leader*. New York, NY: Addison-Wesley.

Bennis, W. (1993). *An invented life* (p. 107). Reading, MA: Addison- Wesley.

Bennis, W. G. (1997). *Managing people is like herding cats: Warren Bennis on leadership*. Provo, UT: Executive Excellence Publishing.

Bolton, R. (1998). Managing conversations: the medium for achieving "breakthrough" results. *Career Development International*, *3*(6), 233–237. doi:10.1108/13620439810234482

Brandt, R., Brandt, J., Emmert, P., & Emmert, V. (1992, March). *The development of the listening practices feedback report*. Paper presented at the 13th Annual Convention of the International Listening Association. Seattle, Washington

Brandt Management Group. (1999). *The complete guide: Listening practices feedback report—360 degrees*. Virginia: J. Brandt.

Brannick, T., de Búrca, S., Fynes, B., Roche, A., & Ennis, S. (2002). Service management practice performance model: A focus on training and listening practices. *Journal of European Industrial Training*, *26*(8), 394–403. doi:10.1108/03090590210444973

Burley-Allen, M. (1995). *Listening: The forgotten skill*. New York, NY: Wiley and Sons.

Cahavet, S. R. (1995). *Words that change minds*. Kendall/Hunt Publishing Company.

Chaudron, D. (1995). The "far out" success of teleworking. *Supervisory Management*, *40*(1), 6–9.

Checkland, P. (1999). *Soft systems methodology. A 30-year retrospective*. Chichester, UK: John Wiley.

Courtright, J. A., Fairhurst, G. T., & Rogers, L. E. (1989). Interaction patterns in organic and mechanistic systems. *Academy of Management Journal*, *32*(4), 773–802. doi:10.2307/256568

Covey, S. (1989). *The 7 habits of highly effective people*. New York, NY: Simon and Schuster Publishing.

Covey, S. R. (1989). *The 7 habits of highly effective people* (p. 237). New York, NY: Fireside.

DePree, M. (1992). *Leadership is an art*. New York, NY: Doubleday.

Drucker, P. (2002). *The effective executive*. New York, NY: Harper-Collins.

Drucker, P. F. (1992). *Managing for the future: The 1990s and beyond*. New York, NY: Truman Talley Books/Dutton.

Eccles, R. G., & Nohria, N. (1991). *Beyond the hype: Rediscovering the essence of management*. Boston, MA: Harvard School Press.

Emmert, P. (1996). President's perspective. *ILA Listening Post*, *56*, 2–3.

Forsyth, D. R. (1990). *Group dynamics* (2nd ed.). Pacific Grove, CA: Brooks/Cole.

Forza, C. (1995). The impact of information systems on quality performance. *International Journal of Operations & Production Management*, *15*(5), 66–83.

Fritz, M. B. W., Narasimhan, S., & Rhee, H. (1998). Communication and coordination in the virtual office. *Journal of Management Information Systems*, *14*(4), 7–28.

Geneen, H. (1984). *Managing*. New York, NY: Doubleday.

Giddens, A. (1990). *The consequences of modernity*. Stanford, CA: Stanford University Press.

Goss, T. (1996). *The last word on power*. London, UK: Piatkus.

Hamilton, C.-A. (1987). Telecommuting. *The Personnel Journal*, *66*, 91–101.

House, R. J. (2004). *Culture, leadership, and organizations: The GLOBE study of 62 societies*. Thousand Oaks, CA: SAGE Publications.

Huws, U. (1984). New technology homeworkers. *Employment Gazette*, *92*, 13–17.

Johnson, S. D., & Bechler, C. (1998). Examining the relationship between listening effectiveness and leadership emergence: Perceptions, behaviors, and recall. *Small Group Research*, *29*, 452–471. doi:10.1177/1046496498294003

Katz, D., & Kahn, R. L. (1978). *The social psychology of organizations* (2nd ed.). New York, NY: Wiley.

Kouzes, J., & Posner, B. (1987). *Leadership challenge*. San Francisco, CA: Jossey-Bass.

Kozolowski, S. W. J., & Klein, K. J. (2000). A multilevel approach to theory and research in organizations: Contextual, temporal and emergent processes. In *Multilevel theory, research, and methods in organizations: Foundations, extensions, and new directions* (pp. 3–90). San Francisco, CA: Jossey-Bass.

Kramer, R. (1997). Leading by listening: An empirical test of Carl Rogers's theory of human relationship using interpersonal assessments of leaders by followers. *Dissertation Abstracts International Section A: Humanities & Social Sciences, 58,* 514.

Kraut, R., Lundmark, V., Patterson, M., Kiessler, S., Mukopadhyay, T., & Sherlis, W. (1988). Internet paradox: A social technology that reduces social involvement and psychological well-being? *The American Psychologist*, *53*, 1017–1031. doi:10.1037/0003-066X.53.9.1017

Kurland, N. B., & Cooper, C. D. (2002). Manager control and employee isolation in telecommuting environments. *The Journal of High Technology Management Research*, *13*, 107–126. doi:10.1016/S1047-8310(01)00051-7

Landauer, T. K. (1996). *The trouble with computers: Usefulness, usability, and productivity*. Cambridge, MA: MIT Press.

Landy, F. J., Barnes, J. L., & Murphy, K. R. (1978). Correlates of perceived fairness and accuracy of performance evaluation. *The Journal of Applied Psychology*, *63*(6), 751–754. doi:10.1037/0021-9010.63.6.751

McKewan, E. (2003). *Ten traits of highly effective principals*. Thousand Oaks, CA: Corwin Press.

Messmer, M. (2004). Closing the communication gap. In *Handbook of business strategy*, (pp. 87-92).

Mintzberg, H. (1973). *The nature of management work*. New York, NY: Harper & Row.

Nichols, R., & Stevens, L. (1957). *Are you listening?*New York, NY: McGraw-Hill.

Norman, P., Collins, S., Conner, M., Martin, R., & Rance, J. (1995). Attributions, cognitions, and coping styles: Teleworkers′ reactions to work-related problems. *Journal of Applied Social Psychology*, *25*(2), 117–128. doi:10.1111/j.1559-1816.1995.tb01587.x

Orick, L. M. (2002). *Listening practices of leaders*. Doctoral Dissertation, The University of New Mexico, 2002. (UMI No. 3056912).

Osterman, K. F. (1993). *Communication skills: A key to caring, collaboration, and change*. A paper presented at the Annual Conference of the University of Council for Educational Administration, Houston, Texas, October 29-31, 1993. ED 363 973.

Resnick, P. (2002). Beyond bowling together: Sociotechnical capital. In Carroll, J. M. (Ed.), *HCI in the new millennium* (pp. 247–272). Boston, MA: Addison-Wesley.

Robbins, S. P. (1989). *Organizational behavior* (p. 274). Englewood Cliffs, NJ: Prentice-Hall.

Sayles, L. R. (1979). *Leadership: What effective managers really do... and how they do it*. New York, NY: McGraw-Hill.

Scarnati, J. T. (1998). Beyond technical competence: Learning to listen. *Career Development International*, *3*(2), 79–81. doi:10.1108/13620439810207590

Shojo, A. R., Woods, J., & Smith, P. (2006). *Is anyone listening? The dichotomous perspective of school administrator listening skills*. University Council for Educational Administration 2006 Annual Conference.

Spears, L. C. (1995). *Reflections on leadership: How Robert K. Greenleaf's theory of servant leadership influenced today's top management thinkers*. New York, NY: John Wiley.

Spinks, N., & Wells, B. (1995). Quality communication: A key to quality leadership. *Training for Quality*, *3*(2), 14–19. doi:10.1108/09684879510087486

Weick, K. E. (1987). Theorizing about organizational communication. In Jablin, F. M., Putnam, L. L., Roberts, K. H., & Porter, L. W. (Eds.), *Handbook of organizational communication* (pp. 97–122). Newbury Park, CA: Sage.

Weiner, S. P., & Hill, E. J. (1995). *Effective leadership in a telework environment: A training needs analysis*. Presented at the Annual Conference of The American Psychological Society, New York, New York.

Wells, B., & Spinks, N. (1992). *Organizational communication: A leadership approach* (3rd ed.). Houston, TX: Dame Publications.

KEY TERMS AND DEFINITIONS

Active Listening: Provides a climate for decision making, genuine dialogue and motivation to pursue new initiatives.

Communication: The transmission of meaning from one person to another or to a group of people, whether verbally or non verbally. It's a crucial skill since the ability to communicate well is critical for effectiveness.

Leader: Person who make people feel as though they are accomplishing something and not just "putting time"; a leader must work with people and not against them. Leaders are able to express themselves fully.

Leadership: The ability of an individual to influence, motivate and enable others to contribute toward the effectiveness and success of the organizations of which they are members.

Listening Barriers: Exist when someone hears what he wants to hear not what is really communicated.

Quality Communication: The avenue by which leaders clarify their visions and foster participative management within the organization.

Chapter 18
Analysis of Gender Equality in Higher Management Levels: A Study Model

Nuria Calvo
University of A Coruña, Spain

María Bastida
University of Santiago de Compostela, Spain

Jacobo Feás
University of Santiago de Compostela, Spain

ABSTRACT

The main goal of this chapter is to undertake a critical analysis of the current situation concerning the equal treatment of female managers in Spain. In this chapter, the authors analysed the dynamics of business behaviour in order to understand why inequality of women managers for gender reasons persists in spite of the anti-discrimination measures recommended by the legislative framework in place. This analysis has allowed proposals for measures to be drawn up to be taken into account in designing human resources strategies, based on systems of management by competencies and assessment of managerial performance.

1. INTRODUCTION

The economic participation of women is essential in shaping the economy of a country, not just for reasons of equity and equality, but also as a strategic question. According to the resources and capability based view (Wernerfelt, 1984; Barney, 1985; Rumelt, 1987; Conner, 1991), the human capital of business organizations, comprised of a combination of people's skills, knowledge, and reasoning and decision-making abilities, constitutes the main competitive value of firms (Grant 1996). Societies that do not use the talent of half the population carry out an allocation of resources that is not efficient, grow less and put their competitiveness at risk.

However, the inequalities between men and women continue in the business world, related

DOI: 10.4018/978-1-61350-207-5.ch018

to their access to employment, vertical segregation ("the glass ceiling effect"), (Hymowitz & Schellhardt, 1986; Morrison & Von Glinow, 1990), differences in pay or inequality in working contidions.

In Spain, Law 3/2007 establishes a regulatory framework that obliges organisations to adopt equality of opportunity as an objective to be pursued. Nevertheless, in spite of this initiative, statistics show that the problem persists. Paradoxically, since this law came into force, there are some cases where the percentage of women has stabilised or even diminished. In addition, three new matters for concern emerge:

- According to Amuedo (2007), it is becoming extremely difficult to find women willing to belong to Boards of Directors, for fear of being regarded as "quotas"[1]. This phenomenon could spread to the rest of the organisation, self-exclusion appearing for certain promotions in level, because of identical reservations.
- The confusion between "equality" and "positive discrimination" has popularised the idea that the current context offers opportunities to women *at the cost of* men. In the words of Gómez-Mejía the "*beaten man syndrome*" (Gómez-Mejía et al., 2007). As a consequence of this situation resentment arises, anxiety increases and prejudice grows amongst those who feel threatened. A greater proportion of men perceive these measures in terms of a threat than do women. They believe that currently they are discriminated against by measures favouring women, that they have fewer employment opportunities and that they are paying for the consequences of a situation in which women suffered from discrimination in the past.

In these conditions it seems clear that the Law, even though it may help to correct certain deficiencies, is not as effective as it aims to be. It is necessary, therefore, to understand what the dynamics of business behaviour are that can cause the phenomenon of the "glass ceiling" to appear in promotion to management and, on the basis of this understanding, propose policies for action that allow organisations to exploit all their managerial talent, independently of whether this talent is possessed by men or women.

2. DIAGNOSIS OF THE SITUATION

The inclusion of women in paid employment is one of the great achievements of the 20th century. Nevertheless, the situation in which women have been included on equal conditions is yet to be reached.

A longitudinal analysis shows the progressive inclusion of women into the labour market. In the following table (Table 1) it can be seen that the female employment rate has increased by more than 20 percentage points in recent years, in a clear trend towards convergence with the male rate.

Nevertheless, trends and scrutinising data are different things. It is true that, in percentage terms, women have been entering the labour market in an increasing proportion, but it is no less true that their numbers continue to be below those of men (Table 2).

It can be seen from the data in Table 2 that the activity rate as well as the occupation rate is lower in the case of women, while the only figure that is higher is the unemployment rate.

This difference between trend and reality is a common denominator in the analysis of the labour market at the moment. Generally, it can be said that different inequalities persist: (1) at entry level (2) in promotion and (3) in salaries.

1. Inequalities at entry level, or horizontal segregation. Data gathered by the latest survey of the active population warn us of the persistence of this inequality at the

Table 1. Employment Rate Development (Source: Institute for Women, 2010)

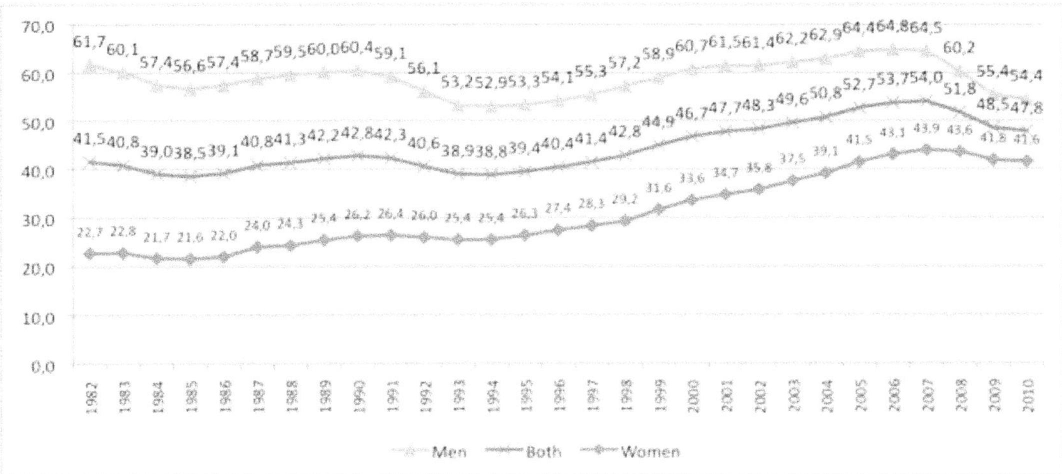

Table 2. Active Population Devlopment (Source: EPA 2010)

		2,001	2,002	2,003	2,004	2005	2006	2,007	2,008	2,009	IQ 2010
Both sexes	ACTIVITY RATE	53,41	54,63	55,91	56,74	57,72	58,58	59,12	60,13	59,76	59,83
	EMPLOYMENT RATE	47,74	48,29	49,56	50,75	52,70	53,72	54,03	51,77	48,50	47,84
	UNEMPLOYMENT RATE	10,63	11,62	11,37	10,56	8,70	8,30	8,60	13,91	18,83	20,05
Women	ACTIVITY RATE	40,96	42,72	44,47	45,79	46,95	48,56	49,37	51,38	51,70	52,05
	EMPLOYMENT RATE	34,71	35,79	37,54	39,13	41,50	43,05	43,94	43,6	41,84	41,55
	UNEMPLOYMENT RATE	15,25	16,22	15,60	14,55	11,61	11,36	11,00	15,14	19,07	20,16
Men	ACTIVITY RATE	66,55	67,16	67,92	68,19	68,95	69,00	69,23	69,21	68,14	67,95
	EMPLOYMENT RATE	61,47	61,42	62,17	62,90	64,37	64,81	64,51	60,24	55,44	54,39
	UNEMPLOYMENT RATE	7,62	8,54	8,46	7,76	6,64	6,06	6,83	12,96	18,64	19,96

level of access to employment (Table 3). As already remarked upon, in spite of the fact that women constitute approximately half of the labour force in the country, the number who are actually working is noticeably below that of men.

In addition, when analysing this apparent "barrier" to access to employment for women, another piece of data should be taken into account: the *quality* of the employment of those who do have work. As can be seen from the following figures (Table 4), the level of "precarity" or "temporariness" of women is significant.

Table 3. Source: Active Population 2008 (Source: EPA Q4 2008)

MEN	
Population from 16 to 64 years	15636.2
Activity rate (16-64)	82.72
Unemployment rate (16-64)	13.04
Employment rate (16-64)	71.93
WOMEN	
Population from 16 to 64 years	15265.2
Activity rate (16-64)	65.29
Unemployment rate (16-64)	15.21
Employment rate (16-64)	55.36

Table 4. Types of contract (Source: Institute for Women 2008)

	1987		1,997		2,007	
	BOTH SEXES	% WOMEN	BOTH SEXES	% WOMEN	BOTH SEXES	% WOMEN
Contract of indefinite duration	6,841.60	28.21	6,950.70	35.68	11,658.20	42.50
Permanent over time	6,696.60	28.08	6,806.90	35.14	11,423.70	42.03
Discontinuous	145.10	34.25	143.70	61.52	234.50	65.37
Temporary	1,656.50	35.03	3,436.60	38.43	5,218.40	46.06
Short-term for reasons of production	65.40	35.32	-	-	923.30	51.44
Learning, training or work placement	65.40	35.32	127.30	40.61	177.10	42.41
Seasonal or fixed period	557.50	33.42	267.80	41.37	276.10	52.01
In probationary period	-	-	31.50	42.54	68.50	45.99
Substituting another worker in their absence	-	-	151.20	72.95	431.30	72.50
For a fixed task or service	-	-	796.70	19.52	2,017.20	31.86
Verbal not included in previous options	-	-	-	-	465.40	67.10
Other type	1,033.70	35.87	2,062.10	42.63	178.20	59.48
Do not know	-	-	14.80	35.14	661.30	44.81
	TOTAL	**TOTAL**	**TOTAL**	**TOTAL**	**TOTAL**	**TOTAL**
	8,511.00	**29.51**	**10,404.10**	**36.59**	**16,876.50**	**43.61**

From among this data, the proportion of women employed on a fixed-term contract basis or, above all, those of a temporary nature in order to substitute a worker, should be noted. In these conditions, and taking the current economic situation into account, it would not be a great risk to suggest that the number of women swelling the unemployment figures in coming months will be significantly higher than that of men.

2. Vertical inequality. In exactly the same way, we can see that women have more difficulty in being promoted to positions of responsibility in organisations. This is a

Table 5. Women and Management (Source: Institute for Women 2009)

	2007	2008
	% Women	% Women
Management of Public Sector and companies	**32.9**	**32.7**
Management of Public Sector and companies with 10 or more employees	**23.4**	**23.4**
Executive and legislative power in public sector, management of organisations	47.8	36.6
Company management	21.9	22.6
Management of companies with less than 10 paid employees	**28.7**	**28.9**
Commerce	35.3	36.3
Hotel and catering	44.6	50.1
Other companies	24.6	20.8
Management of companies without employees	**47.3**	**48.5**
Commerce	55.2	54.0
Hotel and catering	44.6	50.1
Other companies	30.5	33.7

Note: the figures refer to the third quarter in both years

common feature of any statistics we analyse, whatever their sectorial, community, national or international origin.

It seems evident that it is difficult for women to be promoted in enterprises, public as well as private. The following table (Table 5), which is related to the presence of women in managerial positions in public and private enterprises, clearly shows that the proportion of women in these positions is notably below that of men:

The only level on which the numbers seem to be balanced is "management of firms without paid employees", that is to say where women manage because it is their own firm (as in the case of self-employed women, for example) or they are its only employee. This trend is independent of the type of firm, public or private, that is being analysed. In the case of public companies, they do not reach 20% in the best of cases (Table 6).

The situation is not much better in private enterprise. If we look at the situation of firms

Table 6. Women Managers 2008 (Source: Institute for Women 2009)

	Public organisations		State owned firms	
	Both sexes	% Women	Both sexes	% Women
Total	**195**	**18.5**	**1.126**	**13.7**
Presidents	14	14.3	49	6.1
Vice-presidents	6	16.7	33	6.1
Board members	162	18.5	955	13.9
Secretaries	13	23.1	45	28.9
Other members	-	-	44	6.8

Sources: Compiled by the Institute for Women from data provided by the various organisations and firms

Table 7. Women Managers in IBEX firms (Source: Institute for Women 2009)

	Public organisations		State owned firms	
	Both sexes	% Women	Both sexes	% Women
Total Board	**463**	**3,5**	**497**	**6,4**
Presidents	35	2,9	35	2,9
Vice-presidents	41	2,4	53	3,8
Board members	379	3,7	413	7,5
Secretaries	8	0,0	5	0,0

listed in the IBEX 35, the percentage of of women managers is, on average, lower. (Table 7)

In addition, in this final table, it can be seen that change is slow. It is true that, on average, the percentage of women on boards of directors has practically doubled, but due to the amount of inclusion of women in junior management or middle management positions (board members or below), not in the managerial front line (Table 8)

2. Inequality in salaries. The analysis of salaries received by men and women also throws up great inequalities (Table 9). It must be pointed out that we are talking about inequalities, not discrimination. There is a considerable debate about what the correct terminology is. While some authors (Flabbi, 2005; Hernández and Méndez, 2005; Aláez and Uribarri, 1999) maintain that the majority of salary differences between men and women are explained by discrimination, others maintain there is no such discrimination, but inequality (Oliver, 2005). According to these authors, salary differences can be completely explained by a set of objective factors such as women workers having less experience or seniority; less occupation of positions of responsibility; greater presence in the service sector, holding less skilled positions that are more temporary. Others still (Camarero and Vega, 2000; Moreno et al., 1996) believe that both factors combine.

Table 8. Middle Management. Women in managerial tasks. (Source: Institute for Women 2009)

		2004	2005	2006	2007
Both sexes	**Total Board**	**503**	**478**	**463**	**497**
	Presidents	37	37	35	35
	Vice-presidents	39	40	41	53
	Board members	417	388	379	413
	Members/Secretaries	10	13	8	5
% Women	**Total Board**	**2.58**	**2.09**	**3.46**	**6.43**
	Presidents	5.41	0	2.86	2.86
	Vice-presidents	2.56	2.50	2.44	3.77
	Board members	2.88	2.32	3.69	7.51
	Members/Secretaries	0	0	0	0

Table 9. Average Annual Salary (INE, 2009)

	1995	2002	2006	2008
MALES	18.223,29€	22.169,16€	22.051,08€	24.203,33€
FEMALES	12.237,21€	15.767,56€	16.245,17€	18.910,62€
% DIF	32,85%	28,88%	26,33%	21,87%

It can be seen from the Table 9 how the distribution of salaries, on average, is always higher in the case of men. It is true that various reasons can be given, *a prior* objective, for these inequalities: the influence of seniority increments (seniority increments constitute a significant percentage of the salary of workers. Women have recently entered the labour maket, for which reason they probably have less seniority), differing working hours, varying quantities of overtime or productivity bonuses, for example. In fact, researchers disagree when determining what the variables are that have the greatest weight in determining salaries. Aláez and Uribarri (1999) place experience, seniority and level of education first but also concede importance to the type of contract and the sector of activity. For Camarero and Vega (2000) the variables with the greatest influence are seniority (22%), level of education (15%), size of the firm (14%), sex (12%), type of contract (12%), occupation (7%), collective agreement applied (6%), final consumer market (6%) and sector of activity (2%).

From what has been said above it can be deduced that significant inequalities between men and women persist in the labour market. Even considering social pressure in this regard, as well as various political and legal initiatives, these inequalities are maintained through time.

3. THE PROPOSED OBJECTIVE: A CAUSAL ANALYSIS

In spite of the existence of legislation specifically directed towards achieving equality in the workplace with regard to gender, the statistics show a reality that does not correspond to the objectives set out. Inequality between men and women on a professional level has not been reduced at the speed expected. The interaction of factors of a social, cultural or psychological nature, not considered in the legislative framework, continue to make real equality difficult in business organisations.

Therefore organisational designs and action policies are maintained that, far from favouring access to work and promotion for women, create intangible barriers that are difficult to overcome through legislative rules. Currently, the smaller percentage of women that have entered the executive level in firms and institutions permits the existence of organisational behaviour that is not being taken into account to be detected. In recent years neither governments, nor large corporations, banks, academic or scientific institutions, not even international organisations such as the United Nations, have considered assessing the opportunity cost that not using the managerial talents of many women represents. A more balanced distribution of decision-making power in the levels of the hierarchy could not only allow the value of this talent to be exploited, but also favour equality at work at lower levels of the hierarchy.

Gender discrimination in the workplace is such a broad and complex problem that it is impossible to analyse it from a single perspective. For this reason we have limited our study to the phenomenon known as "Discriminated Elites", as we consider that the formation of the organisational conditions that can favour or impede the promotion of women at work to executive positions could affect their future participation in the managerial processes of business organisations.

The starting point for our analysis is based on an initial set of premises that will determine the development of the approach proposed.

- There are no significant differences in managerial competencies in organisations for reasons of gender. On the contrary, levels of competence will depend on the following factors:
    - Specific knowledge: business management, organisational behaviour and psychology and technical knowledge of the area.
    - Management skills: capacity for social relations, negotiating and communication skills, initiative and leadership.
    - Previous executive experience in management or junior management positions.
    - The existence of a broad social network of contacts key for achieving the strategy of the organisation.
- The existence of a smaller number of promotions of female managers compared to male managers, considering the existence of an equal number of candidates of both sexes, could be related to the following:
    - The existence of organisational policies that discourage the promotion of women managers
    - Social, cultural and educational factors that influence the assessors when taking decisions about work promotion in their organisations.
    - Resistance to change or the "status quo" of the current management bodies of the organisation, which are mostly made up of male managers.
- The percentage of women managers who leave is greater than that of men due to life, emotional, and differential behaviour factors. The difficulties female managers have in maintaining themselves as a minority group in top management bodies are related to the increase in the rate they leave.

The traditional view defines women as a problematic aspect of the labour force. Thanks to political concern to encourage measures that reduce inequalities in the workplace between men and women, superficial remedies in practice have multiplied in recent years that neither deal with the strategic causes of inequality nor integrate women at all levels of the labour force. Policies that facilitate leave of absence to take care of children, encourage part-time work, give preferential access to training courses or fiscal benefits for employing women, even if they are very rewarding acts aimed at improving the image of a government or socially responsible firm, are not clearly of any help in really solving the problem of inequality in the workplace, and less so at the level of management.

Research on the advancement (or, more correctly, the "non"-advancement) of women in professional careers (Perry et al., 1994; Tharenou et al., 1994), the reconciliation of personal and professional life (Stroh et al., 1996), female managers (Lyness and Thompson, 1997; Bass and Avolio, 1994), leadership (van Knippenberg, 2004; Hunt, Boal and Dodge, 1999) or gender and organisation (Agars, 2004) indicate that the majority of the "technical" or traditional barriers (access to higher levels of education, a non-discriminatory legal and institutional framework) have been gradually overcome in recent years.

Taking the dynamic role that groups play in determining the form and context of an organisation into account (Ryan and Haslam, 2007; Haslam et al., 2003), we can infer that these barriers are the result of intra-group pressure in the decision-making process, when basically these decisions are taken by men. That is to say that given that the majority of managers in organisations are men, it will be them who have to choose a woman for a management post. Breaking with tradition, prejudices and sexual stereotypes and possible difficulties in ways of dealing with matters because they are different sexes may be some of the reasons that curb male managers from taking this decision. This discrimination could also be interpreted as a way to reduce the status quo and acquired privileges (Schmitt, Ellemers & Branscome, 2003).

From this point of view, because of the internal dynamics of power endogamy (Figure 1), the individuals belonging to a closed group such as a managerial body tend to choose their new members from among those people who are similar to the majority, avoiding the introduction of elements that distort the existing equilibrium and increase the possibility of future conflicts at the heart of the group.

The maintenance of this dynamic through time causes some women members of a management team to leave, because of the isolation created by their minority situation and differences in characteristics from the rest of the group members. The following diagram explains the higher leaving rate that appears amongst women managers compared to their male counterparts. The higher the percentage of men who are at management level, the higher will be the percentage of new male managers recruited. This situation will make the isolated position of the women managers belonging to the same level more acute, which will cause more of them to leave for this reason, which in turn will increase the percentage of male managers with decision-making status.

These dynamics allow the fact that low female representation is maintained at a managerial level in various organisations to be explained. To ameliorate this low representation, a set of policies has been put in place in recent years under the heading "positive discrimination". These are intended to assure a minimum participation of the female minority in managerial bodies. The method advanced by these policies means that if in the selection process there are several equally eligible candidates the selectors will discriminate in favour of the woman. With this method it is intended, in the mid-term, to achieve quantitative equality of both sexes in decision-making bodies. This is a measure that, even though it is obligatory in the

Figure 1. Endogamous Loop (source: made on the premises)

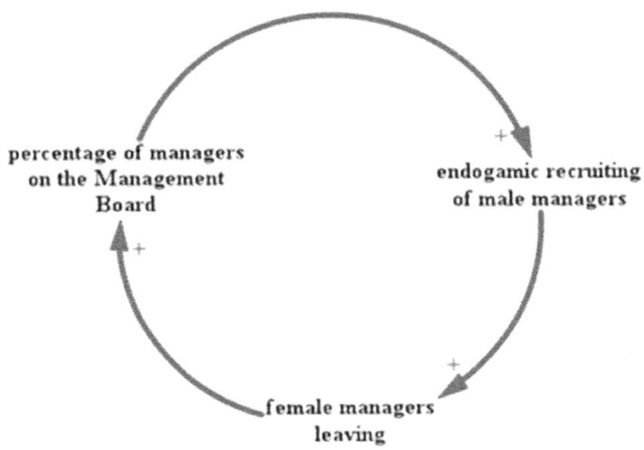

Public Sector, is only a recommendation for good governance in the case of private organisations.

From the proposed approach it is not obvious that this type of policy, recently introduced, can achieve the intended objectives. The complexity of the internal reactions of the organisation itself as a group of individuals, not taken into consideration in the practical application of the proposed action, could mean failure in the real objective of this policy.

The following causal diagram (Figure 2) shows the interrelation between various variables that may influence the situation. The increase in policies promoting equality through positive discrimination creates an attitude of disagreement among the men who occupy the majority of managerial positions. This disagreement increases prejudices towards those new female managers favoured in the selection. This attitude is transmitted to the junior managers, making cooperation difficult between male managers and women junior managers. As a result of this situation the productivity of the female junior managers may decrease in comparison to that of their male colleagues. Afterwards, the assessment of the performance of the female junior managers who are candidates will be inferior to that of their male counterparts. As a consequence the previous condition of equal eligibility will not be met during the promotion process, which will make the promotion of new women managers difficult.

Nevertheless, if this diagram allows the maintenance of female managers in a minority in organisations to be explained theoretically, it does not clarify the situation of inequality between sexes that appears at management level in strategic areas in organisations.

The explanation of this phenomenon can be due, from a comprehensive point of view, to the contradiction that appears between the final objective of the policies put in place (that is to achieve global parity between sexes at a managerial level) and the traditional cultural values that exist in the perception of the majority of the male managers whose job it is to implement these policies. In this way the subjective perception of the lack of

Figure 2. Loop of resistance to change and cession in the power structure (source: made on the premises)

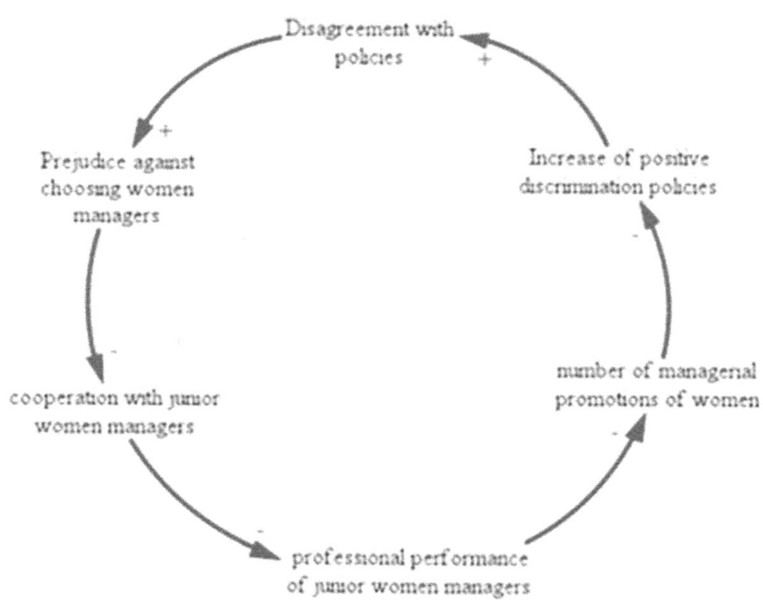

Figure 3. Organisational culture loop (source: made on the premises)

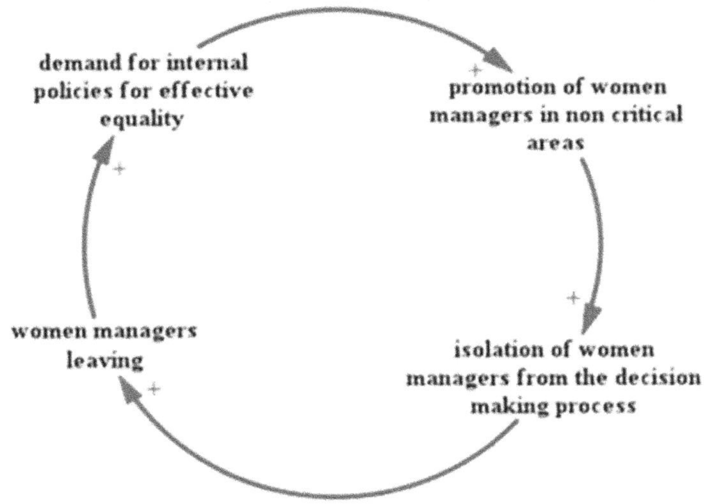

managerial competence of female junior managers favours the promotion of women into managerial jobs that are not critical for the organisation to function well, placing them in areas of activity with lower budgets or considered "feminine" according to gender stereotypes.

This situation, seen from a dynamic point of view (Figure 3), perpetuates a loop able to keep women managers in a minority despite promotion policies put in place to avoid this.

As a result the promotion of the majority of women managers to areas that are not critical for the global functioning of the organisation, which means occupying posts with less responsibility, budget and margin for manoeuvre, causes their isolation in the important decision-making bodies. This situation reduces the support of their male colleagues at the same time as it again stirs up prejudices and gender stereotypes, in time provoking a greater leaving rate among women managers in relation to men.

4. PROPOSED MEASURES

Organisations today are aware of change in the competitive environment. Intangible assets, when they are scarce, valuable and lasting, constitute the basis of sustainable competitive advantage (Barney, 1991; Grant, 1996; Drucker, 2005).

Knowing how to manage and develop the managerial competencies of their employees becomes, therefore, a strategic objective for organisations, key to their differentiation. Renouncing this competence because of a gender question can be manifestly unprofitable. Nevertheless, putting women into managerial positions just because they are women, without them having suitable competencies compared to other candidates, may result in the same situation.

It seems necessary, therefore, to identify a strategy that should guide the behaviour of an organisation that intends to select and maintain the managerial talent of its employees as a strategic resource without gender criteria being able to bias decision-making one way or another.

The measures proposed below require close collaboration between the functions of strategic planning, human resources and organisation.

Without this coordination, as well as the commitment of management to the objective of equality of opportunity, it will not be possible to change the conduct identified in the above analysis as causing the glass ceiling effect.

4.1. Strategic Planning

A firm's business plan is firmly in the sights of partners, investors, creditors and public organisations involved in its going into operation. The fulfilment of each and every one of the declared objectives adequately and in time allows a firm to gain the necessary credibility to stimulate its growth.

For this reason, the management of a company should embody the distribution of tasks to be carried out in the organisational structure. To do so the following is necessary:

A. *A clear definition of the functional and professional job profile of the management positions.* In order to do so the system of assessment of factors and competencies (Orue-Echevarría, J., 2004; Fernández-Ríos, M., 1995) allows the most relevant functions of each position to be identified, as well as the professional qualities required to carry them out successfully. Only when gender is a relevant aspect to be considered in each of these profiles would it be licit to include it as a criterion in the selection of management.

B. *A distribution of objectives through the organisational structure,* which implies a hierarchical distribution related to the level of responsibility (objectives of the company, the area, individuals), as well as a distribution of functions that is coherent with the basis of groupings in the organisational structure (intra-functional objectives, inter-functional objectives). The weighting of objectives within the process of managing performance allows the degree of commitment of every professional to be assessed at the various levels of objectives. Nevertheless, it is the responsibility of management to define the *goals grid* to be accomplished, that will allow the work of the organisation to be aligned with the fulfillment of the objectives in its Business Plan. Ambiguity in responsibilities combined with a changing situation and an organisational structure that is not well-defined are often features that favour biased and opportunistic behaviour, more concerned with individual interests than the accomplishment of business objectives. It is in this context that bias can appear in the selection of management for gender reasons, and endogenous behaviour that tends to eject women from the management team just because they are women.

When defining a strategic process of introducing human resources practices, we have considered the starting point to be the classification hierarchy of Wright and Gardner (2003), that differentiates four levels: (1) guiding principles (2) alternative policies (different practices), (3) products (competencies or behaviour that stimulates practice, and (4) processes of putting into practice (effectiveness in the execution of the practices). It is considered that while some practices such as payment for performance, or selection by competence, could be universally effective, this effectiveness does not really come from the practice itself, but from the *fit* between practice and the result obtained (for example, a salary increase for a good performance level or the selection of suitable professionals in relation to a specific strategy).

The description of management profiles and strategic objectives allows a system of management by competencies to be put in action that will serve as a connecting theme for the measures tending to identify and maintain management talent in the organisation.

4.2. Recruitment and Selection Measures

Considering the non-existence of significant differences in management competencies in organisations because of gender, and taking into account that competence in management is related to matters such as knowledge of administration, communication and negotiation abilities, leadership, previous executive experience and a network of contacts, the following are put forward as *selection criteria* for candidates for management posts.

- *A blind review of merits.* This means the consideration of the curriculum value of the candidates without having any information about their gender.
- *The above assessment of competence connected with systems for managing performance*. In order to do this it will be necessary to design tests specifically directed towards assessing, on a predefined scale, the performance of candidates in solving management problems. Strategic problem simulators allow a complete set of parameters of reasoning to be assessed, at the same time avoiding the assessor's bias. This will make it possible to avoid decisions influenced by personal or cultural factors connected with perception of the gender of the candidate.

4.3. Management Coaching System

One of the reasons identified in the above diagnosis as causing the greater leaving rate of female managers was lack of support from the previous management group. To avoid this situation resulting in organisational behaviour in which the group expels the marginal member, or that this member gives up because of lack of support from equals, *a coaching or tutorial system is proposed that encourages managers to stay*, regardless of their gender, as long as the objective assessment of professional performance advises that they remain.

To do this, managers should have a high level of dynamic vision of the organisational scene in which they carry out their work (Table 10).

The perception of self-efficacy is conditioned by expectancy of control over reinforcements (Rotter, 1966). These expectancies reflect belief in one's own responsibility for the consequences of behaviour, and are influenced, one the one hand, by intellectual and social capabilities and resources, under internal control, and on the other by external reinforcement coming from the perceptions of managerial efficacy that third parties transmit.

The perception of self-efficacy similarly conditions expectancies of results (Bandura, 1986). These refer to the beliefs that, in this case, women managers and junior managers may have about the probability that specific results will be the consequence of controllable behaviour.

Table 10. Adapted from Lewin's Force Field Analysis

Driving forces		Resisting forces
People	→	People
Economic benefits	←	Traditions / previous culture
Situation in the environment	→	Organisational policies
Wants and needs of the management	←	Attitudes of the management team
Availability of technology and tools	?	Costs

Figure 4.Causal diagram of managerial self-exclusion (source: made on the premises)

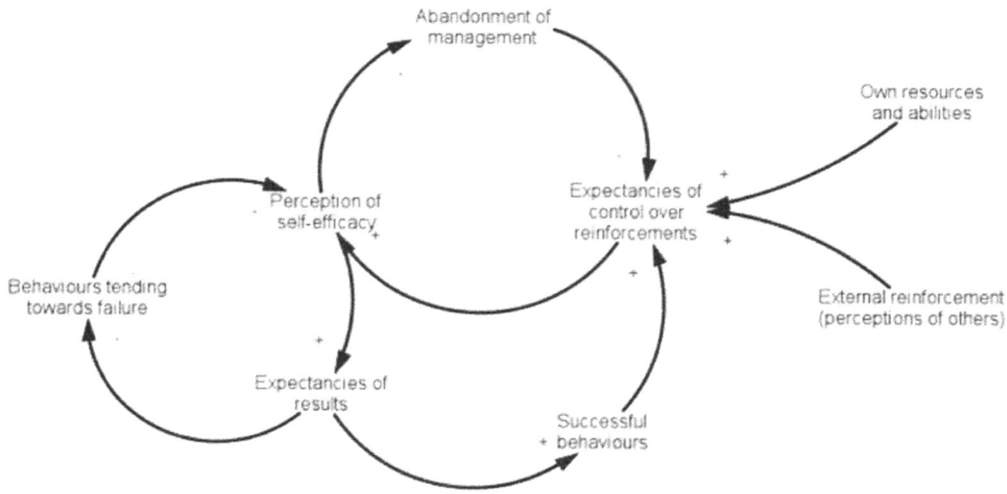

Taking these relationships into account, we have identified three behavioural loops which allow in depth analysis of how self-exclusion works as a barrier to the promotion of women managers.

A. *Prior deterrent loop:* The perception of others, biased by gender stereotypes, lessens the effect of the value of the resources and abilities of women hoping to access managerial positions, lowering their expectancies of control over reinforcements. This reduces their perception of self-efficacy, resulting in abandonment of their promotion to manager, even before this happens.
B. *Positive behavioural loop:* The reduction of the belief in their responsibility for the consequences of their behaviour (expectancies of control over reinforcements) reduces female managers' perception of self-efficacy. This causes a lowering of expectancies of results and reduces those behaviours tending towards achieving managerial success. This situation reinforces the lowering of expectancies of control over reinforcements, creating a dynamic of managerial paralysis (lack of proactive behaviour in the absence of perception of the behaviour-results connection).
C. *Negative behavioural loop:* Reduction of female managers' perception of self-efficacy causes a lowering of expectancies of results which, by contrast, may set in motion behaviours tending towards failure, motivated by stress, poor assessment of alternatives or those motivated by personal interests. All of this creates a negative dynamic which reinforces the reduction of female managers' perception of self-efficacy and may cause abandonment of the position of responsibility.

The proposed analysis suggests a dynamic which is difficult to change. The majority of measures proposed by the Government focus on providing management training for women and establishing quota systems for female representation in positions of responsibility within organisations. Measures which, whilst they contribute in part to altering the current situation, do not contribute to changing the underlying causes for the psychological and social barriers which prevent the smooth promotion of female talent in organisations.

The coaching measure proposed is directed towards encouraging greater external reinforcement of the work of pre-selected female managers. This support, aligned with their own managerial competencies, may influence expectancies of control over reinforcements, activating successful behaviour and avoiding leaving motivated by a perception of low self-efficacy.

5. CONCLUSION

In recent years we have seen the proliferation of legal regulations -national and international- directed towards introducing equality of opportunity between men and women. Simultaneously, awareness and social pressure have been increasing in this regard. Nevertheless, the situation of women in the labour market continues to be of concern: less activity, higher unemployment rate, imbalances in levels of entry, salary and managers.

In Spain, Organic Law 3/2007, on the effective equality of men and women, attempts to ameliorate this situation, completely or in part. This law represents a challenge for many organisations. Among the several obligations that the introduction of this law brings with it figures the requirement to draw up an equality plan in firms of more than 250 workers, which will probably before long be extended to the rest of firms.

All together, the greatest challenge that the law creates is the negative collateral effects in the workforce, above all in groups not benefitted by the measures. We have seen that there is a low level of acceptance of positive discrimination measures, above all among men. We also detect that the perception arises among men that their situation is being impinged on to the benefit of women. Conflicts will appear in justice, legitimacy and authority. Women, for their part, may find themselves with problems of self-perception of their capability, getting to the point of excluding themselves from the possibility of benefitting from these measures.

This situation means bad use of managerial talent in the heart of organisations. If in principle business efficiency requires the best use possible to be made of resources, that is to say to always use the "most suitable" resource, applying this to the specific case of human resources implies allocating *the most suitable person* to a post with regard to different attributes (training, work record, productivity, specific knowledge, contribution to results, etc.) that make a person most suitable, independently of their gender.

In this chapter we have analysed the dynamics of business behaviour that allows it to be understood why the situation of inequality of women managers for gender reasons persists in spite of the positive discrimination measures recommended by the legislative framework in force. This analysis has allowed proposals for measures to be drawn up to be taken into account in designing human resources strategies, based on systems of management by competencies and assessment of managerial performance.

From the proposed approach, these measures, adapted to the needs and conditions of each organisation, will allow the orientation of decision-making towards the criterion of efficiency based on exploiting the managerial talent of professionals who work in an organisation, independently of whether this talent is masculine or feminine.

REFERENCES

Agars, M. D. (2004). Reconsidering the impact of gender stereotypes on the advancement of women in organizations. *Psychology of Women Quarterly*, *28*, 103–111. doi:10.1111/j.1471-6402.2004.00127.x

Aláez, R. A., & Uribarri, M. (2000). Discriminación salarial por sexo. Un análisis del sector privado y sus diferencias regionales en España. *Información Comercial Española, 789*, 117–137.

Amuedo, J. (2007). Un toque femenino a los consejos de Adminstración. *Fundación San Telmo, 27*, 18–21.

Bandura, A. (1978). Reflections on self-efficacy. In Rachman, S. (Ed.), *Advances in behavior research and therapy* (*Vol. 1*, pp. 237–269). Oxford, UK: Pergamon.

Bandura, A. (1986). From thought to action: Mechanisms of personal agency. *New Zealand Journal of Psychology, 15*, 1–17.

Barney, J. B. (1985). Organizational culture: Can it be a source of sustained competitive advantage? *Academy of Management Review, 11*, 656–665.

Bass, B. M., & Avolio, B. J. (1994). Shatter the glass ceiling: Women may make better managers. *Human Resource Management, 33*, 549–560. doi:10.1002/hrm.3930330405

Blau, F., & Devaro, J. (2006). *New evidence on gender differences in promotion rates: An empirical analysis of a sample of new hires*. Working Papers 891. Princeton University, Department of Economics, Industrial Relations Section.

Conner, K. R. (1991). A historical comparison of resource-based theory and five schools of thought within industrial organization economics: Do we have a new theory of the firm? *Journal of Management, 17*(1), 121–154. doi:10.1177/014920639101700109

Drucker, P. (2005). *The effective executive in action*. Harper Collins.

EPA. (4th Quarter2008). *Encuesta de Población Activa*. Instituto Nacional de Estadística (INE).

Flabbi, L. (2010). Gender discrimination estimation in a search model with matching and bargaining. *International Economic Review, 51*(3), 745–783. doi:10.1111/j.1468-2354.2010.00600.x

Gómez-Mejía, L. R., Balkin, D. B., & Carly, R. L. (2007). *Gestión de recursos humanos*. Pearson-Prentice Hall.

Grant, R. M. (1996). Toward a knowledge-based theory of the firm. *Strategic Management Journal, 17*(Winter Special Issue), 109-122.

Haslam, S. A., & Turner, J. C. (1995). Context-dependent variation in social stereotyping 3: Extremism as a self categorical basis for polarized judgement. *European Journal of Social Psychology, 25*, 341–371. doi:10.1002/ejsp.2420250307

Hernández, P., & Méndez, I. (2005). La correlación del sesgo de selección en los análisis de corte transversal de discriminación salarial por sexo: Estudio comparativo en los países de la UE. *Estadística Española, 47*(158), 179–214.

Hunt, J. G., Boal, K. B., & Dodge, G. E. (1999). The effects of visionary and crisis-responsive charisma on followers: An experimental examination of two kinds of charismatic leadership. *The Leadership Quarterly, 10*, 423–448. doi:10.1016/S1048-9843(99)00027-2

Hymowitz, C. &Schellhadt, T. (1986, March 24). The glass ceiling. *The Wall Street Journal: A Special Report, The Corporate Woman*, D1, 4-5.

Instituto de la Mujer. (2007). *Plan estratégico de igualdad de oportunidades* (2008-2011).

Instituto de la Mujer. (2007). *Plan estratégico de igualdad de oportunidades* (2008-2011). Ley 3/2007 de Igualdad de Oportunidades entre hombres y mujeres.

Instituto de la Mujer. (2008). *Mujeres y hombres en España 2008*. Madrid.

Instituto de la Mujer. (2009). *Mujeres y hombres en España 2009*. Madrid.

Lyness, K. S., & Thompson, D. E. (1997). Above the glass ceiling? A comparison of matched samples of female and male executives. *The Journal of Applied Psychology*, *82*, 359–375. doi:10.1037/0021-9010.82.3.359

Moreno, G., Rodríguez, J. M., & Vera, J. (1996). *La participación laboral femenina y la discriminación salarial en España*. Madrid, Spain: CEC-Social.

Morrison, A. M., & Von Glinow, M. A. (1990). Women and minorities in management. *The American Psychologist*, *45*(2), 200–208. doi:10.1037/0003-066X.45.2.200

Oliver, J. (2005). *Desigualdad y discriminación salarial de la mujer en España*. Madrid.

Perry, E. L., Davis-Blake, A., & Kulik, C. T. (1994). Explaining gender-based selection decisions: A synthesis of contextual and cognitive approaches. *Academy of Management Review*, *19*, 786–820.

Rotter, J. (1966). Generalized expectancies for internal versus external control of reinforcements. *Psychological Monographs, 80*, Whole No. 609.

Rumelt, R. P. (1987). Theory, strategy and entrepreneurship. In Lamb, R. B. (Ed.), *The competitive challenge: Strategies for industrial innovation and renewal* (pp. 137–158). Cambridge, MA: Ballinger.

Ryan, M. K., & Haslam, S. A. (2005). The glass cliff: Evidence that women are over-represented in precarious leadership positions. *British Journal of Management*, *16*, 81–90. doi:10.1111/j.1467-8551.2005.00433.x

Smith, N. C. (2003). Corporate social responsibility: Whether or how? *California Management Review*, *45*(4), 52–76.

Stroh, L. K., Brett, J. M., & Reilly, A. H. (1996). Family structure, glass ceiling, and traditional explanations for the differential rate of turnover of female and male managers. *Journal of Vocational Behavior*, *49*, 99–118. doi:10.1006/jvbe.1996.0036

Tharenou, P., Latimer, S., & Conroy, D. (1994). How do you make it to the top? An examination of influences on women and men managerial advancement. *Academy of Management Journal*, *37*, 899–931. doi:10.2307/256604

UE. (2002). *Directive 2002/73/CE*.

UE. (2004). *Directive 2004/113/CE*.

Van Knippenberg, D., & Hogg, M. A. (Eds.). (2004). *Leadership and power: Identity processes in groups and organizations*. London, UK: Sage.

Wernerfelt, B. (1984). A resource based view of the firm. *Strategic Management Journal*, *5*, 171–180. doi:10.1002/smj.4250050207

ENDNOTE

1 Law 3/2007 does not directly mention quotas, but the principle of "balanced presence" (First Additional Provision), according to which a 60-40 (maximum) percent distribution between the sexes should be respected at each organisational level.

Compilation of References

Abad, I. M., & Castillo, A. M. (2004). Desarrollo de competencias directivas. Ajuste de la formación universitaria a la realidad empresarial. *Boletín Económico del ICE, 2795*, 29–41.

Abrahamson, E. (1991). Managerial fads and fashions: The diffusion and rejection of innovations. *Academy of Management Review, 16*, 586–612.

Accenture. (2008). *Beyond cost reduction*. Retrieved July 30, 2010, from http://www.accenture.com/NR/rdonlyres/6FBF58AC-7A26-43F9-BCED-90F6E232B70C/0/177090038HRBPOBeyondCostReductionandInnovationv3.pdf

Accenture. (2010). *Business process outsourcing*. Retrieved from https://microsite.accenture.com/explore_outsourcing/Business_Process_Outsourcing/Pages/Business_Process_Outsourcing.aspx

ACM. (1995). Considerations for the virtual office. *Communications of the ACM, 38*(12), 13–15.

Adam, N. (2007). *Workforce e-maturity-characteristics and development*. Nottingham, UK: NAACE. Retrieved January, 9, 2011, from http://www.naace.co.uk/448

Adams, K. (1991). Externalisation versus specialisation: What is happening to personnel? *Human Resource Management Journal, 1*(4), 40–54. doi:10.1111/j.1748-8583.1991.tb00236.x

Agars, M. D. (2004). Reconsidering the impact of gender stereotypes on the advancement of women in organizations. *Psychology of Women Quarterly, 28*, 103–111. doi:10.1111/j.1471-6402.2004.00127.x

Agencia Espacial Europea. (1998-2003). *ESA-Careers at ESA-Core skills*.

Agnes, M. (2003). *Webster's new world dictionary*. Indianapolis, IA: Wiley Publishing.

Aguilar López, J. (2011). *Gestión del Compromiso: De la continuidad por necesidad a la aportación libre de valor*. Capital Humano, No. 250. Wolters Kluwer, 2011.

Aguirre Barrera, V. (2011). *Engagement, más allá de un mero compromiso*. Capital Humano, No. 250. Walters Kluwer, 2011.

Ahmed, P., Hardaker, G., & Carpenter, M. (1996). Integrated flexibility: A key to competition in a turbulent environment. *Long Range Planning, 29*(4), 562–571. doi:10.1016/0024-6301(96)00048-9

Akhtar, S., & Tan, D. (1994). Reassessing and reconceptualising the multidimensional nature of organizational commitment. *Psychological Reports, 75*, 1379–1390. doi:10.2466/pr0.1994.75.3.1379

Aláez, R. A., & Uribarri, M. (2000). Discriminación salarial por sexo. Un análisis del sector privado y sus diferencias regionales en España. *Información Comercial Española, 789*, 117–137.

Alavi, M., & Leidner, D. E. (1999). Knowledge management systems: Issues, challenges, and benefits. *Communications of the Association for Information Systems, 1*, article 7, (pp. 2-36).

Alewell, D., Hauff, S., Thommes, K., & Weiland, K. (2009). Triggers of HR outsourcing decisions-An empirical analysis of German firms. *International Journal of Human Resource Management, 20*(7), 1599–1617. doi:10.1080/09585190902985228

Allen, A. (2009). Shaping up or shipping out? *People Management*, November, 4-9.

Allen, N. J., & Meyer, J. P. (1996). Affective, continuance and normative commitment to the organization: an examination of construct validity. *Journal of Vocational Behavior*, *49*, 252–276. doi:10.1006/jvbe.1996.0043

Allen, T. (1977). *Managing the flow of information.* Cambridge, MA: MIT Press.

Alles, M. A. (2000). *Dirección estratégica de recursos humanos. Gestión por competencias.* Buenos Aires, Argentina: Ediciones Granica.

Alles, M. A. (2010). *Diccionario de preguntas. La trilogía.* Buenos Aires, Argentina: Ediciones Granica.

Alliger, G. M., Tannenbaum, S. I., Bennett, W., Traver, H., & Shotland, A. (1997). A meta-analysis of the relations among training criteria. *Personnel Psychology*, *50*(2), 341–359. doi:10.1111/j.1744-6570.1997.tb00911.x

Allred, B. B., Snow, C. C., & Miles, R. E. (1996). Characteristics of managerial careers in the 21st century. *The Academy of Management Executive*, *10*(4), 17–27. doi:10.5465/AME.1996.3145316

Altés, J. (2003). El outsourcing en el departamento de recursos humanos. *Capital Humano*, July-August, 23-27.

Álvarez de Mon, S. (2000). *El mito del líder. Profesionales, ciudadanos, personas: La sociedad alternativa.* Madrid, Spain: PriceWaterhuse Coopers.

Alvarez, G. (2008). Determinantes del compromiso organizacional. ¿Están los trabajadores a tiempo parcial menos comprometidos que los trabajadores a tiempo completo? *Cuadernos de Estudios Empresariales*, *18*, 73–78.

Alvesson, M. (2000). Social identity in knowledge-intensive companies. *Journal of Management Studies*, *37*(8), 1101–1123. doi:10.1111/1467-6486.00218

Amabile, T. M., & Gryskiewicz, N. D. (1989). The creative environment inventory. *Creativity Research Journal*, *2*, 231–253. doi:10.1080/10400418909534321

American Society for Training and Development. (1995). *National HRD executive survey.*

Amit, R., & Schoemaker, P. (1993). Strategic assets and organizational rent. *Strategic Management Journal*, *14*, 33–46. doi:10.1002/smj.4250140105

Amuedo, J. (2007). Un toque femenino a los consejos de Adminstración. *Fundación San Telmo*, *27*, 18–21.

Amuedo-Dorantes, C. (2001). From "temp-to-perm": Promoting permanent employment in Spain. *International Journal of Manpower*, *22*, 625–647. doi:10.1108/EUM0000000006234

Anderson, N., & Schalk, R. (1998). The psychological contract in retrospect and prospect. *Journal of Organizational Behavior*, *19*(S1), 637–648. doi:10.1002/(SICI)1099-1379(1998)19:1+<637::AID-JOB986>3.0.CO;2-H

Anderson, N., De Creu, C. K. W., & Nijstad, B. A. (2004). The routinization of innovation research: A constructively critical review of the state-of-the-science. *Journal of Organizational Behavior*, *25*(2), 147–173. doi:10.1002/job.236

Anderson, R. E., Krohn, K., & Sandman, R. (1980). *User's guide for the Minnesota computer literacy and awareness assessment.* St Paul, MN: Minnesota Computing Corporation.

Ang, J. L. (2010). Nómina y seguros sociales. In *M. J. Pérez López (Coord.), Externalización de funciones de recursos humanos.* Madrid, Spain: Pearson.

Ang, S. A., & Joseph, A. D. (1996). *Organizational learning and the learning organization: Trigger events, processes, and structures.* Paper for Academy of Management Meetings, Cincinnati, OH, August. Retrieved August 1, 2011, from www.ntu.edu.sg/home/adjoseph/webpages/publications/aom10.pdf

Ang, S., & Inkpen, A. C. (2008). Cultural intelligence and offshore outsourcing success: A framework of firm-level intercultural capability. *Decision Sciences*, *39*(3), 337–358. doi:10.1111/j.1540-5915.2008.00195.x

Angel, H. L., & Perry, J. L. (1981). Empirical assessment of organizational commitment and organizational effectiveness. *Administrative Science Quarterly*, *26*, 1–14. doi:10.2307/2392596

Ansari, M. A., Kee, D. M. H., & Aafaqi, R. (2000). Fairness of human resource management practices, leader-member exchange, and intention to quit. *Journal of International Business and Entrepreneurship*, *8*, 1–19.

Ansari, M. A., Kee, D. M. H., & Aafaqi, R. (2007). Leader-member exchange and attitudinal outcomes: Role of procedural justice climate. *Leadership and Organization Development Journal*, *28*(8), 690–709. doi:10.1108/01437730710835443

Appelbaum, S. H., & Donia, M. (2000). The realistic downsizing preview: A management intervention in the prevention of survivor syndrome (part I). *Career Development International*, *5*(7), 333–350. doi:10.1108/EUM0000000005384

Appelbaum, S. H., Delage, C., Labib, N., & Gouit, G. (1997). The survivor syndrome: aftermath of downsizing. *Career Development International*, *2*(6), 278–286. doi:10.1108/13620439710178639

Appelbaum, S. H., Patton, E., & Shapiro, B. (2003). The early retirement incentive program: A downsizing strategy. *Journal of European Industrial Training*, *27*(1), 22–35. doi:10.1108/03090590310456500

Appelbaum, S. H., Simpson, R., & Shapiro, B. (1987). Downsizing: The ultimate human resource strategy. *Ivey Business Journal*, *52*(2), 52–60.

Arce, E. (2000). Desarrollo de competencias. *Training & Development Digest*, *24*, 28–29.

Argyris, C. (1960). *Understanding organizational behavior*. London, UK: Tavistock Publication.

Argyris, C. (1999). *On organizational learning* (2nd ed.). Oxford, UK: Blackwell Publishing.

Aristotle,. (1980). *The Nicomachean ethics* (Ross, D., Trans.). Oxford, UK: Oxford University Press. Also published in World's Classics collection. (Original work published 1925)

Arnold, M., Gibbs, M., & Wright, P. (2003). Intranets and the creation of local community: 'Yes, an intranet is all very well, but do we still get free beer and a barbeque? In Huysman, M., Wenger, E., & Wulf, V. (Eds.), *Communities and technologies*. Dordrecht, The Netherlands: Kluwer Academic Publishers.

Arthur, J. B. (1994). Effects of human resource systems on manufacturing performance and turnover. *Academy of Management Journal*, *37*(3), 670–687. doi:10.2307/256705

Arulampalam, W., & Booth, A. L. (1998). Training and labour market flexibility: Is there a trade-off? *British Journal of Industrial Relations*, *36*(4), 521–536. doi:10.1111/1467-8543.00106

Ashmos, D. P., McDaniel, R. R., & Duchon, D. (1990). Differences in perception of strategic decision-making processes: The case of physicians and administrators. *The Journal of Applied Behavioral Science*, *26*, 201–218. doi:10.1177/0021886390262010

Ashton, C., & Morton, L. (2005). Managing talent for competitive advantage. *Strategic HR Review*, *4*(5), 28–31. doi:10.1108/14754390580000819

Assens, J. (2006). *Huevos con beicon. Cómo aumentar el compromiso de los empleados con la empresa y cómo gestionar su propio compromiso como empleados*. Barcelona: Granica.

Atkinson, J. (1984). Manpower strategies for flexible organisations. *Personnel Management*, August, 28-31.

Atkinson, J., & Gregory, D. (1986). Is flexibility just a flash in the pan? *Personnel Management*, September, 26-29.

Atkinson, J., & Meager, N. (1986). *Changing working patterns. How companies achieve flexibility to meet new needs*. London, UK: National Economic Development Office (NEDO).

Aubert, J., & Gilbert, P. (2003). *L'evaluation des competences*. Sprimont, Belgium: Mardaga.

Aubert, P., Caroli, E., & Muriel, R. (2006). New technologies, organisation and age: Firm-level evidence. *The Economic Journal*, *116*, F73–F93. doi:10.1111/j.1468-0297.2006.01065.x

Audretsch, D. B., Thurik, A. R., Verheul, I., & Wennekers, S. (Eds.). (2002). *Entrepreneurship: Determinants and policy in a European US comparison*. Boston, MA: Kluwer Academic Publishers. doi:10.1007/b109395

Auer, P., & Cazes, S. (2000). Fortaleza de la relación de trabajo duradera en los países industrializados. *Revista Internacional del Trabajo*, *119*(4), 419–450.

Autor, D., Levy, F., & Murnane, R. (2003). The skull content of recent technological change: An empirical exploration. *The Quarterly Journal of Economics*, *118*(4), 1279–1333. doi:10.1162/003355303322552801

Ayoko, O. B. (2007). Communication openness, conflict events and reactions to conflict in culturally diverse workgroups. *Cross Cultural Management: An International Journal*, *14*(2), 105–124. doi:10.1108/13527600710745723

Backtrom, L., Huyyrnlovhrt, D., Kleinberg, J., & Lan, X. (2006). Group formation in large social networks: Membership, growth, and evolution. *Proceedings of the 12th International Conference on Knowledge Discovery in Data Mining* (pp. 44-54). New York, NY: ACM Press.

Backus, C., Keegar, K., Gluck, C., & Gulik, V. (2010). Accelerating leadership development via immersive learning and cognitive apprenticeship. *International Journal of Training and Development*, *14*(2), 144–148. doi:10.1111/j.1468-2419.2010.00347.x

Bacq, G. (1996). Décrypter les compétences. In J. M. Peretty (Ed.), *Tout DRH* (pp. 94-98). Paris, France: Editions D'Organisation.

Bailey, T. (1993). *Discretionary effort and the organization of work, employee participation and work reform since Hawthorne*. Discussion Paper, Teacher College, University of Columbia, New York.

Baker, J. G., & Baker, R. D. F. (1999). Perceived ideological differences, job satisfaction and organizational commitment among psychiatrists in a community mental health center. *Community Mental Health Journal*, *35*, 85–95. doi:10.1023/A:1018704327272

Baker, W. K. (1995). Allen & Meyer's 1990 longitudinal study: A reanalysis and reinterpretation using structural equation modelling. *Human Relations*, *48*, 169–186. doi:10.1177/001872679504800204

Bakhtin, M. M. (1993). *Toward a philosophy of the act*. Austin, TX: University of Texas Press.

Bal, P. M., De Lange, A. H., Cansen, P. G. W., & Van Der Velde, M. E. G. (2008). Psychological contract breach and job attitudes: A meta-analysis of age as a moderator. *Journal of Vocational Behavior*, *72*(1), 143–158. doi:10.1016/j.jvb.2007.10.005

Band, D. C., & Tustin, M. (1995). Strategic downsizing. *Management Decision*, *33*(8), 36–45. doi:10.1108/00251749510093923

Bandura, A. (1977). *Social learning theory*. Englewood Cliffs, NJ: Prentice Hall.

Bandura, A. (1978). Reflections on self-efficacy. In Rachman, S. (Ed.), *Advances in behavior research and therapy* (*Vol. 1*, pp. 237–269). Oxford, UK: Pergamon.

Bandura, A. (1986). From thought to action: Mechanisms of personal agency. *New Zealand Journal of Psychology*, *15*, 1–17.

Banker, R. D., Field, J. M., Schroeder, R. G., & Shina, K. K. (1996). Impact of work teams on manufacturing performance: A longitudinal field study. *Academy of Management Journal*, *39*, 867–890. doi:10.2307/256715

Barber, A. E. (1998). *Recruiting employees*. Thousand Oaks, CA: SAGE.

Barney, J. B. (1985). Organizational culture: Can it be a source of sustained competitive advantage? *Academy of Management Review*, *11*, 656–665.

Barney, J. B. (1991). Firm resources and sustained competitive advantage. *Journal of Management*, *17*(1), 99–120. doi:10.1177/014920639101700108

Baron, J. N., & Kreps, D. M. (1999). Consistent human resource practices. *California Management Review*, *41*(3), 29–53.

Barraza, M. A., & Ortega, M. F. (2009). *Satisfacción laboral en instituciones formadoras de docentes. Un primer acercamiento*. Communication presented at the X Congreso de Investigación Educativa, Veracruz, México.

Barret, D. J. (2006). A leadership communication: A communication approach for senior level managers. In *Handbook of business strategy* (pp. 385–390). Emerald Group Publishing.

Barrett, D. J. (2000). Strong communication skills a must for today's leaders. In *Handbook of business strategy*, (pp. 385-390).

Barrett, R., & Mayson, S. (2007). Human resource management in growing small firms. *Journal of Small Business and Enterprise Development*, *14*(2), 307–320. doi:10.1108/14626000710746727

Barroso, C., 7 Casillas, J. C. (1998). Análisis del efecto de la reestructuración en las empresas españolas durante el quinquenio 90-94. *Revista Europea de Dirección y Economía de la Empresa*, *7*(3), 53–64.

Bartlett, C. A., & Goshal, S. (1997). The myth of the generic manager: New personal competencies for new management roles. *California Management Review*, *40*(1), 92–116.

Bartram, D. (2000). Internet recruitment and selection: Kissing frogs to find princes. *International Journal of Selection and Assessment*, *8*(4), 261–274. doi:10.1111/1468-2389.00155

Baruch, Y., & Nicholson, N. (1997). Home 82Sweet work: Requirements for effective homeworking. *Journal of General Management*, *23*(2), 15–30.

Basadur, M., Runco, M. A., & Vega, L. A. (2000). Understanding how creative thinking skills, attitudes, and behaviors work together: A casual process model. *The Journal of Creative Behavior*, *34*, 77–100.

Bashford, S. (2004). The survivor syndrome. *Human Resources (E-Journal of London)*, June, 43-44.

Bass, B. (1990). *Bass and Stogdill's handbook of leadership: Theory, research, and applications* (3rd ed.). New York, NY: Free Press.

Bass, B. M. (1985). *Leadership and performance beyond expectations*. Nueva York, NY: Free Press.

Bass, B. M. (1998). *Transformational leadership: Industry, military and educational impact*. Hillsdale, NJ: L Erlbaum.

Bass, B. M., & Avolio, B. J. (1994). *Improving organizational effectiveness through transformational leadership*. Thousand Oaks, CA: Sage.

Bass, B. M., & Avolio, B. J. (1994). Shatter the glass ceiling: Women may make better managers. *Human Resource Management*, *33*, 549–560. doi:10.1002/hrm.3930330405

Bassi, L. J., Cheney, S., & Van Buren, M. (1997). Training industry trends 1997. *Training & Development*, (November): 46–59.

Bassi, L., Benson, G., & Cheney, S. (1996). The top ten trends. *Training & Development*, *50*, 28–42.

Bateman, T. S., & Strasser, S. (1984). A longitudinal analysis of the antecedents of organizational commitment. *Academy of Management Journal*, *27*, 95–112. doi:10.2307/255959

Bathmaker, S. (1999). So, what's the deal? The state of the psychological contract in a new university. *Journal of Vocational Education and Training*, *51*(2), 265–282. doi:10.1080/13636829900200084

Bauman, Z. (2007). *Liquid times: Living in an age of uncertainty*. Cambridge, MA: Polity.

Bauman, Z. (2007). *Tiempos líquidos*. Barcelona, Spain: Tusquets.

Bayo, A., & Merino, J. (2001). Quality management and high performance work practices: Do they coexist. *International Journal of Production Economics*, *73*(3), 251–260. doi:10.1016/S0925-5273(01)00089-5

Bayón, F. (1994). *Gestión de recursos humanos. Manual para técnicos en empresas turísticas* (2nd ed.). Madrid, Spain: Editorial Síntesis.

Bayón, F., Cubeiro, J., Romo, M., & Sanz, J. (2006). *Coaching realmente*. Madrid, Spain: Pearson Education.

Beatson, M. (1995). *Labour market flexibility. Research Series, 48*. Sheffield: Employment Department.

Bechler, C., & Johnson, S. D. (1995). Leadership and listening: A study of member perceptions. *Small Group Research*, *26*, 77–85. doi:10.1177/1046496495261004

Becker, B. E., & Huselid, M. A. (1998). High performance work systems and firm performance: a synthesis of research and managerial implications. In Rowland, K. M., & Ferris, G. R. (Eds.), *Research in personnel and human resource management* (pp. 53–101). Greenwich, CT: JAI Press.

Becker, B. E., & Huselid, M. A. (2006). Strategic human resource management: Where do we go from here? *Journal of Management*, *32*(6), 898–925. doi:10.1177/0149206306293668

Becker, B., & Gerhart, B. (1996). The impact of human resource management on organizational performance: Progress and prospects. *Academy of Management Journal*, *39*, 779–801. doi:10.2307/256712

Becker, H. S. (1960). Notes on the concept of commitment. *American Journal of Sociology*, *66*, 33–42. doi:10.1086/222820

Begley, T., & Boyd, D. (1986). Psychological characteristics associated with performance in entrepreneurial firms and smaller business. *Journal of Business Venturing*, *2*, 79–93. doi:10.1016/0883-9026(87)90020-6

Bell, B. S., & Kozlowski, S. W. (2002). A typology of virtual team. *Group & Organization Management*, *27*(1), 14–49. doi:10.1177/1059601102027001003

Bell, D. (1973). *The coming of post-industrial society: A venture in social forecasting*. New York, NY: Basic Book.

Beltrán, I., Roca, V., Escrig, A., & Bou, J. C. (2008). Human resource flexibility as a mediating variable between high performance work systems and performance. *Journal of Management*, *34*(5), 1009–1044. doi:10.1177/0149206308318616

Beltrán-Martín, I., Roca-Puig, V., Escrig-Tena, A., & Bou-Llusar, J. C. (2009). Internal labour flexibility from a resource-based view approach: Definition and proposal of a measurement scale. *International Journal of Human Resource Management*, *20*(7), 1576–1598. doi:10.1080/09585190902985194

Bennet, R. (2002). Employer's demands for personal transferable skills in graduates: A content analysis of 1000 job advertisements and an associated empirical study. *Journal of Vocational Education and Training*, *54*(2), 457–475. doi:10.1080/13636820200200209

Bennis, W. (1989). *On becoming a leader*. New York, NY: Addison-Wesley.

Bennis, W. (1993). *An invented life* (p. 107). Reading, MA: Addison- Wesley.

Bennis, W. G. (1997). *Managing people is like herding cats: Warren Bennis on leadership*. Provo, UT: Executive Excellence Publishing.

Bergman, M. E. (2006). The relationship between affective and normative commitment: Review and research agenda. *Journal of Organizational Behavior*, *27*, 645–663. doi:10.1002/job.372

Bernard, R. (1996). *The corporate Intranet*. New York, NY: John Wiley & Sons.

Berry, J. (1990). Linking management development to business strategies. *Training & Development*, (August): 20–21.

Bhatnagar, J. (2007). Talent management strategy of employee engagement in Indian ITES employees: Key to retention. *Employee Relations*, *29*, 640–663. doi:10.1108/01425450710826122

Bhattacharya, M., Gibson, D. E., & Doty, D. H. (2005). The effects of flexibility in employee skills, employee behaviors, and HR practices on firm performance. *Journal of Management*, *31*, 622–640. doi:10.1177/0149206304272347

Bhidé, A. (1994). How entrepreneurs craft strategies that work. *Harvard Business Review*, *72*(2), 150–161.

Bhidé, A. (1999). Developing start-up strategies. In Sahlman, W. A., Stevenson, H. H., Roberts, M. J., & Bhidé, A. (Eds.), *The entrepreneurial venture* (pp. 121–137). Boston, MA: Harvard Business School Press.

Biech, E. (2008). *American Society for Training and Development handbook for workplace learning professionals*. Alexandria, VA: ASTD Press.

Birley, S., & Stockley, S. (2000). Entrepreneurial teams and venture growth. In Sexton, D. L., & Landström, H. (Eds.), *The Blackwell handbook of entrepreneurship* (pp. 287–308). Oxford, UK: Blackwell.

Björkman, I., & Smale, A. (2010). La gestión global del talento: Retos y soluciones. *Universia Business Review*, *27*, 30–43.

Black, J. S., & Mendenhall, M. (1989). A practical but theory-based framework for selecting cross-cultural training methods. *Human Resource Management*, *28*(4), 511–539. doi:10.1002/hrm.3930280406

Black, J. S., & Mendenhall, M. (1990). Cross-cultural training effectiveness: A review and theoretical framework for future research. *Academy of Management Review*, *15*(1), 113–136.

Blanco Prieto, A. (2007). *Trabajadores competentes. Introducción y reflexiones sobre la gestión de recursos humanos por competencias*. Madrid, Spain: ESIC.

Blau, F., & Devaro, J. (2006). *New evidence on gender differences in promotion rates: An empirical analysis of a sample of new hires*. Working Papers 891. Princeton University, Department of Economics, Industrial Relations Section.

Blau, G. J., & Boal, K. R. (1987). Conceptualizing how job involvement and organizational commitment affect turnover and absenteeism. *Academy of Management Review*, 290.

Blyton, P., & Morris, J. (1992). HRM and the limits of flexibility. In Blyton, B., & Turnbull, P. (Eds.), *Reassessing human resource management*. London, UK: Sage Publications.

Bolleu, R., Bowers, D., Boyatzis, R., & Kolb, D. (1999). Fellowship in lifelong learning: An executive development program for advanced professionals. *Journal of Management Education*, *23*(4), 338–354. doi:10.1177/105256299902300403

Bolton, R. (1998). Managing conversations: the medium for achieving "breakthrough" results. *Career Development International*, *3*(6), 233–237. doi:10.1108/13620439810234482

Bondarouk, T. V., & Ruël, H. (2009). Electronic human resource management: Challenges in the digital era. *International Journal of Human Resource Management*, *20*(3), 505–514. doi:10.1080/09585190802707235

Boone, L. E., & Kurtz, D. L. (2010). *Contemporary business*. Retrieved from http://books.google.es

Borstorff, P. C., Marker, M. B., & Bennett, D. S. (2007). Online recruitment: Attitudes and behaviours of job seekers. *Journal of Strategic E-commerce*, *5*(1&2), 1–23.

Bose, K. (2003). An e-learning experience. *Campus-Wide Information Systems*, *20*(5), 193–199. doi:10.1108/10650740310507399

Boshoff, C., & Mels, G. (2000). The impact of multiple commitments on intentions to resign: An empirical assessment. *British Journal of Management*, *11*, 255–272. doi:10.1111/1467-8551.00165

Boudreau, J. W., & Ramstad, P. M. (2005). Talentship and the new paradigm for human resource management: From professional practises to strategic talent decision science. *Human Resource Planning*, *28*(2), 17–26.

Boudreau, J., & Ramstad, P. (1997). Measuring intellectual capital: Learning from financial history. *Human Resource Management*, *36*(3), 343–356. doi:10.1002/(SICI)1099-050X(199723)36:3<343::AID-HRM6>3.0.CO;2-W

Boudreau, K. J., & Lakhani, K. R. (2009). How to manage outside innovation. *MIT Sloan Management Review*, *50*(4), 69–77.

Boundreau, J. W. (1991). Utility analysis in human resource management decisions. In Dunette, M. D., & Hough, L. M. (Eds.), *Handbook of industrial and organizational psychology* (2nd ed., *Vol. 2*, pp. 271–326). Palo Alto, CA: Consulting Psychologists Press.

Bowen, D. E., & Ostroff, C. (2004). Understanding HRM-firm performance linkages: The role of the "strength" of the HRM system. *Academy of Management Review*, *29*, 203–221.

Bowman, E. H., & Singh, H. (1993). Corporate restructuring: Reconfiguring the firm. *Strategic Management Journal, 14*(Summer Special Edition), 5-14.

Box, K. (1999). *Human interaction during teacher training courses delivered via the Internet*. San Antonio, TX: Society for Information Technology & Teacher Education Conference.

Boxall, P. F. (1996). The strategic HRM debate and the resource-based view of the firm. *Human Resource Management Journal*, *6*(3), 59–75. doi:10.1111/j.1748-8583.1996.tb00412.x

Boyatzis, R. (2006). Intentional change from a complexity perspective. *Journal of Management Development*, *25*(7), 607–623. doi:10.1108/02621710610678445

Boyatzis, R. E. (1982). *The competent manager: A model for effective performance*. New York, NY: John Wiley & Sons.

Boyatzis, R. E., & Saatcioglu, A. (2008). A 20-year view of trying to develop emotional, social and cognitive intelligence competencies in graduate management education. *Journal of Management Development*, *27*(1), 92–108. doi:10.1108/02621710810840785

Boyatzis, R., & Van Oosten, E. (2002) *Developing emotionally intelligent organizations*. Retrieved April 10, 2006, from http://www.eiconsortium.org

Boyatzis, R., Goleman, D., & Rhee, K. (2000). Clustering competence in emotional intelligence: Insights from the emotional competence inventory (ECI). In Bar-On, R., & Parker, J. D. A. (Eds.), *Handbook of emotional intelligence*. San Francisco, CA: Jossey-Bass.

Boyd, D., & Ellison, N. (2007). Social network sites: Definition, history, and scholarship. *Journal of Computer-Mediated Communication*, *13*(1). Retrieved from http://jcmc.indiana.edu/vol13/issue1/boyd.ellison.htmldoi:10.1111/j.1083-6101.2007.00393.x

Bradbury, M. D., & Waechter, G. D. (2009). Extreme outsourcing in local government: At the top and all but the top. *Review of Public Personnel Administration*, *29*(3), 230–248. doi:10.1177/0734371X09332572

Brandt Management Group. (1999). *The complete guide: Listening practices feedback report—360 degrees*. Virginia: J. Brandt.

Brandt, R., Brandt, J., Emmert, P., & Emmert, V. (1992, March). *The development of the listening practices feedback report*. Paper presented at the 13th Annual Convention of the International Listening Association. Seattle, Washington

Brannick, T., de Búrca, S., Fynes, B., Roche, A., & Ennis, S. (2002). Service management practice performance model: A focus on training and listening practices. *Journal of European Industrial Training*, *26*(8), 394–403. doi:10.1108/03090590210444973

Bravard, J. L., & Morgan, R. (2006). *Smarter outsourcing: An executive guide to understanding, planning and exploiting successful outsourcing relationships*. Dorchester, MA: Pearson Education.

Bravo, M. I., Peiró, J. M., & Rodríguez, I. (1996). Satisfacción laboral. In J. M. Peiró & F. Prieto (Eds.), *Tratado de psicología del trabajo, vol. I: La actividad laboral en su contexto*. Madrid, Spain: Síntesis

Brenner, M. (2010). Group dynamics can enhance leadership development. In *SHRM's 2010 HR Trend book. Special supplement to HR Magazine* (pp. 2–33). Alexandria, VA: SHRM.

Breu, K., Hemingway, C., & Strathern, M. (2001). Workforce agility: The new employee strategy for the knowledge economy. *Journal of Information Technology*, *17*, 21–31. doi:10.1080/02683960110132070

Brewster, C., Gunnigle, P., & Morley, M. (1994). Continuity and change in European industrial relations: Evidence from a 14 country survey. *Personnel Review*, *23*(3), 4–20. doi:10.1108/00483489410064531

Brockhaus, R. H., & Horwitz, P. S. (1986). The psychology of the entrepreneur. In Sexton, D. L., & Smilor, W. (Eds.), *The art and science of entrepreneurship* (pp. 25–49). Cambridge, MA: Ballinguer Publishing Co.

Brookes, B. (2009). *Leadership development using online group executive coaching*. UMUC Working Paper Series. Number 2009-001. University of Maryland.

Brungardt, C. (1996). The making of leaders: A review of the research in leadership development and education. *The Journal of Leadership Studies*, *3*(3), 81–95. doi:10.1177/107179199700300309

Bruyat, C. (1993). *Création d'entreprise: Contribution epistémologique et modélisation*. Doctoral dissertation, École Superieure de Affaires, Université Pierre Mendès France, Grenobles, France.

Bruyat, C., & Julien, P. A. (1999). Defining the field of research in entrepreneurship. *Journal of Business Venturing*, *16*(2), 165–180. doi:10.1016/S0883-9026(99)00043-9

Buckingham, M., & Coffman, C. (1999). *First break all the rules: What the world's greatest managers do differently*. Sydney, Australia: Simon & Schuster.

Budros, A. (1999). A conceptual framework for analyzing why organizations downsize. *Organization Science*, *10*(1), 69–82. doi:10.1287/orsc.10.1.69

Bueno Campos, E. (1998). El capital intangible como clave estratégica en la competencia actual. *Boletín de Estudios Económicos*, *53*(164), 207–229.

Bueno, E. (1996). *Organización de empresas. Estructura, procesos y modelos*. Madrid, Spain: Pirámide.

Bunk, G. P. (1994). Teaching competence in initial and continuing vocational training in the Federal Republic of Germany. *Vocational Training European Journal*, *1*, 8–14.

Burgelman, R. (1994). Fading memories: A process theory of strategic business exit in dynamic business environments. *Administrative Science Quarterly, 39*, 24–56. doi:10.2307/2393493

Burke, M., & Day, R. (1986). A cumulative study of the effectiveness of managerial training. *The Journal of Applied Psychology, 71*(2), 232–245. doi:10.1037/0021-9010.71.2.232

Burke, W., & Litwin, G. (1992). A causal model of organizational performance and change. *Journal of Management, 18*(3), 532–545. doi:10.1177/014920639201800306

Burley-Allen, M. (1995). *Listening: The forgotten skill.* New York, NY: Wiley and Sons.

Burns, J. M. (1978). *Leadership.* Nueva York, NY: Harper & Row.

Burt, R. (1992). *Structural holes.* Cambridge, MA: Cambridge University Press.

Bygrave, W., & Hofer, C. (1991). Theorizing about entrepreneurship. *Entrepreneurship Theory and Practice, 16*(2), 13–22.

Byrne, C. L., Mumford, M. D., Barrett, J. D., & Vessey, W. B. (2009). Examining the leaders of creative efforts: What do they do, and what do they think about? *Creativity and Innovation Management, 18*(4), 256–268. doi:10.1111/j.1467-8691.2009.00532.x

Byrne, J. A. (1993). The horizontal corporation. *Business Week*, 76-81.

C. E. O. E. (2002). La gestión por competencias en España. *Informe para el Observatorio Europeo de Gestión por Competencias.* Retrieved from www.ceoe.es/webceoe/ceoe/obseu/informe.pdf

Cable, D. A. J. (2008). *The psychological contract: The development and validation of a managerial measure.* Unpublished Master Doctoral Thesis of University of Waikato.

Cabrera, A. (2002). La gestión del conocimiento. In Bonache, J., & Cabrera, A. (Eds.), *Dirección estratégica de personas: Evidencias y perspectivas para el siglo XXI.* Madrid, Spain: Prentice Hall.

Cadrain, D. (2010). A leave law that just won't go away. *HRMagazine*, (July): 49–52.

Cahavet, S. R. (1995). *Words that change minds.* Kendall/Hunt Publishing Company.

Caiyun, G., Zhiqiang, L., & Bing, L. (2008). *Research on human resource management outsourcing.* Wireless Communications, Networking and Mobile Computing, 2008. WiCOM '08. 4th International Conference. IEEE.

Calhoun, C. A. (1989). *BIVOPROB: A computer program for maximum-likelihood estimation of bivariate ordered-probit models for censored data.* Working Paper, 38, Laxenburg, Austria: International Institute for Applied Systems Analysis.

Calhoun, C. A. (1991). Desired and excess fertility in Europe and the United States: Indirect estimates from world fertility survey data. *European Journal of Population, 7*, 29–57. doi:10.1007/BF01796615

Calhoun, C. A. (1994). The impact of children on the labour supply of married women: Comparative estimates from European and U.S. Data. *European Journal of Population, 10*, 293–318. doi:10.1007/BF01266566

Caligiuri, P. M. (2006). Developing global leaders. *Human Resource Management Review, 16*, 219–228. doi:10.1016/j.hrmr.2006.03.009

Calvo Babío, N., & García Rodríguez, R. (2010). Talent management in professional services firms: A HR issue? *The International Journal of Organizational Analysis, 18*(4), 392–411. doi:10.1108/19348831011081877

Cameron, K. S. (1994). Strategies for successful organizacional downsizing. *Human Resource Management, 33*(2), 189–211. doi:10.1002/hrm.3930330204

Campbell, D. J. (2000). The proactive employee: Managing workplace initiative. *The Academy of Management Executive, 14*(3), 52–66. doi:10.5465/AME.2000.4468066

Camps, J., & Luna-Arocas, R. (2009). High involvement work practices and firm performance. *International Journal of Human Resource Management, 20*(5), 1056–1077. doi:10.1080/09585190902850273

Canós, L., Valdés, J., & Zaragoza, P. (2003). La gestión por competencias como pieza fundamental para la gestión del conocimiento. *Boletín de Estudios Económicos, 48*(180), 445–463.

Cantera, F. J., García-Morán, R., & Gómez, G. (1996). Ingeniería por competencias. *Capital Humano, 95*, 36–41.

Capelli, P., & Neumark, D. (2001). Do high-performance work practices improve establishment level outcomes? *Industrial & Labor Relations Review, 54*(4), 737–775. doi:10.2307/2696111

Capital Humano. (2007). *RPO o como rentabilizar los procesos de selección* (p. 215).

Capital Humano. (2008). *Outsourcing de RR.HH. Para reducir costes en tiempos de crisis*, (p. 226).

Capital Humano. (2009). *El outsourcing ayuda a afrontar la crisis al 33,1 por ciento de las empresas*, (p. 235).

Capital Humano. (2009). *Los servicios de outsourcing mantienen su crecimiento a pesar de la crisis* (p. 238).

Cappelli, P. (2001). Making the most of on-line recruiting. *Harvard Business Review, 79*(3), 139–146.

Cappelli, P., & Crocker-Hefter, A. (1996). Distinctive human resources are firm's core competencies. *Organizational Dynamics, 24*(3), 7–22. doi:10.1016/S0090-2616(96)90002-9

Cappellli, P. (2008). Talent management for the twenty-first century. *Harvard Business Review, 86*(3), 74, 76–81.

Carazo, J. A. (2007). *Bancaja fideliza a sus empleados a través del compromiso interno*. Capital Humano, No. 205. Wolters Kluwer, 2007.

Cardona, P. (2000). El liderazgo relacional. In *Raíces éticas del liderazgo*. Pamplona, Spain: Ediciones Universidad de Navarra.

Cardona, P., & García Lombardía, P. (2005). *Cómo desarrollar las competencias de liderazgo*. Pamplona, Spain: EUNSA.

Carison, D., Upton, N., & Seaman, P. (2006). The impact of human resources practices and compensation design performance. An analysis of family owned SMES. *Journal of Small Business Management, 44*(4), 531–543. doi:10.1111/j.1540-627X.2006.00188.x

Carland, J. A., & Carland, J. W. (1991). An empirical investigation into the distinctions between male and female entrepreneurs and managers. *International Small Business Journal, 9*(3), 62–72. doi:10.1177/026624269100900304

Carley, K. (1992). Organizational learning and personnel turnover. *Organization Science, 3*(1), 20–40. doi:10.1287/orsc.3.1.20

Carstensen, L. L., Fung, H. H., & Charles, S. T. (2003). Socioemotional selectivity theory and the regulation of emotion in the second half of life. *Motivation and Emotion, 27*(2), 103–123. doi:10.1023/A:1024569803230

Carstensen, L. L., Isaacowitz, D. M., & Charles, S. T. (1999). Taking time seriously: A theory of socioemotional selectivity. *The American Psychologist, 54*(3), 165–181. doi:10.1037/0003-066X.54.3.165

Carvin, B. N. (2009). *Effective exit-interview strategies*. Human Resource Executive Online, September. Retrieved from http://www.hreonline.com/HRE/story.jsp?storyId=250192808

Casado, J. M. (2003). *El valor de la persona*. Madrid, Spain: Prentice Hall.

Casado, J. M. (2009). *El compromiso del miedo. Capital Humano, No. 237*. Madrid, Spain: Wolters Kluwer.

Casado, J. M. (2011). *El mantra del talento*. Deusto.

Cascio, W. (1995). Whither industrial and organizational psychology in a changing world of work? *The American Psychologist, 50*, 928–939. doi:10.1037/0003-066X.50.11.928

Cassar, V., & Briner, R. B. (2005). Psychological contract "breach": A multiple component perspective to an over-researched construct? *Revista de Psicología Social, 25*(1), 125–136. doi:10.1174/0213474052871079

Casson, M., & Wadeson, N. (2007). The discovery of opportunities: Extending the economic theory of the entrepreneur. *Small Business Economics, 28*(4), 285–300. doi:10.1007/s11187-006-9037-7

CEDEFOP. (2006). *ICT skills certification in Europe. Dossier. European Center for the Development of Vocational Training*. Luxembourg: European Communities.

Celaya, J. (2008). *La empresa en la web 2.0: El impacto de las redes sociales en la estrategia empresarial*. Barcelona, Spain: Ediciones Gestión 2000.

Céspedes, J. J., Jerez, P., & Valle, R. (2005). Las prácticas de recursos humanos de alto rendimiento y la capacidad de aprendizaje organizativo: incidencia e implicaciones. *Cuadernos de Economía y Dirección de la Empresa*, *24*, 29–56.

Chadwick, C., & Cappelli, P. (1999). Alternatives to generic strategy typologies in strategic human resource management. In Wright, P., Dyer, L., Boudreau, J., & Milkovich, G. (Eds.), *Research in personnel and human resources management, supplement 4*. Greenwich, CT: JAI Press, Inc.

Chadwick, C., & Cappelli, P. (2002). *Functional or numerical flexibility? Which pays off for organizations?* Discussion Paper, Institute for Labor and Industrial Relations, University of Illinois.

Chambers, E., Handfield-Jones, H., Hanking, S., & Michaels, E. III (1998). Win the war for top talent. *Workforce*, December, 50-56.

Chaney, L. H., & Martin, J. S. (2007). *Intercultural business communication*. New Jersey: Pearson, Prentice Hall.

Chaston, I. (1998). Self-managed teams: Assessing the benefits for small service-sector firms. *British Journal of Management*, *9*, 1–12. doi:10.1111/1467-8551.00071

Chaudron, D. (1995). The "far out" success of teleworking. *Supervisory Management*, *40*(1), 6–9.

Checkland, P. (1999). *Soft systems methodology. A 30-year retrospective*. Chichester, UK: John Wiley.

Chell, E., Haworth, J. M., & Brearley, S. (1991). *The entrepreneurial personality: Concepts, cases and categories*. London, UK: Routledge.

Chen, Y. J. (2007). Relationships among service-orientations, job satisfaction, and organizational commitment in the international tourist hotel industry. *Journal of American Academy of Business*, *11*, 71–82.

Cherniss, C., & Goleman, D. (2001). *The emotionally intelligence workplace*. New York, NY: Wiley & Sons.

Chesbrough, H. (2006). *Open business models: How to drive in the new innovation landscape*. Boston, MA: Harvard Business School Press.

Chesbrough, H., Vanhaverbeke, W., & West, J. (Eds.). (2006). *Open innovation: Researching a new paradigm*. Oxford University Press.

Chiavenato, I. (2002). *Gestión del talento humano*. McGraw-Hill.

Chow, I. H. S. (1998). The impact of rules and regulations on workforce flexibility in Hong Kong. *International Journal of Human Resource Management*, *9*(3), 494–505. doi:10.1080/095851998341035

Chua, J. H., Chrisman, J. J., & Chang, E. P. C. (2004). Are family firms born or made? An exploratory investigation. *Family Business Review*, *17*(1), 37–54. doi:10.1111/j.1741-6248.2004.00002.x

Churh, A. H., & Bracken, D. W. (Eds.). (1997). Special issue: 360-degree feedback systems. *Group & Organization Management*, *22*(2).

CIPD. (2009). *HR outsourcing and the HR function: Threat or opportunity?* London, UK: Chartered Institute of Personnel and Development.

Clark, A. E. (2001). What really matters in a job? Hedonic measurement using quit data. *Labour Economics*, *8*, 223–242. doi:10.1016/S0927-5371(01)00031-8

Clark, A., Oswald, A., & Warr, P. (1996). Is job satisfaction u-shaped in age? *Journal of Occupational and Organizational Psychology*, *69*(1), 57–81. doi:10.1111/j.2044-8325.1996.tb00600.x

Click, R. L., & Duening, T. N. (2005). *Business process outsourcing: The competitive advantage*. London, UK: Wiley.

Cobo Romani, J. C. (2009). *Strategies to promote the development of e-competencies in the next generation of professionals: European and international trends*. ERC Centre on Skills, Knowledge and Organisational Performance, Cardiff and Oxford Universities. SKOPE Monograph No. 13 ISSN 1466-1535. Retrieved December 27, 2010 from http://www.cardiff.ac.uk/socsi/research/researchcentres/skope/publications/monographs/monograph13.pdf

Coe, N., & Hess, M. (2006). *Global production networks: Debates and challenges*. Presented at GPERG Workshop, University of Manchester.

Coggburn, J. D. (2007). Outsourcing human resources: The case of the Texas Health and Human Services Commission. *Review of Public Personnel Administration*, *27*(4), 315–335. doi:10.1177/0734371X07301976

Cohen, A. (1993). Work commitment in relation to withdrawal intentions and union effectiveness. *Journal of Business Research*, *26*, 75–90. doi:10.1016/0148-2963(93)90044-P

Cohen, A. (2003). *Multiple commitments at work: An integrative approach*. Hillsdale, New York: Lawrence Erlbaum.

Cohen, S. G., & Bailey, D. E. (1997). What makes teams work: Group effectiveness research from the shop floor to the executive suite. *Journal of Management*, *23*, 239–290. doi:10.1177/014920639702300303

Cohen, S. G., & Ledford, G. E. Jr. (1994). The effectiveness of self-managing teams: A quasi-experimental. *Human Relations*, *47*, 13–43. doi:10.1177/001872679404700102

Colbert, B. A. (2004). The complex resource-based view: Implications for theory and practice in strategic human resource management. *Academy of Management Review*, *29*(3), 341–358.

Coleman, A. (2006). Does e-recruitment work? *Director (Cincinnati, Ohio)*, *59*(12), 27.

Collier, J. (1995). The virtuous organization. *Business Ethics (Oxford, England)*, *4*(3), 143–149. doi:10.1111/j.1467-8608.1995.tb00245.x

Collins, C. J., & Clark, K. D. (2003). Strategic human resource practices, top management team social networks, and firm performance: The role of human resource practices in creating organizational competitive advantage. *Academy of Management Journal*, *46*(6), 740–751. doi:10.2307/30040665

Collins, C. J., & Smith, K. G. (2006). Knowledge exchange and combination: The role of human resource practices in the performance of high-technology firms. *Academy of Management Journal*, *49*, 544–560. doi:10.5465/AMJ.2006.21794671

Collins, D. (2002). *The effectiveness of managerial leadership development programs: a meta-analysis of studies from 1982-2001*. Dissertation Submitted to the Graduate Faculty of the Louisiana State University and Agricultural and Mechanical College in partial fulfillment of the requirements for the degree of Doctor of Philosophy in The School of Human Resource Education and Workforce Development, West Virginia University.

Collins, D., & Holton, E. (2004). The effectiveness of managerial leadership development: A meta-analysis of studies from 1982 to 2001. *Human Resource Development Quarterly*, *15*(2). doi:10.1002/hrdq.1099

Collins, D., Lowe, J., & Arnett, C. (2000). In Holton, E. F., & Naquin, S. S. (Eds.), *Developing high-performance leadership competency* (*Vol. 6*, pp. 18–46). Baton Rouge, LA: Academy of Human Resource Development.

Collins, J. (2001). *Good to great: Why some companies make the lead and other don't*. New York, NY: Harper Collins.

Collins, O. F., & Moore, D. G. (1979). *The organizational makers: A behavioral study of independent entrepreneurs*. New York, NY: Meredith.

Colvin, A., & Boswell, W. (2007). The problem of action and interest alignment: Beyond job requirements and incentive compensation. *Human Resource Management Review*, *17*(1), 38–51. doi:10.1016/j.hrmr.2006.11.003

Colvin, G. (2008). *Talent is overrated*. USA: Penguin Group.

Colvin, R. (2008). *Building expertise. Cognitive methods for training and performance improvement*. San Francisco, CA: Pfeiffer.

Combs, J., Liu, Y., Hall, A., & Ketchen, D. (2006). How much do high-performance work practices matter? A meta-analysis of their effects on organizational performance. *Personnel Psychology*, *59*(3), 501–528. doi:10.1111/j.1744-6570.2006.00045.x

Commeiras, N., & Fournier, C. (2001). Critical evaluation of Porter et al.'s organizational commitment questionnaire: Implications for researchers. *Journal of Personal Selling & Sales Management*, *21*, 239–245.

Conger, J. (1992). *Learning to lead: The art of transforming managers into leaders*. San Francisco, CA: Jossey Bass.

Connely, M., Gilbert, J., Zaccaro, S., Threlfall, K., Marks, M., & Mumford, M. (2000). Exploring the relationship of leadership skills and knowledge to leader performance. *The Leadership Quarterly*, *11*, 65–86. doi:10.1016/S1048-9843(99)00043-0

Conner, K. R. (1991). A historical comparison of resource-based theory and five schools of thought within industrial organization economics: Do we have a new theory of the firm? *Journal of Management*, *17*(1), 121–154. doi:10.1177/014920639101700109

Consulting Business Intelligence, S. R. I. (2003). *Quality and effectiveness of e-learning*. Learning on Demand.

Conway, N., & Briner, R. B. (2005). *Understanding psychological contracts at work -A critical evaluation of theory and research*. Oxford University Press. doi:10.1093/acprof:oso/9780199280643.001.0001

Cook, M. F. (1999). *Outsourcing human resources functions. Strategies for providing enhanced HR services at lower cost*. New York, NY: AMACOM.

Cooke, F. L., Shen, J., & Mcbride, A. (2005). Outsourcing HR as a competitive strategy? A literature review and an assessment of implications. *Human Resource Management*, *44*(4), 413–432. doi:10.1002/hrm.20082

Cooper, A., Woo, C., & Dunkelberg, W. (1989). Entrepreneurship and the initial size of firms. *Journal of Business Venturing*, *4*, 317–332. doi:10.1016/0883-9026(89)90004-9

Corbin, J., & Strauss, A. (2008). *Basics of qualitative research: Techniques and procedures for developing grounded theory* (3rd ed.). Thousand Oaks, CA: SAGE.

Cordery, J. (1989). Multi-skilling: A discussion of proposed benefits of new approaches to labour flexibility within enterprises. *Personnel Review*, *18*(3), 13–22. doi:10.1108/EUM0000000000772

Cordery, J. L., Mueller, W. S., & Smith, L. M. (1991). Attitudinal and behavioral effects of autonomous group working: A longitudinal field study. *Academy of Management Journal*, *34*(2), 464–476. doi:10.2307/256452

Cordery, J., Sevastos, P., Mueller, W., & Parker, S. (1993). Correlates of employee attitudes toward functional flexibility. *Human Relations*, *46*(6), 705–723. doi:10.1177/001872679304600602

Córdoba, A. (2004). *El reto de la gestión empresarial*. Barcelona, Spain: Deusto.

Corso, M., Giacobbe, A., & Martini, A. (2009). Rethinking knowledge management: The role of ICT and the rise of the virtual workspace. *International Journal of Learning and Intellectual Capital*, *6*(3), 272–292. doi:10.1504/IJLIC.2009.025045

Coughlin, E. C., & Lemke, C. (1999). *Professional competency continuum*. Retrieved form http://www.mff.org/pubs/ME159.pdf

Courtright, J. A., Fairhurst, G. T., & Rogers, L. E. (1989). Interaction patterns in organic and mechanistic systems. *Academy of Management Journal*, *32*(4), 773–802. doi:10.2307/256568

Covey, S. (1989). *The 7 habits of highly effective people*. New York, NY: Simon and Schuster Publishing.

Coyle, D. (2009). *The talent code*.

Coyle-Shapiro, J., & Kessler, I. (2000). Consequences of the psychological contract for the employment relationship: A large scale survey. *Journal of Management Studies*, *37*(7), 903–927. doi:10.1111/1467-6486.00210

Creagh, M., & Brewster, C. (1998). Identifying good practice in flexible working. *Employee Relations*, *20*(5), 490–503. doi:10.1108/01425459810238792

Credit Union Management. (2004, November 7). Many managers lack management skills. www. Cumanagement.org.

Cross, T., Bazron, B., Dennis, K., & Isaacs, M. (1989). *Towards a culturally competent system of care* (*Vol. 1*). Washington, DC: Georgetown University Child Development Center, CASSP Technical Asistance Center.

Csoka, L. S. (1995). A new employer-employee contract? *Employment Relations Today*, *22*(2), 21–31. doi:10.1002/ert.3910220204

Cubeiro, J. C. (2004). *En un lugar del talento*. Pearson Education.

Cullen, B. (2001). e-Recruiting is driving HR systems integration. *Strategic Finance*, *83*(1), 22–26.

Curral, L. A., Forrester, R. H., Dawson, J. F., & West, M. A. (2001). It's what you do and the way you do it: Team task, team size, and innovation-related group processes. *European Journal of Work and Organizational Psychology*, *10*, 187–204. doi:10.1080/13594320143000627

Currivan, D. (1999). The causal order of job satisfaction and organizational commitment in models of employee turnover. *Human Resource Management Review*, *9*, 495–524. doi:10.1016/S1053-4822(99)00031-5

Curry, J. P., Wakefield, D. S., Price, J. L., & Mueller, C. W. (1986). On the causal ordering of job satisfaction and organizational commitment. *Academy of Management Journal*, *29*, 847–858. doi:10.2307/255951

Cutcher-Gershenfeld, J. (1991). The impact on economic performance of a transformation in industrial relations. *Industrial & Labor Relations Review*, *44*, 241–260. doi:10.2307/2524806

D'annunzio-Green, N., & Francis, H. (2005). Human resource development and the psychological contract: Great expectations or false hopes? *Human Resource Development International*, *8*(3), 327–344. doi:10.1080/13678860500199725

Daft, R. (2000). *Management* (5th ed.). Fort Worth, TX: Dryden Press.

Dalmau, J. I., & Baixauli, J. J. (1989). *La productividad*. Valencia, Spain: Servicio de publicaciones de la Universidad Politécnica de Valencia.

Damanpour, F. (1998). Innovation type, radicalness, and the adoption process. *Communication Research*, *15*, 545–567. doi:10.1177/009365088015005003

Daniels, A. (1989). *Bringing out the best in people*. New York, NY: McGraw-Hill.

Darden, W. R., Hampton, R., & Howell, R. D. (1989). Career versus organizational commitment: Antecedents and consequences of retail salespeople's commitment. *Journal of Retailing*, *65*, 80–106.

Datta, D., Guthrie, J., & Wright, P. (2005). Human resource management and labor productivity: does Industry matter? *Academy of Management Journal*, *48*(1), 135–145. doi:10.5465/AMJ.2005.15993158

Davenport, T. H., Delong, D. W., & Beers, M. C. (1998). Successful knowledge management projects. *Sloan Management Review*, *39*(2), 43–57.

David, P., & Foray, D. (2003). Economic fundamentals of the knowledge society. *Policies Futures in Education*, *1*(1), 20–49. doi:10.2304/pfie.2003.1.1.7

Davis, F. D. (1989). Perceived usefulness, perceived ease of use, and user acceptance of Information Technology. *Management Information Systems Quarterly*, *13*(3), 319–340. doi:10.2307/249008

Davis, R., & Wong, D. (2007). Conceptualizing and measuring the optimal experience of the e-learning environment. *Decision Sciences Journal of Innovative Education*, *5*(1), 97–125. doi:10.1111/j.1540-4609.2007.00129.x

Davis-Blake, A., Broschak, J. P., & George, E. (2003). Happy together? How using non-standard workers affects exit, voice, and loyalty among standard employees. *Academy of Management Journal*, *46*, 475–485. doi:10.2307/30040639

Dawson, P., & Dobson, S. (2010). The influence of social pressure and nationality on individual decisions: Evidence from the behaviour of referees. *Journal of Economic Psychology*, *31*(2), 181–191. doi:10.1016/j.joep.2009.06.001

Day, D. (2001). Leadership development: a review in context. *The Leadership Quarterly*, *11*(4), 581–613. doi:10.1016/S1048-9843(00)00061-8

Day, D., & Halpin, S. (2001). *Leadership development: A review of industry best practices*. United States Technical report by Army Research Institute for the behavioural and Social Sciences, Alexandria.

De Andrés, E., & Primo, D. (2007). *Hacia una cultura de compromiso*. Capital Humano, No. 213. Wolters Kluwer, 2007.

de Ansonera, A. (1996). *15 pasos para la selección de personal con éxito: método e instrumentos*. Barcelona, Spain: Paidós.

De Benito, C. (2005). *Reflexiones sobre lealtad y compromiso en la empresa*. Wolters Kluwer. Capital Humano, No. 188. May, 2005.

De Cieri, H., & Kramar, R. (2005). *Human resource management in Australia: Strategy, people, performance* (2nd ed.). Macquarie Park, Australia: McGraw-Hill.

De Geus, A. P. (1997). The living company. *Harvard Business Review*, *75*(2), 51–59.

De Haro, J. (2005). El cuadro de mando competencial, una herramienta para el desarrollo de los recursos humanos en las organizaciones. *Revista del Centro de Estudios Financieros*, *6*(262), 245–280.

De Haro, J. (2009a). La evaluación de la formación, ¿un problema sin resolver? *Training and Development Digest*, *75*, 28–36.

De Haro, J. (2009b). ¿Podemos predecir el talento? *Revista del Centro de Estudios Financieros*, *320*, 137–170.

de Jong, J. P. J., & den Hartog, D. N. (2010). Measuring innovative work behaviour. *Creativity and Innovation Management*, *19*(1), 23–36. doi:10.1111/j.1467-8691.2010.00547.x

De Jong, J. P. J., Vanhaverbeke, T., Chalet, T., & Chesbrough, H. (2008). *Policies for open innovation: theory, framework and cases*. Helsinki, Finland: Research project funded by Vision Era-Net.

de la Florida, R. (2002). *The rise of the creative class*. New York, NY: Basic Books.

De la Rosa, M. D., & Carmona, A. (2010). Cómo afecta la relación del empleado con su líder al compromiso con la organización. *Universia Business Review*, *2*, 112–132.

De Miguel Fernández, E. (2001). *Introducción a la gestión (management)*. Valencia: Universidad Politécnica de Valencia.

De Rose, C., & Tichi, N. (2009). *Leadership judgement: Without it nothing else matters*. Leader to Leader Institute.

De Vos, A., Buyens, D., & Schalk, R. (2003). Psychological contract development during organizational socialization: Adaptation to reality and the role of reciprocity. *Journal of Organizational Behavior*, *24*(5), 537–559. doi:10.1002/job.205

Delery, J. E., & Doty, D. H. (1996). Modes of theorizing in strategic human resource management: Tests of universalistic, contingency, and configurational performance predictions. *Academy of Management Journal*, *39*(4), 802–836. doi:10.2307/256713

Delery, J. E., & Shaw, J. D. (2001). The strategic management of people in work organizations: Review, synthesis and extension. *Research in Personnel and Human Resources Management*, *20*, 165–197. doi:10.1016/S0742-7301(01)20003-6

Delmar, F., & Davidsson, P. (2000). Where do they come from? Prevalence and characteristics of nascent entrepreneurs. *Entrepreneurship & Regional Development*, *12*, 1–23. doi:10.1080/089856200283063

Delmotte, J., & Sels, L. (2008). HR outsourcing: threat or opportunity? *Personnel Review*, *37*(5), 543–563. doi:10.1108/00483480810891673

Denton, K. (2007). Corporate Intranets: How can they give a new meaning to training and development? *Development and Learning in Organizations*, *21*(6), 12–14. doi:10.1108/14777280710828576

DePree, M. (1992). *Leadership is an art*. New York, NY: Doubleday.

Dess, G. G., & Lumpkin, G. T. (2005). The role of entrepreneurial orientation in stimulating effective corporate entrepreneurship. *The Academy of Management Executive*, *19*(1), 147–156. doi:10.5465/AME.2005.15841975

Dess, G. G., & Pickens, J. C. (2000). Changing roles: Leadership in the 21st century. *Organizational Dynamics*, *28*, 18–34. doi:10.1016/S0090-2616(00)88447-8

Detlor, B. (2010). Information management. *International Journal of Information Management*, *30*(2), 103–108. doi:10.1016/j.ijinfomgt.2009.12.001

Devesa, J. E., Devesa, M., & Vidal, C. (1997). Viabilidad de los planes de prejubilación. *Dirección y Organización*, *18*, 5–13.

Devine, D. J., Clayton, L. D., Philips, J. L., Dunford, B. B., & Melner, B. S. (1999). Teams in organizations: Prevalence, characteristics, and effectiveness. *Small Group Research*, *30*, 678–711. doi:10.1177/104649649903000602

Diaz-Mayans, M. A., & Sánchez, R. (2004). Temporary employment and technical efficiency in Spain. *International Journal of Manpower*, *25*, 181–194. doi:10.1108/01437720410535981

DiBella, A. J., Nevis, E. C., & Gould, J. M. (1996). Understanding organizacional learning capability. *Journal of Management Studies*, *33*(3), 361–379. doi:10.1111/j.1467-6486.1996.tb00806.x

Dionne, P. (1996). The evaluation of training activities: A complex issue involving different stakes. *Human Resource Development Quarterly*, *7*(3), 279–286. doi:10.1002/hrdq.3920070308

Dirube, J. L. (1999). Competencias ¿impuestas o inducidas? *Training & Development Digest*, *17*, 40–42.

Dirube, J. L. (2004). *Un modelo de gestión por competencias. Lecciones aprendidas*. Barcelona, Spain. *Gestion*, ▪▪▪, 2000.

Dodgson, M. (1993). Organizational learning: A review of some literature. *Organization Studies*, *14*(3), 375–394. doi:10.1177/017084069301400303

Donovan-Wright, M. A. (2002). How to capitalize on competencies. *Workforce online*. Retrieved from www.workforce.com/archive/feature/23/31/50/index.php

Dotlich, D. L., & Noel, J. L. (1998). *Action learning: How the world's top companies are re-creating their leaders and themselves*. San Francisco, CA: Jossey-Bass Inc.

Dougherty, T. W., Bluedorn, A. C., & Keon, T. L. (1985). Precursors of employee turnover: A multiple-sample causal analysis. *Journal of Occupational Behaviour*, *6*, 259–271. doi:10.1002/job.4030060404

Drach-Zahavy, A., & Somech, A. (2001). Understanding team innovation: The role of team processes and structures. *Group Dynamics*, *5*(2), 111–123. doi:10.1037/1089-2699.5.2.111

Drath, W. (1998). Approaching the future of leadership development. In McCauley, C., Moxley, R. S., & Van Velsor, E. (Eds.), *Handbook of leadership development* (pp. 403–432). San Francisco, CA: Jossey-Bass Publishers.

Drath, W., & Palus, C. (1994). *Making common sense: Leadership as meaning making in a community of practice*. Greensboro, NC: Center for Creative Leadership.

Drazin, R., & Schoonhoven, C. B. (1996). Community population, and organizational effects on innovation: A multilevel perspective. *Academy of Management Journal*, *39*, 1065–1083. doi:10.2307/256992

Drew, S. (1994). Downsizing to improve strategic position. *Management Decision*, *32*(1), 4–11. doi:10.1108/00251749410050624

Drucker, P. (1968). *The age of discontinuity: Guidelines to our changing society*. New York, NY: Harper & Row.

Drucker, P. (1999). *Sobre la profesión del management*. Barcelona, Spain: Apóstrofe.

Drucker, P. (2002). *The effective executive*. New York, NY: Harper-Collins.

Drucker, P. (2005). *The effective executive in action*. Harper Collins.

Drucker, P. F. (1992). *Managing for the future: The 1990s and beyond*. New York, NY: Truman Talley Books/Dutton.

Druker, P. (1993). *The rise of knowledge society*. Boston, MA: Butterworth & Heinemann.

Ducharme, M. (2004). The cognitive-behavioral approach to executive coaching. *Consulting Psychology Journal: Practice and Research*, *56*(4), 214–224. doi:10.1037/1065-9293.56.4.214

Duffy, M. N. (2001). Outsourcing a 401(K) plan. *Journal of Accountancy*, *191*(5), 30–35.

Dunahee, M. H., & Wangler, L. A. (1974). The psychological contract: A conceptual structure for management/employee relations. *The Personnel Journal*, *53*(7), 518–530.

Dunham, R. B., Grube, J. A., & Castaneda, M. B. (1994). Organizational commitment: The utility of an integrative definition. *The Journal of Applied Psychology*, *79*, 370–380. doi:10.1037/0021-9010.79.3.370

Dweck, C. (2006). *Mindset, the new psychology of success*. New York, NY: Ballantine Books.

Dyer, L., & Ericksen, J. (2005). In pursuit of marketplace agility: Applying precepts of self-organizing systems to optimize human resource scalability. *Human Resource Management*, *44*(2), 183–188. doi:10.1002/hrm.20062

Dyer, L., & Shafer, R. A. (1999). From human resource strategy to organizational effectiveness: Lessons from research on organizational agility. In Wright, P., Dyer, L., Boudreau, J., & Milkovich, G. (Eds.), *Research in personnel and human resources management, supplement 4*. Greenwich, CT: JAI Press, Inc.

Dyer, L., & Shafer, R. A. (2002). Dynamic organizations: Achieving marketplace and organizational agility with people. In Peterson, R., & Mannix, E. (Eds.), *Understanding the dynamic organization*. Mahwah, NJ: Lawrence Erlbaum Associates.

Earley, P. C., & Ang, S. (2003). *Cultural intelligence: Individual interactions across cultures*. Palo Alto, CA: Stanford University Press.

Easterby-Smith, M., Crossan, M., & Nicolini, D. (2000). Organizational learning: Debates past, present and future. *Journal of Management Studies*, *37*(6), 783–796. doi:10.1111/1467-6486.00203

Eccles, R. G., & Nohria, N. (1991). *Beyond the hype: Rediscovering the essence of management*. Boston, MA: Harvard School Press.

Edgard, F., & Geare, A. (2005). HRM practice and employee attitudes: Different measures – Different results. *Personnel Review*, *34*, 534–549. doi:10.1108/00483480510612503

Eisenberger, R., Cummings, J., Armeli, S., & Lynch, P. (1997). Perceived organizational support, discretionary treatment, and job satisfaction. *The Journal of Applied Psychology*, *82*(5), 812–820. doi:10.1037/0021-9010.82.5.812

E-Learning 2.0. (2010). *Wikipedia, la enciclopedia libre*. Retrieved August 16, 2010, from http://en.wikipedia.org/wiki/E-learning#E-Learning_2.0

Emmert, P. (1996). President's perspective. *ILA Listening Post*, *56*, 2–3.

Engle, A. D., Mendenhall, M. E., Powers, R. L., & Stedham, Y. (2001). Conceptualizing the global competency cube: A transnational model of human resource. *Journal of European Industrial Training*, *25*(6), 346–353.

EPA. (4th Quarter2008). *Encuesta de Población Activa*. Instituto Nacional de Estadística (INE).

Ericsson, A. (2006). The influence of experience and deliberate practice on the development of superior expert performance. In Ericsson, A., Charness, N., Feltovich, P., & Hoffman, R. (Eds.), *The Cambridge handbook of expertise and expert performance*. New York, NY: Cambridge University Press.

Ericsson, A., & Charness, N. (1994). Expert performance: Its structure and acquisition. *The American Psychologist*, *49*, 725–747. doi:10.1037/0003-066X.49.8.725

Ericsson, A., & Smith, J. (Eds.). (1991). *Toward a general theory of expertise: Prospects and limits*. New York, NY: Cambridge University Press.

Ericsson, A., Charness, N., Feltovich, P., & Hoffman, R. (Eds.). (2006). *The Cambridge handbook of expertise and expert performance*. New York, NY: Cambridge University Press.

Ericsson, A., Krampe, R., & Tesch-Romer, C. (1993). The role of deliberate practice in the acquisition of expert performance. *Psychological Review*, *100*(3), 363–406. doi:10.1037/0033-295X.100.3.363

Ermel, L., & Bohl, D. (1997). Responding to a tight labor market: Using incentives to attract and retain talented workers. *Compensation and Benefits Review*, *29*(6), 25–29. doi:10.1177/088636879702900605

Eskildsen, J. K., Kristensen, K., & Westlund, A. H. (2002). Measuring employee assets. *Business Process Management Journal*, *10*(5).

Espasa-Calpe. (2005). *Diccionario de la lengua Española*. Madrid, Spain: Author.

Ettinger, A., Holton, V., & Blass, E. (2002). E-learner experiences: Key questions to ask when considering implementing e-learning. *Industrial and Comercial Training*, *38*, 143–147. doi:10.1108/00197850610659409

European Employee Index. (2002). *A benchmarking model for increasing the value of human capital*. Retrieved from http://www.ennovaconsulting.com/Default.aspx?ID=362

Evans, J. S. (1991). Strategic flexibility for high technology manoeuvres: A conceptual framework. *Journal of Management Studies*, *28*(1), 69–89. doi:10.1111/j.1467-6486.1991.tb00271.x

Fan, J. P. H., & Lang, L. H. P. (2000). The measurement of relatedness: An application to corporate diversification. *The Journal of Business, 73*, 629–660. doi:10.1086/209657

Fang, C. H., Kang, Y. C., Chen, G. L., & Cheng, Y. S. (2008). Study on personality traits and key competence of R&D staff an empirical analysis in Taiwan. *Proceedings of the 2008 IEEE IEEM.*

Farkas, A. J., & Tetrick, L. E. (1989). A three –wave longitudinal analysis of the causal ordering of satisfaction and commitment on turnover decisions. *The Journal of Applied Psychology, 74*, 855–868. doi:10.1037/0021-9010.74.6.855

Farndale, E., Scullion, H., & Sparrow, P. (2010). The role of the corporate HR function in global talent management. *Journal of World Business, 45*(2), 161–168. doi:10.1016/j.jwb.2009.09.012

Fay, D., Borrill, C., Amir, Z., Haward, R., & West, M. A. (2006). Getting the most out of multidisciplinary teams: A multi-sample study of team innovation in health care. *Journal of Occupational and Organizational Psychology, 79*, 553–567. doi:10.1348/096317905X72128

Feist, G. J., & Gorman, M. E. (1998). Psychology of science: Review and integration of a nascent discipline. *Review of General Psychology, 2*, 3–47. doi:10.1037/1089-2680.2.1.3

Fernández Díez de Lastra, R. (2001). *La formación on-line y sus mitos*. Boletín Learnet, Marzo 2001. Instituto Universitario Euroforum Escorial. Retrieved from http://euroforum.cicei.ulpgc.es/learnet/bolMar_01/boletin.htm

Fernández López, J. (2005). *Gestión por competencias. Un modelo estratégico para la gestión de recursos humanos.* Madrid, Spain: Prentice Hall.

Fernández Sánchez, J. A. (2011a). *Uso de estrategias y sistemas de información en recursos humanos*. EAE.

Fernández Sánchez, J. A. (2011b). *Procesos de contratación y reestructuración de plantillas en España.* EAE.

Fernández, E., Montes, J., & Vázquez, C. (1998). Los recursos intangibles como factores de competitividad de la empresa. *Dirección y Organización, 20*, 83–98.

Ferris, G. R., Arthur, M. M., Berkson, H. M., Kaplan, D. M., Harrell-Cook, G., & Frink, D. W. (1998). Toward a social context theory of the human resource management–organization effectiveness relationship. *Human Resource Management Review, 8*, 235–264. doi:10.1016/S1053-4822(98)90004-3

Ferris, G. R., Hochwarter, W. A., Buckley, M. R., Harrell-Cook, G., & Frink, D. D. (1999). Human resources management: Some new directions. *Journal of Management, 25*, 385–415. doi:10.1177/014920639902500306

Fiol, C. M., & Lyles, M. A. (1985). Organizational learning. *Academy of Management Review, 10*(4), 803–813.

Fitz-enz, J. (2009). Disruptive technology for human resources. *Employment Relations Today*, (Winter): 1–10. doi:10.1002/ert.20219

Flabbi, L. (2010). Gender discrimination estimation in a search model with matching and bargaining. *International Economic Review, 51*(3), 745–783. doi:10.1111/j.1468-2354.2010.00600.x

Fletcher, C., & Williams, R. (1996). Performance management, job satisfaction and organizational commitment. *British Journal of Management, 7*, 169–179. doi:10.1111/j.1467-8551.1996.tb00112.x

Forsyth, D. R. (1990). *Group dynamics* (2nd ed.). Pacific Grove, CA: Brooks/Cole.

Forsythe, S. (1997). Human factors in agile manufacturing: A brief overview with emphasis on communication and information infrastructure. *Human Factors and Ergonomics in Manufacturing, 7*(1), 3–10. doi:10.1002/(SICI)1520-6564(199724)7:1<3::AID-HFM1>3.0.CO;2-7

Forza, C. (1995). The impact of information systems on quality performance. *International Journal of Operations & Production Management, 15*(5), 66–83.

Frank, F. D., & Taylor, C. R. (2004). Talent management: Trends that will shape the future. *Human Resource Planning, 27*(1), 33–41.

Freeman, S. J. (1994). Organizational downsizing as convergence or reorientation: Implications for human resource management. *Human Resource Management, 33*(2), 213–238. doi:10.1002/hrm.3930330205

Frese, M., Kring, W., Soose, A., & Zempel, J. (1996). Personal initiative at work: Differences between East and West Germany. *Academy of Management Journal*, *39*(1), 37–63. doi:10.2307/256630

Friedli, T., Billiger, S., Kickuth, M., & Fleisch, E. (2004). Managing flexibility strategically: A case study on repositioning. *Proceedings of the Production and Operation Management Society, World Conference 2004,* Cancún (Mexico).

Friedman, T. (2005). *The world is flat: A brief history of the twenty-first century*. New York, CA: Farrar, Straus and Giroux.

Friedrich, A., Kabst, R., Weber, W., & Rodehuth, M. (1998). Functional flexibility: Merely reacting or acting strategically. *Employee Relations*, *20*(5), 504–523. doi:10.1108/01425459810238800

Fritz, M. B. W., Narasimhan, S., & Rhee, H. (1998). Communication and coordination in the virtual office. *Journal of Management Information Systems*, *14*(4), 7–28.

Frutos, B., Ruiz, M. A., & San Martín, R. (1998). Análisis factorial confirmatorio de las dimensiones del compromiso de la organización. *Psicológica (Valencia)*, *19*, 345–366.

Fulmer, R., & Warner, S. (1999). Leadership: Lessons from the best. *Training & Development*, *53*(3), 28–33.

Gaertner, S. (1999). Structural determinants of job satisfaction and organizational commitment in turnover models. *Human Resource Management Review*, *9*, 479–493. doi:10.1016/S1053-4822(99)00030-3

Gagné, F. (1991). Toward a differentiated model of giftedness and talent. In Colangelo, N., & Davis, G. A. (Eds.), *Handbook of gifted education*. Boston, MA: Allyn and Bacon.

Gagné, F. (2004). Transforming gifts into talents: The DMGT as a developmental theory. *High Ability Studies*, *15*(2). doi:10.1080/1359813042000314682

Gainey, T. W., & Klaas, B. S. (2005). Outsourcing relationships between firms and their training providers: The role of trust. *Human Resource Development Quarterly*, *16*(1), 7–25. doi:10.1002/hrdq.1121

Gakovic, A., & Tetrick, L. E. (2003). Psychological contract breach as a source of strain for employees. *Journal of Business and Psychology*, *18*(2), 235–246. doi:10.1023/A:1027301232116

Galagan, P. (2002). Mission e-possible: The CISCO e-leaarining story. In Rossett, A. (Ed.), *The ASTD e-learning handbook*. New York, NY: Mc Graw-Hill.

Galanaki, E., & Papalexandris, N. (2005). Outsourcing of human resource management services in Greece. *International Journal of Manpower*, *26*(4), 382–396. doi:10.1108/01437720510609564

Galanaki, E., & Papalexandris, N. (2007). Internationalization as a determining factor of HRM outsourcing. *International Journal of Human Resource Management*, *18*(8), 1557–1567. doi:10.1080/09585190701502687

Galanaki, E., Bourantas, D., & Papalexandris, N. (2008). A decision model for outsourcing training functions: Distinguishing between generic and firm-job-specific training content. *International Journal of Human Resource Management*, *19*(12), 2332–2351. doi:10.1080/09585190802479579

Gallardo, E. (2008). Evolución en el estudio y medida del compromiso organizativo: Problemáticas y soluciones. In Pindado, J., & Payne, G. (Eds.), *Estableciendo puentes en una economía global* (p. 17).

Galunic, D., & Rodan, S. (1998). Resource recombinations in the firm: Knowledge structures and the potential for Schumpeterian innovation. *Strategic Management Journal*, *19*, 1193–1201. doi:10.1002/(SICI)1097-0266(1998120)19:12<1193::AID-SMJ5>3.0.CO;2-F

Gamboa, E. J. (2010). Satisfacción laboral: Descripción teórica de sus determinantes. *Psicología Científica,* 1-7.

García del Junco, J., & Bras, J. M. (2008). Satisfacción profesional y compromiso organizativo: Un meta-análisis. *Revista Europea de Dirección y Economía de la Empresa*, *17*(3), 61–78.

Garcia Urbón, R. (2008). *El compromiso te hace libre: expectativa vs. compromiso*. Capital Humano, No. 232. Wolters Kluwer, 2008.

García, E., Garza, R., Sáenz, L., & Sepúlveda, L. (2002). *Formación de emprendedores*. México: CECSA.

Gartner, W. (1990). What are we talking about when we talk about entrepreneurship? *Journal of Business Venturing*, *5*(1), 15–29. doi:10.1016/0883-9026(90)90023-M

Garud, R., & Kotha, S. (1994). Using the brain as a metaphor to model flexible production systems. *Academy of Management Review*, *19*, 671–698.

Garud, R., & Shapira, Z. (1997). Aligning the residuals: Risk, return, responsibility and authority. In Shapira, Z. (Ed.), *Organizational decision making* (pp. 238–256). Cambridge, UK: Cambridge University Press.

Gasse, Y. (1982). Elaborations on the psychology of the entrepreneur. In Kent, C., Sexton, D., & Vesper, K. (Eds.), *Encyclopedia of entrepreneurship* (pp. 57–71). Englewood Cliffs, NJ: Prentice-Hall.

Gellaty, I. (1995). Individual and group determinants of employee absenteeism: Test of a causal model. *Journal of Organizational Behavior*, *16*, 469–485. doi:10.1002/job.4030160507

Geneen, H. (1984). *Managing*. New York, NY: Doubleday.

Gereffi, G., & Kaplinsky, R. (2001). The value of value chains. [Special Issue]. *IDS Bulletin*, *32*(3).

Gesteland, R. R. (2005). *Cross-cultural business behavior*. Liber, Denmark: Copenhagen Business School Press.

Ghemawat, P., & Del Sol, P. (1998). Commitment versus flexibility. *California Management Review*, *40*(4), 26–42.

Ghoshal, S. (2005). Bad management theories are destroying good management practices. *Academy of Management Learning & Education*, *4*(1), 75–91. doi:10.5465/AMLE.2005.16132558

Giacalone, R., & Beard, J. W. (1994). Impression management, diversity and international management. *The American Behavioral Scientist*, *37*(5), 621–636. doi:10.1177/0002764294037005004

Gibb, A. (1990). Enterpreneurship and intrapreneurship-exploring the differences. New findings and perspectives. In Donckels, R., & Miettinen, A. (Eds.), *Entrepreneurship*. Aldershot, UK: Gower Publishing Company Limited.

Gibbs, G. (2007). *Analyzing qualitative data*. London, UK: SAGE.

Gick, A., & Pallares, T. (1997). Las competencias: ¿Qué está pasando en Europa? *Capital Humano*, *99*, 43–48.

Giddens, A. (1990). *The consequences of modernity*. Stanford, CA: Stanford University Press.

Gil, F., & García, M. (1996). *Grupos en las organizaciones*. Madrid, Spain: Ediciones Pirámide.

Gilbert, T. (1978). *Human competence: Engineering worthy performance*. New York, NY: McGraw-Hill.

Gilley, K. M., & Rasheed, A. (2000). Making more by doing less: an analysis of outsourcing and its effects on firm performance. *Journal of Management*, *26*(4), 763–790.

Goffee, R. E., & Jones, G. (2007). Leading clever people. *Harvard Business Review*, *85*(3), 72–79.

Goldsmith, M., & Morgan, H. (2003). *Increasing leadership effectiveness: A review of leadership practices development of 8 large corporations*. Retrieved from http://www.personal.marshallgoldsmith.com/html/articulos/increasing.html

Goldsmith, M., Govindarajan, V., Kaye, B., & Albert, A. (2002). *The many facets of leadership*. New York, NY: Financial Times Prentice Hall.

Goldsmith, M., Lyons, L., & Freas, A. (Eds.). (2000). *Coaching for leadership*. San Francisco, CA: Jossey-Bass.

Goldstein, A., & Sorcher, M. (1974). *Changing supervisory behaviour*. New York, NY: Pergamon.

Goleman, D. (1998). What makes a leader? *Harvard Business Review*, *76*(6).

Goleman, D. (2000). An EI-based theory of performance. In Goleman, D., & Cherniss, C. (Eds.), *The emotionally intelligent workplace: How to select for, measure, and improve emotional intelligence in individuals, groups, and organizations*. San Francisco, CA: Jossey-Bass.

Goleman, D., Boyatzis, R., & Mckee, A. (2002). *Primal leadership: Realizing the power of emotional intelligence*. Boston, MA: Harvard Business School Press.

Gollwitzer, P., & Brandstäterr, V. (1997). Implementation intentions and effective goal pursuit. *Journal of Personality and Social Psychology*, *73*, 186–199. doi:10.1037/0022-3514.73.1.186

Gómez, S., & Martí, C. (2010). Externalización y expatriados. Proceso de expatriación. In *M. J. Pérez López (Coord.), Externalización de funciones de recursos humanos*. Madrid, Spain: Pearson.

Gómez-Mejía, L. R., Balkin, D. B., & Cardy, R. L. (1998). *Gestión de recursos humanos*. Madrid, Spain: Prentice Hall.

González-Roma, V. (2008). La innovación en los equipos de trabajo. *Papeles del Psicólogo*, *29*(1), 32–40.

Gooderham, P. N., & Nordhaug, O. (1997). Flexibility in Norwegian and UK firms: Competitive pressure and institutional embeddedness. *Employee Relations*, *19*(6), 568–580. doi:10.1108/01425459710367916

Gospel, H., & Sako, M. (2010). The unbundling of corporate functions: The evolution of shared services and outsourcing in human resource management. *Industrial and Corporate Change*, *19*(5). doi:10.1093/icc/dtq002

Goss, T. (1996). *The last word on power*. London, UK: Piatkus.

Gouswaard, A., Kraan, K. O., & Dhondt, S. (2001). *Flexibility in balance: Flexibility of labour and its consequences for employers and employees*. TNO-Report, R990440.

Govindarajan, V., & Fisher, F. (1990). Strategy, control systems, and resource sharing: Effects on business unit performance. *Academy of Management Journal*, *33*(2), 259–285. doi:10.2307/256325

Gracia, F. J., Silla, I., Peiró, J. M., & Fortes-Ferreira, L. (2007). The state of the psychological contract and its relation to employees' psychological health. *Psychology in Spain*, *11*(1), 33–41.

Grant, R. (1997). The knowledge-based view of the firm: implications for management practise. *Long Range Planning*, *30*(3), 450–454. doi:10.1016/S0024-6301(97)00025-3

Grant, R. (2001). Knowledge and organization. In Nonaka, I., & Teece, D. (Eds.), *Managing industrial knowledge: Creation, transfer and utilization*. London, UK: SAGE Publications Ltd.

Grant, R. M. (1996). Toward a knowledge-based theory of the firm. *Strategic Management Journal*, *17*, 109–122.

Grant, R. M. (2005). *Contemporary strategy analysis* (5th ed.). Blackwell Publishing.

Greene, W. H., & Hensher, D. A. (2009. *Modeling ordered choices*. New York, NY: Mimeo, Stern School of Business. Kirby, E. (1999). *Building interaction in online and distance education course*. TX: Technology& Teacher Education Conference.

Greer, C., Youngblood, S., & Gray, D. (1999). Human resource management outsourcing: The make or buy decision. *The Academy of Management Executive*, *13*(3), 85–96. doi:10.5465/AME.1999.2210317

Griggs, H. E., & Hyland, P. (2003). Strategic downsizing and learning organisations. *Journal of European Industrial Training*, *27*(2-4), 177–187. doi:10.1108/03090590310468994

Gubman, E. (1996). The gauntlet is down. *The Journal of Business Strategy*, (Nov/Dec): 33–35.

Guest, D. (2004). The psychology of employment relationship: An analysis based on the psychological contract. *Applied Psychology: An International Review*, *53*(4), 541–555. doi:10.1111/j.1464-0597.2004.00187.x

Guest, D. E. (1987). Human resource management and industrial relations. *Journal of Management Studies*, *24*(5), 503–521. doi:10.1111/j.1467-6486.1987.tb00460.x

Guest, D. E. (1998). Is the psychological contract worth taking seriously? *Journal of Organizational Behavior*, *19*(S1), 649–664. doi:10.1002/(SICI)1099-1379(1998)19:1+<649::AID-JOB970>3.0.CO;2-T

Guest, D. E., & Conway, N. (1999). Peering into the black hole: The downside of the new employment relations in the UK. *British Journal of Industrial Relations*, *37*(3), 367–389. doi:10.1111/1467-8543.00133

Guest, D. E., & Conway, N. (2002). Communicating the psychological contract: An employer perspective. *Human Resource Management Journal*, *12*(2), 22–38. doi:10.1111/j.1748-8583.2002.tb00062.x

Guillén, M. (2006). *Ética en las organizaciones. Construyendo confianza*. Madrid, Spain: Prentice-Hall (Pearson).

Gunnigle, P. (2012). (in press). Weathering the storm? Multi-national companies (MNCs), human resources management (HRM) and industrial relations through the global financial crisis. *International Journal of Manpower*.

Guo, C., Shim, J. P., & Otondo, R. (2010). Social network services in China: An integrated model of centrality, trust, and technology acceptance. *Journal of Global Information Technology Management*, *13*(2), 76–79.

Gupta, A. (2001). Talent management in crisis. *Mid-American Journal of Business*, *16*(2), 2.

Guth, W. D., & Ginsberg, A. (1990). Guest editors′ introduction: Corporate entrepreneurship. *Strategic Management Journal*, *11*, 5–15.

Guthrie, J. (2001). High involvement work practices, turnover and productivity: Evidence from New Zealand. *Academy of Management Journal*, *44*(1), 180–190. doi:10.2307/3069345

Guthrie, J. P., Spell, C. S., & Nyamori, R. O. (2002). Correlates and consequences of high involvement work practices: the role of competitive strategy. *International Journal of Human Resource Management*, *13*(1), 183–197. doi:10.1080/09585190110085071

Hagan, C. M., Konopaske, R., Bernardin, H. J., & Tyler, C. L. (2006). Predicting assessment center performance with 360-degree, top-down, and customer-based competency assessments. *Human Resource Management*, *45*(3), 357–390. doi:10.1002/hrm.20117

Haiyang, L., & Yan, Z. (2002). *Founding team comprehension and behavioral integration: Evidence from new technology ventures in china*. Academy of Management Proceedings.

Hall, L., & Torrington, D. (1998). *The human resource function. The dynamics of change and development*. London, UK: FT-Pitman Publishing.

Hall, S., & Scraton, P. (1981). Law, class and control. In Fitzgerald, M., McLennan, G., & Pawson, J. (Eds.), *Crime and society*. London, UK: RKP.

Hamel, G., & Prahalad, C. K. (1994). *Competing for the future*. Boston, MA: Harvard Business School Press.

Hamilton, C.-A. (1987). Telecommuting. *The Personnel Journal*, *66*, 91–101.

Harkins, P., Brown, S., & Sullivan, R. (1995). Shining new light on a growing trend. *HRMagazine*, *40*(12), 75–79.

Harless, J. (1989). *Job and analysis, design and development*. Newnan, GA: Harless Performance Guild.

Harper, S. C. (1991). *The McGraw-Hill guide to starting your own business*. New York, NY: McGraw-Hill.

Harris, S. G., Hirsfelch, R. R., Field, H. S., & Mossholder, K. W. (1993). Psychological attachment: Relationships with job characteristics, attitudes, and preferences for newcomer development. *Group & Organization Management*, *18*, 459–481. doi:10.1177/1059601193184005

Harrison, D. A., Newman, D. A., & Roth, P. L. (2006). How important are job attitudes? Meta-analytic comparisons of integrative behavioural. *Academy of Management Journal*, *49*, 305–325. doi:10.5465/AMJ.2006.20786077

Harrison, J. K., & Hubbard, R. (1998). Antecedents to organizational commitment among Mexican employees of a US firm in Mexico. *The Journal of Social Psychology*, *138*(5), 609–623. doi:10.1080/00224549809600416

Haslam, S. A., & Turner, J. C. (1995). Context-dependent variation in social stereotyping 3: Extremism as a self categorical basis for polarized judgement. *European Journal of Social Psychology*, *25*, 341–371. doi:10.1002/ejsp.2420250307

Hawkins, P., & Smith, N. (2006). *Coaching, mentoring and organizational consultancy*. Open University Press.

Hay Group. (2010). Study "best companies for leadership". In G. Fernandez (Ed.), *Entrevista con Mary Fontaine*, (pp. 3-5). En persona.

Hayek, F. A. (1985). Richard Cantillon. *The Journal of Libertarian Studies*, *7*(2), 217–247.

Hayton, J. C., & McEvoy, G. M. (2006). Competencies in practice: An interview with Hanneke C. Frese. *Human Resource Management*, *45*(3), 495–500. doi:10.1002/hrm.20126

Hedge, J. W., Borman, W. C., & Lammlein, S. E. (2006). *The aging workforce: Realities, myths, and implications for organizations*. Washington, DC: American Psychological Association. doi:10.1037/11325-000

Hedlund, J., Forsythe, G., Horvath, J., Williams, W., Snook, S., & Sterberg, R. (2003). Identifying and assessing tacit knowledge: Understanding the practical intelligence of military leaders. *The Leadership Quarterly*, *14*, 117–140. doi:10.1016/S1048-9843(03)00006-7

Heery, E., & Noon, M. (Eds.). (2001). *Dictionary of human resource management*. Oxford, UK: Oxford University Press.

Heijke, H., Meng, C., & Ris, C. (2003). Fitting to the job: The role of generic and vocational competencies in adjustment and performance. *Labour Economics*, *10*(2), 215–229. doi:10.1016/S0927-5371(03)00013-7

Helfat, C. E., & Raubitschek, R. S. (2000). Product sequencing: Co-evolution of knowledge, capabilities and products. *Strategic Management Journal*, *21*, 961–979. doi:10.1002/1097-0266(200010/11)21:10/11<961::AID-SMJ132>3.0.CO;2-E

Henderson, R. M., & Clark, K. B. (1990). Architectural innovation: The reconfiguration of existing product technologies and the failure of established firms. *Administrative Science Quarterly*, *35*, 9–30. doi:10.2307/2393549

Herath, T., & Kishore, R. (2007). Outsourcing success: Psychological contract perspective. In *AMCIS 2007 Proceedings*, Paper 58. Retrieved from http://aisel.aisnet.org/amcis2007/58

Hernández, P., & Méndez, I. (2005). La correlación del sesgo de selección en los análisis de corte transversal de discriminación salarial por sexo: Estudio comparativo en los países de la UE. *Estadística Española*, *47*(158), 179–214.

Herranz, A., & de la Vega, R. (1999). Las competencias: Pasado y presente. *Capital Humano*, *123*, 58–66.

Herrbach, O. (2006). A matter of feeling? The affective tone of organizational commitment and identification. *Journal of Organizational Behavior*, *27*, 629–643. doi:10.1002/job.362

Herriot, P. (1984). *Down from the ivory tower: Graduates and their jobs*. Chichester, UK: Wiley.

Herriot, P. (1989). Selection as a social process. In Smith, M., & Robertson, I. T. (Eds.), *Advances in selection and assessment* (pp. 171–187). Chichester, UK: Wiley.

Herriot, P., & Pemberton, C. (1995). A new deal for middle managers. *People Management*, *1*(12), 32–34.

Herriot, P., & Pemberton, C. (1997). Facilitating new deals. *Human Resource Management Journal*, *7*(1), 45–56. doi:10.1111/j.1748-8583.1997.tb00273.x

Herriot, P., Manning, W. E. G., & Kidd, J. M. (1997). The content of the psychological contract. *British Journal of Management*, *8*(2), 151–162. doi:10.1111/1467-8551.0047

Hersey, P., & Blanchard, K. H. (1982). *Management of organizational behaviour: Utilizing human resources* (4th ed.). Englewood Cliffs, NY: Prentice-Hall.

Hervás, J. (1999). El outsourcing se debe contratar sólo por razones estratégicas. *AEDIPE*, March, 23-25.

Hesketh, B., & Neal, A. (1999). Technology and performance. In Ilgen, D., & Pulakos, E. (Eds.), *The changing nature of performance. Implications for staffing, motivation, and development* (pp. 21–55). San Francisco, CA: Jossey-Bass.

Heskett, J., Sasser, E., & Schlesinger, L. (1997). *The service-profit-chain: How leading companies link profit to loyalty, satisfaction and value*. New York, NY: The Free Press.

Heslin, P., Vandewalle, D., & Latham, G. (2006). Keen to help? Managers' implicit person theories and their subsequent employee coaching. *Personnel Psychology*, *59*, 871–902. doi:10.1111/j.1744-6570.2006.00057.x

Hiltrop, J. M. (1995). The changing psychological contract: The human resource challenge of the 1990s. *European Management Journal*, *13*(3), 286–294. doi:10.1016/0263-2373(95)00019-H

Hiltrop, J. M. (1999). The quest for the best: Human resource practices to attract and retain talent. *European Management Journal*, *17*(4), 422–430. doi:10.1016/S0263-2373(99)00022-5

Hirsch, W. (2000). *Succession planning de-mystified, report 372*. Brighton: Institute of Employment Studies.

Hislop, D. (2003). Linking human resource management and knowledge management via commitment. A review and research agenda. *Employee Relations*, *25*(2), 182–202. doi:10.1108/01425450310456479

Hitt, M. A., Hoskisson, R. E., Johnson, R. A., & Moesel, D. D. (1996). The market for corporate control and firm innovation. *Academy of Management Journal, 39*, 1084–1196. doi:10.2307/256993

Hitt, M. A., Ireland, R. D., Camp, S. M., & Sexton, D. L. (2002). *Strategic entrepreneurship: Integrating entrepreneurial and strategic management perspectives*. Oxford, UK: Blackwell Publishers.

Hofstede, G. (1980). *Culture's consequences: International differences in work related values*. Beverly Hills, CA: Sage.

Hofstede, G. (2001). *Culture's consequences: Comparing values, behaviours, institutions and organizations across nations* (2nd ed.). London, UK: Sage Publications.

Hogan, R., & Kaiser, R. (2005). What we know about leadership. *Review of General Psychology, 9*(2), 169–180. doi:10.1037/1089-2680.9.2.169

Holland, J. L. (1985). *Making vocational choices: Theory of careers*. Englewood Cliffs, NJ: Prentice-Hall.

Holm, A. B. (2009, May). *Virtual HRM: A case of e-recruitment.* Paper presented at the HRIS 2009 – 3rd International Workshop on Human Resource Information Systems, Milan, Italy.

Holm, A. B., & Ulhøi, J. P. (2007, December). *From physical towards virtual organization of business activities*. Paper presented at the 2007 Annual Conference of The Danish Academy of Management, Aarhus, Denmark.

Holm, A. B., & Ulhøi, J. P. (2010). A framework for conceptualising virtual organising. *International Journal of Business and Systems Research, 4*(3), 293–311. doi:10.1504/IJBSR.2010.032952

Holton, E. III. (1996). The flawed four-level evaluation model. *Human Resource Development Quarterly, 7*(1), 5–21. doi:10.1002/hrdq.3920070103

Hom, P. W., & Griffeth, R. W. (1995). *Employee turnover*. Cincinnati, OH: South Western.

Hortal, P. (n.d.). *Personal blog*. Retrieved from http://pauhortal.net/blog/

Horton, S. (2000). Competency management in the British civil service. *International Journal of Public Sector Management, 13*(4), 354–368. doi:10.1108/09513550010350508

Horton, W., & Horton, K. (2003). *E-learning, tools and technologies*. Indianapolis, IN: Wiley.

Horwitz, F. M., & Townshend, M. (1993). Elements in participation teamwork and flexibility in South Africa. *International Journal of Human Resource Management, 4*(4), 917–932. doi:10.1080/09585199300000064

House, R., Hanges, P., Javidan, M., Dorfman, P., & Gupta, V. (Eds.). (2004). *Culture, leadership, and organizations: The Globe study of 62 societies*. Thousand Oaks, CA: Sage Publications, Inc.

House, R., Javidan, M., Hanges, P., & Dorfman, P. (2002). Understanding cultures and implicit leadership theories across the globe: An introduction to project GLOBE. *Journal of World Business, 37*(1), 3–10. doi:10.1016/S1090-9516(01)00069-4

Houtzagers, G., & Janssen, S. (2008). Shifting from HR outsourcing to business process sourcing. *IHRIM Journal, 7*(1), 30–42.

Huang, H. J., & Cullen, J. B. (2001). Labour flexibility and related HRM practices: A study of large Taiwanese manufacturers. *Canadian Journal of Administrative Sciences, 18*(1), 33–39. doi:10.1111/j.1936-4490.2001.tb00242.x

Huang, J.-H., Yangb, C., Jinb, B.-H., & Chiua, H. (2004). Measuring satisfaction with business-to-employee systems. *Computers in Human Behavior, 20*(1), 17–35. doi:10.1016/S0747-5632(03)00047-5

Hughes, T. (1987). The evolution of large technological systems. In Bijker, W., Hughes, T., & Pinch, T. (Eds.), *The social construction of technological systems* (pp. 51–82). Cambridge, MA: MIT Press.

Huiskamp, R. (1995). Regulating the employment relationship: An analytical framework. In Van Ruysseveldt, J., Huiskamp, R., & Van Hoof, J. (Eds.), *Comparative industrial and employment relations*. London, UK: Sage.

Hunt, J. G., Boal, K. B., & Dodge, G. E. (1999). The effects of visionary and crisis-responsive charisma on followers: An experimental examination of two kinds of charismatic leadership. *The Leadership Quarterly*, *10*, 423–448. doi:10.1016/S1048-9843(99)00027-2

Huselid, M. A. (1995). The impact of human resource management practices on turnover, productivity and corporate financial performance. *Academy of Management Journal*, *38*(3), 635–672. doi:10.2307/256741

Huselid, M. A., Jackson, S. E., & Schuler, R. S. (1997). Technical and strategic human resource management effectiveness as determinants of firm performance. *Academy of Management Journal*, *40*(1), 171–188. doi:10.2307/257025

Hutton, D., & Cummins, R. (1997). Development of the psychological contract inventory (PsyCon). *Australian Journal of Career Development*, *6*(3), 35–41.

Huws, U. (1984). New technology homeworkers. *Employment Gazette*, *92*, 13–17.

Hyatt, D. E., & Ruddy, T. M. (1997). An examination of the relationship between work group and performance: Once more into the breech. *Personnel Psychology*, *50*(3), 555. doi:10.1111/j.1744-6570.1997.tb00703.x

Hymowitz, C. &Schellhadt, T. (1986, March 24). The glass ceiling. *The Wall Street Journal: A Special Report, The Corporate Woman*, D1, 4-5.

Iacobucci, D. (2002). Explaining business groups started by habitual entrepreneurs in the Italian manufacturing sector. *Entrepreunership and Regional Development*, *14*(1), 31–48. doi:10.1080/08985620110096636

Ichniowski, C., Kochan, T. A., Levine, D., Olson, C., & Strauss, G. (1996). What works at work: overview and assessment. *Industrial Relations*, *35*, 352–332. doi:10.1111/j.1468-232X.1996.tb00409.x

Ichniowski, C., Shaw, K., & Prennushi, G. (1993). *The effects of human resource management practices on productivity.* Working paper, Graduate School of Industrial Administration, Carnegie Mellon University.

Ichniowski, C., Shaw, K., & Prennushi, G. (1997). The effects of human resources management practices on productivity: A study of steel finishing lines. *The American Economic Review*, *87*(3), 291–313.

Instituto de la Mujer. (2007). *Plan estratégico de igualdad de oportunidades* (2008-2011). Ley 3/2007 de Igualdad de Oportunidades entre hombres y mujeres.

Instituto de la Mujer. (2008). *Mujeres y hombres en España 2008*. Madrid.

Instituto de la Mujer. (2009). *Mujeres y hombres en España 2009*. Madrid.

Isaksen, S. G., Lauer, K. J., Ekwall, G., & Britz, A. (2001). Perceptions of the best and worst climates for creativity: Preliminary validation evidence for the situational outlook questionnaire. *Creativity Research Journal*, *13*, 171–184. doi:10.1207/S15326934CRJ1302_5

ISR (International Survey Research). (2004). *Survey design and implications for driving job satisfaction and employee commitment*. Case study 901. Retrieved from http://www.isrinsight.com/pdf/insight/CaseStudy901.pdf

Jackson, S. E., & Schuler, R. S. (1995). Understanding human resource management in the context of organizations and their environments. *Annual Review of Psychology*, *46*, 237–264. doi:10.1146/annurev.ps.46.020195.001321

Jackson, S. E., & Schuler, R. S. (2000). *Managing human resources: A partnership perspective 7e*. Cincinnati, OH: South-Western Publishing.

Jackson, S. E., Joshi, A., & Erhardt, N. L. (2003). Recent research on team and organizational diversity: SWOT analysis and implications. *Journal of Management*, *29*(6), 801–830.

Jackson, S. E., Mayo, K. E., & Whitney, K. (1995). Understanding the diversity of dynamics in decision making teams. In Guzzo, R. A., & Salas, E. (Eds.), *Team effectiveness and decision making in organizations* (pp. 204–261). San Francisco, CA: Jossey-Bass.

Jacobs, R. L., & Jones, M. J. (1995). *Structures on-the-job training: Unleashing employee expertise in the workplace*. San Francisco, CA: Berrett-Koehler.

Jaik, A., Tena, J. A., & Villanueva, R. (2010). Satisfacción laboral y compromiso institucional de los docentes de posgrado. *Revista Electrónica Diálogos Educativos*, *19*, 119–130.

Janssen, O., Van de Vliert, E., & West, M. (2004). The bright and dark sides of individual and group innovation: A special issue introduction. *Journal of Organizational Behavior*, *25*(2), 129–145. doi:10.1002/job.242

Janssens, M., Sels, L., & Van Den Brande, I. (2003). Multiple types of psychological contracts: A six-cluster solution. *Human Relations*, *56*(11), 1349–1378. doi:10.1177/00187267035611004

Jennings, T. (1994). *Multiples perspectives on entrepreneurship*. Cincinnati, OH: South-Western Publishing Co.

Jenssen, S., & Kolvereid, L. (1992). The entrepreneurs reasons leading to start-up as determinants of survival and failure among Norwegian new ventures. In Birley, S., & MacMillan, I. C. (Eds.), *International Perspectives on entrepreneurship research*. Londres, UK: Elsevier Science Publishers.

Jerico, P. (2000). *La gestión del talento: Del profesional con talento al talento organizativo*. Pearson Education.

Jericó, P. (2008). *La nueva gestión del talento. Construyendo compromiso*. Madrid, Spain: Prentice Hall.

Johnson, B. (1990). Toward a multidimensional model of entrepreneurship: The case of achievement motivation and the entrepreneur. *Entrepreneurship Theory and Practice*, *14*(3), 39–54.

Johnson, J. L., & O'Leary-Kelly, A. M. (2003). The effects of psychological contract breach and organizational cynicism: Not all social exchange violations are created equal. *Journal of Organizational Behavior*, *24*(5), 627–647. doi:10.1002/job.207

Johnson, J. L., Lee, R. P. W., Saini, A., & Grohmann, B. (2003). Market-focused strategic flexibility: Conceptual advances and an integrative model. *Journal of the Academy of Marketing Science*, *31*(1), 74–89. doi:10.1177/0092070302238603

Johnson, S. D., & Bechler, C. (1998). Examining the relationship between listening effectiveness and leadership emergence: Perceptions, behaviors, and recall. *Small Group Research*, *29*, 452–471. doi:10.1177/1046496498294003

Johnston, M. W., Parasuraman, A., Futrell, C. M., & Black, W. C. A. (1990). Longitudinal assessment of the impact of selected organizational influences on salespeople's organizational commitment during early employment. *JMR, Journal of Marketing Research*, *27*, 333–344. doi:10.2307/3172590

Joinson, C. (2002). Managing virtual teams: Keeping members on the same page without being in the same place poses challenges for managers workplace trends. *Human Resource Magazine*, *47*(6), 68–72.

Jones, G. R., & Wright, P. M. (1992). An economic approach to conceptualizing the utility of human resource management practices. In Rowland, K., & Ferris, G. (Eds.), *Research in personnel and human resources management* (pp. 271–299). Greenwich, CT: JAI Press.

Jones, W. (2008). *Keeping found things found: The study and practice of personal information management*. Burlington, MA: Morgan Kaufman.

Jonkeren, S. (2009). HR's time to shine: Challenging economy spells opportunity to demonstrate value. *IHRIM Journal*, *13*(1), 3–7.

Joras, M. (1996). Utiliser les bilans de compétences. In J. M. Peretty (Ed.), *Tout DRH* (pp. 165-172). Paris, France: Editions D'Organisation.

Juaneda, E., & González, L. (2007). Definición, antecedentes y consecuencias del compromiso organizativo. In Ayala Calvo, J. C. (Eds.), *Universidad de La Rioja: Conocimiento, innovación y emprendedores: Camino al futuro* (pp. 3590–3609).

Judge, T. A., Bono, J. E., Thoresen, C. J., & Patton, G. K. (2001). The job satisfaction-job performance relationship: A qualitative and quantitative review. *Psychological Bulletin*, *127*, 376–407. doi:10.1037/0033-2909.127.3.376

Judge, T., Bono, J., Llies, R., & Gerhardt, M. (2002). Personality and leadership: A qualitative and quantitative review. *The Journal of Applied Psychology*, *87*, 765–780. doi:10.1037/0021-9010.87.4.765

Kakabadse, A. P., & Kakabadse, N. (2000). Outsourcing: A paradigm shift. *Journal of Management Development*, *19*(8), 668–778. doi:10.1108/02621710010377508

Kakabadse, A. P., & Kakabadse, N. (2002). Trends in outsourcing: Contrasting USA and Europe. *European Management Journal*, *20*(2), 189–198. doi:10.1016/S0263-2373(02)00029-4

Kakabadse, A., & Kakabadse, N. (2005). Outsourcing: Current and future trends. *Thunderbird International Business Review*, *47*(2), 183–204. doi:10.1002/tie.20048

Kalleberg, A. L. (2001). Organizing flexibility: The flexible firm in a new century. *British Journal of Industrial Relations*, *39*(4), 479–504. doi:10.1111/1467-8543.00211

Kampa-Kokesch, S., & Anderson, M. (2001). Executive coaching: A comprehensive review of the literature. *Consulting Psychology Journal: Practice and Research*, *53*(4). doi:10.1037/1061-4087.53.4.205

Kanazawa, S. (2003). Why productivity fade with age: The crime-genius connection. *Journal of Research in Personality*, *37*(4), 257–272. doi:10.1016/S0092-6566(02)00538-X

Kanter, R. (1997). *Frontiers of management*. Boston, MA: Harvard Business Review Press.

Kanter, R. (1997). Restoring people to the heart of the organization of the future. In Hesselbein, F., Golesmith, M., & Beckhard, R. (Eds.), *The organization of the future*. San Francisco, CA: Jossey-Bass.

Kara, S., Kayis, B., & O'Kane, S. (2002). The role of human factors in flexibility management: A survey. *Human Factors and Ergonomics in Manufacturing*, *12*(1), 75–119. doi:10.1002/hfm.10004

Karuppan, C. M. (2004). Strategies to foster labor flexibility. *International Journal of Productivity and Performance Management*, *53*(6), 532–547. doi:10.1108/17410400410556192

Katz, D., & Kahn, R. L. (1966). *The social psychology of organizations*. New York, NY: John Wiley & Sons.

Katz, D., & Kahn, R. L. (1978). *The social psychology of organizations* (2nd ed.). New York, NY: Wiley.

Katz, J., & Miller, F. (1996). Coaching leaders through cultural change. *Consulting Psychology Journal: Practice and Research*, *48*, 104–114. doi:10.1037/1061-4087.48.2.104

Katzell, R. A. (1964). Personal values, job satisfaction and job behaviour. In Borow, H. (Ed.), *Man in a world at work*. Boston, MA: Houghton Mifflin.

Katzenbach, J. R., & Smith, D. K. (1993). *The wisdom of teams* (pp. 21, 45–48). Cambridge, MA: Harvard University Press.

Kayes, D. C., Kayes, A. B., & Yamazaki, Y. (2005a). Essential competencies for cross-cultural knowledge absorption. *Journal of Managerial Psychology*, *20*(7), 578–589. doi:10.1108/02683940510623399

Kayes, D. C., Kayes, A. B., & Yamazaki, Y. (2005b). Transferring knowledge across cultures: A learning competencies approach. *Performance Improvement Quality*, *18*(4), 87–100. doi:10.1111/j.1937-8327.2005.tb00352.x

Keeys, G. (1994). Effective leaders need to be good coaches. *Personnel Management*, *26*(11), 52–55.

Kelly, K. (2006). *Next 5000 days of Internet*. Retrieved May 3, 2010, from http://www.ted.com/index.php/talks/kevin_kelly_on_the_next_5_000_days_of_the_web.html

Kepner, C., & Tregoe, B. (1965). *The rational manager: A systematic approach to problem solving and decision making*. New York, NY: McGraw Hill.

Kichuk, S. L., & Weisner, W. H. (1997). The big five personality factors and team performance: Implications for selecting successful product design teams. *Journal of Engineering and Technology Management*, *14*(3-4), 195–222. doi:10.1016/S0923-4748(97)00010-6

Kiffin-Peteron, S. A., & Cordery, J. L. (2003). Trust, individualism, and job characteristics of employee reference for teamwork. *International Journal of Human Resource Management*, *14*(1), 93–116. doi:10.1080/0958519021015853 8

Kilburg, R. (1996). Toward a conceptual understanding and definition of executive coaching. *Consulting Psychology Journal: Practice and Research*, *48*(2). doi:10.1037/1061-4087.48.2.134

Kim, D. H. (1993). The link between individual and organizational learning. *Sloan Management Review*, *34*(1), 37–50.

Kim, S. W., Price, J. L., Mueller, C. W., & Watson, T. W. (1996). The determinants of career intent among physicians at a U.S. Air Force hospital. *Human Relations*, *49*, 947–975. doi:10.1177/001872679604900704

Kim, W. (2003). Economic crisis, downsizing and "layoff survivor's syndrome". *Journal of Contemporary Asia*, *33*(4), 449–464. doi:10.1080/00472330380000281

King, N., & Anderson, N. (2002). *Managing innovation and change: A critical guide for organizations*. London, UK: Thomson.

Kinlaw, D. C. (1991). *Developing superior work teams* (pp. 3–21). Lexington, MA: Lexington Books.

Kinnie, N., Purcell, J., Hutchinson, S., Terry, M., Collinson, M., & Scarbrough, H. (1999). Employment relations in SMEs: Market-driven or customer-shaped? *Employee Relations*, *21*(3), 218–236. doi:10.1108/01425459910273071

Kirkman, B. L., & Rosen, B. (1999). Beyond self-management: Antecedents and consequences of team empowerment. *Academy of Management Journal*, *42*(1), 58–74. doi:10.2307/256874

Kirkman, B. L., & Shapiro, D. L. (1997). The impact of cultural values on employee resistance to teams: toward a model of globalize self-managing work team effectiveness. *Academy of Management Review*, *22*(3), 730–757.

Kirkman, B. L., Gibson, C. B., & Shapiro, D. L. (2001). Exporting teams: Enhancing the implementation and effectiveness of work teams in global affiliates. *Organizational Dynamics*, *30*(1), 12–29. doi:10.1016/S0090-2616(01)00038-9

Kirkman, B. L., Rosen, B., Gibson, C. B., Tesluk, P. E., & McPherson, S. O. (2002). Five challenges to virtual team success: Lessons from Sabre, Inc. *The Academy of Management Executive*, *6*(3), 67. doi:10.5465/AME.2002.8540322

Klaas, B. S. (2008). Outsourcing and the HR function: An examination of trends and developments within North American firms. *International Journal of Human Resource Management*, *19*(8), 1500–1514. doi:10.1080/09585190802200280

Klaas, B. S., Klimchak, M., Semadeni, M., & Holmes, J. J. (2010). The adoption of human capital services by small and medium enterprises: A diffusion of innovation perspective. *Journal of Business Venturing*, *25*, 349–360. doi:10.1016/j.jbusvent.2008.12.002

Klaas, B. S., Mcclendon, J., & Gainey, T. (1999). HR outsourcing and its impact: The role of transaction costs. *Personnel Psychology*, *52*(1), 113–136. doi:10.1111/j.1744-6570.1999.tb01816.x

Klaas, B. S., Mcclendon, J., & Gainey, T. (2001). Outsourcing HR: The impact of organizational characteristics. *Human Resource Management*, *40*(2), 125–138. doi:10.1002/hrm.1003

Klaas, B. S., Yang, H., Gainey, T., & McClendon, J. A. (2005). HR in the small business enterprise: Assessing the impact of PEO utilization. *Human Resource Management*, *44*(4), 433–448. doi:10.1002/hrm.20083

Kleiner, M. M. (1990). The role of industrial relations in firm performance. In Fossum, J. A., & Mattson, J. (Eds.), *Employee and labour relations* (pp. 4.23–4.43). Washington, DC: BNA Press.

Kluytmans, F., & Ott, M. (1999). Management of employability in The Netherlands. *European Journal of Work and Organizational Psychology*, *8*(2), 261–272. doi:10.1080/135943299398357

Knowles, M. (1988). *Adult education. What works at work*. Minneapolis, MN: Lake- wood Books.

Koch, M., & McGrath, R. (1996). Improving labour productivity: Human resource management policies do matter. *Strategic Management Journal*, *17*, 335–354. doi:10.1002/(SICI)1097-0266(199605)17:5<335::AID-SMJ814>3.0.CO;2-R

Kochanski, J., & Ruse, D. (1996). Designing a competence-based human resources organization. *Human Resource Management*, *35*, 19–34. doi:10.1002/(SICI)1099-050X(199621)35:1<19::AID-HRM2>3.0.CO;2-#

Kofman, F. (2007). *Metamanagement*. Barcelona, Spain: Granica.

Kolb, D. A. (1984). *Experiential learning: Experience as the source of learning and development*. Englewood Cliffs, NJ: Prentice Hall.

Köllermeier, T. (1992). The East German transition. In Abell, D. F., & Köllermeier, T. (Eds.), *Dynamic entrepreneurship in Central and Eastern Europe* (pp. 65–86). Amsterdam, The Netherlands: European Foundation for Entrepreneurship Research.

Komaki, J. (1998). *Leadership from an operant perspective*. New York, NY: Routledge.

Kosnik, T., Wong-MingJi, J. D., & Hoover, K. (2006). Outsourcing vs insourcing in the human resource supply chain: A comparison of five generic models. *Personnel Review*, *35*(6), 671–683. doi:10.1108/00483480610702728

Kotey, B., & Sheridan, A. (2004). Changing HRM practices with firm growth. *Journal of Small Business and Enterprise Development*, *11*(4), 474–485. doi:10.1108/14626000410567125

Kotter, J. (1990). What leaders really do. *Harvard Business Review*, (May-June): 37–60.

Kotter, J. P. (1973). The psychological contract: Managing the joining-up process. *California Management Review*, *15*(3), 91–99.

Kouzes, J., & Posner, B. (1987). *Leadership challenge*. San Francisco, CA: Jossey-Bass.

Kouzes, J., & Posner, B. (1995). *The leadership challenge: How to keep getting extraordinary things done in organizations*. San Francisco, CA: Jossey-Bass.

Kouzes, J., & Posner, B. (2003). *Credibility: How leaders gain and lose it, why people demand it*. San Francisco, CA: Jossey-Bass.

Kozolowski, S. W. J., & Klein, K. J. (2000). A multilevel approach to theory and research in organizations: Contextual, temporal and emergent processes. In *Multilevel theory, research, and methods in organizations: Foundations, extensions, and new directions* (pp. 3–90). San Francisco, CA: Jossey-Bass.

Kram, K. (1986). Mentoring in the workplace. In Hall, D. T. (Ed.), *Career development in organizations*. San Francisco, CA: Jossey–Bass.

Kram, K., & Isabella, L. (1985). Alternatives to mentoring: The role of peer relationships in career development. *Academy of Management Journal*, *28*, 110–132. doi:10.2307/256064

Kramer, R. (1997). Leading by listening: An empirical test of Carl Rogers's theory of human relationship using interpersonal assessments of leaders by followers. *Dissertation Abstracts International Section A: Humanities & Social Sciences*, *58*, 514.

Kraut, R., Lundmark, V., Patterson, M., Kiessler, S., Mukopadhyay, T., & Sherlis, W. (1988). Internet paradox: A social technology that reduces social involvement and psychological well-being? *The American Psychologist*, *53*, 1017–1031. doi:10.1037/0003-066X.53.9.1017

Krueger, A. R. (1988). *El grupo de discusión. Guía práctica para la investigación aplicada*. Madrid, Spain: Ediciones Pirámide.

Kuchinke, K. (2000). The role of feedback in management training settings. *Human Resource Development Quarterly*, *11*, 381–401. doi:10.1002/1532-1096(200024)11:4<381::AID-HRDQ5>3.0.CO;2-3

Kumar, R., Novak, J., & Tomkins, A. (2006). *Structure and evolution of online social networks*. In ACM SIGKDD, August 20-23, Philadelphia.

Kurland, N. B., & Cooper, C. D. (2002). Manager control and employee isolation in telecommuting environments. *The Journal of High Technology Management Research*, *13*, 107–126. doi:10.1016/S1047-8310(01)00051-7

Kuvaas, B. (2008). An exploration of how the employee-organization relationship affects the linkage between perception of developmental human resource practices and employee outcomes. *Journal of Management Studies*, *45*(1), 1–25.

Kydd, C. T., & Oppenheim, L. (1990). Using human resource management to enhance competitiveness: Lessons from four excellent companies. *Human Resource Management*, *29*(2), 145–166. doi:10.1002/hrm.3930290203

Lado, A. A., & Wilson, M. C. (1994). Human resource systems and sustained competitive advantage: A competency-based perspective. *Academy of Management Review*, *19*(4), 699–727.

Lam, L., & White, L. (1998). Human resource orientation and corporate performance. *Human Resource Development Quarterly*, *9*(4), 351–364. doi:10.1002/hrdq.3920090406

Lambing, P., & Kuelh, Ch. (1997). *Entrepreneurship*. London, UK: Prentice-Hall International (UK) limited.

Lance, C. E. (1991). Evaluation of a structural model relating job satisfaction, organizational commitment, and precursors to voluntary turnover. *Multivariate Behavioral Research*, *26*, 137–162. doi:10.1207/s15327906mbr2601_7

Landauer, T. K. (1996). *The trouble with computers: Usefulness, usability, and productivity*. Cambridge, MA: MIT Press.

Landis, D., & Brislin, R. (1983). *Handbook of intercultural training* (*Vol. 1*). Elmsford, NY: Pergamon Press.

Landy, F. J., Barnes, J. L., & Murphy, K. R. (1978). Correlates of perceived fairness and accuracy of performance evaluation. *The Journal of Applied Psychology*, *63*(6), 751–754. doi:10.1037/0021-9010.63.6.751

Larrainzar, M. J., Miñarro, J., Molinos, F. J., & Verdú, M. A. (2001). La satisfacción laboral en el Hospital de Sant Cugat de Asepeyo. *Capital Humano*, *143*, 46–50.

Lau, R. S. M. (1999). Critical factors for achieving manufacturing flexibility. *International Journal of Operations & Production Management*, *19*(3), 328–341. doi:10.1108/01443579910249741

Lawler, E. E. (1992). *The ultimate advantage: Creating the high-involvement organization*. San Francisco, CA: Jossey- Bass.

Lawler, E. E. III. (2005). From human resource management to organizational effectiveness. *Human Resource Management*, *44*(2), 165–169. doi:10.1002/hrm.20059

Lawler, E. E., & Mohrman, S. A. (2003). *Creating a strategic human resources organisation*. Palo Alto, CA: Stanford University Press.

Lawler, E. E., Mohrman, S. A., & Ledford, G. (1995). *Creating high performance organizations: Practises and results of employee involvement and total quality management: Practice and results in Fortune 1000 companies*. San Francisco, CA: Jossey-Bass.

Lawler, E. J. (1992). Affective attachments to nested groups: A choice-process theory. *American Sociological Review*, *57*, 327–339. doi:10.2307/2096239

Lazear, E. P. (1998). *Personnel economics for managers*. Jonh Wiley & Sons, Inc.

Le Boterf, G. (2000). *Ingeniería de las competencias*. Barcelona, Spain: Ediciones Gestión.

Leal, A. (Coord.), Alfaro de Prado, A., Rodríguez, L., & Román, M. (1999). *El factor humano en las relaciones laborales*. Madrid, Spain: Pirámide.

LeBlanc, P., & Mclnerney, P. (1994). Need a change? Jump on the banding wagon. *The Personnel Journal*, *73*(1), 72–76.

Lee, C., & Bruvold, N. T. (2003). Creating value for employees: Investment in employee development. *International Journal of Human Resource Management*, *14*, 981–1000. doi:10.1080/0958519032000106173

Lee, I. (2005). The evolution of e-recruiting: A content analysis of fortune 100 career websites. *Journal of Electronic Commerce in Organizations*, *3*(3), 57–68. doi:10.4018/jeco.2005070104

Lee, J. (2005). Effects of leadership and leader-member exchange on commitment. *Leadership and Organization Development Journal*, *26*(8), 655–672. doi:10.1108/01437730510633728

Lee, J., & Miller, D. (1999). People matter: Commitment to employees, strategy and performance in Korean firms. *Strategic Management Journal*, *20*(6), 579–593. doi:10.1002/(SICI)1097-0266(199906)20:6<579::AID-SMJ37>3.0.CO;2-C

Lee, K. S., & Gao, T. (2005). Studying organizational commitment with the OCQ in the Korean retail context: Its dimensionality and relationships with satisfaction and work outcomes. *International Review of Retail, Distribution and Consumer Research*, *15*(4), 375–399. doi:10.1080/09593960500197495

Lee, Y., & Chang, H. (2008). Relations between team work and innovation in organizations and the job satisfaction of employees: A factor analytic study. *International Journal of Management*, *25*(3), 732–739.

Leite, E. (2000). *O Fenómeno do Empreendedorismo* (2nd ed.). Brasil: Edicoes Bagaco.

Lengnick-Hall, M. L., & Andrade, L. (2008). Talent staffing systems for effective knowledge management. In Vlaiman, V., & Vance, C. (Eds.), *Smart talent management: Building knowledge capital for competitive advantage*. Edward Elgar Publishing.

Leonard, D., & Sensiper, S. (1998). The role of tacit knowledge in group innovation. *California Management Review*, *40*(3), 112–131.

Leong, C. S., Furnham, A., & Cooper, C. L. (1996). The moderating effect of organizational commitment on the occupational stress outcome relationship. *Human Relations*, *49*, 1345–1363. doi:10.1177/001872679604901004

Lepak, D. P., & Snell, S. A. (1998). Virtual HR: Strategic human resource management in the 21st century. *Human Resource Management Review*, *8*(3), 215–234. doi:10.1016/S1053-4822(98)90003-1

Lepak, D. P., & Snell, S. A. (2002). Examining the human resource architecture: The relationships among human capital, employment, and human resource configurations. *Journal of Management*, *28*, 517–543.

Lepak, D. P., Liao, H., Chung, Y., & Harden, E. H. (2006). A conceptual review of human resource management systems in strategic human resource management research. In Martocchio, J. J. (Ed.), *Research in personnel and human resources management* (*Vol. 25*, pp. 217–271). Oxford, UK: Elsevier. doi:10.1016/S0742-7301(06)25006-0

Lepak, D., Takeuchi, R., & Snell, S. A. (2003). Employment flexibility and firm performance: Examining the interaction effects of employment mode, environmental dynamism, and technological intensity. *Journal of Management*, *29*(5), 681–703.

Lermusiaux, Y., & Snell, A. (2001). *Best practices for European 500 career web site recruiting*. Taleo Research.

Lever, S. (1997). An analysis of managerial motivations behind outsourcing practices in human resources. *Human Resource Planning*, *20*(2), 37–47.

Levie, J., & Muzyka, D. (1996). *New venture growth in France: Do local government policies make a difference? Frontiers of Entrepreneurship Research*. Wellesley, MA: Babson College, Babson Park.

Levinson, H. (1996). Executive coaching. *Consulting Psychology Journal*, *48*(2), 115–123. doi:10.1037/1061-4087.48.2.115

Levinson, H., Price, C., Munden, K., Mandl, H., & Solley, C. (1962). *Men management and mental health*. Cambridge, MA: Harvard University Press.

Levy–Leboyer, C. (1997). *Gestión de las competencias. Cómo analizarlas. Cómo evaluarlas. Cómo desarrollarlas,* 1st ed. Barcelona, Spain: Ediciones Gestión 2000.

Levy-Leboyer, C. (1997). *Gestión de las competencias: Cómo analizarlas, cómo evaluarlas, cómo desarrollarlas*. Barcelona, Spain: Ediciones Gestión 2000.

Levy-Leboyer, C. (2002). *La gestion des competences*. Paris, France: Editions d'Organisation.

Lewis, R. E., & Heckman, R. J. (2006). Talent management: A critical review. *Human Resource Management Review*, *16*, 139–154. doi:10.1016/j.hrmr.2006.03.001

Liaw, S. (2008). Investigating students′ perceived satisfaction, behavioral intention, and effectiveness of e-learning: A case of study of the blackboard system. *Computers & Education*, *51*, 864–873. doi:10.1016/j.compedu.2007.09.005

Libert, B., & Faulk, R. (2008). *Barack, Inc: Winning business lessons of the Obama campaign*. New York, NY: Prentice Hall.

Lichtenstein, B., Dooley, K. J., & Lumpkin, G. T. (2006). Measuring emergence in the dynamics of new venture creation. *Journal of Business Venturing*, *21*(2), 153–175. doi:10.1016/j.jbusvent.2005.04.002

Liden, R. C., Wayne, S. J., & Sparrowe, R. T. (2000). An examination of the mediating role of psychological empowerment on the relations between the job, interpersonal relationships and work outcomes. *The Journal of Applied Psychology*, *85*, 407–416. doi:10.1037/0021-9010.85.3.407

Liebeskind, J. P. (1996). Knowledge, strategy, and the theory of the firm. *Strategic Management Journal*, *17*, 93–107.

Lievens, F., & De Corte, W. (2008). Development and test of a model of external organizational commitment in human resources outsourcing. *Human Resource Management*, *47*(3), 559–579. doi:10.1002/hrm.20232

Lievens, F., & Harris, M. (2003). Research on internet recruiting and testing: Current status and future directions. In Cooper, C. L., & Robertson, I. T. (Eds.), *International review of industrial and organizational psychology* (*Vol. 16*, pp. 131–165). Chichester, UK: John Wiley & Sons, Ltd.

Limam, A. (2009). Enabling HR transformation in the European public sector: e-HR and outsourcing. *IHRIM Journal*, *13*(1), 41–48.

Lin, C. P., & Ma, H. C. (2004). Career commitment as a moderator of the relationship among procedural justice, perceived organizational support, organizational commitment, and turnover intentions. *Asia Pacific Management Review*, *9*, 519–538.

Lincoln, J., & Kalleberg, A. L. (1990). *Culture, control and commitment: A study of work organization and work orientations in the United States and Japan*. Cambridge, UK: Cambridge University Press.

Linz, S. J. (2003). Job satisfaction among Russian workers. *Journal of Manpower*, *24*, 626–645. doi:10.1108/01437720310496139

Lo, S., & Aryee, S. (2003). Psychological contract breach in a Chinese context: An integrative approach. *Journal of Management Studies*, *40*(4), 1005–1020. doi:10.1111/1467-6486.00368

Loasby, B. J. (2007). A cognitive perspective on entrepreneurship and the firm. *Journal of Management Studies*, *44*(7), 1078–1106. doi:10.1111/j.1467-6486.2007.00729.x

Locke, E. A. (1969). What is job satisfaction? *Organizational Behavior and Human Performance*, *4*, 309–336. doi:10.1016/0030-5073(69)90013-0

Logan, M. S., & Ganster, D. C. (2007). The effects of empowerment on attitudes and performance: The role of social support and empowerment beliefs. *Journal of Management Studies*, *44*(8), 1523–1550.

Lombardo, M., & Eicheinger, R. (1989*). Eighty eight assignments for development in place: Enhancing the development challenger of existing jobs*. Technical report, 136, Greensboro, NC. Center for creative leadership.

Lombardo, M., & Eichinger, R. (1997). The HR function that consists in creating leaders with advantage on competency. In Ulrich, D., Losey, M. R., & Lake, G. (Eds.), *Tomorrow's HR management*. Wiley and sons.

Lombardo, M., & Eichinger, R. (2002). *The leadership machine*. Minneapolis, MN: Lominger.

Longenecker, C., & Neubert, M. (2005). The practices of effective managerial coaches. *Business Horizons*, *48*, 493–500. doi:10.1016/j.bushor.2005.04.004

Looise, J. C., van Riemsdijk, M., & de Lange, F. (1998). Company labour flexibility strategies in The Netherlands: An institutional perspective. *Employee Relations*, *20*(5), 461–482. doi:10.1108/01425459810238747

Lookwood, N. R. (2007). *Leveraging employee engagement for competitive advantage: HR's strategic role*. Society for Human Resource Management Research Quarterly.

López Mena, L. (1989). *Intervención psicológica en la empresa*. Barcelona, Spain: Martinez Roca.

Lorenzo, R. (2006). ¿A qué se denomina talento? Estado del arte acerca de su conceptualización. *Intangible Capital*, *11*(2), 72–163.

Losey, M. R. (1999). Mastering the competencies of HR management. *Human Resource Management*, *38*(2), 99–102. doi:10.1002/(SICI)1099-050X(199922)38:2<99::AID-HRM3>3.0.CO;2-T

Lowe, K., Kroeck, K. G., & Sivasubrabramanian, N. (1996). Effective correlates of transformational and transactional leadership: A meta-analytic review. *The Leadership Quarterly*, *7*, 385–425. doi:10.1016/S1048-9843(96)90027-2

Lumpkin, G. T., & Dess, G. G. (2001). Linking two dimensions of entrepreneurial orientation to firm performance: The moderating role of environment and industry life cycle. *Journal of Business Venturing*, *16*(5), 429–451. doi:10.1016/S0883-9026(00)00048-3

Luna, R., & Morley, M. (2011). *Talent management and organizational performance: The mediator role of innovation*. Communication at the VIII International Workshop on Human Resource Management, 12-13th May, Sevilla (Spain).

Luna-Arocas, R., & Camps, J. (2008). A model of high performance work practices and turnover intentions. *Personnel Review*, *37*(1), 26–46. doi:10.1108/00483480810839950

Luna-Arocas, R., & Morley, M. (2010). *Strategic management and job performance: The mediating role of job satisfaction*. Communication at the 11th International Human Resource Management Conference, 9-12th June, Birmingham (UK).

Luthans, F., & Kreitner, R. (1985). *Organizational behavior modification and beyond: An operant and social learning-approach*. Glenview, IL: Scott & Foresman.

Lyham, S. (2000). Theory building in the human resource development profession. *Human Resource Development Quarterly*, *11*(2), 159–178. doi:10.1002/1532-1096(200022)11:2<159::AID-HRDQ5>3.0.CO;2-E

Lyness, K. S., & Thompson, D. E. (1997). Above the glass ceiling? A comparison of matched samples of female and male executives. *The Journal of Applied Psychology*, *82*, 359–375. doi:10.1037/0021-9010.82.3.359

Lyon, D. W., Lumpkin, G. T., & Dess, G. G. (2000). Enhancing entrepreneurial orientation research: Operationalizing and measuring a key strategic decision making process. *Journal of Management*, *26*(5), 1055–1085. doi:10.1177/014920630002600503

Macaulay, M., & Lawton, A. (2006). From virtue to competence: Changing the principles of public service. *Public Administration Review*, (September- October): 702–710. doi:10.1111/j.1540-6210.2006.00635.x

MacDuffie, J. (1995). Human resource bundles and manufacturing performance: Organizational logic and flexible production systems in the world auto industry. *Industrial & Labor Relations Review*, *48*, 197–221. doi:10.2307/2524483

Makin, P., & Cox, C. (2004). *Changing behaviour at work: A practical guide*. Oxfordshire, UK: Routledge.

Malhotra, A., Majchrzak, A., & Rosen, B. (2007). Leading virtual teams. *The Academy of Management Perspectives*, (February): 60–70. doi:10.5465/AMP.2007.24286164

Malhotra, Y. (1996). *Organizational learning and learning organizations: An overview* [WWW document]. Retrieved May 1, 2011, from http://www.brint.com/papers/orglrng.htm

Malone, T. W., & Crowston, K. (1994). The interdisciplinary study of coordination. *ACM Computing Surveys*, *26*(1). doi:10.1145/174666.174668

Man, D. C., & Lam, S. K. S. (2003). The effects of job complexity and autonomy on cohesiveness in collectivistic and individualistic work groups: A cross-cultural analysis. *Journal of Organizational Behavior*, *24*, 979–1001. doi:10.1002/job.227

Management, H. R. (2006). Critical issues in HR drive 2006 priorities: N1 in talent management. *HRFocus*, *83*(1).

Mansfield, R. S. (1996). Building competency models: Approaches for HR professionals. *Human Resource Management. Special Issue on Human Resource Competencies*, *35*(1), 7–18.

March, J. G., & Simon, H. A. (1958). *Organizations*. New York, NY: Wiley.

Marchetti, C., & Wilson, A. (2006). Outsourcing: Letting go to gain control. *CriticalEYE*, *15*, 1–6.

Marchington, M., & Grugulis, I. (2000). Best practice human resource management: Perfect opportunity or dangerous illusion? *International Journal of Human Resource Management*, *11*(6), 1104–1121. doi:10.1080/09585190050177184

Marr, R., & García, S. (1997). *La dirección corporativa de los recursos humanos*. Madrid, Spain: Díaz de Santos.

Marsden, D. (2004). The 'network economy' and models of the employment contract. *British Journal of Industrial Relations*, *42*(4), 659–684. doi:10.1111/j.1467-8543.2004.00335.x

Martí, N. (2008). *High commitment human resource practices in Spanish call centres. Nature, Sources and effects on voluntary turnover*. Thesis Doctoral Universitat Rovira i Virgili.

Martin, G., & Pear, J. (1996). *Behaviour modification: What it is and how to do it* (5th ed.). Englewood Cliffs, NJ: Prentice Hall.

Martínez-León, I. M., Ruiz Mercader, J., & Martínez-León, J. A. (2008). The effect of organisational learning tools on business results. *International Journal of Knowledge and Learning*, *4*(6), 539–552. doi:10.1504/IJKL.2008.022887

Mathews, J. A. (2002). A resource-based view of Schumpeterian economic dynamics. *Journal of Evolutionary Economics*, *12*, 29–54. doi:10.1007/s00191-002-0106-z

Mathieu, J. E. (1991). A cross-level nonrecursive model of the antecedents of organizational commitment and satisfaction. *The Journal of Applied Psychology*, *76*, 607–618. doi:10.1037/0021-9010.76.5.607

Mathieu, J. E., & Zajac, D. M. (1990). A review and meta-analysis of the antecedents, correlates, and consequences of organizational commitment. *Psychological Bulletin*, *108*, 171–194. doi:10.1037/0033-2909.108.2.171

Matveev, A. V., & Nelson, P. E. (2004). Cross cultural communication competence and multicultural team performance. *International Journal of Cross Cultural Management*, *4*(2), 253–270. doi:10.1177/1470595804044752

Maurer, R., & Mobley, N. (1998). Outsourcing: Is it the HR department of the future? *HRFocus*, (November): 9–10.

Maurer, T. J., Wrenn, K. A., Pierce, H. R., Tross, S. A., & Collins, W. C. (2003). Beliefs about "improvability" of career-relevant skills: Relevance to job/task analysis, competency modelling, and learning orientation. *Journal of Organizational Behavior*, *24*, 107–131. doi:10.1002/job.182

Mayer, J., & Caruso, D. (2002). *The effective leader: Understanding and applying emotional intelligence*. The University of Ontario Ivey Business Journal, November/December.

Mayer, J., Salovey, P., & Caruso, D. (2000b). Emotional Intelligence. In *R. J. Sternberg (2000), Handbook of Intelligence* (pp. 396–421). New York: Cambridge University Press.

Mayne, L., Tregaskis, O., & Brewster, C. (1996). A comparative analysis of the link between flexibility and HRM strategy. *Employee Relations*, *18*(3), 5–24. doi:10.1108/01425459610116447

McCall, M. (1998). *High flyers: Developing the next generation of leaders*. Boston, MA: Harvard Business School.

McCall, M., Lombardo, M., & Morrison, A. (1988). *The lessons of experience: How successful executives develop on the job*. New York, NY: The Free Press.

McCauley, C., & Brutus, S. (1998). *Management development through job experiences*. Greensboro, NC: Center for Creative Leadership.

McCauley, C., & Hughes-James, M. (1994). *An evaluation of the outcomes of a leadership development program*. Greensboro, NC: Center for Creative leadership.

McCauley, C., & Wakefield, M. (2006). Talent management in the 21st century. *Journal for Quality and Participation*, *29*(4).

McCauley, C., Rudermar, M., Ohlott, P., & Morrow, J. (1994). Assessing the developmental components of managerial jobs. *The Journal of Applied Psychology*, *79*, 544–560. doi:10.1037/0021-9010.79.4.544

McClelland, D. (1961). *The achieving society*. D Priceton van Nostrand.

McClelland, D. (1973). Testing for competencies rather than intelligence. *The American Psychologist*, *28*(1), 1–14. doi:10.1037/h0034092

McClelland, D. (1985). How motives, skills and values determine what people do. *The American Psychologist*, *40*, 812–825. doi:10.1037/0003-066X.40.7.812

McClelland, J. L. (1989). Parallel distributed processing: Implications for cognition and development. In Morris, R. (Ed.), *Parallel distributed processing: Implications for psychology and neurobiology* (pp. 8–45). New York, NY: Oxford University Press.

McGee, G. W., & Ford, R. C. (1987). Two (or more) dimensions of organizational commitement: Re-examination of the affective and continuance commitment scales. *The Journal of Applied Psychology*, *72*, 638–642. doi:10.1037/0021-9010.72.4.638

McGill, M. E., Slocum, J. W., & Lei, D. (1992). Management practices in learning organizations. *Organizational Dynamics*, *21*(Summer), 5–17. doi:10.1016/0090-2616(92)90082-X

McHugh, D., Groves, D., & Alker, A. (1998). Managing learning: What do we learn from a learning organization? *The Learning Organization*, *5*(5), 209–220. doi:10.1108/09696479810238215

McKewan, E. (2003). *Ten traits of highly effective principals*. Thousand Oaks, CA: Corwin Press.

McKinley, W., Sánchez, C. M., & Schick, A. G. (1995). Organizational downsizing: Constraining, cloning, learning. *The Academy of Management Executive*, *9*(3), 32–44. doi:10.5465/AME.1995.9509210276

McLagan, P. (2008). Competencies and the changing world of work. In Biech, E. (Ed.), *ASTD handbook for workplace learning professionals*. Virginia: ASTD Press.

Mclean Parks, J., Kidder, D. L., & Gallagher, D. G. (1998). Fitting square pegs into round holes: Mapping the domain of contingent work arrangements onto the psychological contract. *Journal of Organizational Behavior*, *19*(S1), 697–730. doi:10.1002/(SICI)1099-1379(1998)19:1+<697::AID-JOB974>3.0.CO;2-I

Mcnaughton, R. B., Quickenden, P., Matear, S., & Gray, B. (1999). Intranet adoption and inter-functional co-ordination. *Journal of Marketing Management*, *15*(5), 387–403. doi:10.1362/026725799784870270

McPherson, A., Homan, G., & Wilkinson, K. (2005). The implementation and use of e-learning in the corporate university. *Journal of Workplace Learning*, *17*, 33–48. doi:10.1108/13665620510574441

Meisel, S., & Fearon, D. (1996). Leading learning. In Calvieri, S., & Fearon, D. (Eds.), *Managing in organisations that learn* (pp. 180–209). Cambridge, MA: Blackwell.

Menchero, A., & Vale, V. (2002). Técnicas para cerrar el grifo. *Revista Emprendedores*, *55*, 60–66.

Mendelson, H. (2000). Organizational architecture and success in the IT industry. *Management Science*, *46*, 513–529. doi:10.1287/mnsc.46.4.513.12060

Mendenhall, M., & Oddou, G. (1985). The dimensions of expatriate acculturation: A review. *Academy of Management Review*, *10*(1), 39–47.

Mendenhall, M., & Oddou, G. (1986). Acculturation profiles of expatriate managers: Implications for cross-cultural training programs. *The Columbia Journal of World Business*, *21*(4), 73–79.

Mendoza, M. I., Orgambídez, A., & Carrasco, A. M. (2010). Orientación a la calidad total, satisfacción laboral, comunicación y compromiso en establecimientos de turismo rural de la Sierra de Huelva. *Pasos. Revista de Turismo y Patrimonio Cultural*, *8*(2), 351–361.

Menkes, J. (2005). *Executive intelligence*. New York, NY: HarperCollins Publisher.

Meso, P., & Smith, R. (2000). A resource-based view of organisational knowledge management systems. *Journal of Knowledge Management*, *4*(3), 224–234. doi:10.1108/13673270010350020

Messmer, M. (2004). Closing the communication gap. In *Handbook of business strategy*, (pp. 87-92).

Meyer, J. P., & Allen, N. J. (1991). A three component conceptualization or organizational commitment. *Human Resource Management Review*, *1*, 61–89. doi:10.1016/1053-4822(91)90011-Z

Meyer, J. P., & Herscovich, L. (2001). Commitment in the workplace: toward a general model. *Human Resource Management Review*, *11*, 299–326. doi:10.1016/S1053-4822(00)00053-X

Meyer, J. P., Stanley, D. J., Herscovitch, L., & Topolnytsky, L. (2002). Affective, continuance and normative commitment to the organization: A meta-analysis of antecedents, correlates and consequences. *Journal of Vocational Behavior*, *61*, 20–52. doi:10.1006/jvbe.2001.1842

Michaels, E., Handfield, H., & Axelrod, B. (2001). *The war for talent*. Boston, MA: Harvard Business School Press.

Michalos, A. C. (1986). Job satisfaction, marital satisfaction and the quality of life: A review and preview. In F. M Andrews (Ed.), *Research on the quality of life*. Ann Arbor, MI: Survey Research Center Institute.

Michie, J., & Sheehan-Quinn, M. (2001). Labour market flexibility, human resource management and corporate performance. *British Journal of Management*, *12*, 287–306. doi:10.1111/1467-8551.00211

Miles, M. B., & Huberman, M. A. (1994). *Qualitative data analysis* (2nd ed.). Thousand Oaks, CA: SAGE.

Minton-Eversole, T. (2007). E-recruitment comes of age, survey says. *HRMagazine*, *52*(8), 34–34.

Mintzberg, H. (1973). *The nature of management work*. New York, NY: Harper & Row.

Mintzberg, H. (1979). *The structuring of organizations*. Englewood Cliffs, NJ: Prentice-Hall.

Mise, I. J. (2001). Payroll and HR outsourcing. *CMA Management*, *75*(1), 30–32.

Mohr, R. D., & Zoghi, C. (2008). The high-involvement work design and job satisfaction. *Industrial & Labor Relations Review*, *61*(3), 275–296.

Molleman, E., & Slomp, J. (1999). Functional flexibility and team performance. *International Journal of Production Research*, *37*(8), 1837–1858. doi:10.1080/002075499191021

Montes, J. L., Moreno, A. R., & Morales, V. G. (2005). Influence of support leadership and teamwork cohesion on organizational learning, innovations and performance: An empirical examination. *Technovation*, *25*, 1159–1172. doi:10.1016/j.technovation.2004.05.002

Moorman, R. H., Blakely, G. L., & Niehoff, B. P. (1998). Does perceived organizational support mediate the relationship between procedural justice and organizational citizenship behaviour? *Academy of Management Journal*, *41*(3), 351–357. doi:10.2307/256913

Moreno, G., Rodríguez, J. M., & Vera, J. (1996). *La participación laboral femenina y la discriminación salarial en España*. Madrid, Spain: CEC-Social.

Morris, D. (2010). Are teachers technophobes? Investigating professional competency in the use of ICT to support teaching and learning. *Procedia - Social and Behavioral Science, 2*(2), 4010-4015.

Morrison, A. M., & Von Glinow, M. A. (1990). Women and minorities in management. *The American Psychologist*, *45*(2), 200–208. doi:10.1037/0003-066X.45.2.200

Morrison, E. W., & Phelps, C. C. (1999). Taking charge at work: Extra-role efforts to initiate workplace change. *Academy of Management Journal*, *42*(4), 403–419. doi:10.2307/257011

Morrison, E., & Robinson, S. (1997). When the employers feel betrayed: A model of how the psychological contract violation develops. *Academy of Management Review*, *22*(1), 226–256.

Motowidlo, S. J., & Schmit, M. J. (1999). Performance assessment in unique jobs. In Ilgen, D., & Pulakos, E. (Eds.), *The changing nature of performance. Implications for staffing, motivation, and development* (pp. 56–86). San Francisco, CA: Jossey-Bass.

Mottaz, C. J. (1988). Determinants of organizational commitment. *Human Relations*, *41*, 467–482. doi:10.1177/001872678804100604

Mowday, R. T., Porter, L. W., & Steers, R. M. (1982). *Employee-organization linkages: The psychology of commitment, absenteeism and turnover*. New York, NY: Academic Press.

Mowday, R. T., Steers, R. M., & Porter, L. W. (1979). The measure of organizational commitment. *Journal of Vocational Behavior*, *14*, 224–247. doi:10.1016/0001-8791(79)90072-1

Moynihan, D., & Pandey, S. (2007). Finding workable levers over work motivation: Comparing job Satisfaction, job involvement and organizational commitment. *Administration Camp Society*, *39*, 803–832. doi:10.1177/0095399707305546

Mueller, C. W., & Lawler, E. J. (1996). *Commitment to different foci: The case of nested organizational units*. Iowa: Department of Sociology, the University of Iowa.

Mueller, C. W., Boyer, E. M., Price, J. L., & Iverson, R. D. (1994). Employee attachment and non-coercive conditions of work: The case of dental hygienists. *Work and Occupations*, *21*, 179–212. doi:10.1177/0730888494021002002

Mullin, R. (1996). Managing the outsourced enterprise. *The Journal of Business Strategy*, *17*(4), 28–32. doi:10.1108/eb039792

Mullins, L. (2007). *Management and organizational behaviour*. New York, NY: Prentice Hall.

Mumford, E. (1976). *Work design and job satisfaction*. Manchester, UK: Business School.

Mumford, M. D., & Gustafson, S. B. (1988). Creativity syndrome: Integration, application, and innovation. *Psychological Bulletin*, *103*, 27–43. doi:10.1037/0033-2909.103.1.27

Mumford, M. D., & Hunter, S. T. (2005). Innovation in organizations: A multi-level perspective on creativity. In Yamarino, F. J., & Dansereau, F. (Eds.), *Research in multi-levels issues* (*Vol. 4*). Oxford, UK: Elsevier. doi:10.1016/S1475-9144(05)04001-4

Mumford, M. D., Eubanks, D. L., & Murphy, S. T. (2007). Creating the conditions for success: Best practices in leading for innovation. In Conger, J. A., & Riggio, R. E. (Eds.), *The practice of leadership* (pp. 107–197). San Francisco, CA: Jossey-Bass.

Mumford, M. D., Feldman, J. M., Hein, M. B., & Nago, D. J. (2001). Tradeoffs between ideas and structure: Individual versus group performance in creative problem-solving. *The Journal of Creative Behavior*, *35*, 1–23.

Mumford, M. D., Marks, M. A., Connelly, M. S., Zaccaro, S. J., & Johnson, T. F. (1998). Domain based scoring of divergent thinking tests: Validation evidence in an occupational sample. *Creativity Research Journal*, *11*, 151–164. doi:10.1207/s15326934crj1102_5

Mumford, M. D., Marks, M. A., Connelly, M. S., Zaccaro, S. J., & Reiter-Palmon, R. (2000). Development of leadership skills: Experience, timing, and growth. *The Leadership Quarterly*, *11*, 87–114. doi:10.1016/S1048-9843(99)00044-2

Mumford, M. D., Scott, G. M., Gaddis, B., & Strange, J. M. (2002). Leading creative people: Orchestrating expertise and relationships. *The Leadership Quarterly*, *13*, 705–750. doi:10.1016/S1048-9843(02)00158-3

Mumford, M., Friedrich, T., Caughron, J., & Antes, A. (2009). Leadership development and assessment, describing and rethinking the state of the art. In Ericsson, A. (Ed.), *Development of professional expertise. Toward measurement of expert performance and design of optimal learning environments*. New York, NY: Cambridge University Press.

Murphy, K. R. (1999). The challenge of staffing a post-industrial workplace. In Howard, A. (Ed.), *The changing nature of work* (pp. 295–324). San Francisco, CA: Jossey-Bass.

Murphy, P. E., & Jackson, S. E. (1999). Managing work role performance: Challenges for twenty-first century organizations and their employees. In Ilgen, D., & Pulakos, E. (Eds.), *The changing nature of performance. implications for staffing, motivation, and development*. San Francisco, CA: Jossey-Bass.

Naquin, S. S., & Elwood, F. H. (2003). Redefining state leadership and management development: A process for competence-based development. *Public Personnel Management*, *32*(1), 23–46.

NCCC National Center for Cultural Competence, Georgetowm University Center for Child and Human Development, Centers for Excellence in Developmental Disabilities. (1998). *Definitions of cultural competences*. Washington, DC: Georgetown University. Retrieved 10th January, 2011, from http://www.nccccurricula.info/culturalcompetence.html

Netteland, G., Wasson, B., & Morch, A. I. (2007). E-learning in a large organization. *Journal of Workplace Learning*, *19*(6), 392–411. doi:10.1108/13665620710777129

Nichols, R., & Stevens, L. (1957). *Are you listening?*New York, NY: McGraw-Hill.

Nicholson, N. (1996). Career systems in crisis: Change and opportunity in the Information Age. *The Academy of Management Executive*, *10*(4), 40–51. doi:10.5465/AME.1996.3145318

Nicolini, D., & Meznar, M. B. (1995). The social construction of organizational learning: Concepts and practical issues in the field. *Human Relations*, *48*(7), 727–746. doi:10.1177/001872679504800701

Nonaka, I. (1991). The knowledge-creating company. *Harvard Business Review*, *69*(6), 96–104.

Nonaka, I. (1994). A dynamic theory of organizational knowledge. *Organization Science*, *5*(1), 14–37. doi:10.1287/orsc.5.1.14

Nonaka, I., & Takeuchi, H. (1995). *The knowledge-creating company*. New York, NY: Oxford University Press.

Nonaka, I., Toyama, R., & Konno, N. (2000). SECI, Ba and leadership: A unified model of dynamic knowledge creation. *Long Range Planning*, *33*(1), 5–34. doi:10.1016/S0024-6301(99)00115-6

Norman, P., Collins, S., Conner, M., Martin, R., & Rance, J. (1995). Attributions, cognitions, and coping styles: Teleworkers′ reactions to work-related problems. *Journal of Applied Social Psychology*, *25*(2), 117–128. doi:10.1111/j.1559-1816.1995.tb01587.x

Nuñez, I. (2009). Outsourcing occupational safety and health: An analysis of the make or buy decision. *Human Resource Management, 48*(6), 941–958. doi:10.1002/hrm.20323

O'Donnel, M., & Shields, J. (2002). Performance management and the psychological contract in the australian federal public sector. *The Journal of Industrial Relations, 44*(3), 435–457. doi:10.1111/0022185602044003008

O'Keeffe, T. (2002). Organizational learning: A new perspective. *Journal of European Industrial Training, 26*(2), 130–141. doi:10.1108/03090590210422012

O'Really, C. A., & Chatman, J. (1986). Organizational commitment and psychological attachment: The effect of compliance, identification, and internationalization on prosocial behaviour. *The Journal of Applied Psychology, 71*, 492–499. doi:10.1037/0021-9010.71.3.492

O'Reilly, C. A., & Pfeffer, J. (2000). Unlocking the hidden valued in organizations. *Employment Relations Today, 27*(2), 63–80. doi:10.1002/ert.3910270208

O'ReillyY. T. (2005) *What is Web 2.0? Design patterns and business models for the next generation of software.* Retrieved July 3, 2010, from http://oreilly.com/web2/archive/what-is-web-20.html

Oakley, J., & Dean, C. (2001). *Virtue ethics and professional roles*. Cambridge, UK: Cambridge University Press. doi:10.1017/CBO9780511487118

Observatorio De Comunicación Interna E Identidad Corporativa. (2009). *Estudio sobre la comunicación interna y la gestión del cambio en las empresas privadas y las Administraciones Públicas*. Retrieved from www.observatoriocomunicacioninterna.es

OECD. (2005). *New perspectives on ICT skills and employment. Working paper on the information economy.* Directorate for Science, Technology and Industry, Committee for Information, Computer and Communications Policy. Paris, France: Organisation for Economic Co-operation and Development. DSTI/UCCP/IE. (2004). *10/Final.*

Oettingen, G., & Mayer, D. (2002). The motivation function of thinking about the future: Expectations vs. fantasies. *Journal of Personality and Social Psychology, 87*, 1198–1212. doi:10.1037/0022-3514.83.5.1198

Ohtott, P. J. (2004). Job assignments. In McCauley, C. D., & Van Velsor, E. (Eds.), *The Center for Creative Leadership handbook of leadership development* (pp. 151–182). San Francisco, CA: Jossey-Bass.

Oliver, J. (2005). *Desigualdad y discriminación salarial de la mujer en España*. Madrid.

Olivero, G., Bane, K., & Kopelman, R. (1997). Executive coaching as a transfer of training tool: Effects on productivity in a public agency. *Public Personnel Management, 26*, 461–469.

Oliveros, L. (2006). Identificación de competencias: Una estrategia para la formación en el Espacio Europeo de Educación Superior. *Revista Complutense de Educación, 17*(1), 101–118.

Olson, M. (1965). *The logic of collective action*. Cambridge, MA: Harvard University Press.

Olson, M. L. (2001). *E-Coaching, ASTD's source for e-learning* [online]. Retrieved March 3, 2008, from http://www.learningcircuits.org/2001/sep2001/olson.html

Orbea, T. (2008). *UNE 66181. La calidad en la formación virtual*. Symposium 4th March, Ministry of Industry, Commerce and Tourism. Spanish Government.

Ordiz, M. (2002). Prácticas de alto rendimiento en recursos humanos: Concepto y factores que motivan su adopción. *Cuadernos de Economía y Dirección de la Empresa, 12*, 247–265.

Ordiz, M., & Fernández, E. (2003). High-involvement practices in human resource management: Concept and factors that motive their adoption. *International Journal of Human Resource Management, 14*(4), 511–529. doi:10.1080/0958519032000057565

Ordóñez, M. (1995). *La nueva gestión de los recursos humanos*. Barcelona, Spain: Ediciones Gestión.

Organ, D. W. (1988). *O.C.B: The good soldier syndrome*. Lexington, MA: Lexington Books.

Orick, L. M. (2002). *Listening practices of leaders*. Doctoral Dissertation, The University of New Mexico, 2002. (UMI No. 3056912).

Osterman, K. F. (1993). *Communication skills: A key to caring, collaboration, and change*. A paper presented at the Annual Conference of the University of Council for Educational Administration, Houston, Texas, October 29-31, 1993. ED 363 973.

Ostroff, C. (2000). *Human resource management and firm performance*. Working paper, Arizona State University.

Otter, T., & Holincheck, J. (2009). *Magic quadrant for e-recruitment software*. Gartner, Inc.

Parker, S. (2000). From passive to proactive motivation: The importance of flexible role orientations and role breath self-efficacy. *Applied Psychology: An International Review*, *49*(3), 447–469. doi:10.1111/1464-0597.00025

Parker, S. K., Wall, T. D., & Jackson, P. R. (1997). "That's not my job": Developing flexible employee work orientations. *Academy of Management Journal*, *40*(4), 899–929. doi:10.2307/256952

Parnell, J. A., Lester, D. L., & Menefee, M. L. (2000). Strategy as a response to organizational uncertainty: An alternative perspective on the strength-performance relationship. *Management Decision*, *38*, 301–392. doi:10.1108/00251740010352811

Parry, E., & Tyson, S. (2008). An analysis of the use and success of online recruitment methods in the UK. *Human Resource Management Journal*, *18*(3), 257–257. doi:10.1111/j.1748-8583.2008.00070.x

Parry, S. B. (1996). The quest for competencies. *Training (New York, N.Y.)*, *33*(7), 48–56.

Parsons, G. D., & Pascale, R. T. (2007). Crisis at the summit. *Harvard Business Review*, *85*(3), 80–89.

Parthasarthy, R., & Sethi, S. P. (1993). Relating strategy and structure to flexible automation: A test of fit and performance implications. *Strategic Management Journal*, *14*, 529–549. doi:10.1002/smj.4250140704

Pastor, J. (2002). El liderazgo de personas en la organización. In Bonache, J., & Cabrera, A. (Eds.), *Dirección estratégica de personas*. Madrid, Spain: Prentice Hall.

Pavlou, P. A., & El Sawy, O. (2004). *From IT competence to competitive advantage in turbulent environments: A dynamic capabilities model*. Working Paper, University of California.

Pazzani, M. J. (2000). Representation of electronic mail filtering profiles: A user study. In *Proceedings of the 5th International Conference of Intelligent User Interfaces*. New York, NY: ACM Press.

Pearce, J. H. II, & Ravlin, E. C. (1987). The design and activation of self-regulating work groups. *Human Relations*, *40*, 751–782. doi:10.1177/001872678704001104

Pearn, M. A., & Kandole, R. S. (1988). *Job analysis: A practical guide for managers*. Londres, UK: Institute for Personnel Management.

Peccei, R., & Rosenthal, P. (2001). Delivering customer-oriented behaviour through empowerment: An empirical test of HRM assumptions. *Journal of Management Studies*, *38*, 831–857. doi:10.1111/1467-6486.00261

Pedler, M., Burgogyne, J., & Boydell, T. (1997). *The learning company: A strategy for sustainable development* (2nd ed.). London, UK: McGraw-Hill.

Peiró, J. M., García-Montalvo, J., & Gracia, F. (2002). How do young people cope with job flexibility? Demographic and psychological antecedents of the resistance to accept a job with non-preferred flexibility features. *Applied Psychology: An International Review*, *51*(1), 43–66. doi:10.1111/1464-0597.0078z

Penley, L. E., & Gould, S. (1988). Etzioni's model of organizational involvement: A perspective for understanding commitment to organizations. *Journal of Organizational Behavior*, *9*, 43–59. doi:10.1002/job.4030090105

Penrose, E. T. (1959). *The theory of growth of the firm*. London, UK: Basil Blackwell.

Pereda, S., & Berrocal, F. (1999). *Gestión de recursos humanos por competencias*. Madrid, Spain: Editorial Centro de Estudios Ramón Areces.

Pereda, S., & Berrocal, F. (2001). *Técnicas de gestión de recursos humanos por competencias*. Madrid, Spain: Editorial Centro de Estudios Ramón Areces.

Pereda, S., Berrocal, F., & López, M. (2002). Gestión por competencias y gestión del conocimiento. *Revista de Dirección. Organización y Administración de Empresas*, *28*, 43–54.

Perretti, F., & Negro, G. (2007). Mixing genres and matching people: A study in innovation and team composition in Hollywood. *Journal of Organizational Behavior*, *28*(5), 563–586. doi:10.1002/job.464

Perry, E. L., Davis-Blake, A., & Kulik, C. T. (1994). Explaining gender-based selection decisions: A synthesis of contextual and cognitive approaches. *Academy of Management Review*, *19*, 786–820.

Peteraf, M. A. (1993). The cornerstones of competitive advantage: A resource-based view. *Strategic Management Journal*, *14*, 179–191. doi:10.1002/smj.4250140303

Peters, T. (2005). *Tom Peters essentials: Talent*. Dorling Kindersley.

Pfeffer, J. (1994). Competitive advantage through people. *California Management Review*, *36*(2), 9–28.

Pfeffer, J., & Sutton, R. I. (1999). Knowing "what" to do is not enough: Turning knowledge into action. *California Management Review*, *42*(1), 83–108.

Phillips, J. (1997a). *Return on investment in training and performance improvement programs*. Houston, TX: Gulf Publishing.

Phillips, J., & Phillips, P. (2005). *ROI at work: Best practice case studies from the real world*. Alexandria, VA: ASTD Press.

Pichault, F., Warnotte, G., & Wilkin, L. (Dirs.). (1998). *La fonction ressources humaines face aux restructurations. Trois cas de downsizing*. Paris, France: L'Harmattan.

Pierce, J., & Delbecq, L. A. (1977). Organizational structure, individual attitudes, and innovation. *Academy of Management Review*, *2*, 27–37.

Pieters, Z. (2009). *Expatriation as a career experience*. Master's Thesis. University of Stellenbosch. Retrieved April 15, 2011, from http://hdl.handle.net/10019.1/2343

Pin Arboledas, J. R. (2002). *Libro Blanco. Las mejores prácticas (best practices) en los procesos de reestructuración de plantillas*. Centre Telemàtic Editorial.

Pin, J. R., & Sáenz-Díez, I. (2002). *Outsourcing de recursos humanos*. Barcelona, Spain: IESE Business School. IRCO (International Research Center on Organizations), OP 03/4.

Pinch, S., Mason, C., & Witt, S. (1991). Flexible employment strategies in British industry: Evidence from the UK Sunbelt. *Regional Studies*, *25*(3), 207–218. doi:10.1080/00343409112331346427

Pinchot, G. (1985). *Intrapreneuring: Why you don't have to leave the organization to become an entrepreneur*. New York, NY: Harper & Row.

Ployhart, R. E. (2006). Staffing in the 21st century: New challenges and strategic opportunities. *Journal of Management*, *32*, 868–897. doi:10.1177/0149206306293625

Ployhart, R. E., Schneider, B., & Schmitt, N. (2006). *Staffing organizations: Contemporary practice and theory* (3rd ed.). Mahwah, NJ: Lawrence Erlbaum Associates, Publishers.

Plunket, L., & Hale, G. (1982). *The proactive manager*. New York, NY: Wiley.

Pool, S., & Pool, B. (2007). A management development model. *Journal of Management Development*, *26*, 353–369. doi:10.1108/02621710710740101

Porter, L. W., Crampon, W. J., & Smith, F. J. (1976). Organizational commitment and managerial turnover: A longitudinal study. *Organizational Behavior and Human Performance*, *15*, 87–89. doi:10.1016/0030-5073(76)90030-1

Porter, L. W., Steers, R. M., & Mowday, R. T. (2005). Do employee attitudes towards organizations matter? In Smith, K., & Hitt, M. (Eds.), *Great minds in management* (pp. 171–189). Great Britain: Oxford University Press.

Porter, L. W., Steers, R. M., Mowday, R. T., & Boulian, P. V. (1974). Organizational commitment, job satisfaction and turnover among psychiatric technicians. *The Journal of Applied Psychology*, *59*, 603–609. doi:10.1037/h0037335

Powell, D., & Meyer, J. P. (2003). Side-bet theory and the three component model of organizational commitment. *Journal of Vocational Behavior*, *65*, 157–177. doi:10.1016/S0001-8791(03)00050-2

Prahalad, C. K., & Hamel, G. (1990). The core competence of the corporation. *Harvard Business Review*, *68*(3), 79–91.

Presnky, M. (2001). Digital natives, digital immigrants. *On the Horizon, 9*(5). Retrieved June 5, 2010, from http://www.marcprensky.com/writing/Prensky%20-%20Digital%20Natives,%20Digital%20Immigrants%20-%20Part1.pdf

Price, J. L. (1997). Handbook of organizational measurement. *International Journal of Manpower, 18*, 303–558. doi:10.1108/01437729710182260

Price, J. L., & Mueller, C. W. (1986). *Absenteeism and turnover among hospital employees*. Greenwich, CT: Jai Press.

Priem, R. L., Harrison, D. A., & Muir, N. K. (1995). Structured conflict and consensus outcomes in group decision making. *Journal of Management, 11*, 691–710. doi:10.1177/014920639502100406

Pulakos, E. D., Arad, S., Donovan, M. A., & Plondon, K. E. (2000). Adaptability in the workplace: Development of a taxonomy of adaptive performance. *The Journal of Applied Psychology, 85*, 612–624. doi:10.1037/0021-9010.85.4.612

Pulignano, V. (2011). Bringing labour markets "back in": Restructuring international businesses in Europe. *Economic and Industrial Democracy*. Retrieved June 4, 2011, from http://eid.sagepub.com

Quinn, J. B. (1980). *Strategies for change: Logical incrementalism*. Homewood, IL: Richard D. Irwin.

Rábago López, E. (2010). *Gestión por competencias. Un enfoque para mejorar el rendimiento personal y empresarial*. La Coruña, Spain: Netbiblo.

Rackman, N., & Morgan, T. (1977). *Behavior analysis in training*. New York, NY: Mc Graw-Hill.

Rahmandad, H., Repenning, N., & Stearman, J. (2009). Effects of feedback delay on learning. *System Dynamics Review, 25*(4), 309–338. doi:10.1002/sdr.427

Rainnie, A., Barrett, R., Burgess, J., & Connell, J. (2008). Introduction: Call centers, the networked economy and the value chain. *The Journal of Industrial Relations, 50*(2), 195–208. doi:10.1177/0022185607087897

Ramírez, A. (1996). La gestión de las competencias. *Estudios Financieros, 160*, 87–100.

Randall, M. D., & Cote, J. A. (1991). Interrelationships of work commitment constructs. *Work and Occupations, 18*, 194–211. doi:10.1177/0730888491018002004

Rankin, N. (Ed.). (2001). *The IRS handbook on competencies. Law and practice*. London, UK: IRS LexisNexis Group.

Rausch, E., Sherman, H., & Washbush, J. B. (2001). Defining and assessing competencies for competency-based, outcome-focused management development. *Journal of Management Development, 21*(3), 184–200. doi:10.1108/02621710210420264

Ray, G., Barney, J. B., & Muhanna, W. A. (2004). Capabilities, business processes, and competitive advantage: Choosing the dependent variable in empirical tests of the resource-based view. *Strategic Management Journal, 25*, 23–37. doi:10.1002/smj.366

Redding, W. C. (1972). *Communication within the organization: An interpretive review of theory and research*. New York, NY: Industrial Communication Council.

Redmon, J., & Salopek, J. J. (2000). A year in the life of an e-learning project. *Training & Development*, (September): 36–41.

Renzulli, J. (1986). The three-ring conception of giftedness: A developmental model for creative productivity. In Sternberg, R. (Ed.), *Conceptions of giftedness*. Cambridge, UK: Cambridge University Press. doi:10.1017/CBO9780511610455.015

Requejo, A. (1996). Reducción de plantilla y problemas de viabilidad financiera: El papel de la estructura de capital. *Investigacion Economica, 20*(1), 43–70.

Resnick, P. (2002). Beyond bowling together: Sociotechnical capital. In Carroll, J. M. (Ed.), *HCI in the new millennium* (pp. 247–272). Boston, MA: Addison-Wesley.

Revans, R. (1980). *Action learning: New techniques for managers*. London, UK: Blond & Briggs.

Reyes Contreras, Y., & Martínez León, I. M. (2007). Los contratos psicológicos: Sus efectos en los resultados de las organizaciones. In *V Encuentro Iberoamericano de Finanzas y Sistemas de Información*. EFSI.

Richard, O. C., & Johnson, N. B. (2001). Strategic human resource management effectiveness and firm performance. *International Journal of Human Resource Management, 12*, 299–310.

Rifkin, J. (2004). *The end of work*. New York, NY: Penguin.

Riley, M., & Lockwood, A. (1997). Strategies and measurement for workforce flexibility: An application of functional flexibility in a service setting. *International Journal of Operations & Production Management, 17*(4), 413–419. doi:10.1108/01443579710159996

Ring, P. S., & Van De Ven, A. H. (1994). Developmental processes of cooperative interorganisational relationships. *Academy of Management Review, 19*(1), 90–118.

Rippin, S., & Dawson, G. (2001, September 27). How to outsource the HR function. *People Management*, 42-43.

Ripsas, S. (1998). Toward an interdisciplinary theory of entrepreneurship. *Small Business Economics, 10*(2), 103–115. doi:10.1023/A:1007975330428

Robbins, S. (1983). *Organization theory: The structure and design of organizations*. Englewood Cliffs, NJ: Prentice Hall Inc.

Robbins, S. P. (1989). *Organizational behavior* (p. 274). Englewood Cliffs, NJ: Prentice-Hall.

Robbins, S. P. (1994). *Comportamiento organizacional*. México: Prentice Hall.

Robbins, S. P., & Judje, T. A. (2009). *Comportamiento organizacional* (13th ed.). México: Pearson Educación.

Roberts, B. (2009). Outsourcing in turbulent times. *HR-Magazine, 54*(11), 42–47.

Robinson, S. L. (1996). Trust and breach of the psychological contract. *Administrative Science Quarterly, 41*(4), 574–599. doi:10.2307/2393868

Robinson, S. L., & Morrison, E. W. (2000). The development of psychological contract breach and violation: A longitudinal study. *Journal of Organizational Behavior, 21*(5), 525–546. doi:10.1002/1099-1379(200008)21:5<525::AID-JOB40>3.0.CO;2-T

Robinson, S. L., & Rousseau, D. M. (1994). Violating the psychological contract: Not the exception but the norm. *Journal of Organizational Behavior, 15*(3), 245–259. doi:10.1002/job.4030150306

Roca, V., Escrig, A. B., & Bou, J. C. (2002). Compromiso con los empleados y estrategia competitiva: un análisis intersectorial de su repercusión en los resultados. *Cuadernos de Economía y Dirección de la Empresa, 12*, 267–289.

Roche, W. K. (1999). In search of commitment-oriented human resource management practices and the conditions that sustain them. *Journal of Management Studies, 36*, 653–678. doi:10.1111/1467-6486.00153

Rochford, L., & Rudelius, W. (1992). How involving more functional areas within a firm affects the new product process. *Journal of Product Innovation Management, 9*, 287–299. doi:10.1111/1540-5885.940287

Rodgers, D. (1998b). *Business communications. international case studies in English*. Cambridge, UK: Cambridge University Press.

Rodríguez, L. F. (2007). Emprendimiento y asociatividad como herramienta de desarrollo social para la educación media. *Cuadernos Latinoamericanos de Administración, 4*(2).

Rogers, D. (1987). The development of a measure of perceived communication openness. *Journal of Business Communication, 24*(4), 53–61. doi:10.1177/002194368702400404

Rosenblatt, Z., & Inbal, B. (1999). Skill flexibility among school teachers: Operationalization and organizational implications. *Journal of Educational Administration, 37*, 345–366. doi:10.1108/09578239910285589

Rossett, A., & Marino, G. (2005). If coaching is good, then e-coaching is.... *Training & Development, 59*, 11.

Rotter, J. (1966). Generalized expectancies for internal versus external control of reinforcements. *Psychological Monographs, 80*, Whole No. 609.

Roujol, B. (n.d.). *Definición de e-coaching*. Retrieved 22 July, 2010, from http://www.coachingavenue.com/2/view_news.php?id_news=472

Rousseau, D. M. (1989). Psychological and implied contract in organizations. *Employee Responsibilities and Rights Journal*, *2*, 121–139. doi:10.1007/BF01384942

Rousseau, D. M. (1990). New hire perception of their own and their employer's obligations: A study of psychological contracts. *Journal of Organizational Behavior*, *11*(5), 389–400. doi:10.1002/job.4030110506

Rousseau, D. M. (1995). *Psychological contracts in organizations: Understanding written and unwritten agreements*. Thousand Oaks, CA: Sage.

Rousseau, D. M. (2001). Schema, promise and mutuality: The building blocks of the psychological contract. *Journal of Occupational and Organizational Psychology*, *74*(4), 511–541. doi:10.1348/096317901167505

Rousseau, D. M., & Greller, M. M. (1994). Human resource practices: Administrative contract makers. *Human Resource Management*, *33*(3), 385–401. doi:10.1002/hrm.3930330308

Rousseau, D. M., & Mclean Parks, J. (1993). The contracts of individuals and organizations. In Cummings, L. L., & Staw, B. M. (Eds.), *Research in organizational behavior* (pp. 1–47). Greenwich, CT: JAI Press.

Rousseau, D. M., & Shperling, Z. (2003). Pieces of the action: Ownership and the changing employment relationship. *Academy of Management Review*, *28*(4), 553–570.

Rousseau, D. M., & Tijoriwala, S. A. (1998). Assessing psychological contracts: Issues, alternatives and measures. *Journal of Organizational Behavior*, *19*(S1), 679–695. doi:10.1002/(SICI)1099-1379(1998)19:1+<679::AID-JOB971>3.0.CO;2-N

Rousseau, D., & Parks, J. (1993). The contract of individuals and organizations. *Research in Organizational Behavior*, *15*(1), 1–43.

Roy, K. (2006). The impact of learning styles on interactivity in asynchronous e-learning. *Performance Improvement Journal*, *45*(10), 21–26. doi:10.1002/pfi.4930451026

Ruël, H., Bondarouk, T., & Looise, J. K. (2004). E-HRM: Innovation or irritation. An explorative empirical study in five large companies on web-based HRM. *Management Review*, *15*(3), 364–380.

Ruggles, R. (1998). The state of the notion: Knowledge management in practice. *California Management Review*, *40*(3), 80–89.

Rumelt, R. P. (1987). Theory, strategy and entrepreneurship. In Lamb, R. B. (Ed.), *The competitive challenge: Strategies for industrial innovation and renewal* (pp. 137–158). Cambridge, MA: Ballinger.

Rummler, G., & Brache, A. (1995). *Improving performance: How to manage the white space on the organization chart*. San Francisco, CA: Jossey- Bass.

Ryan, M. K., & Haslam, S. A. (2005). The glass cliff: Evidence that women are over-represented in precarious leadership positions. *British Journal of Management*, *16*, 81–90. doi:10.1111/j.1467-8551.2005.00433.x

Saari, L., Jonhson, T., McLaughlin, S., & Zimmerle, D. (1988). A survey of management training and education practices in U.S. companies. *Personnel Psychology*, *41*, 731–743. doi:10.1111/j.1744-6570.1988.tb00650.x

Sáenz Blanco, M. T. (2001). Schneider Electric: El valor exponencial de las tecnologías aplicadas a la gestión de RR.HH. *Capital Humano*, *148*, 28–36.

Sagi-Vela, L. (2004). *Gestión por competencias. El reto compartido del crecimiento personas y de la organización*. Madrid, Spain: ESIC.

Sajaia, Z. (2008). Maximum likelihood estimation of a bivariate ordered probit model: Implementation and Monte Carlo simulations. *The Stata Journal*, *3*(2), 311–328.

Sala, G. (2006). La gestión por competencias en el sector bancario español. *Trayectorias*, *8*(22), 21–33.

Sánchez de Dios, L., & Valldeperes, P. (1998). Metodología de evaluación de competencias directivas. *Capital Humano*, *110*, 20–30.

Sánchez, I. (2004). The creation of employee commitment in Spanish hotels through cultural practices. *Management Research*, *2*(2), 175–192.

Sanchez, R. (1995). Strategic flexibility in product competition. *Strategic Management Journal*, *16*, 135–159. doi:10.1002/smj.4250160921

Sanchez, R. (1997). Preparing for an uncertain future. Managing organizations for strategic flexibility. *International Studies of Management and Organizations*, *27*, 71–94.

Sanchez, R. (2004). Understanding competence-based management. Identifying and managing five modes of competence. *Journal of Business Research*, *57*, 518–532. doi:10.1016/S0148-2963(02)00318-1

Sarasvathy, S. D. (2001). *What makes entrepreneurs entrepreneurial? Schumpeter, J. (2000). The theory of economic development*. New Brunswick, NJ: Transaction Publishers.

Sashkin, M. (1984). Participative management is an ethical imperative. *Organizational Dynamics*, *12*(4), 5–22. doi:10.1016/0090-2616(84)90008-1

Sayles, L. R. (1979). *Leadership: What effective managers really do... and how they do it*. New York, NY: McGraw-Hill.

Scacchi, W., & Noll, J. (1997). Process-driven Intranets: Life-cycle support for process reengineering. *IEEE Internet Computing*, *1*(5), 42–49. doi:10.1109/4236.623967

Scarnati, J. T. (1998). Beyond technical competence: Learning to listen. *Career Development International*, *3*(2), 79–81. doi:10.1108/13620439810207590

Schaubroeck, J., Lam, S. S. K., & Cha, S. E. (2007). Embracing transformational leadership: Team values and the impact of leader behaviour on team performance. *The Journal of Applied Psychology*, *92*(4), 1020–1030. doi:10.1037/0021-9010.92.4.1020

Schein, E. H. (1965). *Organizational psychology*. Englewood Cliffs, NY: Prentice-Hall.

Schein, E. H. (1978). *Career dynamics: Matching individual and organizational needs*. Reading, MA: Addison- Wesley.

Schein, E. H. (1996). Career anchors revisited: Implications for career development in the 21st century. *The Academy of Management Executive*, *10*(4), 80–88. doi:10.5465/AME.1996.3145321

Schiller, S. Z., & Cui, J. (2010). Communication openness in the workplace: The effects of medium (F2F and IM) and culture (U.S. and China). *Journal of Global Information Technology Management*, *13*(2), 37–75.

Schippmann, J. S., Ash, R. A., Battista, M., Carr, L., Eyde, L. D., & Hesketh, B. (2000). The practice of competency modeling. *Personnel Psychology*, *53*, 703–740. doi:10.1111/j.1744-6570.2000.tb00220.x

Schlemenson, A. (2003). *The strategy of the talent*. Paidos.

Schuler, R. S., & Jackson, S. E. (1987a). Linking competitive strategy with human resource management practices. *The Academy of Management Executive*, *1*(3), 207–219. doi:10.5465/AME.1987.4275740

Schuler, R. S., & Jackson, S. E. (1987b). Organizational strategy and organization level as determinants of human resource management practices. *Human Resource Planning*, *10*(3), 125–141.

Seligman, M. (1993). *What you can change and what you can't*. New York, NY: Knopf.

Senge, P. M. (1990). *The fifth discipline*. London, UK: Century Business.

Senge, P. M. (1990). *The fifth discipline: The art and practice of the learning organization*. New York, NY: Doubleday.

Senge, P. M. (1993). Transforming the practice of management. *Human Resource Development Quarterly*, *4*, 5–32. doi:10.1002/hrdq.3920040103

Serenko, A., Bontis, N., & Hardie, T. (2007). Organizational size and knowledge flow: A proposed theoretical link. *Journal of Intellectual Capital*, *8*(4), 610–627. doi:10.1108/14691930710830783

Servage, L. (2005). Strategizing for workplace e-learning: Some critical considerations. *Journal of Workplace Learning*, *17*(5), 304–317. doi:10.1108/13665620510606733

Shafer, R. A., Dyer, L., Kitty, J., Amos, J., & Ericksen, G. A. (2001). Crafting a human resource strategy to foster organizational agility: A case study. *Human Resource Management*, *40*(3), 197–211. doi:10.1002/hrm.1011

Shafritz, J. M., Ott, J. S., & Jang, Y. S. (2005). *Classics of organization theory* (6th ed.). Belmont, CA: Thomson Wadsworth.

Shamir, B., & Eilan, G. (2005). What's your story? A life-stories approach to authentic leadership development. *The Leadership Quarterly*, *16*, 395–417. doi:10.1016/j.leaqua.2005.03.005

Shane, S., & Venkatamaran, S. (2000). The promise of entrepreneurship as a field of research. *Academy of Management Review*, *25*(1), 217–226.

Shane, S., & Venkataraman, S. (2007). The promise of entrepreneurship as a field of research. *Entrepreneurship*, *2*, 171–184.

Shapero, A. (1975). *Entrepreneurship and economic development. Entrepreneurship and enterprise development: A worldwide perspective*. Milwaukee, WI: Proceedings of Project ISEED.

Shapero, A., & Sokol, L. (1982). The social dimension of entrepreneurship. In Kent, C., Sexton, D., & Vesper, K. (Eds.), *Encyclopedia of entrepreneurship* (pp. 72–90). Englewood Cliffs, NJ: Prentice Hall.

Sharma, P., & Chrisman, J. J. (1999). Toward a reconciliation of the definitional sigues in the field of corporate entrepreneurship. *Entrepreneurship Theory and Practice*, Spring, 11-27.

Sharma, R., & Bhatnagar, J. (2009). Talent management - Competency development: Key to global leadership. *Industrial and Commercial Training*, *41*(3), 118–132. doi:10.1108/00197850910950907

Shaver, K. G., & Scott, L. R. (1991). Person, process, choice: The psychology of new venture creation. *Entrepreneurship Theory and Practice*, *16*(2), 23–45.

Sheehan, C. (2009). Outsourcing HRM activities in Australian organisations. *Asia Pacific Journal of Human Resources*, *47*(2), 236–253. doi:10.1177/1038411109105444

Shen, J. (2005). Human resource outsourcing: 1990-2004. *Journal of Organisational Transformation and Social Change*, *2*(3), 275–296. doi:10.1386/jots.2.3.275/1

Shih, H., Chiang, Y., & Hsu, C. (2006). Can high performance work systems really lead to better performance? *International Journal of Manpower*, *27*(8), 741–763. doi:10.1108/01437720610713530

Shojo, A. R., Woods, J., & Smith, P. (2006). *Is anyone listening? The dichotomous perspective of school administrator listening skills.* University Council for Educational Administration 2006 Annual Conference.

Shore, L. M., & Barksdale, K. (1998). Examining degree of balance and level of obligation in the employment relationship: A social exchange approach. *Journal of Organizational Behavior*, *19*, 731–744. doi:10.1002/(SICI)1099-1379(1998)19:1+<731::AID-JOB969>3.0.CO;2-P

Shore, L. M., & Tetrick, L. E. (1994). The psychological contract as an explanatory framework in the employment relationship. In Cooper, C. L., & Rousseau, D. M. (Eds.), *Trends in organizational behavior* (*Vol. 1*, pp. 91–109). Chichester, UK: John Wiley & Sons.

Shore, L. M., & Wayne, S. J. (1993). Commitment and employee behavior: Comparison of affective commitment and continuance commitment with perceived organizational support. *The Journal of Applied Psychology*, *78*, 774–780. doi:10.1037/0021-9010.78.5.774

SHRM. (2008a). *Online technologies and their impact on recruitment strategies: Using social networking websites to attract talent*. Society for Human Resource Management.

SHRM. (2008b). *SHRM staffing research: Online technologies and their impact on recruitment strategies*. Society for Human Resource Management.

Siebdrat, F., Hoegl, M., & Ernst, H. (2009). How to manage virtual teams. *MIT Sloan Management Review*, *50*(4), 62–69.

Siegel, G. B. (2000). Outsourcing personnel functions. *Public Personnel Management*, *29*(2), 225–236.

Silla, I., Gracia, F. J., & Peiró, J. M. (2005). Diferencias en el contenido del contrato psicológico en función del tipo de contrato y de la gestión empresarial pública o privada. *Revista de Psicología Social*, *20*(1), 61–72. doi:10.1174/0213474052871097

Simmonds, P. G., Dawley, D. D., Ritchie, W. J., & Anthony, W. P. (2001). An exploratory examination of the knowledge transfer of strategic management concepts from the academic environment to practicing managers. *Journal of Managerial Issues*, *13*(3), 360–375.

Simon, H. A. (1991). Bounded rationality and organizational learning. *Organization Science*, *2*(1), 125–134. doi:10.1287/orsc.2.1.125

Simon, H. A. (1993). Altruism and economics. *The American Economic Review*, *83*, 156–161.

Sims, H., & Manz, C. H. (1982). Social learning-theory: The role of modelling in the exercise of leadership. *Journal of Organizational Behavior Management*, *3*, 55–63. doi:10.1300/J075v03n04_06

Sims, R. R. (1994). Human resource management's role in clarifying the new psychological contract. *Human Resource Management*, *33*(3), 373–382. doi:10.1002/hrm.3930330306

Şimşek, Ö., Altum, E., & Ateş, A. (2010). Developing ICT skills of visually impaired learners. *Procedia-Social and Behavioural Sciences*, *2*(2), 4655–4661. doi:10.1016/j.sbspro.2010.03.745

Singh, H., & Reed, C. (2002). Demystifying e-learning standards. *Industrial and Commercial Training*, *34*(2), 62–65. doi:10.1108/00197850210417546

Sisson, K., & Storey, J. (2000). *The reality of human resource management*. Oxford, UK: Oxford University Press.

Sivera, S. (2008). *Marketing viral*. Barcelona, Spain: Editorial UOC.

Smith, N. C. (2003). Corporate social responsibility: Whether or how? *California Management Review*, *45*(4), 52–76.

Snook, S., & Khurana, R. (2004). Developing "leaders of character": Lessons from West Point. In Gandossy, R., & Sonnenfeld, J. (Eds.), *Leadership and governance from the inside out* (pp. 213–233). New York, NY: John Wiley & Sons.

Snyder, K., Farrell, R., & Baker, N. (2000). Online mentoring: A case study involving cognitive apprenticeship and a technology-enabled learning environment. In J. Bourdeau & R. Heller (Eds.), *Proceedings of World Conference on Educational Multimedia, Hypermedia and Telecommunications 2000*, (pp. 1742-1743). Chesapeake, VA: AACE

Solís, E., & Corral, A. (2010). Reclutamiento y selección de personal. Externalización. In *M. J. Pérez López (Coord.), Externalización de funciones de recursos humanos*. Madrid, Spain: Pearson.

Solomon, R. C. (1992). *Ethics and excellence: Cooperation and integrity in business*. New York, NY: Oxford University Press.

Solomon, R. C. (2004). Aristotle, ethics and business organizations. *Organization Studies*, *25*(6), 1021–1043. doi:10.1177/0170840604042409

Somech, A., & Drach-Zahavy, A. (2004). Exploring organizational citizenship behaviour from an organizational perspective: The relationships between organizational learning and organizational citizenship behaviour. *Journal of Occupational and Organizational Psychology*, *77*, 281–298. doi:10.1348/0963179041752709

Sparrow, P. (1998). The pursuit of multiple and parallel organizational flexibilities: Reconstituting jobs. *European Journal of Work and Organizational Psychology*, *7*(1), 79–95. doi:10.1080/135943298398970

Spears, L. C. (1995). *Reflections on leadership: How Robert K. Greenleaf's theory of servant leadership influenced today's top management thinkers*. New York, NY: John Wiley.

Spencer, L. M., & Spencer, S. M. (1993). *Competence at work. Models for superior performance*. New York, NY: John Wiley & Sons.

Spinks, N., & Wells, B. (1995). Quality communication: A key to quality leadership. *Training for Quality*, *3*(2), 14–19. doi:10.1108/09684879510087486

Stahl, B. C., & Elbeltagi, I. (2004). Cultural universality versus particularity in CMC. *Journal of Global Information Technology Management*, *7*(4), 47–65.

Stebbins, R. A. (2001). *Exploratory research in the social sciences*. Thousand Oaks, CA: SAGE.

Sternberg, R. (1996). *Successful intelligence: How practical and creative intelligence determine success in life*. New York, NY: Plume.

Sternberg, R. (1999). The theory of successful intelligence. *Review of General Psychology*, *3*(4), 292–316. doi:10.1037/1089-2680.3.4.292

Stogdill, R. M. (1948). Personal factor associated with leadership: A survey of the literature. *The Journal of Psychology, 25*, 35–71. doi:10.1080/00223980.1948.9917362

Stogdill, R. M. (1974). *Handbook of leadership: A survey of the literature*. New York, NY: Free Press.

Stoner, J. A. R., & Freeman, R. E. (1992). *Management* (5th ed.). New Delhi, India: Prentice-Hall of India.

Stopford, J. M., & Baden-Fuller, C. W. F. (1994). Creating corporate entrepreneurship. *Strategic Management Journal, 15*(7), 521–536. doi:10.1002/smj.4250150703

Storey, J., Quintas, P., Taylor, P., & Fowle, W. (2002). Flexible employment contracts and their implications for product and process innovation. *International Journal of Human Resource Management, 13*(1), 1–18. doi:10.1080/09585190110092758

Stroh, L. K., Brett, J. M., & Reilly, A. H. (1996). Family structure, glass ceiling, and traditional explanations for the differential rate of turnover of female and male managers. *Journal of Vocational Behavior, 49*, 99–118. doi:10.1006/jvbe.1996.0036

Strohmeier, S. (2007). Research in e-HRM: Review and implications. *Human Resource Management Review, 17*(1), 19–37. doi:10.1016/j.hrmr.2006.11.002

Suárez González, I. (1999). Decisiones de reducción de plantilla en la gran empresa española. *Investigaciones Europeas de Dirección y Economía de la Empresa, 5*(1), 79–100.

Sun, L. Y., Aryee, S., & Law, K. S. (2007). High performance human resources practices, citizenship behavior and organizational performance: A relational perspective. *Academy of Management Journal, 50*, 558–577. doi:10.5465/AMJ.2007.25525821

Sun, P., Tsai, R. J., Chen, Y., & Yeh, D. (2008). What drives a successful e-learning? An empirical investigation of the critical factors influencing learner satisfaction. *Computers & Education, 50*, 1183–1202. doi:10.1016/j.compedu.2006.11.007

Sutton, G., & Griffin, M. A. (2004). Integrating expectations, experiences, and psychological contract violations: A longitudinal study of new professionals. *Journal of Occupational and Organizational Psychology, 77*(4), 493–514. doi:10.1348/0963179042596487

Svetlik, I., & Stavrou-Costea, E. (2007). Connecting human resources management and knowledge management. *International Journal of Manpower, 28*(3/4), 197–206. doi:10.1108/01437720710755209

Svyantek, D., & Deshon, R. (1993). Organizational attractors: A chaos theory explanation of why cultural change efforts often fail. *Public Administration Quarterly, 17*(3), 339–355.

Swailes, S. (2002). Organizational commitment: A critique of the construct and measures. *International Journal of Management Reviews, 4*(2), 155–178. doi:10.1111/1468-2370.00082

Swanson, R., & Gradous, D. (1988). *Forecasting financial benefits of human resource development*. San Francisco, CA: Jossey-Bass Publishers.

Swanson, R., & Holton, E. (1999). *Results: How to assess performance, learning, and perceptions in organizations*. San Francisco, CA: Berrett-Koehler.

Syedain, H. (2009). Out of the ordinary. *People Management, 15*(23), 20–22.

Szulanski, G. (1996). Exploring stickiness: Impediments to the transfer of best practice within the firm. *Strategic Management Journal, 17*(Winter Special Issue), 27-43.

Tabernero, C., Arenas, A., & Briones, E. (2005). Impacto del feedback negativo sobre los procesos de autorregulación del grupo. La ruptura del contrato grupal relacional. *Revista de Psicología Social, 20*(1), 93–108. doi:10.1174/0213474052871033

Takeuchi, R., Lepak, D. P., Wang, H., & Takeuchi, K. (2007). An empirical examination of the mechanisms mediating between high performance work systems and the performance of Japanese organizations. *The Journal of Applied Psychology, 92*(4), 1069–1083. doi:10.1037/0021-9010.92.4.1069

Tam, Y. M., Korczynski, M., & Frenkel, S. J. (2002). Organisational and occupational commitment: Knowledge workers in large corporations. *Journal of Management Studies, 39*(6), 775–801. doi:10.1111/1467-6486.00311

Tannenbaum, A. (1993). Giftedness: A psychological approach. In Sternberg, R. (Ed.), *Conceptions of giftedness*. Cambridge, UK: Cambridge University Press.

Tannenbaum, S. (1997). Enhancing continuous learning: Diagnostic findings from multiple companies. *Human Resource Management*, *36*, 437–452. doi:10.1002/(SICI)1099-050X(199724)36:4<437::AID-HRM7>3.0.CO;2-W

Tansley, C., Turner, P., & Foster, C. (2007). *Talent: Strategy, management, measurement. Research into practice*. London, UK: Chartered Institute of Personnel and Development.

Tassabehji, R., Wallace, J., & Cornelius, N. (2007). E-technology and the emergent environment: Implications for organizational form and function. *The Journal of High Technology Management Research*, *18*, 15–30. doi:10.1016/j.hitech.2007.03.007

Taylor, C. (2001, May). E-recruitment is powerful weapon in war for talent. *People Management,* 10.

Taylor, S. M., & Tekleab, A. G. (2004). Taking stock of psychological contract research: Assessing progress, addressing troublesome issues, and setting research priorities. In Coyle-Shapiro, J., Taylor, M. S., & Tetrick, L. E. (Eds.), *The employment relationship. Examining psychological and contextual perspectives* (pp. 253–283). Oxford University Press.

Teece, D. J., Pisano, G., & Shuen, A. (1997). Dynamic capabilities and strategic management. *Strategic Management Journal*, *18*, 509–533. doi:10.1002/(SICI)1097-0266(199708)18:7<509::AID-SMJ882>3.0.CO;2-Z

Tekleab, A. G., & Taylor, M. S. (2003). Aren't there two parties in an employment relationship? Antecedents and consequences of organization-employee agreement on contract obligations and violations. *Journal of Organizational Behavior*, *24*(5), 585–608. doi:10.1002/job.204

Teng, B. (2007). Corporate entrepreneurship activities through strategic alliances: A resource-based approach toward competitive strategy. *Journal of Management Studies*, *44*(1), 119–142. doi:10.1111/j.1467-6486.2006.00645.x

Testa, M. R. (2001). Organizational commitment, job satisfaction and effort in the service environment. *The Journal of Psychology*, *135*, 226–236. doi:10.1080/00223980109603693

Tharenou, P., Latimer, S., & Conroy, D. (1994). How do you make it to the top? An examination of influences on women and men managerial advancement. *Academy of Management Journal*, *37*, 899–931. doi:10.2307/256604

Thomas, H., & Anderson, N. (1998). Changes in newcomers' psychological contracts during organizational socialization: A study of recruits entering the British Army. *Journal of Organizational Behavior*, *19*(S1), 745–767. doi:10.1002/(SICI)1099-1379(1998)19:1+<745::AID-JOB967>3.0.CO;2-I

Thompson, L. F., Braddy, P. W., & Wuensch, K. L. (2008). E-recruitment and the benefits of organizational Web appeal. *Computers in Human Behavior*, *24*(5), 2384–2398. doi:10.1016/j.chb.2008.02.014

Tierney, P., Farmer, S. M., & Graen, G. B. (1999). An examination of leadership and employee creativity: The relevance of traits and relationships. *Personnel Psychology*, *52*, 591–620. doi:10.1111/j.1744-6570.1999.tb00173.x

Timmons, J. (1997). *New venture creation. Entrepreneurship for the 21st century* (revised 4th ed.). McGraw-Hill.

Tobin, D. R., & Pettingell, M. S. (2008). *The AMA guide to management development*. Retrieved from http://books.google.es/

Toffier, A. (1990). *Powershift: Knowledge, wealth and violence at the edge of 21st century*. New York, NY: Bantam Books.

Toffler, A. (1980). *The third wave*. New York, NY: Bantam Books.

Toffler, A. (1998). *Interview on Viva Newsletter*. Retrieved 24 July, 2010, from http://www.cuentometro.com.ar/221.htm

Tolich. (2005). Phillip Brown and Anthony Hesketh: The mismanagement of talent: Employability and jobs in the knowledge economy. *Administrative Science Quarterly, 50*(2), 306-307

Tomasko, R. M. (1996). *Repensar la empresa. La arquitectura del cambio*. Barcelona, Spain: Paidós Ibérica.

Topa, G., & Palací, F. (2004). Psychological contract breach or unfulfilment? A meta-analytic review of empiric research. *Acción Psicológica*, *3*, 155–171.

Topa, G., & Palací, F. (2005). La ruptura del contrato psicológico entre los Soldados Profesionales Españoles y su relación con la decisión de permanecer o abandonar. *Revista de Psicología Social*, *20*(1), 45–60. doi:10.1174/0213474052871088

Topa, G., Alonso, E., & Palací, F. (2002). Una medida de socialización. Aplicación de un estudio longitudinal con Soldados Profesionales Españoles. *Metodología de las Ciencias del Comportamiento*, 544-549.

Torrington, D., & Mackay, L. (1986). Will consultants take over the personnel function? *Personnel Management*, February, 34-37.

Touron, J. (2001). *Igualdad, eficacia y excelencia. Retos del sistema educativo ante la sociedad del conocimiento.* Actas del IV Congreso de economía Navarra. Pamplona, Spain: Departamento de Economía y Hacienda.

Townsend, A. M., DeMarie, S. M., & Hendrickson, A. R. (1998). Virtual teams: Technology and the workplace of the future. *The Academy of Management Executive*, *12*(3), 17–29. doi:10.5465/AME.1998.1109047

Tremblay, M., Patry, M., & Lanoie, P. (2008). Human resources outsourcing in Canadian organizations: An empirical analysis of the role of organizational characteristics, transaction costs and risks. *International Journal of Human Resource Management*, *19*(4), 683–715. doi:10.1080/09585190801953699

Truss, C., Gratton, L., Hope-Hailey, V., McGovern, P., & Stiles, P. (1997). Soft and hard models of human management: A reappraisal. *Journal of Management Studies*, *34*, 53–73. doi:10.1111/1467-6486.00042

Tsui, A. S., Pearce, J. L., Porter, L. W., & Tripoli, A. M. (1997). Alternative approaches to the employee-organization relationship: Does investment in employees pay off? *Academy of Management Journal*, *40*(5), 1089–1121. doi:10.2307/256928

Tung, R. (1981). Selecting and training of personnel for overseas assignments. *The Columbia Journal of World Business*, *16*(1), 68–78.

Tung, R. (1982). Selection and training procedures of U.S., European & Japanese multinationals. *California Management Review*, *25*(1), 57–71.

Turban, E., Leidner, D., Mclean, E., & Wetherbe, J. (2006). *Information Technology for management: Transforming organizations in the digital age* (5th ed.). John Wiley & Sons.

Turnbull, J. (2002). Inside outsourcing. *People Management*, 10–11.

Turnley, W. H., & Feldman, D. C. (1999). A discrepancy model of psychological contract violations. *Human Resource Management Review*, *9*(3), 367–386. doi:10.1016/S1053-4822(99)00025-X

Turnley, W. H., & Feldman, D. C. (2000). Re-examining the effects of psychological contract violations: Unmet expectations and job dissatisfaction as mediators. *Journal of Organizational Behavior*, *21*(1), 25–42. doi:10.1002/(SICI)1099-1379(200002)21:1<25::AID-JOB2>3.0.CO;2-Z

Turnley, W. H., Bolino, M. C., Lester, S. W., & Bloodgood, J. M. (2003). The impact of psychological contract fulfillment on the performance of in-role and organizational citizenship behaviors. *Journal of Management*, *29*(2), 187–206. doi:10.1177/014920630302900204

Tüselmann, H. J. (1996). Progress towards greater labour flexibility in Germany. The impact of recent reforms. *Employee Relations*, *18*(1), 50–67. doi:10.1108/01425459610110236

Tushman, M. L., & O'Reilly, C. A. (1997). *Winning through innovation.* Cambridge, MA: Harvard Business School Press.

Tynjälä, P., & Häkkinen, P. (2005). E-learning at work: Theoretical underpinnings and pedagogical challenges. *Journal of Workplace Learning*, *17*(5), 318–336. doi:10.1108/13665620510606742

UE. (2002). *Directive 2002/73/CE.*

UE. (2004). *Directive 2004/113/CE.*

Uhalde, R., & Strohl, J. (2006). *America in the global economy: A background paper for the new commission on the skills of the American workforce*. Washington, DC: National Center on Education and the Economy.

Ulrich, D. (1997). Measuring human resources: An overview of practices and a prescription for results. *Human Resource Management*, *36*(3), 303–320. doi:10.1002/(SICI)1099-050X(199723)36:3<303::AID-HRM3>3.0.CO;2-#

Ulrich, D. (1998). A new mandate for human resources. *Harvard Business Review*, (January-February): 124–134.

Ulrich, D. (1998). Intellectual capital = Competence x commitment. *Sloan Management Review*, *39*(2), 15–26.

Ulrich, D., & Samllwood, N. (2004). Capitalizing on capabilities. *Harvard Business Review*, *82*(6), 119–127.

Ulrich, D., & Smallwood, N. (2007). *Leadership brand, developing customer-focused leaders to drive performance and build lasting value*. Cambridge, MA: Harvard Business School Press.

Ulrich, D., Brockbank, W., Yeung, A., & Lake, D. G. (1995). Human resource competencies: An empirical assessment. *Human Resource Management*, *34*(4), 473–495. doi:10.1002/hrm.3930340402

Upton, D. M. (1995). Flexibility as process mobility: The management of plant capabilities for quick response manufacturing. *Journal of Operations Management*, *12*(3-4), 205–224. doi:10.1016/0272-6963(95)00004-C

Urch, V., Sala, F., & Mount, G. (2006). *Linking emotional intelligence and performance at work*. London, UK: Lawrence Erlbaum Associates.

Uzzi, B. (1997). Social structure and competition in interfirm networks: The paradox of embeddedness. *Administrative Science Quarterly*, *42*, 35–67. doi:10.2307/2393808

Valverde, M., Quijano, S., & Soler, C. (2003). *HRM as a shared function: An exploratory study of responsibilities and agents in the managing of people*. Paper presented at the IV International Workshop on Human Resources, Cadiz, May.

Valverde, M., Ryan, G., & Soler, C. (2006). Distributing HRM responsibilities: A classification of organisations. *Personnel Review*, *35*(6), 618–636. doi:10.1108/00483480610702692

Valverde, M., Tregaskis, O., & Brewster, C. (2000). Labor flexibility and firm performance. *International Advances in Economic Research*, *6*, 649–661. doi:10.1007/BF02295375

Van Breukelen, W., Van Der Vlist, R., & Steensma, H. (2004). Voluntary employee turnover combining variables from traditional turnover literature with the theory of planned behaviour. *Journal of Organizational Behavior*, *25*, 893–914. doi:10.1002/job.281

van de Ven, C. (2004). *The psychological contract: A big deal!* Behavioural Sciences Service Centre. Ministry of Defence. The Hague, The Netherlands. Retrieved April 18, 2011, from http://www.internationalmta.org/Documents/2004/2004024P.pdf

van den Berg, P. T., & van der Velde, M. E. G. (2005). Relationships of functional flexibility with individual and work factors. *Journal of Business and Psychology*, *20*(1), 111–129. doi:10.1007/s10869-005-6994-9

Van der Heijde, C., & van der Heijden, B. (2006). A competence-based and multidimensional operationalization and measurement of employability. *Human Resource Management*, *45*(3), 449–476. doi:10.1002/hrm.20119

Van Dyne, L., Vandewalle, D., Kostova, T., Latham, M. E., & Cummings, L. L. (2000). Collectivism, propensity to trust and self-esteem as predictors of organizational citizenship in a non-work setting. *Journal of Organizational Behavior*, *21*(3), 3–23. doi:10.1002/(SICI)1099-1379(200002)21:1<3::AID-JOB47>3.0.CO;2-6

Van Knippenberg, D., & Hogg, M. A. (Eds.). (2004). *Leadership and power: Identity processes in groups and organizations*. London, UK: Sage.

van Knippenberg, D., & Schippers, M. C. (2007). Work group diversity. *Annual Review of Psychology*, *58*, 515–541. doi:10.1146/annurev.psych.58.110405.085546

Vanderberg, R. J., & Lance, C. E. (1992). Examining the causal order of job satisfaction and organizational commitment. *Journal of Management*, *18*, 153–167. doi:10.1177/014920639201800110

Varma, A., Beatty, R. W., Schneier, C. E., & Ulrich, D. O. (1999). High performance work systems: Exciting discovery or passing fad? *Human Resource Planning*, *22*(1), 26–38.

Vázquez-Palacios, I. (2010). Externalización del área de formación. In *M. J. Pérez López (Coord.), Externalización de funciones de recursos humanos*. Madrid, Spain: Pearson.

Veciana, J. M. (1999). Entrepreneurship as a scientific research programme. *Revista Europea de Dirección y Economia de la Empresa*, *8*, 2–10.

Venkatesh, V., Morris, M. G., Davis, G. B., & Davis, F. D. (2003). User Acceptance of Information Technology: Toward a unified view. *Management Information Systems Quarterly*, *27*(3), 425–478.

Verdú, A. J. (2002). *Relación entre flexibilidad y desempeño organizativo. Una aproximación desde la perspectiva de la gestión de la calidad total. Elche (Alicante)* (Hernández, U. M., Ed.).

Vernon, P., Philips, J., Brewster, C., & Ommeren, J. (2000). *European trends in HR outsourcing*. Cranfield, UK: Cranfield School of Management & William M. Mercer.

Volberda, H. W. (1996). Toward the flexible form: How to remain vital in hypercompetitive environments. *Organization Science*, *17*(4), 359–374. doi:10.1287/orsc.7.4.359

Volberda, H. W. (1998). *Building the flexible firm. How to remain competitive*. New York, NY: Oxford University Press.

Von Krogh, G. (1998). Care in knowledge creation. *California Management Review*, *40*(3), 133–153.

Vossio, R. (2002). Certificación y normalización de competencias. Orígenes, conceptos y prácticas. *Boletín Cinterfor*, *152*, 51–73.

Vroom, V., & Yetton, P. (1973). *Leadership and decision making*. Pittsburgh, PA: University of Pittsburgh Press.

Walker, A., & Smither, J. (1999). A five-year study of upward feedback: What managers do with their results matters. *Personnel Psychology*, *52*, 393–423. doi:10.1111/j.1744-6570.1999.tb00166.x

Walsh, J. P., & Ungson, G. R. (1991). Organizational memory. *Academy of Management Review*, *16*(1), 57–91.

Walton, R. E. (1985). From control to commitment in the workplace. *Harvard Business Review*, *63*, 77–84.

Wang, C. L., & Ahmed, P. K. (2003). Organizational learning: A critical review. *The Learning Organization*, *10*(1), 8–17. doi:10.1108/09696470310457469

Wang, L., & Wentling, T. (2001). *The relationship between distance coaching and the transfer of training*. Paper presented at the Academy of Human Resource Development, Tulsa, Oklahoma.

Ward, E. A., & Davis, E. (1995). The effect of benefit satisfaction on organizational commitment. *Compensation & Benefits Management*, *11*, 35–40.

Warner, R., & Sternberg, R. (1985). Practical intelligence in real-world pursuits: The role of tacit knowledge. *Journal of Personality and Social Psychology*, *49*, 436–458. doi:10.1037/0022-3514.49.2.436

Watkins, R., & Kaufman, R. (1996). An update on relating needs assessment and needs analysis. *Performance Improvement*, *35*(10), 10–13. doi:10.1002/pfi.4170351005

Watson, W., Kumar, K., & Michaelsen, L. K. (1993). Cultural diversity's impact on interaction process and performance: comparing homogeneous and diverse task groups. *Academy of Management Journal*, *36*, 590–602. doi:10.2307/256593

Way, S. A. (2002). High performance work systems and intermediate indicators of firm performance within the US small business sector. *Journal of Management*, *28*(6), 765.

Weerawardena, J., & Sullivan, M. G. (2006). Investigating social entrepreneurship: A multidimensional model. *Journal of World Business*, *41*(1), 21–35. doi:10.1016/j.jwb.2005.09.001

Weick, K. E. (1987). Theorizing about organizational communication. In Jablin, F. M., Putnam, L. L., Roberts, K. H., & Porter, L. W. (Eds.), *Handbook of organizational communication* (pp. 97–122). Newbury Park, CA: Sage.

Weick, K. E. (1991). The nontraditional quality of organizational learning. *Organization Science*, *2*(1), 116–124. doi:10.1287/orsc.2.1.116

Weick, K. E., & Roberts, K. H. (1993). Collective mind in organizations: Heedful interrelating on flight decks. *Administrative Science Quarterly*, *38*, 357–381. doi:10.2307/2393372

Weiner, S. P., & Hill, E. J. (1995). *Effective leadership in a telework environment: A training needs analysis*. Presented at the Annual Conference of The American Psychological Society, New York, New York.

Wellins, R., Byham, W., & Wilson, J. M. (1991). *Empowerment teams: Creating self-directed work groups that improve quality, productivity and participation*. San Francisco, CA: Jossey-Bass.

Wells, B., & Spinks, N. (1992). *Organizational communication: A leadership approach* (3rd ed.). Houston, TX: Dame Publications.

Wensels, O. (2010). Recolocación. In *M. J. Pérez López (Coord.), Externalización de funciones de recursos humanos*. Madrid, Spain: Pearson.

Wernerfelt, B. (1984). A resource-based view of the firm. *Strategic Management Journal*, *5*, 171–180. doi:10.1002/smj.4250050207

Wernerfelt, B., & Karnani, A. (1987). Competitive strategy under uncertainty. *Strategic Management Journal*, *8*, 187–195. doi:10.1002/smj.4250080209

West, M. A. (2002). Sparkling fountains or stagnant ponds: An integrative model of creativity and innovation implementation in work groups. *Applied Psychology-An International Review-Psychologies Appliquee-Revue Internationale*, *51*(3), 355–387.

West, M. A., & Hirst, G. (2003). Cooperation and teamwork for innovation. In West, M. A., Tjosvold, D., & Smith, K. G. (Eds.), *International handbook of organizational teamwork and cooperative working* (pp. 297–319). Chichester, UK: Wiley. doi:10.1002/9780470696712

Whetten, D. A., Keiser, J. D., & Urban, T. (1995). Implications of organizational downsizing for the human resource management function. In Ferris, G. R., Rosen, S. D., & Barnum, D. T. (Eds.), *Handbook of human resource management* (pp. 282–296). Cambridge, MA: Blackwell Publishers.

Whiddett, S., & Hollyforde, S. (2003). *A practical guide to competencies: How to enhance individual and organisational performance*. London, UK: CIPD (Chartered Institute of Personnel and Development).

Wikipedia. (2010). *Web 2.0*. Retrieved 20 August, 2010, from http://en.wikipedia.org/wiki/Web_2.0

Wilkowska, I. (4-6 June 2003). *Corporate restructuring and human resource management: A case study of the banking sector in Britain*. Paper presented at the 7th Conference on International Human Resource Management, "Exploring the Mosaic: Developing the Discipline." University of Limerick, Ireland.

Williams, L. J., & Hazer, J. T. (1996). Antecedents and consequences of job satisfaction and commitment in turnover models: A reanalysis using latent variable structural equation methods. *The Journal of Applied Psychology*, *71*, 219–231. doi:10.1037/0021-9010.71.2.219

Williams, R., & Edge, D. (1996). The social shaping of technology. *Research Policy*, *25*(6), 865–899. doi:10.1016/0048-7333(96)00885-2

Winterton, J. (19, 20 June, 2003), *Un modèle heuristique de restructuration utilisé pour prévoir les pertes d'emploi et les futurs besoins de compétences*. Paper presented at the congress Responsible restructuring in Europe, Dublin.

Wiseman, R. (2009). *59 seconds, think a little, change a lot*. London, UK: McMillan.

Wood, S., & Albanase, M. T. (1995). Can we speak of a high commitment management on the shop floor? *Journal of Management Studies*, *32*, 215–247. doi:10.1111/j.1467-6486.1995.tb00341.x

Wood, S., & de Menezes, L. M. (2008). Comparing perspectives on high involvement management and organizational performance across the British economy. *International Journal of Human Resource Management*, *19*(4), 639–682. doi:10.1080/09585190801953673

Woodall, J., Scott-Jackson, W., Newham, T., & Gurney, M. (2009). Making the decision to outsource human resources. *Personnel Review*, *38*(3), 236–252. doi:10.1108/00483480910943313

Wooldridge, B., & Floyd, S. W. (1990). The strategy process, middle management involvement, and organizational performance. *Strategic Management Journal*, *11*, 231–241. doi:10.1002/smj.4250110305

Wooldridge, M., & Jennings, N. R. (1995). Intelligent agents: Theory and practice. *The Knowledge Engineering Review*, *10*(2), 115–152. doi:10.1017/S0269888900008122

Wright, P. M., & Snell, S. A. (1991). Toward an integrative view of strategic human resource management. *Human Resource Management Review*, *1*(3), 203–225. doi:10.1016/1053-4822(91)90015-5

Wright, P. M., & Snell, S. A. (1998). Toward a unifying framework for exploring fit and flexibility in strategic human resource management. *Academy of Management Review*, *23*, 756–772.

Wright, P. M., Dunford, B. B., & Snell, S. A. (2001). Human resources and the resource based view of the firm. *Journal of Management*, *27*(6), 701–721. doi:10.1177/014920630102700607

Wright, P. M., McMahan, G. C., & McWilliams, A. (1994). Human resources and sustained competitive advantage: A resource-based perspective. *International Journal of Human Resource Management*, *5*, 301–326. doi:10.1080/09585199400000020

Wright, P., & McMahan, G. (1992). Theoretical perspectives for strategic resource management. *Journal of Management*, 18.

Wustemann, L. (2003). La comunicación a los empleados y su compromiso en el diseño del modelo de competencias. *Training & Development Digest*, *39*, 64–72.

Yahya, S., & Goh, W. K. (2002). Managing human resources toward achieving knowledge management. *Journal of Knowledge Management*, *6*(5), 457–468. doi:10.1108/13673270210450414

Yamazaki, Y., & Kayes, D. C. (2004). An experiential approach to cross-cultural learning: A review and integration of competencies for successful expatriate adaptation. *Academy of Management Learning & Education*, *3*(4), 362–379. doi:10.5465/AMLE.2004.15112543

Yavas, U., & Bodur, M. (1999). Satisfaction among expatriate managers: Correlates and consequence. *Career Development International*, *4*, 261–269. doi:10.1108/13620439910279743

Youndt, M. A., & Snell, S. A. (2004). Human resource configurations, intellectual capital, and organizational performance. *Journal of Managerial Issues*, *16*(3), 337–360.

Youndt, M. A., Snell, S. A., Dean, J. W., & Lepak, D. P. (1996). Human resource management, manufacturing strategy, and firm performance. *Academy of Management Journal*, *39*, 836–866. doi:10.2307/256714

Yunkl, G. (2006). *Leadership in organizations*. Englewood Cliffs, NJ: Prentice Hall.

Zahra, S. A., Neubaum, D. O., & Huse, M. (2000). Entrepreneurship in medium-size companies: Exploring the effects of ownership and governance systems. *Journal of Management*, *26*(5), 947–976. doi:10.1177/014920630002600509

Zajac, E. J., Golden, B. R., & Shortell, S. M. (1991). New organizational forms for enhancing innovation: The case of internal corporate joint ventures. *Management Science*, *37*(2), 170–185. doi:10.1287/mnsc.37.2.170

Zamagni, S. (2010). *Speech at the 1st Symposium Society, Economy and Values*, IESE Business School, Barcelona, Spain.

Zammuto, R. F., & O'Connor, E. J. (1992). Gaining advanced manufacturing technologies' benefits: The roles of organization design and culture. *Academy of Management Review*, *17*(4), 701–728.

Zarifain (1999). *Objectif Competente, pour une nouvelle logique*. Paris, France: Éditions Liasions.

Zárraga, C., & Bonache, J. (2005). Equipos de trabajo para la gestión del conocimiento: La importancia de un clima adecuado. *Cuadernos de Economía y Dirección de la Empresa*, *22*, 27–48.

Zatzick, C. D., & Iverson, R. D. (2006). High-involvement management and workforce reduction: Competitive advantage or disadvantage? *Academy of Management Journal*, *49*(5), 999–1015. doi:10.5465/AMJ.2006.22798180

Zehir, C., & Keskin, H. (2003). A field research on the effects of MIS on organizational restructuring. *Journal of African Business*, *3*(1/2), 270–287.

Zeigarnick, A. (2007). Bluma Zeigarnick: A memory. *Gestalt Theory*, *29*(3), 256–268.

Zhang, Z., & Jasimuddin, S. M. (2008). Toward a strategic framework of mobile knowledge management. *Knowledge and Process Management*, *15*(2), 87–96. doi:10.1002/kpm.300

Zhao, H., Seibert, S. E., & Lumpkin, G. T. (2010). The relationship of personality to entrepreneurial intentions and performance: A meta-analytic review. *Journal of Management*, *36*(2), 381–404. doi:10.1177/0149206309335187

Zhao, H., Wayne, S. J., Glibkowski, B. C., & Bravo, J. (2007). The impact of psychological contract breach on work-related outcomes: A meta-analysis. *Personnel Psychology*, *60*(3), 647–680. doi:10.1111/j.1744-6570.2007.00087.x

Zimmerman, B., Bonner, S., & Kovach, R. (1996). *Developing self-regulated learners*. Washington, DC: APA. doi:10.1037/10213-000

Zingheim, P. K., & Schuster, J. R. (2003). Competencies and rewards: Substance or just style? *Compensation and Benefits Review*, *35*(5), 40–44. doi:10.1177/0886368703256932

Zuboff, S. (1988). *In the age of the smart machine: The future of work and power*. New York, NY: Basic Books.

About the Contributors

Susana de Juana-Espinosa has been a full time Assistant Professor in the Department of Business Organization of the University of Alicante since 2001. Previously, she worked as a researcher for the University of Murcia and the CEBAS-CSIC, and she has worked in the private sector as a director of personnel and accounting. She holds, besides her PhD in Business Organization, an MA in European Studies and a Degree in Business Administration. Her main teaching areas are Human Resource Management, Information Systems Management, Quality Management, and E-business. Dr. de Juana's research has mainly focused on using information and communications technologies (ICTs) to facilitate transformations in the strategies, business models, organisational designs, and human resource policies in both private and public organizations. She has worked closely with telecommunications organizations and local councils through research and consultancy.

José Antonio Fernández Sánchez is an assistant Professor at the Department of Business Organization at the University of Alicante (Spain) since 2000. He was previously an accounting manager in the private sector. Professor Fernández holds a Spanish Degree in Business Administration (Bachelor in Business Administration, BBA) and a Ph.D. in Business Organization (both at the University of Alicante). His main teaching areas are Human Resource Management and Strategic and Information Systems Management. He is also part of the staff of the Master of Human Resources program at the University of Alicante (Spain). Dr. Fernandez's research focuses mainly on the use of strategies and information systems to facilitate human resource policies in the fields of recruitment and downsizing in private and public organizations.

Encarnación Manresa Marhuenda has been a full time Assistant Professor in the Department of Business Organization of the University of Alicante since 2001. She holds, besides her PhD in Business Organization, a Degree in Economics. Her main teaching areas are Strategic Management and Human Resource Management. Dr. Manresa has focused her research on the analysis of corporate strategy, focusing on the processes of outsourcing, subcontracting, and vertical integration. She combines the subject with the influence of such a strategy on human resources and technology. She has worked in projects with local councils through Research and Consultancy.

Jorge Valdés Conca has been a full time Assistant Professor in the Department of Business Organization of the University of Alicante since 2000. Previously, he has worked in the private sector as teacher in private academies, as a business administrator in a services company, and as a manager of the department of saving accounts in a bank. He holds a Degree in Business Administration and a PhD in Business

Organization. His main teaching areas are Human Resource Management, Business Administration, and Competency-Based Human Resource Models. Dr. Valdés's research has been centrally focused on design and implementation processes of Competency-Based Human Resource Models in companies. He is currently coordinating the OECA (Economic and Commercial Observatory of Alicante), developed for the City Council of Alicante.

***

Carmen Avilés Palacios became PhD in Economics in Madrid (Universidad Rey Juan Carlos). She is currently professor at the Technical University of Madrid teaching Business economics and Organization and Management. She authored several papers about Human resource Management, Corporate Responsibility; and Environmental Economics. Carmen is also very interested on TIC and its influence over teaching and learning processes.

Alexis J. Bañon Gomis is the Director of a training workshop and the Deputy Director of the Institute for Ethics in Communication and Organizations (IECO). He is also Deputy Director the IECO-UNESCO Chair of Management, Trust, and Alterité. Bañón is Associate-Professor of the Polytechnic University of Valencia (Spain). He has attended numerous academic and professional conferences and has been a research visiting scholar at IESE (Barcelona). He got his PhD at the University of Valencia, and is MBA by ESTEMA Business School. His current research interests are focused on organizational behaviour and business ethics.

Maria Bastida Domínguez is an Associate Professor at the University of Santiago de Compostela (Spain), PhD in Business Administration and degree in Economics. She has published more than 20 papers, and her research focuses on two fields Human International Resources management, and the analysis of barriers for the professional career and promotion of women in public and private organizations. In collaboration with Nuria Calvo and Jacobo Feás, she has realized numerous international contributions to this research. Nowadays she makes her educational work compatible with her responsibility as member of the research group in human resources and performance of the University of Santiago de Compostela.

Inmaculada Beltrán-Martín is an Associate Professor at the Department of Business Management and Marketing at Universitat Jaume I (Castellón, Spain). She obtained her PhD in Business Administration. Her current research interests include labor flexibility, initiative, and strategic human resource management. Some of the results of her researches have been presented in national and international conferences and published in journals such as *Journal of Management, International Journal of Human Resource Management,* or *Personnel Review*.

Nuria Calvo Babío is an Associate Professor at the University of A Coruña, PhD in Business Administration and degree in Economics and Industrial Psychology. After twelve years working as consultant in several firms (Accenture, Inditex), now she works as a Researcher in the Economic Analysis and Business Administration Department at University of A Coruña (Spain). Her research focuses on modelling human resources policies in organizations, and the assessment of different strategic approaches in spin-offs. She has published in *System Dynamics Review, International Journal of Organizational*

Analysis, European Research Studies Journal, and she has been visitor Professor in University of Porto (Portugal) and The Dublin Institute for Technology (Ireland).

Lourdes Canós Darós has a degree in Business and Administration and her Ph.D. is about Human Resources Management. Today, she is Vice-Director of Tourism Management Area in Universitat Politècnica de València, where she participates in different commissions. Her research expertise concerns some topics related to human resources management, strategic management, and business management. She has published papers in different international journals with recognized quality in the management field. She has also published some books about business topics. Currently, she is reviewer of top-level journals and she also reviews papers presented in international congresses. She has taught in different courses and seminars around the world. These conferences were about management in touristic companies, management under uncertainty, human resources management, new technologies and information systems, decision making, et cetera.

José Antonio Carazo Muriel is a journalist by the Universidad Complutense de Madrid, PDD by IESE (Universidad de Navarra), and Diploma in Top Management Development Program by Universidad Antonio de Nebrija. He is currently Editor of the magazines *Capital Humano, Gestión Práctica de Riesgos Laborales,* and *Revista de Comunicación*, published by Wolters Kluwer Spain. He is also Academic Director of the Training Division of this company. He has directed the fair Salón Capital Humano since 1996 and in 1997 prompted the creation of Capital Humano Awards for Human Resource Management. He is the Board Director of the Internal Communication and Corporate Identity Observatory, member of the Governing Council of Human Resources Latin-American Forum, and member of the Advisory Council of the Spanish Foundation for the Promotion and Development of Scientific and Professional Psychology (Psicofundación). He is also member of several professional associations (AEDIPE, AIG, APIE, AERCO, Top Ten). In 2000 he was awarded with the Company Values Award, and in 2007 and 2011, he was chosen as one of "The Most Influential People in Human Resources," by the newspaper *RRHHDigital*. He is coauthor of the book "Do you live or work?"

José Antonio Carrión López is Professor at the Department of Business Organization at the University of Alicante. He is an Assistant Professor at the Murcia Business School (ENAE), as well as in the Master of Personnel and Human Resources Management at the University of Alicante Business School (training course on Techniques for Human Resource Management). He has a degree in Psychology from the University of Murcia and Master's in Human Resources Management from the Center of Financial Studies, in Madrid. He has developed his career in the Human Resources field for over 20 years, both as a consultant and director of companies such as Grupo Santander, Adecco, Alexandre Tic Consultants, and BR Group. He has successfully developed and implemented human resources projects, such as: goal-oriented management systems and goal-driven performance assessment; variable salary system, adapted to different groups within the company; writing of a job-description and posts-analysis handbook; drawing-up of procedures for recruitment, selection, and employees handbook (manuals, schedules, and organization); and tailor-made training plans. He has been taking part in many research agreements, developed together with the University of Alicante, Deloitte, Tea Cegos, Sage, Woltters Klower, and Randstand.

Erik Døving is an Associate Professor of Management at the Oslo and Akershus University College of Applied Sciences, located in Oslo, Norway. He holds a PhD in Management from the Norwegian School of Economics (NHH), as well as Master's degrees in Industrial Engineering and in Organization Science. Dr. Døving has headed several contract research projects, including program evaluations on behalf of various national ministries. His teaching is primarily within organizational theory and behavior, human resource management, change management, strategic management, and applied statistics and research methods. Dr. Døving has published his research in edited books, national journals, and international journals, including the *Strategic Management Journal.*

Jacobo Feás, BSc in Economics and Business Administration, MSc, and PhD, and is Associate Professor at the Department of Finance and Accounting at University of Santiago de Compostela (Spain). He worked in a consultant company for several years before starting his academic career. He was granted a one year European Commission scholarship to finish his PhD in Italy and worked at the Fondazione ENI Enrico Mattei as Senior Researcher one year more. He also worked five years as Lecturer at the department of Management and Marketing at the University of Santiago de Compostela. He was involved in several international research projects and collaborated as expert with international organizations as European Commission, FAO and World Bank. His research focuses on decision theory, and in particular, multicriteria decision analysis and decision support systems applied to management policies. Multicriteria analysis, decision tools, portfolio investment management, and glass ceiling are his main research areas.

Fernando J. Fuentes-García (Córdoba, 1965), holds a degree in Business Management from the ICADE (Universidad Pontificia de Comillas, Madrid) and a PhD in Economics and Business at the University of Córdoba. He is currently a Lecturer in the Statistics and Business Administration Department and the Vice-Dean in the Faculty of Law and Business at the University of Cordoba (Spain). Previously, he developed nine years of professional activity in the Economic Department of the SAFA Foundation in Jaén (Spain), in the auditing and consulting area of Arthur Andersen (now Deloitte Touche) in Granada and in financial management and administration in Bodegas Alvear (Córdoba).

Joaquín García-Alfonso has a Finance & HR background at multinational companies in several industries: education, pharmaceutical, tech and direct selling. In the last four years has been Professor lecturing in the Management Unit at the Polytechnic University of Madrid (Forestry Engineers), specialized in Human Resources and Competences Development area. He is currently CFO and HR at FLP Spain and Portugal, subsidiary of the world leader in the manufacture of Aloe Vera, Forever Living Products. He has extensive experience in various sectors such Engineering and innovation consulting (CFO and HR in a subsidiary of French multinational, Altran), pharmaceutical industry (Coordinator SmithKline Beecham Foundation), in a dot com company (CFO and HR at enfemenino.com), etc. He earned a degree in Economics and Business Administration from Autónoma University of Madrid and completed the Executive Development Program in Human Resources Management (MHRM).

Dolores Garzón Benítez is PhD in Business Organization by the Polytechnic University of Valencia and degree in Business Administration by the University of Córdoba. She is professor in the Management Department of the Polytechnic University of Valencia, specialized in subjects related to strategic management, entrepreneurship and companies creation. She has an extensive career in business and experience in supporting the creation and development of small companies.

Jose Gasco is a Full Professor in Business Management and Human Resources at the University of Alicante. His current research interests include Human Resources and Information Systems. He has published articles in several journals, namely, Revue Internationale P.M.E., Direction et Gestion des Entreprises, Corporate Communications: an International Journal, The International Journal of Public Sector Management, Business Process Management Journal, Business Process Management Journal, Information Systems Management, Information Processing and Management, International Journal of Information Management, Information Management & Computer Security, Industrial Management & Data Systems, Total Quality Management, Journal of European Industrial Training, Information Technology and People, and Logistics Information Management.

Ignacio Gil Pechuán is Professor of Management and Information and Technology Systems at Polytechnic University of Valencia, and PhD in Management and Information Systems. Prof. Gil is the Director of Business and Management Department at the Polytechnic University of Valencia, and GEDESTIC Chair of Information, Technology and Management. He is Deputy Director the IECO-UNESCO Chair of Management, Trust and Alterité. His research has been focused in management and Information Systems and technologies. He has a degree in Business Administration, in Information and Library Science, and in Market Research and Techniques. He also earned a Master's in Information Systems, Technology, and Management at IESE Business School (University of Navarra).

Manuel Guillén Parra is Professor of Management and Business Ethics at the University of Valencia (Spain). He is the Director of the Institute for Ethics in Communication and Organizations (IECO); he is also the Director of the IECO-UNESCO Chair of Management, Trust and Alterité, and the General Secretary of the European Business Ethics Network Spanish branch (EBEN-Spain). Guillén earned his PhD in Management and ethics at UV and has been visiting scholar at the University of St. Thomas, the IESE Business School, the University of Notre Dame, the CBE at Bentley University, and the RCC at Harvard University. He is author of several books on business ethics, including "Ética en las Organizaciones: Construyendo Confianza" (Prentice-Hal, 2006).

Santiago Gutierrez-Broncano is professor on Human Resources and Management at Castilla-La Mancha University. He was graduated in Business Administration from the University of Alcalá de Henares, Master in Development of People from the University Autónoma of Madrid, and a PhD from the University San Pablo CEU in 2003. He has collaborated with Spanish and foreign universities. He has directed several doctoral theses in the area of human resources, and he has participated in various national and international conferences. He is author of several papers in journals such as Total Quality Management & Business Excellence, Management, Universia Business Review, Journal of Tourism, International Journal of Small and Medium Enterprises, among others. His main research areas are leadership, high commitment practices in human resources, small and medium enterprises and family businesses and flexible organizational design. Currently, his main area is focused on innovative management teams.

José Manuel de Haro García is a Senior Work and Organizational Psychologist specialized in results-oriented training, talent management, personnel assessment, performance management, competen¬cies, and leadership with more than twenty-four years of experience in HR positions in private and public companies (Agbar-Suez Environnement, National Employment Office-Labour and Social Security

Ministry) and in HR consultant projects (General Electric Plastics Spain). He has been lecturer at Universities and academic institutions (Alicante University, Murcia University, UNED, UMH, ENAE, Fundesem Business School and EEN). As research, he is author of around 30 publications and studies presented at both national and international conferences about HR issues. He was awarded the AEDIPE HUMAN 2004 prize in Innovation in Human Resources for the article: "¿Do somebody really knows what a professional competence is?: From McClelland to the ISO 9000". He currently, holds the position of recruitment, training & development manager at Agbar, Comunidad Valenciana in Spain and is editorial member team of the "International Journal of mentoring and Coaching".

Anna B. Holm is Assistant Professor at the Department of Business Administration, Business and Social Sciences, Aarhus University in Denmark. She holds a PhD (2010) in Organization and Management Theory from Aarhus University and a Master's degree in Business Administration (1995) from Bradford University School of Management, UK. Dr. Holm's research area is new ways of organizing business activities based on information and communication technologies (ICTs) and the Internet. She has presented and published a number of conference papers and articles on the subjects of recruitment, electronic human resource management (e-HRM), virtual organizing, business model innovation, and institutional theory. Prior to her academic career, Dr. Holm worked for over a decade as a business development consultant and in various managerial positions in industry.

Jaime Izquierdo Pereira is CEO at NewMahwah, the Social Xperts Lab. He studied Mathematics in Madrid, and has a working background in IT in American and Spanish companies. He learned about HRM while working as Sales and Marketing Director at Forum Informático, an IT services company which developed a software application for competence management. In 2008 he founded Competencias 2.0TM, a research project nested in the Technological Park of the Balearic Islands (ParcBIT) to identify, systematize, and develop the essential competencies that result in a presence both pleasant and productive in social networks. In March 2011 he started NewMahwah - The Social Xperts Lab together with Proyectos de Inversión Beta (VC) to develop and trade Target-ToolTM, a methodology that helps organizations and individuals to define their optimal presence in Web 2.0. Jaime has an Advanced Management Program by Instituto de Empresa Business School, and speaks five languages.

Manuel Lopez-Quero is a Forestry Engineering specialized in Management. He has an MBA by Instituto de Empresa Business School Manuel has been Vice-Rector and occupied several responsible post at the Technical University of Madrid. At present Manuel manages the Chair "University-Business-Trade Union: Labor, Environment, Health". He is the author of several papers about Human resource Management, Corporate Responsibility; and Environmental Economics.

Roberto Luna-Arocas is an Assistant Professor in the Department of Business Management, Faculty of Economy at the University of Valencia, Spain. His research interest and professional experience is focused on talent management and executive coaching. He has published several books (i.e., Human Resources in Tourism, Ed. Pearson Prentice Hall; The Trip to Talent, 2010, Ed. Obelisco; Wake up the Talent, 2011, LidEditorial) and many journal articles (i.e., *British Journal of Management, Journal of Business Ethics, International Journal of Human Resource Management, Personnel Review, Management Research,* or *International Focus Society for Human Resource Management*). He has reviewed papers in some main journals and has been the principal investigator (PI) of two major three-year Research and Development project concerning strategic human resource management granted by the Spanish Ministry of Education. Also, he is the director of the human resources research group, called People No Limits, and belongs to the national human resource management association, AEDIPE, in Spain. He is the President of the Executive Coaching Association AECOP in Levante Area in Spain (association linked with the European Coaching Association European Mentoring and Coaching Councill, EMCC). He is also Director of a Master's degree program about Talent Management. From a professional side he has done many business consultancies being a usual invited speaker in all around the world.

Irene Martín-Rubio is an Associate Professor of Industrial Organization, Business Economics, and Management at the Industrial Engineering College, Technical University of Madrid since 2002. She received her Ph.D. from Universidad Complutense de Madrid. Previously she held positions at Universidad Carlos III de Madrid, CU Cardenal Cisneros- Universidad Complutense, Universidad Europea de Madrid, and Universidad Oberta de Cataluña. Her research examines organizational learning, technology, quality management, and the impact of human resources on strategic management in a global environment. She has published four books and more than 15 journal articles and book chapters, as well as articles in the mass media. She recently published Manual of Economics and Business Management in Engineering (Thomson Reuters-Cívitas, 2011). Dr. Martín-Rubio visited Oslo University College in 2006; this collaboration was made possible by a grant from the Technical University of Madrid and the Staff Mobility Program of the European Union. She is a member of Editorial Boards in Spanish Mentoring & Coaching Review.

Inocencia María Martínez-León has been a Professor of Business Management and Human Resource at Universidad Politécnica de Cartagena (Spain) since 1999. Before this, she had gathered experience in the business and financial management at G's Group Ltd. From 2005 to 2008 she was a Planning Coordinator of her university. Since 2010 she is a head of the department for Business Management at Universidad Politécnica de Cartagena, Spain. Inocencia's doctoral thesis was in the field of organizational learning and Information Technologies. Her current research interests are organizational learning, psychological contract, organizational structure, corporate reputation, intangibles management, and gender studies. Her papers are publishing in prestigious Spanish and international journals.

Ignacio Moreno Zapatero is an innovation and value-driven professional with a special focus in education and e-learning. He has a Bachelor in Law and Economics by ICADE and MBA by Insead, and he has more than 15 years of professional experience. Ignacio started his professional career as a consultant in Strategy in Accenture, has worked freelance for different consulting firms, and has been Director at the Conference and Training branch of Unidad Editorial, the largest media company in Spain. In the last 4 years, he has specialized in the Spanish and Brazilian Education markets, where he develops strategic and business development projects for leading international education corporations, through his own company, ICARINDI. Through this company, he also develops and manages online training schools and programs for different kinds of institutions, ranging from bar associations, to private limited corporations. Ignacio is also an active investor in innovative e-learning projects in a broad range of subjects as different as online parenting schools and online sales academies, which have been launched in Spain and Brazil.

Miguel R. Olivas-Luján is a full Professor of the Administrative Science department at Clarion University of Pennsylvania (USA) and is affiliated with Tecnológico de Monterrey (México). He is a Series Editor (with T. Bondarouk) for the Advanced Series in Management, an Emerald Group Publishing research book series, and has served as Editor-in-Chief to *The Business Journal of Hispanic Research.* As a researcher and author on business management topics, his scholarly and practitioner-oriented articles have appeared in journals, books and magazines in four languages. He has also been a member of the Society for Human Resource Management (SHRM) Global HR Experts Panel (2008-11) and of the United States Technical Advisory Group (US TAG) on Human Resource Management for the International Standards Organization (ISO). His research has been presented in five continents.

Javier Pagán Castaño is corporate coach, master degree in coaching and development by Portsmouth University Business School, International MBA by Instituto de Empresa and Loyola University Business School, certified senior coach by AECOP and Bachelor of Economics by the University of Valencia. He has worked as team director in Spain and other countries, he is part of the teaching staff of ENAE Business School and Luis Vives Business School. He is certified trainer by Dale Carnegie University and co-author of the book "dynamic of success".

Drew Rodgers taught International Business Communications at Oslo University College, School of Business, until 2010 when he retired. He visited Technical University of Madrid in 2010, made possible by a grant from the Staff Mobility Program of the European Union. Mr. Rodgers has published several books on international business communication for Cambridge University Press following a cross-cultural case study. He has participated in several projects studying international business communication in the global environment.

Sergi Romeu Crusat, born in Tarragona (Spain), holds a Degree in Economics and Business Administration from the University of Barcelona (UB), plus a Grade in Cash Management from the Polytechnic University of Catalonia (UPC) and Escuela de Aministracion de Empresas (EAE). He is currently developing research for his PhD thesis on human resource management at the Universitat Rovira i Virgili (URV, Tarragona, Spain). He started his professional career as a financial auditor, later assuming several roles in accounting and finance in a multinational environment of the industrial sector and taking positions of responsibility in Treasury and Taxes. He has attended several professional seminars and has been involved in developing programs to adapt European legislation at country level. Additionally, he has participated in some scholarship programs to develop business simulations.

Paolo Rungo is European PhD in Economics, and Associate Professor of the Department of Economic Analysis and Business Administration at the University of A Coruña, Spain. His main lines of research deal with the economics of education and health economics. He has authored a number of papers, which have been published in prestigious international journals, with a special interest in the determinants of success in the learning process, and the impact of professional competencies on employability.

Sandra Mª Sánchez-Cañizares (Córdoba, 1976) has a European Doctorate degree in Management Science. She currently works as a Lecturer in the Statistics and Business Administration Department at the University of Cordoba (Spain). She has been a visiting Lecturer at British, French, Chilean, Portuguese, and Italian universities. She has directed several projects about job satisfaction and has participated in different competitive projects about communitarian tourism and economic development. She is a specialist in organizational behavior and has published three books and over 40 scientific papers related to human resources, organizational commitment, entrepreneurship, and the tourist trade.

Cristina Santandreu-Mascarell has a degree in Business and Administration (financial and human resources management) and she is a Ph.D. student at Universidad Politécnica de Valencia. Today, she is a lecturer in Business and Innovation Management. She participates in the research institute IGIC. Her research expertise concerns some topics related with human resources management, strategic management and business management. She has published papers in different journals with recognized quality in the management field. She has also published some book chapters about business topics and a book of business development. Moreover, she is reviewer of some journals. She has taught in different courses and seminars. These conferences were about management in touristic companies, human resources management, new technologies and information systems, decision making, etc., She has recently developed a collaboration with the Federal University of Rio de Janeiro, researching about innovative companies.

Teresa Torres-Coronas has a bachelor's degree in economics (Barcelona University) and a PhD in management (Rovira i Virgili University). She won first prize in the 2000 edition of EADA related management research. She is the author of the book Valuing Brands (Ediciones Gestión 2000, Spain), co-author of the book Retrieve Your Creativity (Septem Ediciones, Spain), and co-editor of the books Changing the way you teach: Creative tools for management education (Septem Ediciones, Spain), e-HRM: Managing knowledge people (Idea Group, USA), Higher creativity for virtual teams: Developing platforms for co-creation (Information Science Reference) and, The Encyclopedia of HRIS: Challenges in e-HRM (Information Science Reference). She is author of many articles and conference papers about intangible management, management education, and applied creativity and IT. She is management professor at the Universitat Rovira i Virgili. She is one of the researchers of the Technology and Governance Network (Royal Holloway, University of London). She is an active member of the Management Education and Development Division (Academy of Management) and the Information Resources Management Association (IRMA).

Mireia Valverde is a senior Lecturer in Human Resource Management at Universitat Rovira i Virgili (URV, Tarragona, Spain). Her university education is from Universitat de Barcelona, University of Limerick, Cranfield School of Management, and URV. She has taught HRM and research methods in Ireland, Spain, and other countries. She was the Assistant Director and later the Director of the Business Management Department at the URV, managing more than 140 personnel. Her research interests relate to the input of different agents into the HR function (the HR department, top and middle managers, HR service providers) as well as to issues on the interface between the HRM and Marketing functions. She is the principal researcher of the research group FHOM (Human Factor, Organisations and Markets), officially recognised by the Catalan Government.

Index

D

E

F

G

H

U

V

W

X

www.ingramcontent.com/pod-product-compliance
Lightning Source LLC
LaVergne TN
LVHW062112300926
849161LV00010B/2437